Matrix Polynomials

This is a volume in
COMPUTER SCIENCE AND APPLIED MATHEMATICS
A Series of Monographs and Textbooks

Editor: WERNER RHEINBOLDT

A complete list of titles in this series appears at the end of this volume.

Matrix Polynomials

I. Gohberg

School of Mathematical Sciences
Tel-Aviv University
Ramat-Aviv, Tel-Aviv, Israel

P. Lancaster

Department of Mathematics and Statistics
University of Calgary
Calgary, Alberta, Canada

L. Rodman

School of Mathematical Sciences
Tel-Aviv University
Ramat-Aviv, Tel-Aviv, Israel

ACADEMIC PRESS 1982
A Subsidiary of Harcourt Brace Jovanovich, Publishers

New York London
Paris San Diego San Francisco São Paulo Sydney Tokyo Toronto

ACADEMIC PRESS, INC.
111 Fifth Avenue, New York, New York 10003

United Kingdom Edition published by
ACADEMIC PRESS, INC. (LONDON) LTD.
24/28 Oval Road, London NW1 7DX

Library of Congress Cataloging in Publication Data

Gohberg, I. TS. (Izrail' TSudikovich), Date.
Matrix polynomials.

(Computer science and applied mathematics series)
Bibliography: p.
Includes index.
1. Matrices. 2. Polynomials. I. Lancaster, Peter, mathematician. II. Rodman, L. III. Title. IV. Series.
QA188.G64 512.9'434 81-22792
ISBN 0-12-287160-X AACR2

PRINTED IN THE UNITED STATES OF AMERICA

82 83 84 85 9 8 7 6 5 4 3 2 1

To the memory of the late
F. R. Gantmacher
in appreciation of his outstanding contributions
as mathematician and expositor

Contents

Chapter 3 Multiplication and Divisibility

Chapter 4 Spectral Divisors and Canonical Factorization

Chapter 5 Perturbation and Stability of Divisors

Chapter 6 Extension Problems

Part II NONMONIC MATRIX POLYNOMIALS 181

Chapter 7 Spectral Properties and Representations

Preface

This book provides a comprehensive treatment of the theory of matrix polynomials. By a matrix polynomial (sometimes known as a λ-matrix) is understood a polynomial of a complex variable with matrix coefficients. Basic matrix theory (including the Jordan form, etc.) may be viewed as a theory of matrix polynomials $I\lambda - A$ of first degree. The theory developed here is a natural extension to polynomials of higher degrees, and forms an important new part of linear algebra for which the main concepts and results have been arrived at during the past five years. The material has important applications in differential equations, boundary value problems, the Wiener–Hopf technique, system theory, analysis of vibrations, network theory, filtering of multiple time series, numerical analysis, and other areas. The mathematical tools employed are accessible even for undergraduate students who have studied matrix theory and complex analysis. Consequently, the book will be useful to a wide audience of engineers, scientists, mathematicians, and students working in the fields mentioned, and it is to this audience that the work is addressed.

Collaboration among the authors on problems concerning matrix polynomials started in early 1976. We came to the subject with quite different backgrounds in operator theory and in applied mathematics, but had in common a desire to understand matrix polynomials better from the point of view of spectral theory. After bringing together our points of view, expertise, and tools, the solution to some problems for monic polynomials could be seen already by the summer of 1976. Then the theory evolved rapidly to include deeper analysis, more general (not monic) problems on the one hand, and more highly structured (self-adjoint) problems on the other. This work, enjoyable and exciting, was initially carried out at Tel-Aviv University, Israel, and the University of Calgary, Canada.

Very soon after active collaboration began, colleagues in Amsterdam, The Netherlands, and Haifa, Israel, were attracted to the subject and began to make substantial contributions. We have in mind H. Bart and M. A. Kaashoek of the Free University, Amsterdam, and L. Lerer of the Technion, Haifa. It is a pleasure to acknowledge their active participation in the development of the work we present and to express our gratitude for thought-provoking discussions. This three-way international traffic of ideas and personalities has been a fruitful and gratifying experience for the authors.

The past four years have shown that, indeed, a theory has evolved with its own structure and applications. The need to present a connected treatment of this material provided the motivation for writing this monograph. However, the material that we present could not be described as closed, or complete. There is related material in the literature which we have not included, and there are still many open questions to be answered.

Many colleagues have given us the benefit of discussion, criticism, or access to unpublished papers. It is a pleasure to express our appreciation for such assistance from E. Bohl, K. Clancey, N. Cohen, M. Cowen, P. Dewilde, R. G. Douglas, H. Dym, C. Foias, P. Fuhrmann, S. Goldberg, B. Gramsch, J. W. Helton, T. Kailath, R. E. Kalman, B. Lawruk, D. C. Lay, J. D. Pincus, A. Ran, B. Rowley, P. Van Dooren, F. Van Schagen, J. Willems, and H. K. Wimmer.

The authors acknowledge financial support for some or all of them from the Natural Sciences and Engineering Research Council of Canada, and the National Science Foundation of the United States. We are also very grateful to our home departments at Tel-Aviv University, the Weizmann Institute, and the University of Calgary for understanding and support. In particular, the second author is grateful for the award of a Killam Resident Fellowship at the University of Calgary. At different times the authors have made extended visits to the Free University, Amsterdam, the State University of New York at Stony Brook, and the University of Münster. These have been important in the development of our work, and the support of these institutions is cordially acknowledged.

Several members of the secretarial staff of the Department of Mathematics and Statistics of the University of Calgary have worked diligently and skillfully on the preparation of drafts and the final typescript. The authors much appreciate their efforts, especially those of Liisa Torrence, whose contributions far exceeded the call of duty.

Matrix Polynomials

Introduction

This is probably the first book to contain a comprehensive theory of matrix polynomials. Although the importance of matrix polynomials is quite clear, books on linear algebra and matrix theory generally present only a modest treatment, if any. The important treatise of Gantmacher [22], for example, gives the subject some emphasis, but mainly as a device for developing the Jordan structure of a square matrix. The authors are aware of only two earlier works devoted primarily to matrix polynomials, both of which are strongly motivated by the theory of vibrating systems: one by Frazer, Duncan, and Collar in 1938 [19], and the other by one of the present authors in 1966 [52b].

By a matrix polynomial, sometimes known as a λ-matrix, we understand a matrix-valued function of a complex variable of the form $L(\lambda) = \sum_{i=0}^{l} A_i \lambda^i$, where $A_0, A_1, \ldots, A_l$ are $n \times n$ matrices of complex numbers. For the time being, we suppose that $A_l = I$, the identity matrix, in which case $L(\lambda)$ is said to be *monic*. Generally, the student first meets with matrix polynomials when studying systems of ordinary differential equations (of order $l > 1$) with constant coefficients, i.e., a system of the form

$$\sum_{i=0}^{l} A_i \left(\frac{d}{dt}\right)^i u(t) = 0.$$

Looking for solutions of the form $u(t) = x_0 e^{\lambda_0 t}$, with x_0, λ_0 independent of t, immediately leads to the eigenvalue–eigenvector problem for a matrix polynomial: $L(\lambda_0)x_0 = 0$.

More generally, the function

$$u(t) = \left\{\frac{t^k}{k!}x_0 + \cdots + \frac{t}{1!}x_{k-1} + x_k\right\}e^{\lambda_0 t}$$

is a solution of the differential equation if and only if the set of vectors $x_0, x_1, \ldots, x_k$, with $x_0 \neq 0$, satisfies the relations

$$\sum_{p=0}^{j} \frac{1}{p!} L^{(p)}(\lambda_0)x_{j-p} = 0, \qquad j = 0, 1, \ldots, k.$$

Such a set of vectors $x_0, x_1, \ldots, x_k$ is called a *Jordan chain* of length $k + 1$ associated with eigenvalue λ_0 and eigenvector x_0.

It is this information on eigenvalues with associated multiplicities and Jordan chains which we refer to as the spectral data for the matrix polynomial, and is first to be organized in a concise and systematic way. The spectral theory we are to develop must include as a special case the classical theory for polynomials of first degree (when we may write $L(\lambda) = I\lambda - A$). Another familiar special case which will be included is that of scalar polynomials, when the A_i are simply complex numbers and, consequently, the analysis of a single (scalar) constant coefficient ordinary differential equation of order l.

Now, what we understand by spectral theory must contain a complete and explicit description of the polynomial itself in terms of the spectral data. When $L(\lambda) = I\lambda - A$, this is obtained when a Jordan form J for A is known together with a transforming matrix X for which $A = XJX^{-1}$, for we then have $L(\lambda) = X(I\lambda - J)X^{-1}$. Furthermore, X can be interpreted explicitly in terms of the eigenvector structure of A (or of $I\lambda - A$ in our terminology). The full generalization of this to matrix polynomials $L(\lambda)$ of degree l is presented here and is, surprisingly, of very recent origin.

The generalization referred to includes a Jordan matrix J of size ln which contains all the information about the eigenvalues of $L(\lambda)$ and their multiplicities. In addition, we organize complete information about Jordan chains in a single $n \times ln$ matrix X. This is done by associating with a typical $k \times k$ Jordan block of J with eigenvalue λ_0 k columns of X in the corresponding (consecutive) positions which consist of the vectors in an associated Jordan chain of length k. When $l = 1$, J and X reduce precisely to the classical case mentioned above. The representation of the coefficients of $L(\lambda)$ is then obtained in terms of this pair of matrices, which we call a *Jordan pair* for $L(\lambda)$. In fact, if we define the $ln \times ln$ matrix

$$Q = \begin{bmatrix} X \\ XJ \\ \vdots \\ XJ^{l-1} \end{bmatrix},$$

then Q is necessarily nonsingular and the coefficient matrices of $L(\lambda)$ are given by

$$[A_0 \quad A_1 \quad \cdots \quad A_{l-1}] = -XJ^l Q^{-1}.$$

Such a representation for the polynomial coefficients gives a solution of the inverse problem: to determine a matrix polynomial in terms of its spectral data. It also suggests further problems. Given only partial spectral data, when can it be extended to complete spectral data for some monic matrix polynomial? Problems of this kind are soluble by the methods we develop.

There is, of course, a close connection between the systems of eigenvectors for $L(\lambda)$ (the *right* eigenvectors) and those of the transposed polynomial $L^{\mathrm{T}}(\lambda) = \sum_{i=0}^{l} A_i^{\mathrm{T}} \lambda^i$ (the *left* eigenvectors). In fact, complete information about the left Jordan chains can be obtained from the $ln \times n$ matrix Y defined in terms of a Jordan pair by

$$\begin{bmatrix} X \\ XJ \\ \vdots \\ XJ^{l-1} \end{bmatrix} Y = \begin{bmatrix} 0 \\ \vdots \\ 0 \\ I \end{bmatrix}.$$

The three matrices X, J, Y are then described as a Jordan triple.

A representation theorem can now be formulated for the *inverse* of a monic matrix polynomial. We have

$$L^{-1}(\lambda) = X(I\lambda - J)^{-1} Y.$$

Results of this kind admit compact closed form solutions of the corresponding differential equations with either initial or two-point boundary conditions.

It is important to note that the role of a Jordan pair X, J can, for many aspects of the theory, be played by any pair, say V, T related to X, J by similarity as follows:

$$V = XS^{-1}, \qquad T = SJS^{-1}.$$

Any such pair V, T is called a *standard pair* for $L(\lambda)$. In particular, if C_1 is the companion matrix for $L(\lambda)$ defined by

$$C_1 = \begin{bmatrix} 0 & I & 0 & \cdots & 0 \\ \vdots & & I & \ddots & \vdots \\ 0 & 0 & & \cdots & I \\ -A_0 & -A_1 & & \cdots & -A_{l-1} \end{bmatrix},$$

then J is also a Jordan form for C_1 and there is an invertible S for which $C_1 = SJS^{-1}$. But more can be said. If U is the $n \times ln$ matrix $[I \quad 0 \quad \cdots \quad 0]$, and Q is the invertible matrix introduced above, then

$$U = XQ^{-1}, \qquad C_1 = QJQ^{-1}.$$

Thus U, C_1 is a standard pair. This particular standard pair admits the formulation of several important results in terms of either spectral data or the coefficients of L. Thus, the theory is not limited by the difficulties associated with the calculation of Jordan normal forms, for example.

A standard triple V, T, W is obtained by adjoining to a standard pair V, T the (unique) $ln \times n$ matrix W for which

$$\begin{bmatrix} V \\ VT \\ \vdots \\ VT^{l-1} \end{bmatrix} W = \begin{bmatrix} 0 \\ \vdots \\ 0 \\ I \end{bmatrix}.$$

The next main feature of the theory is the application of the spectral analysis to factorization problems. In terms of the motivation via differential equations, such results can be seen as replacing an lth-order system by a composition of two lower-order systems, with natural advantages in the analysis of solutions. However, factorizations arise naturally in several other situations. For example, the notion of decoupling in systems theory requires such a factorization, as also the design of filters for multiple time series, and the study of Toeplitz matrices and Wiener–Hopf equations.

An essential part of the present theory of factorization involves a geometrical characterization of right divisors. To introduce this, consider any standard pair V, T of $L(\lambda)$. Let $\mathscr{S}$ be an invariant subspace of T, let V_0, T_0 denote the restrictions of V and T to $\mathscr{S}$, and define a linear transformation $Q_k: \mathscr{S} \to \mathbb{C}^{nk}$ $(1 \le k \le l-1)$ by

$$Q_k = \begin{bmatrix} V_0 \\ V_0 T_0 \\ \vdots \\ V_0 T_0^{k-1} \end{bmatrix}.$$

Then $\mathscr{S}$ is said to be a *supporting subspace* (with respect to T) if Q_k is invertible. It is found that to each such supporting subspace corresponds a monic right divisor of degree k and, conversely, each monic right divisor has an associated supporting subspace. Furthermore, when $\mathscr{S}$ is a supporting subspace with respect to T, the pair V_0, T_0 is a standard pair for the associated right divisor.

This characterization allows one to write down explicit formulas for a right divisor and the corresponding quotient. Thus, if the coefficients of the monic right divisor of degree k are $B_0, B_1, \ldots, B_{k-1}$, then

$$[B_0 \quad B_1 \quad \cdots \quad B_{k-1}] = -V_0 T_0^k Q_k^{-1}.$$

To obtain the coefficients of the corresponding monic quotient of degree $l - k$, we introduce the third component W of the standard tiple V, T, W and any projector P which acts *along* the supporting subspace $\mathscr{S}$, i.e., $\operatorname{Ker} P = \mathscr{S}$. Then define a linear transformation $R_{l-k}\colon \mathbb{C}^{n(l-k)} \to \operatorname{Im} P$ by

$$R_{l-k} = [PW \quad PTPW \quad \cdots \quad PT^{l-k-1}PW].$$

This operator is found to be invertible and, if $C_0, C_1, \ldots, C_{l-k-1}$ are the coefficients of the quotient, then

$$\begin{bmatrix} C_0 \\ C_1 \\ \vdots \\ C_{l-k-1} \end{bmatrix} = R_{l-k}^{-1}.$$

Several applications concern the existence of divisors which have spectrum localized in the complex plane in some way (in a half-plane, or the unit circle, for example). For such cases the geometrical approach admits the transformation of the problem to the construction of invariant subspaces with associated properties. This interpretation of the problem gives new insights into the theory of factorization and admits a comprehensive treatment of questions concerning existence, perturbations, stability, and explicit representations, for example. In addition, extensions of the theory to the study of *systems* of matrix polynomials provide a useful geometrical approach to least common multiples and greatest common divisors.

The spectral theory developed here is also extended to include matrix polynomials for which the leading coefficient is not necessarily the identity matrix, more exactly, the case in which $\det L(\lambda) \not\equiv 0$. In this case it is necessary to include the possibility of a point of spectrum at infinity, and this needs careful treatment. It is easily seen that if $\det L(\lambda_0) \neq 0$, and we define $\mu = \lambda + \lambda_0$, then $M(\mu) = \mu^l L^{-1}(\lambda_0)L(\mu^{-1})$ is monic. Such a transformation is basic, but it is still necessary to analyze the effect of this transformation on the original problem.

Several important problems give rise to *self-adjoint* polynomials, i.e., in which all coefficients $A_0, A_1, \ldots, A_l$ are hermitian matrices. The prototype problem of this kind is that of damped vibrations, when $l = 2$ and A_2 is positive definite. The more familiar special cases which a theory for the self-adjoint case must include are $L(\lambda) = I\lambda - A$ with $A^* = A$ and the case of a scalar polynomial with *real* coefficients. The general theory developed leads to the introduction of a new invariant for a self-adjoint polynomial called the *sign characteristic*.

For self-adjoint matrix polynomials it is natural to seek the corresponding symmetry in a Jordan triple. What is going to be the simplest relationship between X and Y in a Jordan triple X, J, Y? It turns out that knowledge of the sign characteristic is vital in providing an answer to this question. Also, when a factorization involves a real eigenvalue common to both divisor and quotient, then the sign characteristic plays an important part.

The account given to this point serves to clarify the intent of the theory presented and to describe, in broad terms, the class of problems to which it is applied. This is the content of the spectral theory of matrix polynomials, as understood by the authors. Now we would like to mention some related developments which have influenced the authors' thinking.

In the mathematical literature many papers have appeared in the last three decades on operator polynomials and on more general operator-valued functions. Much of this work is concerned with operators acting on infinite-dimensional spaces and it is, perhaps, surprising that the complete theory for the finite-dimensional case has been overlooked. It should also be mentioned, however, that a number of important results discussed in this book are valid, and were first discovered, in the more general case.

On reflection, the authors find that four particular developments in operator theory have provided them with ideas and stimulus. The first is the early work of Keldysh [49a], which motivated, and made attractive, the study of spectral theory for operator polynomials; see also [32b]. Then the work of Krein and Langer [51] led to an appreciation of the importance of monic divisors for spectral theory and the role of the methods of indefinite scalar product spaces. Third, the theory of characteristic functions developed by Brodskii and Livsic [9] gave a clue for the characterization of right divisors by supporting subspaces. Finally, the work of Marcus and Mereutsa [62a] was very helpful in the attack on problems concerning greatest common divisors and least common multiples. In each case, the underlying vector spaces are infinite dimensional, so that the emphasis and objectives may be rather different from those which appear in the theory developed here.

In parallel with the operator theoretic approach, a large body of work on systems theory was evolving, mainly in the engineering literature. The authors learned about this more recently and although it has had its effect on us, the theory presented here is largely independent of systems theory. Even so, several formulas and notions have striking similarities and we know the connection to be very strong. The most obvious manifestation of this is the observation that, from the systems theory point of view, we study here systems for which the transfer function is the inverse of a matrix polynomial. However, this is the tip of the iceberg, and a more complete account of these connections would take us too far afield.

Another influential topic for the authors is the study of Toeplitz matrices and Wiener–Hopf equations. Although we do not pursue the connections in this book, they are strong. In particular, the spectral theory approach admits the calculation of the "partial indices"; a concept of central importance in that theory.

Finally, a short description of the contents of the book is presented. There are 19 chapters which are grouped into four parts. In Part I, consisting of six chapters, the spectral theory for monic matrix polynomials is developed beginning with the basic ideas of linearization and Jordan chains. We then go on to representation theorems for a monic polynomial and its inverse, followed by analysis of divisibility problems. These sections are illustrated by applications to differential and difference equations. The last three chapters are concerned with special varieties of factorization, perturbation and stability of divisors, and extension problems.

Part II consists of three chapters and is devoted to more general problems in which the "monic" condition is relaxed. The necessary extensions to the spectral theory are made and applied to differential and difference equations. Problems concerning least common multiples and greatest common divisors are then discussed in this context.

Further concepts are needed for the analysis of self-adjoint polynomials, which is presented in the four chapters of Part III. After developing the additional spectral theory required, including the introduction of the sign characteristic, some factorization problems are discussed. The last chapter provides illustrations of the theory of Part III in the case of matrix polynomials of second degree arising in the study of damped vibrating systems.

Finally, Part IV consists of six supplementary chapters added to make this work more self-contained. It contains topics not easily found elsewhere, or for which it is useful to have at hand a self-contained development of concepts and terminology.

Part I

Monic Matrix Polynomials

Introduction. If $A_0, A_1, \ldots, A_l$ are $n \times n$ complex matrices and $A_l \neq 0$, the zero matrix, then the matrix-valued function defined on the complex numbers by $L(\lambda) = \sum_{i=0}^{l} A_i \lambda^i$ is called a *matrix polynomial* of *degree* l. When $A_l = I$, the identity matrix, the matrix polynomial is said to be *monic*.

In Part I of this book our objective is the development of a spectral theory for monic matrix polynomials. In particular, this will admit representation of such a polynomial in terms of the spectral data, and also the characterization of divisors, when they exist, in terms of spectral data.

This theory is then put to use in studies of perturbation problems and the inverse problem of admissible extensions to incomplete spectral data. Applications to the solution of differential and difference equations are included.

We begin our analysis with *monic* polynomials for two reasons. First, matrix polynomials frequently occur in analysis and applications which are already in this form. Second, the study of monic polynomials allows one to see more clearly the main features of the spectral theory.

Chapter 1

Linearization and Standard Pairs

The fundamental notions in the spectral theory of matrix polynomials are "linearization" and "Jordan chains." We start this chapter with the detailed analysis of linearization and its applications to differential and difference equations. The definition of a Jordan chain will be seen to be a natural extension of the notion of a chain of generalized eigenvectors associated with a Jordan block of a matrix.

One of the objectives here is to analyze the connection between linearization and Jordan chains.

1.1. Linearization

Let $L(\lambda) = I\lambda^l + \sum_{j=0}^{l-1} A_j\lambda^j$ be a monic $n \times n$ matrix polynomial of degree l, where the A_j are $n \times n$ matrices with complex entries. Consider the following differential equation with constant coefficients:

$$\frac{d^l x}{dt^l} + \sum_{j=0}^{l-1} A_j \frac{d^j x}{dt^j} = f(t), \qquad -\infty < t < \infty, \tag{1.1}$$

where $f(t)$ is a given n-dimensional vector function and $x = x(t)$ is an n-dimensional vector function to be found. Let us study the transformation of the matrix polynomial $L(\lambda)$, analogous to the linearization of the differential equation (1.1). By the linearization of (1.1) we mean the reduction to a first-order equation, using the substitution

$$x_0 = x, \qquad x_1 = \frac{dx_0}{dt}, \qquad \cdots \qquad x_{l-1} = \frac{dx_{l-2}}{dt},$$

in which case the equation takes the form

$$\frac{dx_{l-1}}{dt} + A_{l-1}x_{l-1} + \cdots + A_1 x_1 + A_0 x_0 = f.$$

Note that we increased the dimension of the space containing the unknown function, which is now an ln-dimensional vector

$$\begin{bmatrix} x_0 \\ x_1 \\ \vdots \\ x_{l-1} \end{bmatrix}.$$

The corresponding transformation in the case of monic matrix polynomials resolves the following problem: Find a linear matrix polynomial $I\lambda - A$ which is "equivalent" to a given monic matrix polynomial $L(\lambda)$. It is clear that there is no such linearization if attention is confined to $n \times n$ matrices (the size of $L(\lambda)$ itself). To make the equivalence of $L(\lambda)$ to a linear polynomial $I\lambda - A$ possible, we have to extend the size of our matrices, and choose A and B as matrices of size $(n + p) \times (n + p)$, where $p \geq 0$ is some integer. In this case, instead of $L(\lambda)$ we consider the matrix polynomial

$$\begin{bmatrix} L(\lambda) & 0 \\ 0 & I \end{bmatrix}$$

of size $(n + p) \times (n + p)$, where I is the $p \times p$ unit matrix. So we are led to the following definition.

A linear matrix polynomial $I\lambda - A$ of size $(n + p) \times (n + p)$ is called a *linearization* of the monic matrix polynomial $L(\lambda)$ if

$$I\lambda - A \sim \begin{bmatrix} L(\lambda) & 0 \\ 0 & I \end{bmatrix} \tag{1.2}$$

where $\sim$ means equivalence of matrix polynomials. Recall (see also Section S1.6) that matrix polynomials $M_1(\lambda)$ and $M_2(\lambda)$ of size $m \times m$ are called *equivalent* if

$$M_1(\lambda) = E(\lambda)M_2(\lambda)F(\lambda)$$

for some $m \times m$ matrix polynomials $E(\lambda)$ and $F(\lambda)$ with constant nonzero determinants. Admitting a small abuse of language we shall also call matrix A

from (1.2) a linearization of $L(\lambda)$. Comparing determinants on both sides of (1.2) we conclude that $\det(I\lambda - A)$ is a polynomial of degree nl, where l is the degree of $L(\lambda)$. So the size of a linearization A of $L(\lambda)$ is necessarily nl.

As an illustration of the notion of linearization, consider the linearizations of a scalar polynomial ($n = 1$). Let $L(\lambda) = \prod_{i=1}^{k} (\lambda - \lambda_i)^{\alpha_i}$ be a scalar polynomial having different zeros $\lambda_1, \ldots, \lambda_k$ with multiplicities $\alpha_1, \ldots, \alpha_k$, respectively. To construct a linearization of $L(\lambda)$, let J_i $(i = 1, \ldots, k)$ be the Jordan block of size α_i with eigenvalue λ_i, and consider the linear polynomial $\lambda I - J$ of size $\sum_{i=1}^{k} \alpha_i$, where $J = \operatorname{diag}[J_i]_{i=1}^{k}$. Then

$$\lambda I - J \sim \begin{bmatrix} L(\lambda) & 0 \\ 0 & I \end{bmatrix},$$

because both have the same elementary divisors.

The following theorem describes a linearization of a monic matrix in terms of its coefficients.

Theorem 1.1. *For a monic matrix polynomial of size $n \times n$, $L(\lambda) = I\lambda^l + \sum_{j=0}^{l-1} A_j \lambda^j$, define the $nl \times nl$ matrix*

$$C_1 = \begin{bmatrix} 0 & I & 0 & \cdots & 0 \\ 0 & 0 & I & \cdots & 0 \\ \vdots & \vdots & \vdots & \ddots & \vdots \\ & & & \cdots & I \\ -A_0 & -A_1 & & \cdots & -A_{l-1} \end{bmatrix}.$$

Then

$$I\lambda - C_1 \sim \begin{bmatrix} L(\lambda) & 0 \\ 0 & I \end{bmatrix}.$$

Proof. Define $nl \times nl$ matrix polynomials $E(\lambda)$ and $F(\lambda)$ as follows:

$$F(\lambda) = \begin{bmatrix} I & 0 & \cdots & 0 & 0 \\ -\lambda I & I & \cdots & 0 & 0 \\ \vdots & \vdots & \cdots & \vdots & \vdots \\ 0 & 0 & \cdots & I & 0 \\ 0 & 0 & \cdots & -\lambda I & I \end{bmatrix},$$

$$E(\lambda) = \begin{bmatrix} B_{l-1}(\lambda) & B_{l-2}(\lambda) & \cdots & & B_0(\lambda) \\ -I & 0 & \cdots & & 0 \\ 0 & -I & \cdots & & \vdots \\ \vdots & \vdots & \ddots & & \\ 0 & & & -I & 0 \end{bmatrix},$$

where $B_0(\lambda) = I$ and $B_{r+1}(\lambda) = \lambda B_r(\lambda) + A_{l-r-1}$ for $r = 0, 1, \ldots, l-2$. It is immediately seen that $\det F(\lambda) \equiv 1$ and $\det E(\lambda) \equiv \pm 1$. Direct multiplication on both sides shows that

$$E(\lambda)(\lambda I - C_1) = \begin{bmatrix} L(\lambda) & 0 \\ 0 & I \end{bmatrix} F(\lambda), \tag{1.3}$$

and Theorem 1.1 follows. □

The matrix C_1 from the Theorem 1.1 will be called the (first) *companion matrix* of $L(\lambda)$, and will play an important role in the sequel. From the definition of C_1 it is clear that

$$\det(I\lambda - C_1) = \det L(\lambda).$$

In particular, the *eigenvalues* of $L(\lambda)$, i.e., zeros of the scalar polynomial $\det L(\lambda)$, and the eigenvalues of $I\lambda - C_1$ are the same. In fact, we can say more: since

$$I\lambda - C_1 \sim \begin{bmatrix} L(\lambda) & 0 \\ 0 & I \end{bmatrix},$$

it follows that the elementary divisors (and therefore also the partial multiplicities in every eigenvalue) of $I\lambda - C_1$ and $L(\lambda)$ are the same.

Let us prove here also the following important formula.

Proposition 1.2. *For every $\lambda \in \mathbb{C}$ which is not an eigenvalue of $L(\lambda)$, the following equality holds:*

$$(L(\lambda))^{-1} = P_1(I\lambda - C_1)^{-1}R_1, \tag{1.4}$$

where

$$P_1 = [I \quad 0 \quad \cdots \quad 0]$$

is an $n \times nl$ matrix and

$$R_1 = \begin{bmatrix} 0 \\ \vdots \\ 0 \\ I \end{bmatrix} \tag{1.5}$$

is an $nl \times n$ matrix.

Proof. Consider the equality (1.3) used in the proof of Theorem 1.1. We have

$$\begin{bmatrix} (L(\lambda))^{-1} & 0 \\ 0 & I \end{bmatrix} = F(\lambda)(I\lambda - C_1)^{-1}(E(\lambda))^{-1}. \tag{1.6}$$

It is easy to see that the first n columns of the matrix $(E(\lambda))^{-1}$ have the form (1.5). Now, multiplying the equality (1.6) on the left by P_1 and on the right by P_1^T and using the relation

$$P_1 F(\lambda) = [I \quad 0 \quad \cdots \quad 0] = P_1,$$

we obtain the desired formula (1.4). □

The formula (1.4) will be referred to as a *resolvent form* of the monic matrix polynomial $L(\lambda)$. We shall study the resolvent form in Chapter 2 more thoroughly. The following proposition gives another useful linearization of a monic matrix polynomial.

Proposition 1.3. *The matrix*

$$C_2 = \begin{bmatrix} 0 & \cdots & 0 & -A_0 \\ I & \cdots & & -A_1 \\ \vdots & \cdots & \vdots & \vdots \\ 0 & \cdots & I & -A_{l-1} \end{bmatrix}$$

is a linearization of a matrix polynomial $L(\lambda) = I\lambda^l + \sum_{j=0}^{l-1} A_j \lambda^j$.

The proof is left to the reader as an exercise.

The matrix C_2 from Proposition 1.3 will be called the *second companion matrix* of $L(\lambda)$.

The following result follows directly from the definition of a linearization and Theorem S1.12.

Proposition 1.4. *Any two linearizations of a monic matrix polynomial* $L(\lambda)$ *are similar. Conversely, if a matrix* T *is a linearization of* $L(\lambda)$ *and matrix* S *is similar to* T, *then* S *is also a linearization of* $L(\lambda)$.

1.2. Application to Differential and Difference Equations

In this section we shall use the companion matrix C_1 to solve the homogeneous and nonhomogeneous differential and difference equations connected with the monic matrix polynomial $L(\lambda) = I\lambda^l + \sum_{j=0}^{l-1} A_j \lambda^j$ of size $n \times n$.

We start with the differential equation

$$L\left(\frac{d}{dt}\right)x(t) = \frac{d^l x}{dt^l} + \sum_{j=0}^{l-1} A_j \frac{d^j x}{dt^j} = f(t), \tag{1.7}$$

where $f(t)$ is a given piecewise continuous n-dimensional vector function on $t \in (-\infty, \infty)$. A general solution $x(t)$ of (1.7) can be represented as a sum $x_0(t) + x_1(t)$, where $x_0(t)$ is a general solution of the homogeneous equation

$$L\left(\frac{d}{dt}\right)x = 0, \tag{1.8}$$

and $x_1(t)$ is some particular solution of the nonhomogeneous equation (1.7). The solutions of the homogeneous equation (1.8) form a linear space, which we denote by $\mathscr{S}(L)$, and we have

$$\dim \mathscr{S}(L) = nl. \tag{1.9}$$

In other words, the general solution of (1.8) depends on nl independent complex variables. This statement follows at once from Theorem S1.6, bearing in mind that $\det L(\lambda)$ is a scalar polynomial of degree nl.

We now give a formula for the general solution of (1.7).

In the proof of Theorem 1.5 and elsewhere we shall employ, when convenient, the notation $\operatorname{col}(Z_i)_{i=m}^{p}$ to denote the block column matrix

$$\begin{bmatrix} Z_m \\ Z_{m+1} \\ \vdots \\ Z_p \end{bmatrix}.$$

Theorem 1.5. *The general solution of equation* (1.7) *is given by the formula*

$$x(t) = P_1 e^{tC_1} x_0 + P_1 \int_0^t e^{(t-s)C_1} R_1 f(s)\, ds, \tag{1.10}$$

where

$$P_1 = [I \quad 0 \quad \cdots \quad 0], \qquad R_1 = \begin{bmatrix} 0 \\ \vdots \\ 0 \\ I \end{bmatrix},$$

and $x_0 \in \mathbb{C}^{nl}$ is arbitrary. In particular, the general solution of the homogeneous equation (1.8) *is given by the formula*

$$x(t) = P_1 e^{tC_1} x_0. \tag{1.11}$$

Proof. Let us first prove (1.11). We shall need the following simple observation:

$$P_1 C_1^k = \begin{cases} [0 \quad \cdots \quad 0 \quad I \quad 0 \quad \cdots \quad 0] & \text{with } I \text{ in the } (k+1)\text{th place for } k = 0, \ldots, l-1 \\ [-A_0 \quad -A_1 \quad \cdots \quad -A_{l-1}] & \text{for } k = l. \end{cases} \tag{1.12}$$

The verification of (1.12) we leave to the reader. Using these results, it is not hard to check that

$$x(t) = P_1 e^{tC_1} x_0$$

is a solution of (1.8) for any $x_0 \in \mathbb{C}^n$. Indeed,

$$\frac{d^i x(t)}{dt^i} = P_1 C_1^i e^{tC_1} x_0, \qquad i = 0, 1, \ldots, l.$$

So using (1.12),

$$\begin{aligned} L\left(\frac{d}{dt}\right)x(t) &= P_1 C_1^l e^{tC_1} x_0 + (A_{l-1} P_1 C_1^{l-1} + \cdots + A_0 P_1) e^{tC_1} x_0 \\ &= [-A_0 \quad -A_1 \quad \cdots \quad -A_{l-1}] e^{tC_1} x_0 \\ &\quad + ([0 \quad \cdots \quad 0 \quad A_{l-1}] + \cdots + [A_0 \quad 0 \quad \cdots \quad 0]) e^{tC_1} x_0 = 0. \end{aligned}$$

We show now that (1.11) gives all the solutions of the homogeneous equations (1.8). In view of (1.9) it is sufficient to prove that if

$$P_1 e^{tC_1} x_0 \equiv 0, \tag{1.13}$$

then $x_0 = 0$. Indeed, taking derivatives of the left-hand side of (1.13) we see that

$$P_1 C_1^i e^{tC_1} x_0 \equiv 0 \qquad \text{for every} \quad i = 0, 1, \ldots.$$

In particular, $[\operatorname{col}(P_1 C_1^i)_{i=0}^{l-1}] e^{tC_1} x_0 \equiv 0$. But (1.12) implies that $\operatorname{col}(P_1 C_1^i)_{i=0}^{l-1} = I$ (identity matrix of size $nl \times nl$); so the equality $e^{tC_1} x_0 \equiv 0$ follows. Since e^{tC_1} is nonsingular, $x_0 = 0$.

To complete the proof of Theorem 1.5 it remains to prove that

$$x(t) = P_1 \int_0^t e^{(t-s)C_1} R_1 f(s)\, ds \tag{1.14}$$

is a solution of the nonhomogeneous equation (1.7). Using the familiar substitution $x_0 = x$, $x_i = dx_{i-1}/dt$, $i = 1, \ldots, l-1$, we reduce the system (1.7) to the form

$$\frac{d\tilde{x}}{dt} = C_1 \tilde{x} + g(t), \tag{1.15}$$

where

$$\tilde{x} = \begin{bmatrix} x_0 \\ \vdots \\ x_{l-1} \end{bmatrix} \quad \text{and} \quad g(t) = \begin{bmatrix} 0 \\ \vdots \\ 0 \\ f(t) \end{bmatrix}.$$

The formula

$$\tilde{x}(t) = \int_0^t e^{(t-s)C_1} g(s)\, ds$$

gives a solution of (1.15), as one can check by direct verification. Multiplying the last formula by P_1 on the left and bearing in mind that

$$g(t) = R_1 f(t),$$

we find that (1.14) gives a solution of (1.7). □

We pass now to the difference equation

$$x_{j+l} + A_{l-1}x_{j+l-1} + \cdots + A_0 x_j = y_j, \qquad j = 0, 1, \ldots. \tag{1.16}$$

where $(y_0, y_1, \ldots)$ is a given sequence of vectors on n-dimensional space and $(x_0, x_1, \ldots)$ is a sequence to be found.

We regard the set of all sequences $(x_0, x_1, \ldots)$, $x_m \in \mathbb{C}^n$ as a complex linear space where the sum, and multiplication by a complex number, are defined in the natural way, i.e., coordinatewise. We claim that a general solution $x = (x_0, x_1, \ldots)$ can be represented in the following form (a complete analog of the case of differential equations)

$$x = x^{(0)} + x^{(1)},$$

where $x^{(1)}$ is a fixed particular solution of (1.16) and $x^{(0)}$ is a general solution of the homogeneous equation

$$x_{j+l} + A_{l-1}x_{j+l-1} + \cdots + A_0 x_j = 0, \qquad j = 0, 1, \ldots. \tag{1.17}$$

Indeed, if $x^{(1)}$ is the fixed solution of (1.16) and x is any solution, then the difference $x^{(0)} = x - x^{(1)}$ satisfies the homogeneous equation (1.17). Conversely, for every solution $x^{(0)}$ of (1.17), the sum $x = x^{(0)} + x^{(1)}$ is a solution of (1.16).

Let us focus first on the homogeneous equation (1.17). Clearly, its set of solutions $\tilde{\mathcal{S}}(L)$ is a linear space. Further, $\dim \tilde{\mathcal{S}}(L) = nl$. Indeed, we can choose $x_0, \ldots, x_{l-1}$ arbitrarily, and then compute $x_l, x_{l+1}, \ldots$ sequentially using the relation

$$x_{j+l} = -\sum_{k=0}^{l-1} A_k x_{j-k}.$$

Thus, the general solution of (1.17) depends on nl independent complex variables, i.e., $\dim \tilde{\mathcal{S}}(L) = nl$.

It is not hard to give a formula for the general solution of (1.17); namely,

$$x_i = P_1 C_1^i z_0, \qquad i = 0, 1, \ldots, \tag{1.18}$$

where C_1 and P_1 are as in Theorem 1.5 and $z_0 \in \mathbb{C}^{nl}$ is an arbitrary vector. The proof of this formula is by substitution and using (1.12). As for differential equations, one proves that (1.18) gives *all* the solutions of (1.17).

More generally, the following result holds.

Theorem 1.6. *The general solution of the nonhomogeneous equation* (1.16) *is given by the formula*:

$$x_i = P_1 C_1^i z_0 + P_1 \sum_{k=0}^{i-1} C_1^{i-k-1} R_1 y_k, \qquad i = 0, 1, \ldots, \tag{1.19}$$

where R_1 is as in Theorem 1.5.

Proof. We have only to check that

$$x_0^{(1)} = 0, \qquad x_i^{(1)} = P_1 \sum_{k=0}^{i-1} C_1^{i-k-1} R_1 y_k, \qquad i = 1, 2, \ldots, \tag{1.20}$$

is a particular solution of (1.16). Thus,

$$\begin{aligned} x_{i+l}^{(1)} &+ A_{l-1} x_{i+l-1}^{(1)} + \cdots + A_0 x_i^{(1)} \\ &= P_1 \left(\sum_{k=0}^{i+l-1} C_1^{i+l-k-1} R_1 y_k \right) + A_{l-1} P_1 \sum_{k=0}^{i+l-2} C_1^{i+l-2-k} R_1 y_k \\ &\quad + \cdots + A_0 P_1 \sum_{k=0}^{i-1} C_1^{i-k-1} R_1 y_k. \end{aligned}$$

Put together the summands containing y_0, then the summands containing y_1, and so on, to obtain the following expression (here C_1^j for $j < 0$ will be considered as zero):

$$\begin{aligned} &(P_1 C_1^{i+l-1} R_1 + A_{l-1} P_1 C_1^{i+l-2} R_1 + \cdots + A_0 P_1 C_1^{i-1} R_1) y_0 \\ &\quad + (P_1 C_1^{i+l-2} R_1 + A_{l-1} P_1 C_1^{i+l-3} R_1 + \cdots + A_0 P_1 C_1^{i-2} R_1) y_1 \\ &\quad + \cdots + (P_1 C_1^{l-1} R_1 + A_{l-1} P_1 C_1^{l-2} R_1 + \cdots + A_0 P_1 C_1^{-1} R_1) y_i \\ &\quad + (P_1 C_1^{l-2} R_1 + A_{l-1} P_1 C_1^{l-3} R_1 + \cdots + A_0 P_1 C_1^{-2} R_1) y_{i+1} \\ &\quad + \cdots + P_1 C_1^0 R_1 y_{i+l-1}. \end{aligned} \tag{1.21}$$

From (1.12) it follows that

$$P_1 C_1^l + A_{l-1} P_1 C_1^{l-1} + \cdots + A_0 P_1 C_1^0 = 0, \tag{1.22}$$

and

$$P_1 C_1^k R_1 = \begin{cases} 0 & \text{for} \quad k = 0, \ldots, l-2 \\ I & \text{for} \quad k = l-1. \end{cases} \tag{1.23}$$

Equalities (1.22) and (1.23) ensure that expression (1.21) is just y_i. So (1.20) is a particular solution of the nonhomogeneous equation (1.16). □

1.3. The Inverse Problem for Linearization

Consider the following problem: Let T be a matrix of size $m \times m$; when is T a linearization of some monic matrix polynomial $L(\lambda)$ of degree l and size $n \times n$?

Clearly, a necessary condition for existence of such an $L(\lambda)$ is that $m = nl$. But this is not all. We can discover the additional necessary condition if we examine dim Ker $L(\lambda)$ for different λ. Evidently, dim Ker $L(\lambda) \leq n$ for any $\lambda \in \mathbb{C}$. On the other hand, since $I\lambda - T$ is similar to

$$\begin{bmatrix} L(\lambda) & 0 \\ 0 & I \end{bmatrix},$$

we have

$$\dim \operatorname{Ker}(I\lambda - T) = \dim \operatorname{Ker} L(\lambda), \qquad \lambda \in \mathbb{C}.$$

Thus, the additional necessary condition is that

$$\dim \operatorname{Ker}(I\lambda - T) \leq n \qquad \text{for all} \quad \lambda \in \mathbb{C}.$$

As the following result shows, both necessary conditions are also sufficient.

Theorem 1.7. *The $m \times m$ matrix T is a linearization of some monic matrix polynomial of degree l and size $n \times n$ if and only if the two following conditions hold:*

(1) $m = nl$;
(2) $\max_{\lambda \in \mathbb{C}} \dim \operatorname{Ker}(I\lambda - T) \leq n$.

The second condition means that for every eigenvalue λ_0 of $I\lambda - T$, the number of elementary divisors of $I\lambda - T$ which correspond to λ_0 does not exceed n.

For the scalar case ($n = 1$), Theorem 1.7 looks as follows:

Corollary 1.8. *The $m \times m$ matrix T is a linearization of a monic scalar polynomial of degree m if and only if*

$$\max_{\lambda \in \mathbb{C}} \dim \operatorname{Ker}(I\lambda - T) \leq 1.$$

This condition is equivalent to the following: There exist complex numbers

$a_{m-1}, a_{m-2}, \ldots, a_0$ such that T is similar to a matrix of the form

$$\begin{bmatrix} 0 & 1 & 0 & \cdots & 0 \\ 0 & 0 & 1 & \cdots & 0 \\ \vdots & \vdots & \vdots & \cdots & \vdots \\ 0 & 0 & 0 & \cdots & 1 \\ -a_0 & -a_1 & -a_2 & \cdots & -a_{m-1} \end{bmatrix}.$$

In this case, the matrix T is also said to be *nonderogatory*.

Proof of Theorem 1.7. The necessity of both conditions was explained above. Let us prove their sufficiency. Suppose that T is given, satisfying both conditions (1) and (2). Consider $I\lambda - T$ and its Smith form $D(\lambda)$:

$$I\lambda - T = E(\lambda)D(\lambda)F(\lambda),$$

where $(E(\lambda))^{\pm 1}$ and $(F(\lambda))^{\pm 1}$ are matrix polynomials (see Section S1.1). It is sufficient to prove that

$$D(\lambda) \sim \begin{bmatrix} I_{n(l-1)} & 0 \\ 0 & L(\lambda) \end{bmatrix} \tag{1.24}$$

for some monic matrix polynomial $L(\lambda)$ of degree l. Let

$$D(\lambda) = \begin{bmatrix} I_{nl-k} & 0 \\ 0 & \operatorname{diag}[d_j(\lambda)]_{j=1}^{k} \end{bmatrix},$$

where $d_i(\lambda)$ are the nonconstant invariant polynomials of $D(\lambda)$, and, as usual, $d_{i+1}(\lambda)$ is divisible by $d_i(\lambda)$. Note that $D(\lambda)$ does not contain zeros on its main diagonal because $\det D(\lambda) = \det(I\lambda - T) \not\equiv 0$.

The condition (2) ensures that $k \leq n$. Indeed, if λ_0 is a zero of $d_1(\lambda)$, then

$$\dim \operatorname{Ker} D(\lambda_0) = k,$$

and, in view of (2), we have $k \leq n$. Now represent

$$D(\lambda) = \begin{bmatrix} I_{n(l-1)} & 0 \\ 0 & \operatorname{diag}[\tilde{d}_j(\lambda)]_{j=1}^{n} \end{bmatrix},$$

where $\tilde{d}_1(\lambda) \equiv \cdots \equiv \tilde{d}_{n-k}(\lambda) \equiv 1$ and $\tilde{d}_{n-k+j}(\lambda) = d_j(\lambda)$ for $j = 1, \ldots, k$.

Observe that (since m is the degree of $\det(I\lambda - T) = \det D(\lambda)$)

$$\sum_{j=1}^{n} l_j = m,$$

where l_j is the degree of $\tilde{d}_j(\lambda)$. If all the degrees l_j are equal, then $\operatorname{diag}[\tilde{d}_j(\lambda)]_{j=1}^{n}$ is a monic polynomial of degree $l = m/n$, and (1.24) is trivial. Suppose not all the degrees l_j are equal. Then $l_1 < l$ and $l_n > l$. Now there is a j such that $l_{j-1} < l_j$ and $l_{j-1} \leq l_1 + (l_n - l) < l_j$. So there is a monic scalar polynominal $s(\lambda)$ of degree $l_n - l$ such that $\tilde{d}_1(\lambda)s(\lambda)$ divides $\tilde{d}_j(\lambda)$ and $\tilde{d}_{j-1}(\lambda)$ divides $\tilde{d}_1(\lambda)s(\lambda)$. Then let $q(\lambda)$ be a polynomial for which $d_1(\lambda)s(\lambda)q(\lambda) = -d_n(\lambda)$, and observe that the degree of $d_1(\lambda)q(\lambda)$ is l.

Now perform the following sequence of elementary transformations on the matrix $\tilde{D}(\lambda) = \operatorname{diag}[\tilde{d}_i(\lambda)]_{i=1}^{n}$:

(a) Add s times the first column to the last column.
(b) Add q times the first row to the last row.
(c) Interchange the first and last columns.
(d) Row k goes to row $k + 1$ for $k = 1, 2, \ldots, j - 2$ and row $j - 1$ goes to row one.
(e) Permute the first $j - 1$ columns as the rows are permuted in (d).

The resulting matrix, which is equivalent to $\tilde{D}(\lambda)$, may be written in the form $D' + D''$ where

$$D' = \operatorname{diag}\{d_2, \ldots, d_{j-1}, d_1 s, d_j, \ldots, d_{n-1}, d_1 q\}$$

and D'' has all its elements zero except that in position $(j - 1, n)$ which is d_1.

Now consider the leading $(n - 1) \times (n - 1)$ partition of D':

$$D_1 = \operatorname{diag}\{d_2, \ldots, d_{j-1}, d_1 s, d_j, \ldots, d_{n-1}\}.$$

Since the degree of $d_1 q$ is l, the determinant of D_1 has degree $(n - 1)l$. Furthermore, by the choice of s, D_1 is in the Smith canonical form. This enables us to apply the argument to D_1 already developed for $\tilde{D}$. After a finite number of repetitions of this argument it is found that $\tilde{D}$ is equivalent to an upper triangular matrix polynomial B whose diagonal elements all have degree l. It is easily seen that such a matrix $\tilde{B}$ is equivalent to a monic matrix polynominal, say $L(\lambda)$. As required in (1.24) we now have

$$D(\lambda) \sim \begin{bmatrix} I_{n(l-1)} & 0 \\ 0 & \tilde{D}(\lambda) \end{bmatrix} \sim \begin{bmatrix} I_{n(l-1)} & 0 \\ 0 & L(\lambda) \end{bmatrix},$$

and the theorem is proved. $\square$

1.4. Jordan Chains and Solutions of Differential Equations

Consider a monic $n \times n$ matrix polynomial $L(\lambda) = I\lambda^l + \sum_{j=0}^{l-1} A_j\lambda^j$ and the associated homogeneous differential equation

$$L\left(\frac{d}{dt}\right)u(t) = 0. \tag{1.25}$$

Here $u(t)$ is an n-dimensional vector-valued function to be found. We already have a formula (namely, (1.11)) for the general solution of (1.25), but now we shall express it through elementary vector-valued functions of the real variable t. It turns out that such an expression is closely related to the spectral properties of $L(\lambda)$.

It is well known that in the scalar case ($n = 1$) the solutions of (1.25) are linear combinations of the simple solutions of the form

$$u(t) = t^j e^{\lambda_0 t}, \qquad j = 0, \dots, s-1,$$

where λ_0 is a complex number and $s = s(\lambda_0)$ is a positive integer. It turns out that λ_0 must be a root of the scalar polynomial $L(\lambda)$, and s is just the multiplicity of λ_0 as a root of $L(\lambda)$.

We now wish to generalize this remark for monic matrix polynomials. So let us seek for a solution of (1.25) in the form

$$u(t) = p(t)e^{\lambda_0 t}, \tag{1.26}$$

where $p(t)$ is an n-dimensional vector-valued polynomial in t, and λ_0 is a complex number. It will be convenient for our purposes to write the representation (1.26) as follows:

$$u(t) = \left(\frac{t^k}{k!}x_0 + \frac{t^{k-1}}{(k-1)!}x_1 + \cdots + x_k\right)e^{\lambda_0 t} \tag{1.27}$$

with $x_j \in \mathbb{C}^n$ and $x_0 \neq 0$.

Proposition 1.9. *The vector function $u(t)$ given by* (1.27) *is a solution of equation* (1.25) *if and only if the following equalities hold:*

$$\sum_{p=0}^{i} \frac{1}{p!} L^{(p)}(\lambda_0)x_{i-p} = 0, \qquad i = 0, \dots, k. \tag{1.28}$$

Here and elsewhere in this work $L^{(p)}(\lambda)$ denotes the pth derivative of $L(\lambda)$ with respect to λ.

Proof. Let $u(t)$ be given by (1.27). Computation shows that

$$\left(\frac{d}{dt} - \lambda_0 I\right)u(t) = \left(\frac{t^{k-1}}{(k-1)!}x_0 + \cdots + x_{k-1}\right)e^{\lambda_0 t}.$$

More generally,

$$\left(\frac{d}{dt} - \lambda_0 I\right)^j u(t) = \left(\frac{t^{k-j}}{(k-j)!} x_0 + \frac{t^{k-j-1}}{(k-j-1)!} x_1 + \cdots + x_{k-j}\right) e^{\lambda_0 t} \quad (1.29)$$

for $j = 0, \ldots, k$, and

$$\left(\frac{d}{dt} - \lambda_0 I\right)^j u(t) = 0 \qquad \text{for} \quad j = k+1, k+2, \ldots. \quad (1.30)$$

Write also the Taylor series for $L(\lambda)$:

$$L(\lambda) = L(\lambda_0) + \frac{1}{1!} L'(\lambda_0)(\lambda - \lambda_0) + \cdots + \frac{1}{l!} L^{(l)}(\lambda_0)(\lambda - \lambda_0)^l.$$

Then, replacing here λ by d/dt, we obtain:

$$L\left(\frac{d}{dt}\right)x(t) = L(\lambda_0)u(t) + \frac{1}{1!} L'(\lambda_0)\left(\frac{d}{dt} - \lambda_0 I\right)u(t) + \cdots$$
$$+ \frac{1}{l!} L^{(l)}(\lambda_0)\left(\frac{d}{dt} - \lambda_0\right)^l u(t). \quad (1.31)$$

Now substitution of (1.29) and (1.30) in the formula (1.31) leads to Proposition 1.9. □

The sequence of n-dimensional vectors $x_0, x_1, \ldots, x_k$ ($x_0 \neq 0$) for which equalities (1.28) hold is called a *Jordan chain* of *length* $k + 1$ for $L(\lambda)$ corresponding to the complex number λ_0. Its leading vector x_0 ($\neq 0$) is an *eigenvector*, and the subsequent vectors $x_1, \ldots, x_k$ are sometimes known as *generalized eigenvectors*. A number λ_0 for which a Jordan chain exists is called an *eigenvalue* of $L(\lambda)$, and the set

$$\sigma(L) = \{\lambda_0 \in \mathbb{C} \mid \lambda_0 \text{ is an eigenvalue of } L(\lambda)\}$$

is the *spectrum* of $L(\lambda)$.

This definition of a Jordan chain is a generalization of the well-known notion of a Jordan chain for a square matrix A. Indeed, let $x_0, x_1, \ldots, x_k$ be the Jordan chain of A, i.e.,

$$Ax_0 = \lambda_0 x_0, \qquad Ax_1 = \lambda_0 x_1 + x_0, \qquad \ldots, \qquad Ax_k = \lambda_0 x_k + x_{k-1}.$$

Then these equalities mean exactly that $x_0, x_1, \ldots, x_k$ is a Jordan chain of the matrix polynomial $I\lambda - A$ in the above sense. We stress that for a matrix polynomial of degree greater than one, the vectors in its Jordan chains need not be linearly independent, in contrast to the linear matrix polynomials of type $I\lambda - A$ (with square matrix A). Indeed, the zero vector is admissible as a *generalized* eigenvector. Examples 1.2 and 1.4 below provide such a matrix polynomial.

Note that from the definition of a Jordan chain $x_0, x_1, \ldots, x_k$ it follows that $L(\lambda_0)x_0 = 0$, i.e., $x_0 \in \operatorname{Ker} L(\lambda_0)$. Hence λ_0 is an eigenvalue of $L(\lambda)$ if and only if $\operatorname{Ker} L(\lambda_0) \neq \{0\}$.

We shall sometimes use the evident fact that the solutions of the linear system

$$\begin{bmatrix} L(\lambda_0) & 0 & \cdots & 0 \\ L'(\lambda_0) & L(\lambda_0) & \cdots & 0 \\ \vdots & \vdots & \cdots & \vdots \\ \frac{1}{\alpha!}L^{(\alpha)}(\lambda_0) & \frac{1}{(\alpha-1)!}L^{(\alpha-1)}(\lambda_0) & \cdots & L(\lambda_0) \end{bmatrix} \begin{bmatrix} x_0 \\ x_1 \\ \vdots \\ x_\alpha \end{bmatrix} = 0$$

form the set of all Jordan chains $x_0, x_1, \ldots, x_\alpha$ of $L(\lambda)$ with lengths not exceeding $\alpha + 1$ corresponding to λ_0 (after we drop the first zero vectors $x_0 = \cdots = x_i = 0$).

Consider now some examples to illustrate the notion of a Jordan chain.

EXAMPLE 1.1. Let

$$L(\lambda) = \begin{bmatrix} \lambda^2 & -\lambda \\ 0 & \lambda^2 \end{bmatrix}$$

Since $\det L(\lambda) = \lambda^4$, there exists one eigenvalue of $L(\lambda)$, namely, $\lambda_0 = 0$. Every nonzero vector in $\mathbb{C}^2$ is an eigenvector of $L(\lambda)$. Let us compute the Jordan chains which begin with an eigenvector $x_0 = \left[\begin{smallmatrix} x_{01} \\ x_{02} \end{smallmatrix}\right] \in \mathbb{C}^2 \backslash \{0\}$. For the first generalized eigenvector $x_1 = \left[\begin{smallmatrix} x_{11} \\ x_{12} \end{smallmatrix}\right] \in \mathbb{C}^2$ we have the following equation:

$$L'(0)x_0 + L(0)x_1 = 0,$$

which amounts to

$$\begin{bmatrix} 0 & -1 \\ 0 & 0 \end{bmatrix} \begin{bmatrix} x_{01} \\ x_{02} \end{bmatrix} = 0.$$

So x_1 exists if and only if $x_{02} = 0$, and in this case x_{01} can be taken completely arbitrary. For the second generalized eigenvector $x_2 = \left[\begin{smallmatrix} x_{21} \\ x_{22} \end{smallmatrix}\right]$ we have

$$\tfrac{1}{2}L''(0)x_0 + L'(0)x_1 + L(0)x_2 = 0,$$

which amounts to

$$\begin{bmatrix} 1 & 0 \\ 0 & 1 \end{bmatrix} \begin{bmatrix} x_{01} \\ 0 \end{bmatrix} + \begin{bmatrix} 0 & -1 \\ 0 & 0 \end{bmatrix} \begin{bmatrix} x_{11} \\ x_{12} \end{bmatrix} = 0,$$

or $x_{01} = x_{12}$. If this equality is satisfied, the vector x_2 can be arbitrary. Analogous consideration of the third generalized eigenvector $x_3 = \left[\begin{smallmatrix} x_{31} \\ x_{32} \end{smallmatrix}\right]$ gives

$$\begin{bmatrix} 1 & 0 \\ 0 & 1 \end{bmatrix} \begin{bmatrix} x_{11} \\ x_{12} \end{bmatrix} + \begin{bmatrix} 0 & -1 \\ 0 & 0 \end{bmatrix} \begin{bmatrix} x_{21} \\ x_{22} \end{bmatrix} = 0,$$

which is contradictory, because it implies that $x_{12} = 0$; but as we have seen before, $x_{12} = x_{01}$, and therefore $x_{12} \neq 0$ (since $x_0 \neq 0$).

Summarizing the investigation of this example, we obtain the following: the length of a Jordan chain $x_0, x_1, \ldots, x_{k-1}$ of $L(\lambda)$ cannot exceed 3 (i.e., $k \leq 3$). All Jordan chains can be described as follows:

(1) Jordan chains of length 1 are $x_0 = \left[\begin{smallmatrix} x_{01} \\ x_{02} \end{smallmatrix}\right]$, where $x_{01}, x_{02} \in \mathbb{C}$ are not both zero;
(2) Jordan chains of length 2 are $x_0 = \left[\begin{smallmatrix} x_{01} \\ 0 \end{smallmatrix}\right], x_1$, where $x_{01} \neq 0$ and x_1 is arbitrary;
(3) Jordan chains of length 3 are $x_0 = \left[\begin{smallmatrix} x_{01} \\ 0 \end{smallmatrix}\right]$, $x_1 = \left[\begin{smallmatrix} x_{11} \\ x_{01} \end{smallmatrix}\right]$, x_2, where $x_{01} \neq 0$, and x_{11}, x_2 are arbitrary. □

The next example is somewhat more complicated.

EXAMPLE 1.2. Let

$$L(\lambda) = \begin{bmatrix} \lambda^2(\lambda - 1)(\lambda^2 + 1) & \lambda^3(\lambda - 1) \\ \lambda^2(\lambda - 1)^2 & \lambda^3(\lambda - 1)^2 \end{bmatrix}.$$

Compute the determinant: $\det L(\lambda) = \lambda^7(\lambda - 1)^3$. Thus, there exist only two eigenvalues of $L(\lambda)$: $\lambda_1 = 0$ and $\lambda_2 = 1$. Since $L(0) = L(1) = 0$, every nonzero two-dimensional vector is an eigenvector of $L(\lambda)$ corresponding to the eigenvalue zero, as well as an eigenvector of $L(\lambda)$ corresponding to the eigenvalue $\lambda_2 = 1$.

Let us find the Jordan chains of $L(\lambda)$ corresponding to $\lambda_1 = 0$. Since $L'(0) = 0$, every pair of vectors $x_0, x_1 \in \mathbb{C}$ with $x_0 \neq 0$ forms a Jordan chain of $L(\lambda)$ corresponding to $\lambda_1 = 0$. An analysis shows that this chain cannot be prolonged unless the first coordinate of x_0 is zero. If this is the case, then one can construct the following Jordan chain of length 5:

$$\begin{bmatrix} 0 \\ x_{02} \end{bmatrix}, \quad \begin{bmatrix} -x_{02} \\ x_{12} \end{bmatrix}, \quad \begin{bmatrix} -x_{12} \\ x_{22} \end{bmatrix}, \quad \begin{bmatrix} x_{31} \\ x_{32} \end{bmatrix}, \quad \begin{bmatrix} x_{41} \\ x_{42} \end{bmatrix}, \tag{1.32}$$

where $x_{02}, x_{12}, x_{22}, x_{31}, x_{32}, x_{41}, x_{42}$ are arbitrary complex numbers and $x_{02} \neq 0$. The Jordan chain (1.32) is of maximal length.

Consider now the eigenvalue $\lambda_2 = 1$. It turns out that the Jordan chains of length 2 corresponding to λ_2 are of the following form:

$$\begin{bmatrix} x_{01} \\ -2x_{01} \end{bmatrix}, \quad \begin{bmatrix} x_{11} \\ x_{12} \end{bmatrix} \tag{1.33}$$

where $x_{01} \neq 0$, and x_{11}, x_{12} are arbitrary complex numbers. These Jordan chains cannot prolonged.

Using Proposition 1.9, we can find solutions of the system of differential equations:

$$L\left(\frac{d}{dt}\right)u(t) = \begin{bmatrix} \dfrac{d^5}{dt^5} - \dfrac{d^4}{dt^4} + \dfrac{d^3}{dt^3} - \dfrac{d^2}{dt^2} & \dfrac{d^4}{dt^4} - \dfrac{d^3}{dt^3} \\ \dfrac{d^4}{dt^4} - 2\dfrac{d^3}{dt^3} + \dfrac{d^2}{dt^2} & \dfrac{d^5}{dt^5} - 2\dfrac{d^4}{dt^4} + \dfrac{d^3}{dt^3} \end{bmatrix} \begin{bmatrix} u_1(t) \\ u_2(t) \end{bmatrix} = 0.$$

For instance, the solution corresponding to the Jordan chain (1.32) is

$$\begin{bmatrix} u_1(t) \\ u_2(t) \end{bmatrix} = \begin{bmatrix} -\dfrac{t^3}{6}x_{02} - \dfrac{t^2}{2}x_{12} + tx_{31} + x_{41} \\ \dfrac{t^4}{24}x_{02} + \dfrac{t^3}{6}x_{12} + \dfrac{t^2}{2}x_{22} + tx_{32} + x_{42} \end{bmatrix},$$

and the solution corresponding to the Jordan chain (1.33) is

$$\begin{bmatrix} u_1(t) \\ u_2(t) \end{bmatrix} = \begin{bmatrix} (tx_{01} + x_{11})e^t \\ (-2tx_{01} + x_{12})e^t \end{bmatrix}. \qquad \square$$

These examples show that the structure of Jordan chains for matrix polynomials can be quite complicated. To understand this structure better, it is useful to define canonical systems, which may be regarded as basic Jordan chains in the set of all Jordan chains corresponding to a fixed eigenvalue, and they play a role analogous to a basis in a finite-dimensional linear space. This will be done in Section 1.6.

A convenient way of writing a Jordan chain is given by the following proposition.

Proposition 1.10. *The vectors $x_0, \ldots, x_{k-1}$ form a Jordan chain of the matrix polynomial $L(\lambda) = \sum_{j=0}^{l-1} A_j\lambda^j + I\lambda^l$ corresponding to λ_0 if and only if $x_0 \neq 0$ and*

$$A_0X_0 + A_1X_0J_0 + \cdots + A_{l-1}X_0J_0^{l-1} + X_0J_0^l = 0, \qquad (1.34)$$

where

$$X_0 = [x_0 \quad \cdots \quad x_{k-1}]$$

is an $n \times k$ matrix, and J_0 is the Jordan block of size $k \times k$ with λ_0 on the main diagonal.

Proof. By Proposition 1.9, $x_0, \ldots, x_{k-1}$ is a Jordan chain of $L(\lambda)$ if and only if the vector function

$$u_0(t) = \left(\sum_{p=0}^{k-1} \frac{t^p}{p!} x_{k-p-1}\right)e^{\lambda_0 t}$$

satisfies the equation

$$L\left(\frac{d}{dt}\right)u_0(t) = 0.$$

But then also

$$0 = \left(\frac{d}{dt} - \lambda_0 I\right)^j L\left(\frac{d}{dt}\right)u_0(t) = L\left(\frac{d}{dt}\right)u_j(t), \qquad j = 1, 2, \ldots, k-1,$$

where

$$u_j(t) = \left(\frac{d}{dt} - \lambda_0 I\right)^j u_0(t) = \left(\sum_{p=0}^{k-j-1} \frac{t^p}{p!} x_{k-j-p-1}\right) e^{\lambda_0 t}.$$

Consider the $n \times k$ matrix

$$U(t) = [u_{k-1}(t) \quad \cdots \quad u_0(t)].$$

From the definition of $U(t)$ it follows that

$$U(t) = X_0 \begin{bmatrix} 1 & t & \cdots & \frac{1}{(k-1)!} t^{k-1} \\ & 1 & & \vdots \\ & & \ddots & t \\ 0 & & & 1 \end{bmatrix} e^{\lambda_0 t}.$$

Using the definition of a function of a matrix (see Section S1.9), it is easy to see that

$$U(t) = X_0 e^{J_0 t}. \tag{1.35}$$

Now write

$$L\left(\frac{d}{dt}\right) U(t) = 0,$$

and substitute here formula (1.35) to obtain $[A_0 X_0 + A_1 X_0 J_0 + \cdots + A_{l-1} X_0 J_0^{l-1} + X_0 J_0]e^{J_0 t} = 0$. Since $e^{J_0 t}$ is nonsingular, (1.34) follows.

Reversing the arguments above, we obtain that the converse is also true: formula (1.34) implies that $[x_0, \ldots, x_{k-1}]$ is a Jordan chain of $L(\lambda)$ corresponding to λ_0. □

Observe that in this section, as well as in the next two sections, monicity of the matrix polynomial $L(\lambda)$ is not essential (and it is not used in the proofs). In fact, Propositions 1.9 and 1.10 hold for a matrix polynomial of the type $\sum_{j=0}^{l} A_j \lambda^j$, where A_j are $m \times n$ rectangular matrices (in formula (1.34) the summand $X_0 J_0^l$ should be replaced in this case by $A_l X_0 J_0^l$).

We conclude this section with a remark concerning the notion of a left Jordan chain of a matrix polynomial $L(\lambda)$. The n-dimensional row vectors $y_0, \ldots, y_k$ form a *left Jordan chain* of $L(\lambda)$ corresponding to the eigenvalue λ_0 if the equalities

$$\sum_{p=0}^{i} \frac{1}{p!} y_{i-p} L^{(p)}(\lambda_0) = 0,$$

hold for $i = 0, 1, \ldots, k$. The analysis of left Jordan chains is completely similar to that of usual Jordan chains, since $y_0, \ldots, y_k$ is a left Jordan chain of

$L(\lambda)$ if and only if the transposed vectors $y_0^T, \ldots, y_k^T$ form a usual Jordan chain for the transposed polynomial $(L(\lambda))^T$, corresponding to the same eigenvalue. Thus, we shall deal mostly with the usual Jordan chains, while the left Jordan chains will appear only occasionally.

1.5. Root Polynomials

In order to construct a canonical system of Jordan chains of a monic matrix polynomial $L(\lambda)$ corresponding to a given eigenvalue λ_0, and for other purposes too, it is convenient to describe the Jordan chains in terms of root polynomials, which we now define.

Given an $n \times n$ matrix polynomial $L(\lambda)$ (not necessarily monic), an n-dimensional vector polynomial $\varphi(\lambda)$, such that $\varphi(\lambda_0) \neq 0$ and $L(\lambda_0)\varphi(\lambda_0) = 0$, is called a *root polynomial* of $L(\lambda)$ corresponding to λ_0. The order of λ_0 as a zero of $L(\lambda)\varphi(\lambda)$ is called the *order* of the root polynomial $\varphi(\lambda)$. Develop the root polynomial in powers of $\lambda - \lambda_0$:

$$\varphi(\lambda) = \sum_{j=0}^{q} (\lambda - \lambda_0)^j \varphi_j,$$

and it follows that the vectors $\varphi_0, \varphi_1, \ldots, \varphi_{k-1}$ (where k is the order of $\varphi(\lambda)$) form a Jordan chain of $L(\lambda)$ corresponding to λ_0. To see that, write

$$L(\lambda)\varphi(\lambda) = \left[\sum_{j=0}^{l} \frac{1}{j!} L^{(j)}(\lambda_0)(\lambda - \lambda_0)^j\right]\left[\sum_{j=0}^{q} (\lambda - \lambda_0)^j \varphi_j\right]$$

and equate to zero the coefficients of $(\lambda - \lambda_0)^j$ for $j = 0, \ldots, k-1$, in this product. The converse is also true: if $\varphi_0, \varphi_1, \ldots, \varphi_{k-1}$ is a Jordan chain of $L(\lambda)$ corresponding to λ_0, then the vector polynomial of type

$$\varphi(\lambda) = \sum_{j=0}^{k-1} (\lambda - \lambda_0)^j \varphi_j + (\lambda - \lambda_0)^k \psi(\lambda),$$

where $\psi(\lambda)$ is some vector polynomial, is a root polynomial of $L(\lambda)$ of order k corresponding to the same λ_0.

The notion of a root polynomial makes it easy to prove the following proposition, which will be frequently used in what follows.

Proposition 1.11. *Let $L(\lambda)$ be an $n \times n$ matrix polynomial (not necessarily monic) and let $A(\lambda)$ and $B(\lambda)$ be $n \times n$ matrix polynomials such that $A(\lambda_0)$ and $B(\lambda_0)$ are nonsingular for some $\lambda_0 \in \mathbb{C}$. Then $y_0, \ldots, y_k$ is a Jordan chain of the matrix polynomial $A(\lambda)L(\lambda)B(\lambda)$ corresponding to λ_0 if and only if the vectors*

$$z_j = \sum_{i=0}^{j} \frac{1}{i!} B^{(i)}(\lambda_0) y_{j-i}, \qquad j = 0, \ldots, k \tag{1.36}$$

form a Jordan chain of $L(\lambda)$ corresponding to λ_0.

Proof. Suppose that $z_0, \ldots, z_k$ given by (1.36) form a Jordan chain of $L(\lambda)$ corresponding to λ_0. Then

$$\varphi(\lambda) = \sum_{j=0}^{k} (\lambda - \lambda_0)^j z_j$$

is a root polynomial of $L(\lambda)$ and $L(\lambda)\varphi(\lambda) = (\lambda - \lambda_0)^{k+1}\theta_1(\lambda)$ for some vector polynomial $\theta_1(\lambda)$. From (1.36) it follows that

$$\varphi(\lambda) = B(\lambda)\psi(\lambda),$$

where $\psi(\lambda) = \sum_{j=0}^{k} (\lambda - \lambda_0)^j y_j$. So

$$[A(\lambda)L(\lambda)B(\lambda)]\psi(\lambda) = A(\lambda)[L(\lambda)\varphi(\lambda)] = (\lambda - \lambda_0)^{k+1}A(\lambda)\theta_1(\lambda),$$

so $\psi(\lambda)$ is a root polynomial of $A(\lambda)L(\lambda)B(\lambda)$ of order not less than $k + 1$, and, consequently, $y_0, \ldots, y_k$ is a Jordan chain of $A(\lambda)L(\lambda)B(\lambda)$ corresponding to λ_0.

Conversely, let $y_0, \ldots, y_k$ be a Jordan chain of $A(\lambda)L(\lambda)B(\lambda)$ corresponding to λ_0. Then

$$A(\lambda)L(\lambda)B(\lambda)\psi(\lambda) = (\lambda - \lambda_0)^{k+1}\theta_2(\lambda)$$

for some vector polynomial $\theta_2(\lambda)$ (here $\psi(\lambda) = \sum_{j=0}^{k} (\lambda - \lambda_0)^j y_j$). So

$$L(\lambda)\varphi(\lambda) = (\lambda - \lambda_0)^{k+1}A^{-1}(\lambda)\theta_2(\lambda). \tag{1.37}$$

Since $A(\lambda_0)$ is nonsingular, the right-hand part in (1.37) is a vector polynomial for which $\lambda = \lambda_0$ is a zero of multiplicity not less than $k + 1$. Thus $\varphi(\lambda)$ is a root polynomial of $L(\lambda)$ of order not less than $k + 1$, and $z_0, \ldots, z_k$ is a Jordan chain of $L(\lambda)$ (corresponding to λ_0). □

In particular, it follows from Proposition 1.11 that the matrix polynomials $L(\lambda)$ and $A(\lambda)L(\lambda)$ (where $A(\lambda)$ is square and $\det A(\lambda_0) \neq 0$) have the same set of Jordan chains corresponding to λ_0.

In many cases, Proposition 1.11 allows us to reduce the proofs of results concerning Jordan chains for monic matrix polynomials to the diagonal matrix

$$L(\lambda) = \operatorname{diag}[(\lambda - \lambda_0)^{\kappa_1}, \ldots, (\lambda - \lambda_0)^{\kappa_n}], \tag{1.38}$$

by using the local Smith form (see Theorem S1.10). Here we rely on the fact that $\det L(\lambda) \not\equiv 0$ for a monic matrix polynomial $L(\lambda)$ (so Theorem S1.10 is applicable).

As a demonstration of this reduction method, we shall prove the next result. It also helps to clarify the construction of a canonical set of Jordan chains given in the next section.

We shall say that an eigenvector x_0 of a monic matrix polynomial $L(\lambda)$ corresponding to the eigenvalue λ_0 is of *rank* k if the maximal order of a root polynomial $\varphi(\lambda)$ with $\varphi(\lambda_0) = x_0$ is k. This notion can be applied to nonmonic matrix polynomials as well.

Theorem 1.12. *Let $L(\lambda)$ be a monic matrix polynomial, and let λ_0 be an eigenvalue. Denote by $\mathscr{X}_k \subset \mathbb{C}^n$ $(k = 1, 2, \ldots)$ the maximal linear subspace in $\mathbb{C}^n$ with the property that all its nonzero elements are eigenvectors of $L(\lambda)$ of rank k corresponding to λ_0. Then*

(i) *there exists a finite set $\{k_1, \ldots, k_p\} \subset \{1, 2, \ldots\}$ such that $\mathscr{X}_k \neq \{0\}$ if and only if $k \in \{k_1, \ldots, k_p\}$;*

(ii) *the subspace* $\operatorname{Ker} L(\lambda_0) \subset \mathbb{C}^n$ *is decomposed into a direct sum:*

$$\operatorname{Ker} L(\lambda_0) = \mathscr{X}_{k_1} \dot{+} \cdots \dot{+} \mathscr{X}_{k_p}; \tag{1.39}$$

(iii) *if $x = \sum_{j=1}^{p} x_j \in \operatorname{Ker} L(\lambda_0)$, where $x_j \in \mathscr{X}_{k_j}$, then the rank of x is equal to* $\min\{k_j \mid x_j \neq 0\}$.

Proof. Let

$$L(\lambda) = E_{\lambda_0}(\lambda) D_{\lambda_0}(\lambda) F_{\lambda_0}(\lambda),$$

where

$$D_{\lambda_0}(\lambda) = \operatorname{diag}[(\lambda - \lambda_0)^{\kappa_i}]_{i=1}^{n}, \qquad 0 \le \kappa_1 \le \cdots \le \kappa_n$$

is the local Smith form of $L(\lambda)$ and $E_{\lambda_0}(\lambda)$ and $F_{\lambda_0}(\lambda)$ are matrix polynomials invertible at $\lambda = \lambda_0$ (Theorem S1.10). By Proposition 1.11, y_0 is an eigenvector of $L(\lambda)$ if and only if $z_0 = F_{\lambda_0}(\lambda_0) y_0$ is an eigenvector of $D_{\lambda_0}(\lambda)$ and their ranks coincide. So it is sufficient to prove Theorem 1.12 for $D_{\lambda_0}(\lambda)$.

Denote by e_i the ith coordinate vector in $\mathbb{C}^n$. Then it is easy to see that $e_i, 0, \ldots, 0$ ($\kappa_i - 1$ times zero) is a Jordan chain of $D_{\lambda_0}(\lambda)$, provided $\kappa_i \ge 1$. So the rank of the eigenvector e_i is greater than or equal to κ_i. In fact, this rank is exactly κ_i, because if there were a root polynomial $\varphi(\lambda) = \sum_{j=0}^{q} (\lambda - \lambda_0)^j \varphi_j$ with $\varphi_0 = e_i$ of order greater than κ_i, then

$$(D_{\lambda_0}(\lambda)\varphi(\lambda))^{(\kappa_i)}(\lambda_0) = 0, \tag{1.40}$$

which is contradictory (the ith entry in the left-hand side is different from zero). It follows that $\mathscr{X}_k$ is spanned by the eigenvectors e_i such that $\kappa_i = k$ (if such exist). So (i) and (ii) become evident. To prove (iii), assume that there exists a root polynomial $\varphi(\lambda)$ with $\varphi(\lambda) = x$ and of order greater than $\min\{k_j \mid x_j \neq 0\}$; then we obtain again the contradictory equality (1.15). □

The proof of Theorem 1.12 shows that the integers $k_1, \ldots, k_p$ coincide with the nonzero partial multiplicities of $L(\lambda)$ at λ_0 (see Section S1.5).

1.6. Canonical Set of Jordan Chains

We construct now a canonical set of Jordan chains of a given monic matrix polynomial $L(\lambda)$. Since det $L(\lambda)$ is a scalar polynomial not identically zero, it has only a finite number of roots. It follows that the number of different eigenvalues of $L(\lambda)$ is finite and does not exceed the degree nl of det $L(\lambda)$, where l is the degree of $L(\lambda)$.

Let λ_0 be a fixed eigenvalue of $L(\lambda)$. In the construction of a canonical set of Jordan chains given below, all the root polynomials correspond to the eigenvalue λ_0 of the matrix polynomial $L(\lambda)$. First we remark that the order of any root polynomial does not exceed the multiplicity of λ_0 as a zero of det $L(\lambda)$. Indeed, using the Smith form of $L(\lambda)$ and Proposition 1.11, it is sufficient to prove this statement for a diagonal matrix polynomial; but then it is easily checked. Let now $\varphi_1(\lambda) = \sum_{j=0}^{\kappa_1-1} (\lambda - \lambda_0)^j \varphi_{1j}$ be a root polynomial with the largest order κ_1. It follows from the preceding remark that the orders of the root polynomials are bounded above (by the multiplicity of λ_0 as a zero of det $L(\lambda)$), so such a $\varphi_1(\lambda)$ exists. Further, let $\varphi_2(\lambda) = \sum_{j=0}^{\kappa_2-1} (\lambda - \lambda_0)^j \varphi_{2j}$ be a root polynomial with the largest order among all the root polynomials whose eigenvector is not a scalar multiple of φ_{10}. (In particular, $\kappa_2 \le \kappa_1$.) If $\varphi_1(\lambda), \ldots, \varphi_{s-1}(\lambda)$ are already chosen,

$$\varphi_i(\lambda) = \sum_{j=0}^{\kappa_i - 1} (\lambda - \lambda_0)^j \varphi_{ij}, \qquad i = 1, \ldots, s-1,$$

let $\varphi_s(\lambda) = \sum_{j=0}^{\kappa_s-1} (\lambda - \lambda_0)^j \varphi_{sj}$ be a root polynomial with the largest order κ_s among all the root polynomials whose eigenvector is not in the span of the eigenvectors $\varphi_{10}, \ldots, \varphi_{s-1,0}$. We continue this process until the set Ker $L(\lambda_0)$ of all the eigenvectors of $L(\lambda)$ corresponding to λ_0 is exhausted. Thus, r root polynomials

$$\varphi_i(\lambda) = \sum_{j=0}^{\kappa_i - 1} (\lambda - \lambda_0)^j \varphi_{ij}, \qquad i = 1, \ldots, r,$$

are constructed, where $r = \dim(\operatorname{Ker} L(\lambda_0))$. In this case the Jordan chains

$$\varphi_{10}, \ldots, \varphi_{1,\kappa_1-1}, \qquad \varphi_{20}, \ldots, \varphi_{2,\kappa_2-1}, \qquad \cdots \qquad \varphi_{r0}, \ldots, \varphi_{r,\kappa_r-1}$$

are said to form a *canonical set* of Jordan chains of $L(\lambda)$ corresponding to λ_0. Note that the canonical set is not unique; for instance, we can replace $\varphi_2(\lambda)$ by $\varphi_2(\lambda) + \sum_{j=0}^{\kappa_2-1} (\lambda - \lambda_0)^j \varphi_{1j}$ in the above construction to obtain another canonical set of Jordan chains. As we shall see later, the numbers $\kappa_1, \ldots, \kappa_r$ are uniquely defined (i.e., do not depend on the choice of the canonical set). In fact, $\kappa_1, \ldots, \kappa_r$ are the nonzero partial multiplicities of $L(\lambda)$ at λ_0 (see Section S1.5).

Let us illustrate the notion of a canonical set by examples.

EXAMPLE 1.3. Let

$$L(\lambda) = (\lambda - \lambda_0)^p, \qquad \lambda_0 \in \mathbb{C}$$

be a scalar matrix polynomial, where p is a positive integer. The sole eigenvalue of $L(\lambda)$ is λ_0. From the definition of a Jordan chain it is seen that the sequence $1, 0, \ldots, 0$ ($p - 1$ zeros) forms a Jordan chain of $L(\lambda)$ corresponding to λ_0 (in this case the vectors in a Jordan chain are one-dimensional, i.e., are complex numbers). This single Jordan chain forms also a canonical set. Indeed, we have only to check that there is no Jordan chain of $L(\lambda)$ of length greater than p. If such a chain $y_0, \ldots, y_p$ were to exist, then, in particular,

$$\frac{1}{p!} L^{(p)}(\lambda_0) y_0 + \cdots + L(\lambda_0) y_p = 0.$$

But $L^{(j)}(\lambda_0) = 0$ for $j = 0, \ldots, p - 1$; $L^{(p)}(\lambda_0) = p!$ and $y \neq 0$, so this equality is contradictory (cf. the proof of Theorem 1.12). □

The scalar polynomial

$$L(\lambda) = (\lambda - \lambda_0)^p a(\lambda), \qquad a(\lambda_0) \neq 0 \tag{1.41}$$

has the same Jordan chains, and, consequently, the same canonical set of Jordan chains, as the polynomial $(\lambda - \lambda_0)^p$ (see the remark following after Proposition 1.11). Since any scalar polynomial can be represented in the form (1.41), we have obtained a description of Jordan chains for scalar polynomials, which is quite trivial. However, in view of the local Smith form for matrix polynomials (see Section S1.5), this description plays an important role in the theory.

EXAMPLE 1.4. Let

$$L(\lambda) = \begin{bmatrix} \lambda^2(\lambda - 1)(\lambda^2 + 1) & \lambda^3(\lambda - 1) \\ \lambda^2(\lambda - 1)^2 & \lambda^3(\lambda - 1)^2 \end{bmatrix}$$

be the matrix polynomial from Example 1.2. This polynomial has only two eigenvalues: $\lambda_0 = 0$ and $\lambda_0 = 1$. Let us construct a canonical set of Jordan chains for the eigenvalue $\lambda_0 = 0$. We have seen in Example 1.2 that the order of an eigenvector $\left[\begin{smallmatrix} \varphi_{10} \\ \varphi_{20} \end{smallmatrix}\right]$ is 2 if $\varphi_{10} \neq 0$ and 5 if $\varphi_{10} = 0$, $\varphi_{20} \neq 0$. So, using formula (1.32) we pick the following canonical set of Jordan chains for $\lambda_0 = 0$:

$$\begin{bmatrix} 0 \\ 1 \end{bmatrix}, \quad \begin{bmatrix} -1 \\ 0 \end{bmatrix}, \quad \begin{bmatrix} 0 \\ 0 \end{bmatrix}, \quad \begin{bmatrix} 0 \\ 0 \end{bmatrix}, \quad \begin{bmatrix} 0 \\ 0 \end{bmatrix}; \quad \begin{bmatrix} 1 \\ 0 \end{bmatrix}, \quad \begin{bmatrix} 0 \\ 0 \end{bmatrix}.$$

Another canonical set of Jordan chains would be, for instance, the following one:

$$\begin{bmatrix} 0 \\ 1 \end{bmatrix}, \quad \begin{bmatrix} -1 \\ 100 \end{bmatrix}, \quad \begin{bmatrix} -100 \\ i \end{bmatrix}, \quad \begin{bmatrix} 0 \\ 0 \end{bmatrix}, \quad \begin{bmatrix} 0 \\ 7 \end{bmatrix}; \quad \begin{bmatrix} 2i \\ 12 \end{bmatrix}, \quad \begin{bmatrix} 1 \\ 1 \end{bmatrix}.$$

Analogous considerations for the eigenvalue $\lambda_0 = 1$ gives rise, for instance, to the following canonical set of Jordan chains:

$$\varphi_{10} = \begin{bmatrix} -1 \\ 2 \end{bmatrix}, \qquad \varphi_{11} = 0; \qquad \varphi_{20} = \begin{bmatrix} 1 \\ 0 \end{bmatrix}. \quad \square$$

We shall study now some important properties of a canonical set of Jordan chains.

Proposition 1.13. *Let $L(\lambda)$ be a monic matrix polynomial. Then the lengths $\kappa_1, \ldots, \kappa_r$, of the Jordan chains in a canonical set of Jordan chains of $L(\lambda)$ corresponding to λ_0, are exactly the nonzero partial multiplicities of $L(\lambda)$ at λ_0.*

For the definition of partial multiplicities we refer to Section S1.5.

Proof. We use the local Smith form for $L(\lambda)$:

$$L(\lambda) = E_{\lambda_0}(\lambda) D_{\lambda_0}(\lambda) F_{\lambda_0}(\lambda), \tag{1.42}$$

where $E_{\lambda_0}(\lambda)$, $F_{\lambda_0}(\lambda)$ are matrix polynomials which are nonsingular at λ_0,

$$D_{\lambda_0}(\lambda) = \operatorname{diag}[(\lambda - \lambda_0)^{\nu_i}]_{i=1}^n$$

and $0 \le \nu_1 \le \cdots \le \nu_n$ are the partial multiplicities of $L(\lambda)$ at λ_0. A canonical system of Jordan chains of $D_{\lambda_0}(\lambda)$ (which is defined in the same way as for monic matrix polynomials) is easily constructed: if, for instance, $0 = \nu_1 = \cdots = \nu_{i_0 - 1} < \nu_{i_0}$, then

$$e_n, 0, \ldots, 0, \qquad e_{n-1}, 0, \ldots, 0, \qquad \cdots, \qquad e_{i_0}, 0, \ldots, 0$$

is the canonical system, where $e_j = (0, \ldots, 1, 0, \ldots, 0)^{\mathrm{T}}$ with 1 in the jth place, and the length of the chain $e_j, 0, \ldots, 0$ is ν_j $(j = i_0, i_0 + 1, \ldots, n)$. It follows immediately that Proposition 1.13 holds for $D_{\lambda_0}(\lambda)$.

On the other hand, observe that the system

$$\psi_{i0}, \ldots, \psi_{i, \kappa_i - 1}, \qquad i = 1, \ldots, r,$$

is a canonical set of Jordan chains of $L(\lambda)$ corresponding to λ_0 if and only if the system

$$\varphi_{i0}, \ldots, \varphi_{i, \kappa_i - 1}, \qquad i = 1, \ldots, r$$

is a canonical system of Jordan chains of $D_{\lambda_0}(\lambda)$ corresponding to λ_0, where

$$\varphi_{ij} = \sum_{m=0}^{j} \frac{1}{m!} F_{\lambda_0}^{(m)}(\lambda_0) \psi_{i, j-m}, \qquad j = 0, \ldots, \kappa_i - 1, \quad i = 1, \ldots, r.$$

Indeed, this follows from Proposition 1.11 and the definition of a canonical set of Jordan chains, taking into consideration that $\varphi_{i0} = F_{\lambda_0}(\lambda_0)\psi_{i0}$, $i = 1, \ldots, r$, and therefore $\psi_{10}, \ldots, \psi_{r0}$ are linearly independent if and only if

$\varphi_{10}, \ldots, \varphi_{r0}$ are linearly independent. In particular, the lengths of Jordan chains in a canonical set corresponding to λ_0 for $L(\lambda)$ and $D_{\lambda_0}(\lambda)$ are the same, and Proposition 1.13 follows. $\square$

Corollary 1.14. *The sum $\sum_{i=1}^{r} \kappa_i$ of the lengths of Jordan chains in a canonical set corresponding to an eigenvalue λ_0 of a monic matrix polynomial $L(\lambda)$ coincides with the multiplicity of λ_0 as a zero of* $\det L(\lambda)$.

The next proposition shows that a canonical system plays the role of a basis in the set of all Jordan chains of $L(\lambda)$ corresponding to a given eigenvalue λ_0.

Let μ be the length of the longest possible Jordan chain of $L(\lambda)$ corresponding to λ_0. It will be convenient to introduce into consideration the subspace $\mathcal{N} \subset \mathbb{C}^{n\mu}$ consisting of all sequences $(y_0, \ldots, y_{\mu-1})$ of n-dimensional vectors such that

$$\begin{bmatrix} L(\lambda_0) & 0 & \cdots & 0 \\ L'(\lambda_0) & L(\lambda_0) & \cdots & 0 \\ \vdots & \vdots & \cdots & \vdots \\ \frac{1}{(\mu-1)!} L^{(\mu-1)}(\lambda_0) & \frac{1}{(\mu-2)!} L^{(\mu-2)}(\lambda_0) & \cdots & L(\lambda_0) \end{bmatrix} \begin{bmatrix} y_0 \\ y_1 \\ \vdots \\ y_{\mu-1} \end{bmatrix} = 0. \quad (1.43)$$

We have already mentioned that $\mathcal{N}$ consists of Jordan chains for $L(\lambda)$ corresponding to λ_0, after we drop first zero vectors (if any) in the sequence $(y_0, \ldots, y_{\mu-1}) \in \mathcal{N}$.

Proposition 1.15. *Let*

$$\varphi_{i0}, \ldots, \varphi_{i,\mu_i-1}, \qquad i = 1, \ldots, s \quad (1.44)$$

be a set of Jordan chains of a monic matrix polynomial $L(\lambda)$ corresponding to λ_0. Then the following conditions are equivalent:

(i) *the set* (1.44) *is canonical;*

(ii) *the eigenvectors $\varphi_{10}, \ldots, \varphi_{s0}$ are linearly independent and $\sum_{i=1}^{s} \mu_i = \sigma$, the multiplicity of λ_0 as a zero of* $\det L(\lambda)$;

(iii) *the set of sequences*

$$\gamma_{ij} = (0, \ldots, 0, \varphi_{i0}, \ldots, \varphi_{ij}), \qquad j = 0, \ldots, \mu_i - 1, \quad i = 1, \ldots, s, \quad (1.45)$$

where the number of zero vectors preceding φ_{i0} in γ_{ij} is $\mu - (j + 1)$, form a basis in $\mathcal{N}$.

Proof. Again, we shall use the reduction to a local Smith form. For brevity, write $\boldsymbol{L}$ for the matrix appearing on the left of (1.43). Similarly, $\boldsymbol{A}, \boldsymbol{B}$ will denote the corresponding matrices formed from matrix polynomials

$A(\lambda)$, $B(\lambda)$ at λ_0, respectively. First let us make the following observation: let $A(\lambda)$ and $B(\lambda)$ be matrix polynomials nonsingular at $\lambda = \lambda_0$. Denote by $\tilde{\mathcal{N}} \subset \mathbb{C}^{n\mu}$ the subspace defined by formula (1.43) where $L(\lambda)$ is replaced by $\tilde{L}(\lambda) = A(\lambda)L(\lambda)B(\lambda)$. Then

$$\mathcal{N} = \boldsymbol{B}\tilde{\mathcal{N}}. \tag{1.46}$$

(Note that according to Proposition 1.11 the length of the longest Jordan chain corresponding to λ_0 of $L(\lambda)$ and $\tilde{L}(\lambda)$ is the same μ.) Indeed, (1.46) follows from the following formula (here $\tilde{\boldsymbol{L}}$ is the matrix on the left of (1.43) with $L^{(p)}(\lambda_0)$ replaced by $\tilde{L}^{(p)}(\lambda_0)$, $p = 0, \ldots, \mu - 1$):

$$\tilde{\boldsymbol{L}} = \boldsymbol{ALB},$$

which can be verified by direct computation. Now the reduction to the case where

$$L(\lambda) = \operatorname{diag}[(\lambda - \lambda_0)^{\nu_i}]_{i=1}^{n}, \qquad \nu_1 \geq \cdots \geq \nu_n \geq 0 \tag{1.47}$$

follows from the local Smith form (1.42) of $L(\lambda)$ and formula (1.46).

Thus it remains to prove Proposition 1.15 for $L(\lambda)$ given by (1.47). The part (i) $\Rightarrow$ (ii) follows from the definition of a canonical set and Corollary 1.14. Let us prove (ii) $\Rightarrow$ (iii). Let the chains (1.44) be given such that $\varphi_{10}, \ldots, \varphi_{s0}$ are linearly independent and $\sum_{i=1}^{s} \mu_i = \sigma$. From the linear independence of $\varphi_{10}, \ldots, \varphi_{s0}$ it follows that the vectors

$$\gamma_{ij} = (0, \ldots, 0, \varphi_{i0}, \ldots, \varphi_{ij}), \qquad j = 0, \ldots, \mu_i - 1, \quad i = 1, \ldots, s \tag{1.48}$$

are linearly independent in $\mathcal{N}$. On the other hand, it is easy to see that

$$\dim \mathcal{N} = \sum_{j=1}^{n} \nu_j, \tag{1.49}$$

which coincides with the multiplicity of λ_0 as a zero of $L(\lambda)$, i.e., with $\sigma = \sum_{i=1}^{s} \mu_i$. It follows that the sequences (1.48) form a basis in $\mathcal{N}$, and (iii) is proved.

It remains to prove that (iii) $\Rightarrow$ (i). Without loss of generality we can suppose that μ_i are arranged in the nonincreasing order: $\mu_1 \geq \cdots \geq \mu_s$. Suppose that $\nu_r > \nu_{r+1} = \cdots = \nu_n = 0$ (where ν_i are taken from (1.47)). Then necessarily $s = r$ and $\mu_i = \nu_i$ for $i = 1, \ldots, s$. Indeed, $s = r$ follows from the fact that the dimension of the subspace spanned by all the vectors of the form $(0, \ldots, 0, x)^{\mathrm{T}}$, which belong to $\mathcal{N}$, is just r. Further, from Theorem 1.12 it follows that the dimension of the subspace spanned by all the vectors of the form $(0, \ldots, x_0, \ldots, x_{j-1})^{\mathrm{T}}$, which belong to $\mathcal{N}$ is just

$$\sum_{p \geq j} j \cdot |\{k \mid \nu_k = p\}| + \sum_{p=1}^{j-1} p \cdot |\{k \mid \nu_k = p\}|, \qquad j = 1, \ldots, n, \tag{1.50}$$

where $|A|$ denotes the number of elements in a finite set A. On the other hand, from (iii) it follows that this dimension is

$$\sum_{p\geq j} j\cdot|\{k\,|\,\mu_k = p\}| + \sum_{p=1}^{j-1} p\cdot|\{k\,|\,\mu_k = p\}|, \qquad j = 1,\ldots,n. \tag{1.51}$$

Comparing (1.50) and (1.51), we obtain eventually that $\mu_i = \nu_i$. Now (i) follows from the definition of a canonical set of Jordan chains. □

The following Proposition will be useful sometimes in order to check that a certain set of Jordan chains is in fact a canonical one.

Proposition 1.16. *Let*

$$\varphi_{i0},\ldots,\varphi_{i,\mu_i-1}, \qquad i = 1,\ldots,s \tag{1.52}$$

be a set of Jordan chains of $L(\lambda)$ corresponding to λ_0 such that the eigenvectors $\varphi_{10},\ldots,\varphi_{i0}$ are linearly independent. Then

$$\sum_{i=1}^{s} \mu_i \leq \sigma, \tag{1.53}$$

where σ is the multiplicity of λ_0 as a zero of $\det L(\lambda)$. *The equality in* (1.53) *holds if and only if the set* (1.52) *is canonical.*

The proof of Proposition 1.16 uses the same ideas as the proof of Proposition 1.15 and is left for the reader.

1.7. Jordan Chains and the Singular Part of the Laurent Expansion

Let $L(\lambda)$ be a monic matrix polynomial, and let $\lambda_0 \in \sigma(L)$. Consider the singular part of the Laurent expansion of $L^{-1}(\lambda)$ in a neighborhood of λ_0:

$$SP(L^{-1}(\lambda)) = (\lambda-\lambda_0)^{-\nu}K_1 + (\lambda-\lambda_0)^{-\nu+1}K_2 + \cdots + (\lambda-\lambda_0)^{-1}K_\nu, \tag{1.54}$$

and define $mn \times mn$ matrices S_m $(m = 1,\ldots,\nu)$:

$$S_m = \begin{bmatrix} K_1 & 0 & \cdots & & 0 \\ K_2 & K_1 & \cdots & & 0 \\ \vdots & \vdots & \cdots & \vdots & \vdots \\ K_{m-1} & K_{m-2} & \cdots & K_1 & 0 \\ K_m & K_{m-1} & \cdots & K_2 & K_1 \end{bmatrix}. \tag{1.55}$$

It is convenient to introduce the following notation: Given $n \times n$ matrices $A_1, A_2, \ldots, A_p$, we shall denote by $\Delta(A_1, \ldots, A_p)$ the block triangular $np \times np$ matrix

$$\begin{bmatrix} A_1 & 0 & \cdots & 0 & 0 \\ A_2 & A_1 & \cdots & 0 & 0 \\ \vdots & \vdots & \cdots & \vdots & \vdots \\ A_{p-1} & A_{p-2} & \cdots & A_1 & 0 \\ A_p & A_{p-1} & \cdots & A_2 & A_1 \end{bmatrix}.$$

For example, $S_m = \Delta(K_1, K_2, \ldots, K_m)$.

For an eigenvector $x \in \operatorname{Ker} L(\lambda_0) \backslash \{0\}$, the rank of x (denoted rank x) is the maximal length of a Jordan chain of $L(\lambda)$ corresponding to λ_0 with an eigenvector x.

Proposition 1.17. *Under the notation introduced above,*

$$\operatorname{Ker} L(\lambda_0) = \{x \in \mathbb{C}^n \,|\, \operatorname{col}(0, \ldots, 0, x) \in \operatorname{Im} S_\nu\} \tag{1.56}$$

and for every $x \in \operatorname{Ker} L(\lambda_0)$,

$$\operatorname{rank} x = \max\{r \,|\, \operatorname{col}(0, \ldots, 0, x) \in \operatorname{Im} S_{\nu-r+1}\}. \tag{1.57}$$

If $x \in \operatorname{Ker} L(\lambda_0) \backslash \{0\}$, rank $x = r$ *and* $S_{\nu-r+1} \operatorname{col}(y_1, y_2, \ldots, y_{\nu-r+1}) = \operatorname{col}(0, \ldots, 0, x)$, *then the vectors*

$$\begin{gathered} x_0 = x, \qquad x_1 = K_2 y_{\nu-r+1} + K_3 y_{\nu-r} + \cdots + K_{\nu-r+2} y_1, \quad \ldots, \\ x_{r-1} = K_r y_{\nu-r+1} + K_{r+1} y_{\nu-r} + \cdots + K_\nu y_1 \end{gathered} \tag{1.58}$$

form a Jordan chain of $L(\lambda)$ *corresponding to* λ_0.

Proof. Use the local Smith form of $L(\lambda)$ in a neighborhood of λ_0 (see Section S1.5):

$$L(\lambda) = E(\lambda) D(\lambda) F(\lambda), \tag{1.59}$$

where $D(\lambda) = \operatorname{diag}[(\lambda - \lambda_0)^{\kappa_1}, \ldots, (\lambda - \lambda_0)^{\kappa_n}]$, and $E(\lambda)$, $F(\lambda)$ are matrix polynomials such that $E(\lambda_0)$ and $F(\lambda_0)$ are nonsingular (we shall not use in this proof the property that $\det F(\lambda) \equiv \text{const} \neq 0$). Using (1.58) we shall reduce the proof of Proposition 1.17 to the simple case $L(\lambda) = D(\lambda)$, when it can be checked easily.

Let

$$SP(D^{-1}(\lambda_0)) = (\lambda - \lambda_0)^{-\nu} L_1 + (\lambda - \lambda_0)^{-\nu+1} L_2 + \cdots + (\lambda - \lambda_0)^{-1} L_\nu.$$

It is easy to check that

$$S_m = \Delta(\tilde{F}_0, \tilde{F}_1, \ldots, \tilde{F}_{m-1}) \cdot \Delta(L_1, L_2, \ldots, L_m) \cdot \Delta(\tilde{E}_0, \tilde{E}_1, \ldots, \tilde{E}_{m-1}),$$
$$m = 1, \ldots, \nu,$$

where

$$\tilde{F}_j = \frac{1}{j!}(F^{-1})^{(j)}(\lambda_0), \qquad \tilde{E}_j = \frac{1}{j!}(E^{-1})^{(j)}(\lambda_0), \qquad j = 0, 1, \ldots .$$

Let now

$$\operatorname{col}(0, 0, \ldots, x) \in \operatorname{Im} S_v = \operatorname{Im}[\Delta(\tilde{F}_0, \tilde{F}_1, \ldots, \tilde{F}_{v-1}) \cdot \Delta(L_1, L_2, \ldots, L_v)],$$

so

$$\operatorname{col}(0, 0, \ldots, x) = \Delta(\tilde{F}_0, \tilde{F}_1, \ldots, \tilde{F}_{v-1}) \cdot \Delta(L_1, L_2, \ldots, L_v)\operatorname{col}(z_1, \ldots, z_v)$$

for some n-dimensional vectors $z_1, \ldots, z_v$. Multiplying this equality from the left by $\Delta(F_0, F_1, \ldots, F_{v-1})$, where $F_j = (1/j!)F^{(j)}(\lambda_0)$, we obtain that $\operatorname{col}(0, \ldots, 0, F_0 x) \in \operatorname{Im} \Delta(L_1, L_2, \ldots, L_v)$. Applying Proposition 1.17 for $D(\lambda)$ it then follows that $F_0 x \in \operatorname{Ker} D(\lambda_0)$, or $x \in \operatorname{Ker} L(\lambda_0)$. We have proved inclusion $\supset$ in (1.56). Reversing the line of argument, we obtain the opposite inclusion also, so (1.56) is proved. By an analogous argument one proves (1.57), taking into account that the rank of an eigenvector x of $L(\lambda)$ is equal to the rank of the eigenvector $F_0 x$ of $D(\lambda)$ (see Proposition 1.11).

It remains to prove that the vectors (1.58) form a Jordan chain of $L(\lambda)$ corresponding to λ_0. By the definition of $x_0, \ldots, x_{r-1}$, we have

$$\begin{bmatrix} K_1 & 0 & \cdots & 0 \\ K_2 & K_1 & \cdots & 0 \\ \vdots & \vdots & \ddots & \vdots \\ K_{v-r-1} & K_{v-r} & \cdots & K_1 \\ \vdots & \vdots & & \\ K_v & K_{v-1} & \cdots & K_r \end{bmatrix} \begin{bmatrix} y_1 \\ y_2 \\ \vdots \\ y_{v-r+1} \end{bmatrix} = \begin{bmatrix} 0 \\ 0 \\ \vdots \\ x_0 \\ x_1 \\ \vdots \\ x_{r-1} \end{bmatrix}.$$

Multiplying this equality on the left by $\Delta(A_0, A_1, \ldots, A_{v-1})$, where $A_j = (1/j!)L^{(j)}(\lambda_0)$, we obtain (in view of the equality $L(\lambda) \cdot L^{-1}(\lambda) = I$) that $\Delta(A_0, \ldots, A_{v-1}) \operatorname{col}(0, \ldots, 0, x_0, \ldots, x_{r-1}) = 0$. This equality means, by definition, that $x_0, \ldots, x_{r-1}$ is a Jordan chain of $L(\lambda)$ corresponding to λ_0. □

It is clear that Jordan chains of $L(\lambda)$ do not determine the singular part $SP(L^{-1}(\lambda_0))$ uniquely. For instance, the monic matrix polynomials $L(\lambda)$ and $\lambda L(\lambda)$ have the same Jordan chains for every nonzero eigenvalue, but different singular parts. One can show that the knowledge of Jordan chains of $L(\lambda)$ at λ_0 is equivalent to the knowledge of the subspace $\operatorname{Im} S_v$, where S_v is given by (1.55). However, as we shall see later (Corollary 2.5) by using left Jordan chains as well as the usual Jordan chains, one can recover the singular part entirely, provided it is known how to match Jordan chains with their left counterparts.

1.8. Definition of a Jordan Pair of a Monic Polynomial

Let $L(\lambda) = I\lambda^l + \sum_{j=0}^{l-1} A_j\lambda^j$ be a monic matrix polynomial. In this section, we shall describe spectral properties of $L(\lambda)$ in terms of the Jordan pairs which provide a link between linearizations and Jordan chains of $L(\lambda)$.

Recall that det $L(\lambda)$ is a scalar polynomial of degree nl; so it has exactly nl zeros (counting multiplicities). Let us denote by $\lambda_1, \ldots, \lambda_r$, the different zeros of det $L(\lambda)$, and let α_i be the multiplicity of λ_i, $i = 1, \ldots, r$. So $\sum_{i=1}^{r} \alpha_i = nl$. Clearly, the points $\lambda_1, \ldots, \lambda_r$, form the spectrum of $L(\lambda)$: $\sigma(L) = \{\lambda_1, \ldots, \lambda_r\}$.

Now for every λ_i choose a canonical set of Jordan chains of $L(\lambda)$ corresponding to λ_i:

$$\varphi_{j0}^{(i)}, \ldots, \varphi_{j,\mu_j^{(i)}-1}^{(i)}, \qquad j = 1, \ldots, s_i. \tag{1.60}$$

According to Proposition 1.15 the equality $\sum_{j=1}^{s_i} \mu_j^{(i)} = \alpha_i$ holds. It is convenient for us to write this canonical form in terms of pairs of matrices (X_i, J_i), where

$$X_i = [\varphi_{10}^{(i)} \cdots \varphi_{1,\mu_1^{(i)}-1}^{(i)}, \varphi_{20}^{(i)} \cdots \varphi_{2,\mu_2^{(i)}-1}^{(i)}, \ldots, \varphi_{s_i,0}^{(i)}, \ldots, \varphi_{s_i,\mu_{s_i}^{(i)}-1}^{(i)}]$$

is a matrix of size $n \times (\sum_{j=1}^{s_i} \mu_j^{(i)}) = n \times \alpha_i$ (we just write down the vectors from the Jordan chains (1.60) one after the other to make the columns of X_i); and J_i is block-diagonal

$$J_i = \begin{bmatrix} J_{i1} & & & 0 \\ & J_{i2} & & \\ & & \ddots & \\ 0 & & & J_{i,s_i} \end{bmatrix} = \operatorname{diag}[J_{ij}]_{j=1}^{s_i}, \tag{1.61}$$

where J_{ij} is the Jordan block of size $\mu_j^{(i)} \times \mu_j^{(i)}$ with eigenvalue λ_i.

The pair (X_i, J_i) constructed above will be called a Jordan pair of $L(\lambda)$ *corresponding to* λ_i. Clearly, it is not uniquely defined, because the canonical set of Jordan chains is not uniquely defined. We mention the following important property of the Jordan pair corresponding to λ_i:

$$A_0X_i + A_1X_iJ_i + \cdots + A_{l-1}X_iJ_i^{l-1} + X_iJ_i^l = 0, \tag{1.62}$$

which follows from Proposition 1.10.

Now we are able to give the key definition of a Jordan pair. A pair of matrices (X, J), where X is $n \times nl$ and J is an $nl \times nl$ Jordan matrix is called a *Jordan pair* for the monic matrix polynomial $L(\lambda)$ if the structure of X and J is as follows:

$$X = [X_1 \ ,\ldots,\ X_r], \qquad J = \operatorname{diag}[J_1 \ ,\ldots,\ J_r], \tag{1.63}$$

where (X_i, J_i), $i = 1, \ldots, r$, is a Jordan pair of $L(\lambda)$ corresponding to λ_i.

In the sequel, a block-row matrix

$$[Z_1 \quad \cdots \quad Z_p]$$

will often be written in the form $\operatorname{row}(Z_i)_{i=1}^p$. For example, the matrix X from (1.63) will be written as $\operatorname{row}(X_i)_{i=1}^r$.

For a Jordan pair (X, J) of $L(\lambda)$ an equality analogous to (1.62) holds also:

$$A_0 X + A_1 XJ + \cdots + A_{l-1} XJ^{l-1} + XJ^l = 0. \tag{1.64}$$

Let us illustrate the notion of a Jordan pair by a simple example.

EXAMPLE 1.5. Let

$$L(\lambda) = \begin{bmatrix} \lambda^3 & \sqrt{2}\lambda^2 - \lambda \\ \sqrt{2}\lambda^2 + \lambda & \lambda^3 \end{bmatrix}$$

Compute

$$\det L(\lambda) = \lambda^2(\lambda - 1)^2(\lambda + 1)^2;$$

so $\sigma(L) = \{0, 1, -1\}$.

Let us compute the Jordan pairs (X_0, J_0), (X_1, J_1), and (X_{-1}, J_{-1}) corresponding to the eigenvalues 0, 1, and -1, respectively.

Note that $L(0) = 0$; so there exists a set of two linearly independent eigenvectors $\varphi_1^{(0)}, \varphi_2^{(0)}$ corresponding to $\lambda_0 = 0$. For instance, we can take

$$\varphi_1^{(0)} = \begin{bmatrix} 1 \\ 0 \end{bmatrix}, \qquad \varphi_2^{(0)} = \begin{bmatrix} 0 \\ 1 \end{bmatrix}. \tag{1.65}$$

In view of Proposition 1.15, the set (1.65) is canonical. So

$$X_0 = \begin{bmatrix} 1 & 0 \\ 0 & 1 \end{bmatrix}, \qquad J_0 = \begin{bmatrix} 0 & 0 \\ 0 & 0 \end{bmatrix}$$

is a Jordan pair of $L(\lambda)$ corresponding to the eigenvalue zero.

Consider now the eigenvalue 1.

$$L(1) = \begin{bmatrix} 1 & \sqrt{2} - 1 \\ \sqrt{2} + 1 & 1 \end{bmatrix},$$

and the corresponding eigenvector is

$$\varphi_{10}^{(1)} = \begin{bmatrix} -\sqrt{2} + 1 \\ 1 \end{bmatrix}.$$

To find the first generalized eigenvector $\varphi_{11}^{(1)}$, let us solve the equation

$$L'(1)\varphi_{10}^{(1)} + L(1)\varphi_{11}^{(1)} = 0,$$

or

$$\begin{bmatrix} 3 & 2\sqrt{2} - 1 \\ 2\sqrt{2} + 1 & 3 \end{bmatrix} \begin{bmatrix} -\sqrt{2} + 1 \\ 1 \end{bmatrix} + \begin{bmatrix} 1 & \sqrt{2} - 1 \\ \sqrt{2} + 1 & 1 \end{bmatrix} \varphi_{11}^{(1)} = 0. \tag{1.66}$$

This equation has many solutions (we can expect this fact in advance, since if $\varphi_{11}^{(1)}$ is a solution of (1.66), so is $\varphi_{11}^{(1)} + \alpha\varphi_{10}^{(1)}$, for any $\alpha \in \mathbb{C}$). As the first generalized eigenvector $\varphi_{11}^{(1)}$, we can take any solution of (1.66); let us take

$$\varphi_{11}^{(1)} = \begin{bmatrix} \sqrt{2} - 2 \\ 0 \end{bmatrix}.$$

So,

$$X_1 = \begin{bmatrix} -\sqrt{2} + 1 & \sqrt{2} - 2 \\ 1 & 0 \end{bmatrix}, \qquad J_1 = \begin{bmatrix} 1 & 1 \\ 0 & 1 \end{bmatrix}.$$

Consider now the eigenvalue -1. Since

$$L(-1) = \begin{bmatrix} -1 & \sqrt{2} + 1 \\ \sqrt{2} - 1 & -1 \end{bmatrix},$$

as an eigenvector we can take

$$\varphi_{10}^{(-1)} = \begin{bmatrix} \sqrt{2} + 1 \\ 1 \end{bmatrix}.$$

The first generalized eigenvector $\varphi_{11}^{(-1)}$ is defined from the equation

$$L'(-1)\varphi_{10}^{(-1)} + L(-1)\varphi_{11}^{(-1)} = 0,$$

and a computation shows that we can take

$$\varphi_{11}^{(-1)} = \begin{bmatrix} \sqrt{2} + 2 \\ 0 \end{bmatrix}.$$

So the Jordan pair of $L(\lambda)$ corresponding to the eigenvalue -1 is

$$X_{-1} = \begin{bmatrix} \sqrt{2} + 1 & \sqrt{2} + 2 \\ 1 & 0 \end{bmatrix}, \qquad J_{-1} = \begin{bmatrix} -1 & 1 \\ 0 & -1 \end{bmatrix}.$$

Finally, we write down a Jordan pair (X, J) for $L(\lambda)$:

$$X = \begin{bmatrix} 1 & 0 & -\sqrt{2} + 1 & \sqrt{2} - 2 & \sqrt{2} + 1 & \sqrt{2} + 2 \\ 0 & 1 & 1 & 0 & 1 & 0 \end{bmatrix},$$

$$J = \begin{bmatrix} 0 & & & & & \\ & 0 & & & & \\ & & 1 & 1 & & \\ & & & 1 & & \\ & & & & -1 & 1 \\ & & & & & -1 \end{bmatrix}.$$

One can now check equality (1.64) for Example 1.5 by direct computation. □

Consider now the scalar case, i.e., $n = 1$. In this case the Jordan pair is especially simple and, in fact, does not give any additional information apart from the multiplicity of an eigenvalue.

Proposition 1.18. *Let $L(\lambda)$ be a monic scalar polynomial, and λ_0 be an eigenvalue of $L(\lambda)$ of multiplicity v. Then*

$$X_{\lambda_0} = [1 \quad 0 \quad \cdots \quad 0], \qquad J_{\lambda_0} = \begin{bmatrix} \lambda_0 & 1 & & 0 \\ & \lambda_0 & \ddots & \\ & & \ddots & 1 \\ 0 & & & \lambda_0 \end{bmatrix}$$

is a Jordan pair of $L(\lambda)$ corresponding to λ_0. Here X_{λ_0} is $1 \times v$ and J_{λ_0} is the $v \times v$ Jordan block.

Proof. Let $L(\lambda) = (\lambda - \lambda_0)^v \cdot L_1(\lambda)$, where $L_1(\lambda)$ is a scalar polynomial not divisible by $\lambda - \lambda_0$. It is clear that $1, 0, \ldots, 0$ ($v - 1$ zeros) is a Jordan chain of $L(\lambda)$ corresponding to λ_0. By Proposition 1.15, this single Jordan chain forms a canonical set of Jordan chains corresponding to λ_0. Now Proposition 1.18 follows from the definition of a Jordan pair. □

In general, the columns of the matrix X from a Jordan pair (X, J) of the monic matrix polynomial $L(\lambda)$ are not linearly independent (because of the obvious reason that the number of columns in X is nl, which is greater than the dimension n of each column of X, if the degree l exceeds one). It turns out that the condition $l = 1$ is necessary and sufficient for the columns of X to be linearly independent.

Theorem 1.19. *Let $L(\lambda)$ be a monic matrix polynomial with Jordan pair (X, J). Then the columns of X are linearly independent if and only if the degree l of $L(\lambda)$ is equal to 1.*

Proof. Suppose $L(\lambda) = I\lambda - A$ is of degree 1, and let (X, J) be its Jordan pair. From the definition of a Jordan pair, it is clear that the columns of X form a Jordan basis for A, and therefore they are linearly independent.

Conversely, if the degree of $L(\lambda)$ exceeds 1, then the columns of X cannot be linearly independent, as we have seen above. □

1.9. Properties of a Jordan Pair

The following theorem describes the crucial property of a Jordan pair.

Theorem 1.20. *Let (X, J) be a Jordan pair of a monic matrix polynomial $L(\lambda) = I\lambda^l + \sum_{j=0}^{l-1} A_j \lambda^j$. Then the matrix of size $nl \times nl$*

$$\operatorname{col}(XJ^i)_{i=0}^{l-1} = \begin{bmatrix} X \\ XJ \\ \vdots \\ XJ^{l-1} \end{bmatrix}$$

is nonsingular.

Proof. Let C_1 be the first companion matrix for $L(\lambda)$ (as in Theorem 1.1). Then the equality

$$C_1 \cdot \operatorname{col}(XJ^i)_{i=0}^{l-1} = \operatorname{col}(XJ^i)_{i=0}^{l-1} \cdot J \tag{1.67}$$

holds. Indeed, the equality in all but the last block row in (1.67) is evident; the equality in the last block row follows from (1.64). Equality (1.67) shows that the columns of $\operatorname{col}(XJ^i)_{i=0}^{l-1}$ form Jordan chains for the matrix polynomial $I\lambda - C_1$.

Let us show that these Jordan chains of $I\lambda - C_1$ form a canonical set for every eigenvalue λ_0 of $I\lambda - C_1$. First note that in view of Theorem 1.1, $\sigma(L(\lambda)) = \sigma(I\lambda - C_1)$, and the multiplicity of every eigenvalue λ_0 is the same, when λ_0 is regarded as an eigenvalue of $L(\lambda)$ or of $I\lambda - C_1$. Thus, the number of vectors in the columns of $\operatorname{col}(XJ^i)_{i=0}^{l-1}$ corresponding to some $\lambda_0 \in \sigma(I\lambda - C_1)$ coincides with the multiplicity of λ_0 as a zero of $\det(I\lambda - C_1)$. Let $x_1, \ldots, x_k$ be the eigenvectors of $L(\lambda)$ corresponding to λ_0 and which appear as columns in the matrix X; then $\operatorname{col}(\lambda_0^i x_1)_{i=0}^{l-1}, \ldots, \operatorname{col}(\lambda_0^i x_k)_{i=0}^{l-1}$ will be the corresponding columns in $\operatorname{col}(XJ^i)_{i=0}^{l-1}$. By Proposition 1.15, in order to prove that $\operatorname{col}(XJ^i)_{i=0}^{l-1}$ forms a canonical set of Jordan chains of $I\lambda - C_1$ corresponding to λ_0 we have only to check that the vectors

$$\operatorname{col}(\lambda_0^i x_j)_{i=0}^{l-1} \in \mathbb{C}^{nl}, \qquad j = 1, \ldots, k$$

are linearly independent. But this is evident, since by the construction of a Jordan pair it is clear that the vectors $x_1, \ldots, x_k$, are linearly independent.

So $\operatorname{col}(XJ^i)_{i=0}^{l-1}$ forms a canonical set of Jordan chains for $I\lambda - C_1$, and by Theorem 1.19, the nonsingularity of $\operatorname{col}(XJ^i)_{i=0}^{l-1}$ follows. □

Corollary 1.21. *If (X, J) is a Jordan pair of a monic matrix polynomial $L(\lambda)$, and C_1 is its companion matrix, then J is similar to C_1.*

Proof. Use (1.67) bearing in mind that, according to Theorem 1.20, $\operatorname{col}(XJ^i)_{i=0}^{l-1}$ is invertible. □

Theorem 1.20 allows us to prove the following important property of Jordan pairs.

Theorem 1.22. *Let (X, J) and (X_1, J_1) be Jordan pairs of the monic matrix polynomial $L(\lambda)$. Then they are similar, i.e.,*

$$X = X_1 S, \qquad J = S^{-1} J_1 S, \tag{1.68}$$

where

$$S = [\operatorname{col}(X_1 J_1^i)_{i=0}^{l-1}]^{-1} \operatorname{col}(XJ^i)_{i=0}^{l-1}. \tag{1.69}$$

We already know that (when J is kept fixed) the matrix X from a Jordan pair (X, J) is not defined uniquely, because a canonical set of Jordan chains is not defined uniquely. The matrix J itself is not unique either, but, as Corollary 1.21 indicates, J is similar to C_1. It follows that J and J_1 from Theorem 1.22 are also similar, and therefore one of them can be obtained from the second by a permutation of its Jordan blocks.

Proof. Applying (1.67) for (X, J) and (X_1, J_1), we have

$$C_1 \cdot \operatorname{col}(X_1 J_1^i)_{i=0}^{l-1} = \operatorname{col}(X_1 J_1)_{i=0}^{l-1} \cdot J_1,$$
$$C_1 \cdot \operatorname{col}(XJ^i)_{i=0}^{l-1} = \operatorname{col}(XJ^i)_{i=0}^{l-1} \cdot J.$$

Substituting C_1 from the second equality in the first one, it is found that

$$\operatorname{col}(XJ^i)_{i=0}^{l-1} \cdot J[\operatorname{col}(XJ^i)_{i=0}^{l-1}]^{-1} \operatorname{col}(X_1 J_1^i)_{i=0}^{l-1} = \operatorname{col}(X_1 J_1^i)_{i=0}^{l-1} \cdot J_1.$$

Straightforward calculation shows that for S given by (1.69), the equalities (1.68) hold true. □

Theorem 1.22 shows that a Jordan pair of a monic matrix polynomial is essentially unique (up to similarity). It turns out that the properties described in equality (1.64) and Theorem 1.20 are characteristic for the Jordan pair, as the following result shows.

Theorem 1.23. *Let (X, J) be a pair of matrices, where X is of size $n \times nl$ and J is a Jordan matrix of size $nl \times nl$. Then (X, J) is a Jordan pair of the monic matrix polynomial $L(\lambda) = I\lambda^l + \sum_{j=0}^{l-1} A_j \lambda^j$ if and only if the two following conditions hold*:

(i) $\operatorname{col}(XJ^i)_{i=0}^{l-1}$ *is a nonsingular $nl \times nl$ matrix*;
(ii) $A_0 X + A_1 XJ + \cdots + A_{l-1} XJ^{l-1} + XJ^l = 0.$

Proof. In view of (1.64) and Theorem 1.20, conditions (i) and (ii) are necessary. Let us prove their sufficiency.

Let $J_0 = \operatorname{diag}[J_{01}, \ldots, J_{0k}]$ be the part of J corresponding to some eigenvalue λ_0 with Jordan blocks $J_{01}, \ldots, J_{0k}$. Let $X_0 = [X_{01}, \ldots, X_{0k}]$ be the part of X corresponding to J_0 and partitioned into blocks $X_{01}, \ldots, X_{0k}$ according to the partition of $J_0 = \operatorname{diag}[J_{01}, \ldots, J_{0k}]$ into Jordan blocks $J_{01}, \ldots, J_{0k}$. Then the equalities

$$A_0 X_{0i} + A_1 X_{0i} J_{0i} + \cdots + A_{l-1} X_{0i} J_{0i}^{l-1} + X_{0i} J_{0i}^l = 0$$

hold for $i = 1, 2, \ldots, k$. According to Proposition 1.10, the columns of X_{0i}, $i = 1, \ldots, k$, form a Jordan chain of $L(\lambda)$ corresponding to λ_0.

Let us check that the eigenvectors $f_1, \ldots, f_k$ of the blocks $X_{01}, \ldots, X_{0k}$, respectively, are linearly independent. Suppose not; then

$$\sum_{i=1}^{k} \alpha_i f_i = 0,$$

where the $\alpha_i \in \mathbb{C}$ are not all zero. Then also $\sum_{i=1}^{k} \alpha_i \lambda_0^j f_i = 0$ for $j = 1, \ldots, l-1$, and therefore the columns of the matrix

$$\operatorname{col}(\lambda_0^j f_1, \ldots, \lambda_0^j f_k)_{j=0}^{l-1}$$

are linearly dependent. But this contradicts (i) since $\operatorname{col}(\lambda_0^j f_i)_{j=0}^{l-1}$ is the left-most column in the part $\operatorname{col}(X_{0i} J_{0i}^j)_{j=0}^{l-1}$ of $\operatorname{col}(X J^j)_{j=0}^{l-1}$.

In order to prove that the Jordan chains $X_{01}, \ldots, X_{0k}$ represent a canonical set of Jordan chains corresponding to λ_0, we have to prove only (in view of Proposition 1.15) that $\sum_{i=1}^{k} \mu_{0i} = \sigma$, where $n \times \mu_{0i}$ is the size of X_{0i} and σ is the multiplicity of λ_0 as a zero of $\det L(\lambda)$. This can be done using Proposition 1.16. Indeed, let $\lambda_1, \ldots, \lambda_q$ be all the different eigenvalues of $L(\lambda)$, and let μ_{pi}, $i = 1, \ldots, k_p$, be the sizes of Jordan blocks of J corresponding to λ_p, $p = 1, \ldots, q$. If σ_p is the multiplicity of λ_p as a zero of $\det L(\lambda)$, then

$$\sum_{i=1}^{k_p} \mu_{pi} \leq \sigma_p, \qquad p = 1, \ldots, q \tag{1.70}$$

by Proposition 1.16. On the other hand, $\sum_{p=1}^{q} \sigma_p = \sum_{p=1}^{q} \sum_{i=1}^{k_p} \mu_{pi} = nl$, so that in (1.70) we have equalities for every $p = 1, \ldots, q$. Again, by Proposition 1.16, we deduce that the system $X_{01}, \ldots, X_{0k}$ of Jordan chains corresponding to λ_0 (where λ_0 is an arbitrary eigenvalue) is canonical. □

1.10. Standard Pairs of a Monic Matrix Polynomial

Theorem 1.23 suggests that the spectral properties of a monic matrix polynomial $L(\lambda) = I\lambda^l + \sum_{j=0}^{l-1} A_j \lambda^j$ can be expressed in terms of the pair of matrices (X, J) such that (i) and (ii) of Theorem 1.23 hold. The requirement that J is Jordan is not essential, since we can replace the pair (X, J) by $(XS, S^{-1}JS)$ for any invertible $nl \times nl$ matrix S and still maintain conditions (i) and (ii). Hence, if J is not in Jordan form, by suitable choice of S the matrix $S^{-1}JS$ can be put into Jordan form. This remark leads us to the following important definition.

A pair of matrices (X, T), where X is $n \times nl$ and T is $nl \times nl$, is called a *standard pair* for $L(\lambda)$ if the following conditions are satisfied:

(i) $\operatorname{col}(XT^i)_{i=0}^{l-1}$ is nonsingular;
(ii) $\sum_{i=0}^{l-1} A_i X T^i + XT^l = 0$.

It follows from Theorem 1.20 and formula (1.64), that every Jordan pair is standard. Moreover, Theorem 1.23 implies that any standard pair (X, T), where T is a Jordan matrix, is in fact a Jordan pair.

An important example of a standard pair appears naturally from a linearization of $L(\lambda)$, as the following theorem shows (cf. Theorem 1.1).

Theorem 1.24. *Let* $P_1 = [I \quad 0 \quad \cdots \quad 0]$ *and let*

$$C_1 = \begin{bmatrix} 0 & I & & \cdots & 0 \\ 0 & 0 & I & \cdots & 0 \\ \vdots & \vdots & \vdots & \cdots & \vdots \\ 0 & 0 & 0 & \cdots & I \\ -A_0 & -A_1 & & \cdots & -A_{l-1} \end{bmatrix}$$

be the companion matrix of $L(\lambda)$. *Then* (P_1, C_1) *is a standard pair of* $L(\lambda)$.

Proof. Direct computation shows that

$$P_1 C_1^i = [0 \quad \cdots \quad 0 \quad I \quad 0 \quad \cdots \quad 0], \qquad i = 0, \ldots, l-1$$

with I in the $(i+1)$th place, and

$$P_1 C_1^l = [-A_0 \quad -A_1 \quad \cdots \quad -A_l].$$

So

$$\operatorname{col}(P_1 C_1^i)_{i=0}^{l-1} = I,$$

and condition (i) is trivially satisfied. Further,

$$\sum_{i=0}^{l-1} A_i P_1 C_1^i + P_1 C_1^l = \sum_{i=0}^{l-1} A_i [0 \quad \cdots \quad 0 \quad I \quad 0 \quad \cdots \quad 0]$$
$$+ [-A_0 \quad -A_1 \quad \cdots \quad -A_l] = 0,$$

and (ii) follows also. □

EXAMPLE 1.6. For the matrix $L(\lambda)$ from Example 1.5 the pair (P_1, C_1) looks as follows:

$$P_1 = \begin{bmatrix} 1 & 0 & 0 & 0 & 0 & 0 \\ 0 & 1 & 0 & 0 & 0 & 0 \end{bmatrix}, \qquad C_1 = \begin{bmatrix} 0 & 0 & 1 & 0 & 0 & 0 \\ 0 & 0 & 0 & 1 & 0 & 0 \\ 0 & 0 & 0 & 0 & 1 & 0 \\ 0 & 0 & 0 & 0 & 0 & 1 \\ 0 & 0 & 0 & 1 & 0 & -\sqrt{2} \\ 0 & 0 & -1 & 0 & -\sqrt{2} & 0 \end{bmatrix}. \quad \square$$

Observe the following important formula for a standard pair (X, T) of $L(\lambda)$:

$$C_1 \cdot \operatorname{col}(XT^i)_{i=0}^{l-1} = \operatorname{col}(XT^i)_{i=0}^{l-1} \cdot T. \tag{1.71}$$

So T is similar to C_1. The following result shows in particular that this transformation applies to X also, in an appropriate sense.

Theorem 1.25. *Any two standard pairs (X, T) and (X', T') of $L(\lambda)$ are similar, i.e., there exists an invertible $nl \times nl$ matrix S such that*

$$X' = XS, \qquad T' = S^{-1}TS. \tag{1.72}$$

The matrix S is defined uniquely by (X, T) and (X', T') and is given by the formula

$$S = [\operatorname{col}(XT^i)_{i=0}^{l-1}]^{-1} \cdot \operatorname{col}(X'T'^i)_{i=0}^{l-1}. \tag{1.73}$$

Conversely, if for a given standard pair (X, T) a new pair of matrices (X', T') is defined by (1.72) *(with some invertible matrix S), then (X', T') is also a standard pair for $L(\lambda)$.*

Proof. The second assertion is immediate: If (i) and (ii) are satisfied for some pair (X, T) (X and T of sizes $n \times nl$ and $nl \times nl$, respectively), then (i) and (ii) are satisfied for any pair similar to (X, T).

To prove the first statement, write

$$C_1 \cdot \operatorname{col}(XT^i)_{i=0}^{l-1} = \operatorname{col}(XT^i)_{i=0}^{l-1} \cdot T,$$
$$C_1 \cdot \operatorname{col}(X'T'^i)_{i=0}^{l-1} = \operatorname{col}(X'T'^i)_{i=0}^{l-1} \cdot T',$$

and comparison of these equalities gives

$$T' = S^{-1}TS$$

with S given by (1.73).

From the definition of the inverse matrix it follows that

$$X \cdot [\operatorname{col}(XT^i)_{i=0}^{l-1}]^{-1} = [I \quad 0 \quad \cdots \quad 0];$$

so

$$XS = X \cdot [\operatorname{col}(XT^i)_{i=0}^{l-1}]^{-1} \cdot \operatorname{col}(X'T'^i)_{i=0}^{l-1} = X',$$

and (1.72) follows with S given by (1.73). Uniqueness of S follows now from the relation

$$\operatorname{col}(X'T'^i)_{i=0}^{l-1} = \operatorname{col}(XT^i)_{i=0}^{l-1} \cdot S. \quad \square$$

Until now we have considered polynomials whose coefficients are matrices with complex entries. Sometimes, however, it is more convenient to accept the point of view of linear operators acting in a finite dimensional complex space. In this way we are led to the consideration of operator polynomials $L(\lambda) = I\lambda^l + \sum_{j=0}^{l-1} A_j\lambda^j$, where $A_j: \mathbb{C}^n \to \mathbb{C}^n$ is a linear transformation (or operator). Clearly, there exists a close connection between the matricial and operator points of view: a matrix polynomial can be considered as an operator polynomial, when its coefficients are regarded as linear transformations generated by the matricial coefficients in a fixed basis (usually the standard orthonormal

$(1, 0, \ldots, 0)^T, (0, 1, \ldots, 0)^T, \ldots, (0, \ldots, 0, 1)^T)$. Conversely, every operator polynomial gives rise to a matrix polynomial, when its coefficients are written as a matrix representation in a fixed (usually standard orthogonal) basis.

So in fact every definition and statement concerning matrix polynomials can be given an operator polynomial form and vice versa. In applying this rule we suppose that the powers $(\mathbb{C}^n)^l$ are considered as Euclidean spaces with the usual scalar product:

$$[(x_1, \ldots, x_l), (y_1, \ldots, y_l)] = \sum_{i=1}^{l} (x_i, y_i),$$

where $x_i, y_j \in \mathbb{C}^n$. For example, a pair of linear transformations (X, T), where $X: \mathbb{C}^n \to \mathbb{C}^n$, $T: \mathbb{C}^{nl} \to \mathbb{C}^{nl}$ is a standard pair for the operator polynomial $L(\lambda)$ if $\mathrm{col}(XT^i)_{i=0}^{l-1}$ is an invertible linear transformation and

$$\sum_{i=0}^{l-1} A_i XT^i + XT^l = 0.$$

Theorem 1.25 in the framework of operator polynomials will sound as follows: Any two standard pairs (X, T) and (X', T') of the operator polynomial $L(\lambda)$ are similar, i.e., there exists an invertible linear transformation $S: \mathbb{C}^{nl} \to \mathbb{C}^{nl}$ such that $X' = XS$ and $T' = S^{-1}TS$. In the sequel we shall often give the definitions and statements for matrix polynomials only, bearing in mind that it is a trivial exercise to carry out these definitions and statements for the operator polynomial approach.

Comments

A good source for the theory of linearization of analytic operator functions, for the history, and for further references is [28]. Further developments appear in [7, 66a, 66b]. The contents of Section 1.3 (solution of the inverse problem for linearization) originated in [3a].

The notion of a Jordan chain is basic. For polynomials of type $I\lambda - A$, where A is a square matrix, this notion is well known from linear algebra (see, for instance, [22]). One of the earliest systematic uses of these chains of generalized eigenvectors is by Keldysh [49a, 49b], hence the name Keldysh chains often found in the literature. Keldysh analysis is in the context of infinite-dimensional spaces. Detailed exposition of his results can be found in [32b, Chapter V]. Further development of the theory of chains in the infinite-dimensional case for analytic and meromorphic functions appears in [60, 24, 38]; see also [5]. Proposition 1.17 appears in [24].

The last three sections are taken from the authors' papers [34a, 34b]. Properties of standard pairs (in the context of operator polynomials) are used in [76a] to study certain problems in partial differential equations.

Chapter 2

Representation of Monic Matrix Polynomials

In Chapter 1, a language and formalism have been developed for the full description of eigenvalues, eigenvectors, and Jordan chains of matrix polynomials. In this chapter, triples of matrices will be introduced which determine completely all the spectral information about a matrix polynomial. It will then be shown how these triples can be used to solve the inverse problem, namely, given the spectral data to determine the coefficient matrices of the polynomial. Results of this (and a related) kind are given by the representation theorems. They will lead to important applications to constant coefficient differential and difference equations. In essence, these applications yield closed form solutions to boundary and initial value problems in terms of the spectral properties of an underlying matrix polynomial.

2.1. Standard and Jordan Triples

Let $L(\lambda) = I\lambda^l + \sum_{j=0}^{l-1} A_j\lambda^j$ be a monic matrix polynomial with standard pair (X, T). By definition (Section 1.10) of such a pair a third matrix Y of size

$nl \times n$ can be defined by

$$Y = \begin{bmatrix} X \\ XT \\ \vdots \\ XT^{l-1} \end{bmatrix}^{-1} \begin{bmatrix} 0 \\ \vdots \\ 0 \\ I \end{bmatrix}. \tag{2.1}$$

Then (X, T, Y) is called a *standard triple* for $L(\lambda)$. If $T = J$ is a Jordan matrix (so that the standard pair (X, T) is Jordan), the triple (X, T, Y) will be called *Jordan.*

From Theorem 1.25 and the definition of a standard triple it follows immediately that two standard triples (X, T, Y) and (X', T', Y') are *similar*, i.e., for some $nl \times nl$ invertible matrix S the relations

$$X' = XS, \qquad T' = S^{-1}TS, \qquad Y' = S^{-1}Y \tag{2.2}$$

hold. The converse statement is also true: if (X, T, Y) is a standard triple of $L(\lambda)$, and (X', T', Y') is a triple of matrices, with sizes of X', T', and Y' equal to $n \times nl, nl \times nl$, and $nl \times n$, respectively, such that (2.2) holds, then (X', T', Y') is a standard triple for $L(\lambda)$. This property is very useful since it allows us to reduce the proofs to cases in which the standard triple is chosen to be especially simple. For example, one may choose the triple (based on the companion matrix C_1)

$$X = [I \quad 0 \quad \cdots \quad 0], \qquad T = C_1, \qquad Y = \begin{bmatrix} 0 \\ \vdots \\ 0 \\ I \end{bmatrix}. \tag{2.3}$$

In the next sections we shall develop representations of $L(\lambda)$ via its standard triples, as well as some explicit formulas (in terms of the spectral triple of $L(\lambda)$) for solution of initial and boundary value problems for differential equations with constant coefficients. Here we shall establish some simple but useful properties of standard triples.

First recall the second companion matrix C_2 of $L(\lambda)$ defined by

$$C_2 = \begin{bmatrix} 0 & 0 & \cdots & 0 & -A_0 \\ I & 0 & \cdots & 0 & -A_1 \\ 0 & I & \cdots & 0 & -A_2 \\ \vdots & \vdots & \cdots & \vdots & \vdots \\ 0 & 0 & \cdots & I & -A_{l-1} \end{bmatrix}. \tag{2.4}$$

The following equality is verified by direct multiplication:

$$C_2 = BC_1B^{-1},$$

where

$$B = \begin{bmatrix} A_1 & A_2 & \cdots & A_{l-1} & I \\ A_2 & & \cdot^{\cdot^{\cdot}} & I & 0 \\ \vdots & \vdots & \cdot^{\cdot^{\cdot}} & \vdots & \vdots \\ A_{l-1} & I & & & \\ I & 0 & \cdots & & 0 \end{bmatrix} \tag{2.5}$$

is an invertible $nl \times nl$ matrix. In particular, C_2 is, like C_1, a linearization of $L(\lambda)$ (refer to Proposition 1.3).

Define now an $nl \times nl$ invertible matrix R from the equality $R = (BQ)^{-1}$ with B defined by (2.5) and $Q = \operatorname{col}(XT^i)_{i=0}^{l-1}$ or, what is equivalent,

$$RBQ = I. \tag{2.6}$$

We shall refer to this as the *biorthogonality condition* for R and Q.

We now have

$$C_2 = BC_1B^{-1} = BQTQ^{-1}B^{-1} = R^{-1}TR,$$

whence

$$RC_2 = TR. \tag{2.7}$$

Now represent R as a block row $[R_1 \quad \cdots \quad R_l]$, where R_i is an $nl \times n$ matrix. First observe that $R_1 = Y$, with Y defined by (2.1). Indeed,

$$R_1 = Q^{-1}B^{-1}\begin{bmatrix} I \\ 0 \\ \vdots \\ 0 \end{bmatrix}, \tag{2.8}$$

where the matrix B^{-1} has the following form (as can be checked directly):

$$B^{-1} = \begin{bmatrix} 0 & & \cdots & 0 & B_0 \\ \vdots & & \cdot^{\cdot^{\cdot}} & & B_1 \\ & & \cdot^{\cdot^{\cdot}} & & \vdots \\ 0 & B_0 & \cdot^{\cdot^{\cdot}} & & \\ B_0 & B_1 & & \cdots & B_{l-1} \end{bmatrix} \tag{2.9}$$

with $B_0 = I$ and $B_1, \ldots, B_{l-1}$ defined recursively by

$$B_{r+1} = -(A_{l-1}B_r + A_{l-2}B_{r-1} + \cdots + A_{l-r-1}B_0), \qquad r = 0, \ldots, l-2.$$

Substituting (2.9) in (2.8) yields

$$R_1 = Q^{-1}\begin{bmatrix} 0 \\ \vdots \\ 0 \\ I \end{bmatrix} = Y$$

by definition. Further, (2.7) gives (taking into account the structure of C_2)

$$R_i = T^{i-1}Y \qquad \text{for} \quad i = 1, \ldots, l$$

and

$$YA_0 + TYA_1 + \cdots + T^{l-1}YA_{l-1} + T^lY = 0. \tag{2.10}$$

We have, in particular, that for any standard triple (X, T, Y) the $nl \times nl$ matrix row $(T^iY)_{i=0}^{l-1}$ is invertible.

Note the following useful equalities:

$$\begin{bmatrix} X \\ XT \\ \vdots \\ XT^{l-1} \end{bmatrix} [Y \quad TY \quad \cdots \quad T^{l-1}Y] = B^{-1}, \tag{2.11}$$

and B^{-1} is given by (2.9). In particular,

$$XT^iY = \begin{cases} 0 & \text{for} \quad i = 0, \ldots, l-2 \\ I & \text{for} \quad i = l-1. \end{cases} \tag{2.12}$$

We summarize the main information observed above, in the following statement:

Proposition 2.1. *If (X, T, Y) is a standard triple of $L(\lambda)$, then*:

(i) *row* $(T^jY)_{j=0}^{l-1} = [Y \quad TY \quad \cdots \quad T^{l-1}Y]$ *is an $nl \times nl$ nonsingular matrix*,

(ii) *X is uniquely defined by T and Y*: $X = [0 \quad \cdots \quad 0 \quad I] \cdot [\text{row}\ (T^jY)_{j=0}^{l-1}]^{-1}$,

(iii) $YA_0 + TYA_1 + \cdots + T^{l-1}YA_{l-1} + T^lY = 0.$

Proof. Parts (i) and (iii) were proved above. Part (ii) is just equalities (2.12) written in a different form. □

For the purpose of further reference we define the notion of a left standard pair, which is dual to the notion of the standard pair. A pair of matrices (T, Y), where T is $nl \times nl$ and Y is $nl \times n$, is called a *left standard* pair for the monic matrix polynomial $L(\lambda)$ if conditions (i) and (iii) of Proposition 2.1 hold. An equivalent definition is (T, Y) is a left standard pair of $L(\lambda)$ if (X, T, Y) is a standard triple of $L(\lambda)$, where X is defined by (ii) of Proposition 2.1. Note that (T, Y) is a left standard pair for $L(\lambda)$ iff $(Y^{\mathrm{T}}, T^{\mathrm{T}})$ is a (usual) standard pair for $L^{\mathrm{T}}(\lambda)$, so every statement on standard pairs has its dual statement for the left standard pairs. For this reason we shall use left standard pairs only occasionally and omit proofs of the dual statements, once the proof of the original statement for a standard pair is supplied.

We remark here that the notion of the second companion matrix C_2 allows us to define another simple standard triple (X_0, T_0, Y_0), which is in a certain sense dual to (2.3), as

$$X_0 = [0 \quad \cdots \quad 0 \quad I], \qquad T_0 = C_2, \qquad Y_0 = \begin{bmatrix} I \\ 0 \\ \vdots \\ 0 \end{bmatrix}. \tag{2.13}$$

It is easy to check that the triple (X_0, T_0, Y_0) is similar to (2.3):

$$X_0 = [I \quad 0 \quad \cdots \quad 0]B^{-1}, \qquad C_2 = BC_1B^{-1}, \qquad Y_0 = B\,\mathrm{col}(\delta_{il}I)_{i=1}^{l}.$$

So (X_0, T_0, Y_0) itself is also a standard triple for $L(\lambda)$.

If the triple (X, J, Y) is Jordan (i.e., J is a Jordan matrix), then, as we already know, the columns of X, when decomposed into blocks consistently with the decomposition of J into Jordan blocks, form Jordan chains for $L(\lambda)$. A dual meaning can be associated with the rows of Y, by using Proposition 2.1. Indeed, it follows from Proposition 2.1 that the matrix $\mathrm{col}(Y^{\mathrm{T}}(J^{\mathrm{T}})^j)_{j=0}^{l-1}$ is nonsingular and

$$A_0^{\mathrm{T}}Y^{\mathrm{T}} + A_1^{\mathrm{T}}Y^{\mathrm{T}}J^{\mathrm{T}} + \cdots + A_{l-1}^{\mathrm{T}}Y^{\mathrm{T}}(J^{\mathrm{T}})^{l-1} + Y^{\mathrm{T}}(J^{\mathrm{T}})^l = 0.$$

So $(Y^{\mathrm{T}}, J^{\mathrm{T}})$ is a standard pair for the transposed matrix polynomial $L^{\mathrm{T}}(\lambda)$. Let $J = \mathrm{diag}[J_p]_{p=1}^m$ be the representation of J as a direct sum of its Jordan blocks $J_1, \ldots, J_m$ of sizes $\alpha_1, \ldots, \alpha_m$, respectively, and for $p = 1, 2, \ldots, m$, let

$$K_p = \begin{bmatrix} 0 & & \cdots & 0 & 1 \\ \vdots & & & 1 & 0 \\ & & \cdot^{\cdot^{\cdot}} & & \\ 0 & 1 & & & \vdots \\ 1 & 0 & \cdots & & 0 \end{bmatrix}$$

be a matrix of size $\alpha_p \times \alpha_p$. It is easy to check that

$$K_pJ_p^{\mathrm{T}}K_p = J_p.$$

Therefore, by Theorem 1.22, $(Y^{\mathrm{T}}K, J)$ is a Jordan pair for $L^{\mathrm{T}}(\lambda)$, where $K = \operatorname{diag}[K_p]_{p=1}^m$. So the columns of $Y^{\mathrm{T}}K$, partitioned consistently with the partition $J = \operatorname{diag}[J_p]_{p=1}^m$, form Jordan chains for $L^{\mathrm{T}}(\lambda)$. In other words, the rows of Y, partitioned consistently with the partition of J into its Jordan blocks and taken in each block *in the reverse order* form left Jordan chains for $L(\lambda)$.

More generally, the following result holds. We denote by $L^*(\lambda)$ (not to be confused with $(L(\lambda))^*$) the matrix polynomial whose coefficients are adjoint to those of $L(\lambda)$: if $L(\lambda) = \sum_{j=0}^{l} A_j\lambda^j$, then $L^*(\lambda) = \sum_{j=0}^{l} A_j^*\lambda^j$.

Theorem 2.2. *If (X, T, Y) is a standard triple for $L(\lambda)$, then $(Y^{\mathrm{T}}, T^{\mathrm{T}}, X^{\mathrm{T}})$ is a standard triple for $L^{\mathrm{T}}(\lambda)$ and (Y^*, T^*, X^*) is a standard triple for $L^*(\lambda)$.*

This theorem can be proved using, for instance, Theorem 1.23 and the definitions of standard pair and standard triple. The detailed proof is left to the reader.

We illustrate the constructions and results given in this section by an example.

EXAMPLE 2.1. Let

$$L(\lambda) = \begin{bmatrix} \lambda^3 & \sqrt{2}\lambda^2 - \lambda \\ \sqrt{2}\lambda^2 + \lambda & \lambda^3 \end{bmatrix}.$$

A Jordan pair (X, J) was already constructed in Example 1.5:

$$X = \begin{bmatrix} 1 & 0 & -\sqrt{2}+1 & \sqrt{2}-2 & \sqrt{2}+1 & \sqrt{2}+2 \\ 0 & 1 & 1 & 0 & 1 & 0 \end{bmatrix},$$

$$J = \operatorname{diag}\left\{0, 0, \begin{bmatrix} 1 & 1 \\ 0 & 1 \end{bmatrix} \begin{bmatrix} -1 & 1 \\ 0 & -1 \end{bmatrix}\right\}.$$

Let us now construct Y in such a way that (X, J, Y) is a Jordan triple of $L(\lambda)$. Computation shows that

$$\begin{bmatrix} X \\ XJ \\ XJ^2 \end{bmatrix} = \begin{bmatrix} 1 & 0 & -\sqrt{2}+1 & \sqrt{2}-2 & \sqrt{2}+1 & \sqrt{2}+2 \\ 0 & 1 & 1 & 0 & 1 & 0 \\ 0 & 0 & -\sqrt{2}+1 & -1 & -\sqrt{2}-1 & -1 \\ 0 & 0 & 1 & 1 & -1 & 1 \\ 0 & 0 & -\sqrt{2}+1 & -\sqrt{2} & \sqrt{2}+1 & -\sqrt{2} \\ 0 & 0 & 1 & 2 & 1 & -2 \end{bmatrix}$$

and

$$\begin{bmatrix} X \\ XJ \\ XJ^2 \end{bmatrix}^{-1} = \begin{bmatrix} 1 & 0 & \sqrt{2} & 0 & 0 & 1 \\ 0 & 1 & 0 & -\sqrt{2} & -1 & 0 \\ 0 & 0 & \frac{1}{4} & \frac{2\sqrt{2}+3}{4} & \frac{\sqrt{2}+2}{4} & 0 \\ 0 & 0 & -\frac{1}{4} & \frac{-\sqrt{2}-1}{4} & \frac{-\sqrt{2}-1}{4} & \frac{1}{4} \\ 0 & 0 & -\frac{1}{4} & \frac{2\sqrt{2}-3}{4} & \frac{-\sqrt{2}+2}{4} & 0 \\ 0 & 0 & -\frac{1}{4} & \frac{\sqrt{2}-1}{4} & \frac{-\sqrt{2}+1}{4} & -\frac{1}{4} \end{bmatrix}.$$

So the Jordan triple (X, J, Y) is completed with

$$Y = \begin{bmatrix} X \\ XJ \\ XJ^2 \end{bmatrix}^{-1} \begin{bmatrix} 0 \\ 0 \\ I \end{bmatrix} = \begin{bmatrix} 0 & 1 \\ -1 & 0 \\ \frac{\sqrt{2}+2}{4} & 0 \\ \frac{-\sqrt{2}-1}{4} & \frac{1}{4} \\ \frac{-\sqrt{2}+2}{4} & 0 \\ \frac{-\sqrt{2}+1}{4} & -\frac{1}{4} \end{bmatrix}.$$

It is easy to check that

$$\left(\frac{-\sqrt{2}-1}{4}, \frac{1}{4}\right), \quad \left(\frac{\sqrt{2}+2}{4}, 0\right) \quad \left(\text{resp.} \quad \left(\frac{-\sqrt{2}+1}{4}, -\frac{1}{4}\right), \quad \left(\frac{-\sqrt{2}+2}{4}, 0\right)\right)$$

are left Jordan chains (moreover, a canonical set of left Jordan chains) of $L(\lambda)$ corresponding to the eigenvalue $\lambda_0 = 1$ (resp. $\lambda_0 = -1$), i.e., the equalities

$$\left(\frac{-\sqrt{2}-1}{4}, \frac{1}{4}\right) L(1) = 0,$$

$$\left(\frac{-\sqrt{2}-1}{4}, \frac{1}{4}\right) L'(1) + \left(\frac{\sqrt{2}+2}{4}, 0\right) L(1) = 0$$

hold, and similarly for $\lambda_0 = -1$. □

We conclude this section with a remark concerning products of type XT^iY, where (X, T, Y) is a standard triple of a monic matrix polynomial $L(\lambda)$ of degree l, and i is a nonnegative integer. It follows from the similarity of all standard triples to each other that XT^iY does not depend on the choice of the triple. (In fact, we already know (from the definition of Y) that $XT^iY = 0$ for $i = 0, \ldots, l-2$, and $XT^{l-1}Y = I$.) Consequently, there exist formulas which express XT^iY directly in terms of the coefficients A_j of $L(\lambda)$. Such formulas are given in the next proposition.

Proposition 2.3. *Let $L(\lambda) = I\lambda^l + \sum_{j=0}^{l-1} A_j\lambda^j$ be a monic matrix polynomial with standard triple (X, T, Y). Then for any nonnegative integer ρ,*

$$XT^{l+\rho}Y = \sum_{k=1}^{\rho}\left[\sum_{q=1}^{k}\sum_{\substack{i_1+\cdots+i_q=k\\ i_j>0}}\prod_{j=1}^{q}(-A_{l-i_j})\right](-A_{l+k-\rho-1}) + (-A_{l-\rho-1}),$$

where, by definition, $A_i = 0$ for $i < 0$.

Proof. It is sufficient to prove Proposition 2.3 for the case that $X = [I \quad 0 \quad \cdots \quad 0]$, $T = C_1$ (the companion matrix),

$$Y = \begin{bmatrix} 0 \\ \vdots \\ 0 \\ I \end{bmatrix}.$$

Observe (see also the proof of Theorem 2.4 below) that

$$[I \quad 0 \quad \cdots \quad 0]C_1^l = [-A_0 \quad -A_1 \quad \cdots \quad -A_{l-1}],$$

so Proposition 2.3 is proved for $\rho = 0$ (in which case the sum $\sum_{k=1}^{\rho}$ disappears). For the general case use induction on ρ to prove that the βth block entry in the matrix $[I \quad 0 \quad \cdots \quad 0]C_1^{l+\rho}$ is equal to

$$\sum_{k=1}^{\rho}\left[\sum_{q=1}^{k}\sum_{\substack{i_1+\cdots+i_q=k\\ i_j>0}}\prod_{j=1}^{q}(-A_{l-i_j})\right](-A_{\beta+k-\rho-1}) + (-A_{\beta-\rho-1}),$$

$$\beta = 1, \ldots, l.$$

For a more detailed proof see Theorem 2.2 of [30a].

2.2. Representations of a Monic Matrix Polynomial

The notion of a standard triple introduced in the preceding section is the main tool in the following representation theorem.

Theorem 2.4. *Let $L(\lambda) = I\lambda^l + \sum_{j=0}^{l-1} A_j\lambda^j$ be a monic matrix polynomial of degree l with standard triple (X, T, Y). Then $L(\lambda)$ admits the following representations.*

(1) *right canonical form*:

$$L(\lambda) = I\lambda^l - XT^l(V_1 + V_2\lambda + \cdots + V_l\lambda^{l-1}), \tag{2.14}$$

where V_i are $nl \times n$ matrices such that

$$[V_1 \ \cdots \ V_l] = [\operatorname{col}(XT^i)_{i=0}^{l-1}]^{-1};$$

(2) *left canonical form*:

$$L(\lambda) = \lambda^l I - (W_1 + \lambda W_2 + \cdots + \lambda^{l-1}W_l)T^lY, \tag{2.15}$$

where W_i are $n \times nl$ matrices such that

$$\operatorname{col}(W_i)_{i=1}^{l} = [\operatorname{row}(T^iY)_{i=0}^{l-1}]^{-1};$$

(3) *resolvent form*:

$$(L(\lambda))^{-1} = X(I\lambda - T)^{-1}Y, \qquad \lambda \in \mathbb{C}\backslash\sigma(L). \tag{2.16}$$

Note that only X and T appear in the right canonical form of $L(\lambda)$, while in the left canonical form only T and Y appear. The names "right" and "left" are chosen to stress the fact that the pair (X, T) (resp. (T, Y)) represents right (resp. left) spectral data of $L(\lambda)$.

Proof. Observe that all three forms (2.14), (2.15), (2.16) are independent of the choice of the standard triple (X, T, Y). Let us check this for (2.14), for example. We have to prove that if (X, T, Y) and (X', T', Y') are standard triples of $L(\lambda)$, then

$$XT^l[V_1 \ \cdots \ V_l] = X'(T')^l[V'_1 \ \cdots \ V'_l], \tag{2.17}$$

where

$$[V_1 \ \cdots \ V_l] = [\operatorname{col}(XT^i)_{i=0}^{l-1}]^{-1}, [V'_1 \ \cdots \ V'_l] = [\operatorname{col}(X'T'^i)_{i=0}^{l-1}]^{-1}.$$

But these standard triples are similar:

$$X' = XS, \qquad T' = S^{-1}TS, \qquad Y' = S^{-1}Y.$$

Therefore

$$\begin{aligned}[V'_1 \ \cdots \ V'_l] &= [\operatorname{col}(X'T'^i)_{i=0}^{l-1}]^{-1} = [\operatorname{col}(XT^i)_{i=0}^{l-1}S]^{-1} \\ &= S^{-1}[V_1 \ \cdots \ V_l],\end{aligned}$$

and (2.17) follows.

Thus, it suffices to check (2.14) and (2.16) only for the special standard triple

$$X = [I \quad 0 \quad \cdots \quad 0], \qquad T = C_1, \qquad Y = \operatorname{col}(\delta_{il} I)_{i=1}^{l},$$

and for checking (2.15) we shall choose the dual standard triple defined by (2.13). Then (2.16) is just Proposition 1.2; to prove (2.14) observe that

$$[I \quad 0 \quad \cdots \quad 0]C_1^l = [-A_0 \quad -A_1 \quad \cdots \quad -A_{l-1}]$$

and

$$[V_1 \quad \cdots \quad V_l] = [\operatorname{col}([I \quad 0 \quad \cdots \quad 0]C_1^i)_{i=0}^{l-1}]^{-1} = I,$$

so

$$[I \quad 0 \quad \cdots \quad 0]C_1^l[V_1 \quad V_2 \quad \cdots \quad V_l] = [-A_0 \quad -A_1 \quad \cdots \quad -A_{l-1}],$$

and (2.14) becomes evident. To prove (2.15), note that by direct computation one easily checks that for the standard triple (2.13)

$$C_2^j \operatorname{col}(\delta_{i1} I)_{i=1}^{l} = \operatorname{col}(\delta_{i,j+1} I)_{i=1}^{l}, \qquad j = 0, \ldots, l-1$$

and

$$C_2^j \operatorname{col}(\delta_{i1} I)_{i=1}^{l} = \operatorname{col}(-A_i)_{i=0}^{l-1}.$$

So

$$\operatorname{row}(C_2 \operatorname{col}(\delta_{i1} I)_{i=1}^{l})_{j=0}^{l-1} = I$$

and

$$[\operatorname{row}(C_2^j \operatorname{col}(\delta_{i1} I)_{i=1}^{l})_{j=0}^{l-1}]^{-1} T^l Y = \begin{bmatrix} W_1 \\ \vdots \\ W_l \end{bmatrix} T^l Y = \begin{bmatrix} -A_0 \\ -A_1 \\ \vdots \\ -A_{l-1} \end{bmatrix},$$

so (2.15) follows. □

We obtain the next result on the singular part of the Laurent expansion (refer to Section 1.7) as an immediate consequence of the resolvent form (2.16). But first let us make some observations concerning left Jordan chains. A canonical set of left Jordan chains is defined by matrix Y of a Jordan triple (X, J, Y) provided X is defined by a canonical set of usual Jordan chains. The lengths of left Jordan chains in a canonical set corresponding to λ_0 coincide with the partial multiplicities of $L(\lambda)$ at λ_0, and consequently, coincide with the lengths of the usual Jordan chains in a canonical set (see Proposition 1.13). To verify this fact, let

$$L(\lambda) = E(\lambda)D(\lambda)F(\lambda)$$

be the Smith form of $L(\lambda)$ (see Section S1.1); passing to the transposed matrix, we obtain that $D(\lambda)$ is also the Smith form for $L^T(\lambda)$. Hence the partial multiplicities of $L(\lambda)$ and $L^T(\lambda)$ at every eigenvalue λ_0 are the same.

Corollary 2.5. *Let $L(\lambda)$ be a monic matrix polynomial and $\lambda_0 \in \sigma(L)$. Then for every canonical set $\varphi_0^{(j)}, \varphi_1^{(j)}, \dots, \varphi_{r_j-1}^{(j)}$ $(j = 1, \dots, \alpha)$ of Jordan chains of $L(\lambda)$ corresponding to λ_0 there exists a canonical set $z_0^{(j)}, z_1^{(j)}, \dots, z_{r_j-1}^{(j)}$ $(j = 1, \dots, \alpha)$ of left Jordan chains of $L(\lambda)$ corresponding to λ_0 such that*

$$SP(L^{-1}(\lambda_0)) = \sum_{j=1}^{\alpha} \sum_{k=1}^{r_j} (\lambda - \lambda_0)^{-k} \sum_{r=0}^{r_j-k} \varphi_{r_j-k-r}^{(j)} \cdot z_r^{(j)}. \tag{2.18}$$

Note that $\varphi_k^{(j)}$ are n-dimensional column vectors and $z_k^{(j)}$ are n-dimensional rows. So the product $\varphi_{r_j-k-r}^{(j)} \cdot z_r^{(j)}$ makes sense and is an $n \times n$ matrix.

Proof. Let (X, J, Y) be a Jordan triple for $L(\lambda)$ such that the part X_0 of X corresponding to the eigenvalue λ_0 consists of the given canonical set $\varphi_0^{(j)}, \dots, \varphi_{r_j-1}^{(j)}$ $(j = 1, \dots, \alpha)$. Let J_0 be the part of J with eigenvalue λ_0, and let Y_0 be the corresponding part of Y. Formula (2.16) shows that

$$(L(\lambda))^{-1} = X_0(\lambda I - J_0)^{-1} Y_0 + \cdots,$$

where dots denote a matrix function which is analytic in a neighborhood of λ_0. So

$$SP(L^{-1}(\lambda_0)) = SP(X_0(\lambda I - J_0)^{-1} Y_0)_{\lambda=\lambda_0}. \tag{2.19}$$

A straightforward computation of the right-hand side of (2.19) leads to formula (2.18), where $z_0^{(j)}, \dots, z_{r_j-1}^{(j)}$, for $j = 1, \dots, \alpha$, are the rows of Y_0 taken in the reverse order in each part of Y_0 corresponding to a Jordan block in J_0. But we have already seen that these vectors form left Jordan chains of $L(\lambda)$ corresponding to λ_0. It is easily seen (for instance, using the fact that $[Y \;\; JY \;\; \cdots \;\; J^{l-1}Y]$ is a nonsingular matrix; Proposition 2.1) that the rows $z_0^{(1)}, \dots, z_0^{(\alpha)}$ are linearly independent, and therefore the left Jordan chains $z_0^{(j)}, \dots, z_{r_j-1}^{(j)}$, for $j = 1, \dots, \alpha$ form a canonical set (cf. Proposition 1.16). □

The right canonical form (2.14) allows us to answer the following question: given $n \times nl$ and $nl \times nl$ matrices X and T, respectively, when is (X, T) a standard pair of some monic matrix polynomial of degree l? This happens if and only if $\operatorname{col}(XT^i)_{i=0}^{l-1}$ is nonsingular and in this case the desired monic matrix polynomial is unique and given by (2.14). There is a similar question: when does a pair (T, Y) of matrices T and Y of sizes $nl \times nl$ and $nl \times n$, respectively, form a left standard pair for some monic matrix polynomial of degree l? The answer is similar: when $\operatorname{row}(T^iY)_{i=0}^{l-1}$ is nonsingular, for then the desired matrix polynomial is unique and given by (2.15). The reader can easily supply proofs for these facts.

We state as a theorem the solution of a similar problem for the resolvent form of $L(\lambda)$.

Theorem 2.6. *Let $L(\lambda)$ be a monic matrix polynomial of degree l and assume there exist matrices Q, T, R of sizes $n \times nl$, $nl \times nl$, $nl \times n$, respectively, such that*

$$L^{-1}(\lambda) = Q(I\lambda - T)^{-1}R, \qquad \lambda \notin \sigma(T) \tag{2.20}$$

Then (Q, T, R) is a standard triple for L.

Proof. Note that for $|\lambda|$ sufficiently large, $L^{-1}(\lambda)$ can be developed into a power series

$$L^{-1}(\lambda) = \lambda^{-l}I + \lambda^{-l-1}Z_1 + \lambda^{-l-2}Z_2 + \cdots$$

for some matrices $Z_1, Z_2, \ldots$. It is easily seen from this development that if Γ is a circle in the complex plane having $\sigma(L)$ in its interior, then

$$\frac{1}{2\pi i}\oint_\Gamma \lambda^i L^{-1}(\lambda)\, d\lambda = \begin{cases} 0 & \text{if } i = 0, 1, \ldots, l-2 \\ I & \text{if } i = l-1. \end{cases}$$

But we may also choose Γ large enough so that (see Section S1.8)

$$\frac{1}{2\pi i}\oint_\Gamma \lambda^i (I\lambda - T)^{-1}\, d\lambda = T^i, \qquad i = 0, 1, 2, \ldots .$$

Thus, it follows from (2.20) that

$$\begin{aligned}\operatorname{col}(QT^i)_{i=0}^{l-1}[R \quad TR \quad \cdots \quad T^{l-1}R] &= \frac{1}{2\pi i}\int_\Gamma \begin{bmatrix} 1 & \lambda & \cdots & \lambda^{l-1} \\ \vdots & & & \vdots \\ \lambda^{l-1} & & \cdots & \lambda^{2l-2} \end{bmatrix} L^{-1}(\lambda)\, d\lambda \\ &= \begin{bmatrix} 0 & 0 & \cdots & I \\ \vdots & \vdots & \cdot^{\cdot^{\cdot}} & \vdots \\ 0 & I & & \\ I & & & * \end{bmatrix} \end{aligned} \tag{2.21}$$

and, in particular, both $\operatorname{col}(QT^i)_{i=0}^{l-1}$ and $\operatorname{row}(T^iR)_{i=0}^{l-1}$ are nonsingular.

Then we observe that for $i = 0, \ldots, l-1$,

$$\begin{aligned} 0 = \frac{1}{2\pi i}\oint_\Gamma \lambda^i L(\lambda)L^{-1}(\lambda)\, d\lambda &= \frac{1}{2\pi i}\oint_\Gamma \lambda^i L(\lambda)Q(I\lambda - T)^{-1}R\, d\lambda \\ &= (A_0Q + A_1QT + \cdots + QT^l)T^iR, \end{aligned}$$

and, since $[R \quad TR \quad \cdots \quad T^{l-1}R]$ is nonsingular, we obtain $A_0 Q + \cdots + A_{l-1}QT^{l-1} + QT^l = 0$. This is sufficient to show that, if $S = \operatorname{col}(QT^i)_{i=0}^{l-1}$, then

$$Q = [I \quad 0 \quad \cdots \quad 0]S, \qquad T = S^{-1}C_1 S, \qquad R = S^{-1}\begin{bmatrix} 0 \\ \vdots \\ 0 \\ I \end{bmatrix},$$

i.e., Q, T, R is a standard triple of $L(\lambda)$. □

Example 2.2. For the matrix polynomial

$$L(\lambda) = \begin{bmatrix} \lambda^3 & \sqrt{2}\lambda^2 - \lambda \\ \sqrt{2}\lambda^2 + \lambda & \lambda^3 \end{bmatrix}$$

described in Examples 1.5 and 2.1, the representations (2.14)–(2.16) look as follows, using the Jordan triple (X, J, Y) described in Example 2.1:

(a) right canonical form:

$$L(\lambda) = I\lambda^3 - \begin{bmatrix} 1 & 0 & -\sqrt{2}+1 & \sqrt{2}-2 & \sqrt{2}+1 & \sqrt{2}+2 \\ 0 & 1 & 1 & 0 & 1 & 0 \end{bmatrix}$$

$$\cdot \begin{bmatrix} 0 & 0 & 0 & 0 & 0 & 0 \\ 0 & 0 & 0 & 0 & 0 & 0 \\ 0 & 0 & 1 & 3 & 0 & 0 \\ 0 & 0 & 0 & 1 & 0 & 0 \\ 0 & 0 & 0 & 0 & -1 & 3 \\ 0 & 0 & 0 & 0 & 0 & -1 \end{bmatrix} \left\{ \begin{bmatrix} 1 & 0 \\ 0 & 1 \\ 0 & 0 \\ 0 & 0 \\ 0 & 0 \\ 0 & 0 \end{bmatrix} + \begin{bmatrix} \sqrt{2} & 0 \\ 0 & -\sqrt{2} \\ \frac{1}{4} & \frac{2\sqrt{2}+3}{4} \\ -\frac{1}{4} & \frac{-\sqrt{2}-1}{4} \\ -\frac{1}{4} & \frac{2\sqrt{2}-3}{4} \\ -\frac{1}{4} & \frac{\sqrt{2}-1}{4} \end{bmatrix} \lambda + \begin{bmatrix} 0 & 1 \\ -1 & 0 \\ \frac{\sqrt{2}+2}{4} & 0 \\ \frac{-\sqrt{2}-1}{4} & \frac{1}{4} \\ \frac{-\sqrt{2}+2}{4} & 0 \\ \frac{-\sqrt{2}+1}{4} & -\frac{1}{4} \end{bmatrix} \lambda^2 \right\};$$

(b) left canonical form: we have to compute first

$$[Y \quad JY \quad J^2Y]^{-1}.$$

It is convenient to use the biorthogonality condition (2.6), $RBQ = I$, where

$$R = [Y \quad JY \quad J^2Y], \qquad Q = \begin{bmatrix} X \\ XJ \\ XJ^2 \end{bmatrix},$$

(as given in Example 2.1) and

$$B = \begin{bmatrix} A_1 & A_2 & I \\ A_2 & I & 0 \\ I & 0 & 0 \end{bmatrix}.$$

So it is found that

$$[Y \quad JY \quad J^2Y]^{-1} = BQ = \begin{bmatrix} 0 & -1 & 0 & 0 & 0 & 0 \\ 1 & 0 & 0 & 0 & 0 & 0 \\ 0 & \sqrt{2} & 1 & -1 & -1 & -1 \\ \sqrt{2} & 0 & \sqrt{2}-1 & -2\sqrt{2}+3 & \sqrt{2}+1 & 2\sqrt{2}+3 \\ 1 & 0 & -\sqrt{2}+1 & \sqrt{2}-2 & \sqrt{2}+1 & \sqrt{2}+2 \\ 0 & 1 & 1 & 0 & 1 & 0 \end{bmatrix},$$

and the left canonical form of $L(\lambda)$ is

$$\begin{aligned} L(\lambda) = \lambda^3 I - \left\{ \begin{bmatrix} 0 & -1 & 0 & 0 & 0 & 0 \\ 1 & 0 & 0 & 0 & 0 & 0 \end{bmatrix} \right. \\ + \begin{bmatrix} 0 & \sqrt{2} & 1 & -1 & -1 & -1 \\ \sqrt{2} & 0 & \sqrt{2}-1 & -2\sqrt{2}+3 & \sqrt{2}+1 & 2\sqrt{2}+3 \end{bmatrix} \lambda \\ \left. + \begin{bmatrix} 1 & 0 & -\sqrt{2}+1 & \sqrt{2}-2 & \sqrt{2}+1 & \sqrt{2}+2 \\ 0 & 1 & 1 & 0 & 1 & 0 \end{bmatrix} \lambda^2 \right\} \\ \cdot \begin{bmatrix} 0 & & & & & \\ & 0 & & & & \\ & & 1 & 3 & & \\ & & & 1 & & \\ & & & & -1 & 3 \\ & & & & & -1 \end{bmatrix} \begin{bmatrix} 0 & 1 \\ -1 & 0 \\ \dfrac{\sqrt{2}+2}{4} & 0 \\ \dfrac{-\sqrt{2}-1}{4} & \dfrac{1}{4} \\ \dfrac{-\sqrt{2}+2}{4} & 0 \\ \dfrac{-\sqrt{2}+1}{4} & -\dfrac{1}{4} \end{bmatrix}; \end{aligned}$$

(c) resolvent form:

$$L(\lambda)^{-1} = \begin{bmatrix} 1 & 0 & -\sqrt{2}+1 & \sqrt{2}-2 & \sqrt{2}+1 & \sqrt{2}+2 \\ 0 & 1 & 1 & 0 & 1 & 0 \end{bmatrix}$$

$$\cdot \begin{bmatrix} \lambda & & & & & \\ & \lambda & & & & \\ & & \lambda-1 & -1 & & \\ & & & \lambda-1 & & \\ & & & & \lambda+1 & -1 \\ & & & & & \lambda+1 \end{bmatrix}^{-1} \cdot \begin{bmatrix} 0 & 1 \\ -1 & 0 \\ \frac{\sqrt{2}+2}{4} & 0 \\ \frac{-\sqrt{2}-1}{4} & \frac{1}{4} \\ \frac{-\sqrt{2}+2}{4} & 0 \\ \frac{-\sqrt{2}+1}{4} & -\frac{1}{4} \end{bmatrix}. \quad \square$$

We have already observed in Section 1.7 and Corollary 2.5 that properties of the rational function $L^{-1}(\lambda)$ are closely related to the Jordan structure of $L(\lambda)$. We now study this connection further by using the resolvent form of $L(\lambda)$. Consider first the case when all the elementary divisors of $L(\lambda)$ are linear. Write the resolvent form for $L(\lambda)$ taking a Jordan triple (X, J, Y)

$$L(\lambda)^{-1} = X(I\lambda - J)^{-1}Y.$$

Since the elementary divisors of $L(\lambda)$ are linear, J is a diagonal matrix $J = \operatorname{diag}[\lambda_1 \quad \cdots \quad \lambda_{nl}]$, and therefore $(I\lambda - J)^{-1} = \operatorname{diag}[(\lambda - \lambda_i)^{-1}]_{i=1}^{nl}$. So

$$L(\lambda)^{-1} = X \cdot \operatorname{diag}[(\lambda - \lambda_i)^{-1}]_{i=1}^{nl} \cdot Y. \tag{2.22}$$

Clearly the rational matrix function $L(\lambda)^{-1}$ has a simple pole at every point in $\sigma(L)$. Let now $\lambda_0 \in \sigma(L)$, and let $\lambda_{i_1}, \ldots, \lambda_{i_k}$ be the eigenvalues from (2.22) which are equal to λ_0. Then the singular part of $L(\lambda)^{-1}$ at λ_0 is just

$$X_0 Y_0/(\lambda - \lambda_0),$$

where X_0 (Y_0) is $n \times k$ ($k \times n$) submatrix of X (Y) consisting of the eigenvectors (left eigenvectors) corresponding to λ_0.

Moreover, denoting by X_i (Y_i) the ith column (row) from X (Y), we have the following representation:

$$L(\lambda)^{-1} = \sum_{i=1}^{nl} \frac{X_i Y_i}{\lambda - \lambda_i}. \tag{2.23}$$

In general (when nonlinear elementary divisors of $L(\lambda)$ exist) it is not true that the eigenvalues of $L(\lambda)$ are simple poles of $L(\lambda)^{-1}$. We can infer from Proposition 1.17 that the order of $\lambda_0 \in \sigma(L)$ as a pole of $L^{-1}(\lambda)$ coincides with the maximal size of a Jordan block of J corresponding to λ_0.

Next we note that the resolvent form (2.16) can be generalized as follows:

$$\begin{aligned} \lambda^{r-1}L(\lambda)^{-1} &= XT^{r-1}(I\lambda - T)^{-1}Y, \qquad r = 1, \ldots, l, \\ \lambda^{l}L(\lambda)^{-1} &= XT^{l}(I\lambda - T)^{-1}Y + I. \end{aligned} \tag{2.24}$$

To verify these equalities, write

$$\lambda^{r-1}L(\lambda)^{-1} = \lambda^{r-1}X(I\lambda - T)^{-1}Y$$

according to (2.16), and use the following development of $(I\lambda - T)^{-1}$ into a power series, valid for $|\lambda|$ large enough ($|\lambda| > \|T\|$):

$$(I\lambda - T)^{-1} = \lambda^{-1}I + \lambda^{-2}T + \lambda^{-3}T^2 + \cdots. \tag{2.25}$$

The series $\sum_{i=1}^{\infty} \lambda^{-i}T^{i-1}$ converges absolutely for $|\lambda| > \|T\|$, and multiplying it by $I\lambda - T$, it is easily seen that (2.25) holds. So for $|\lambda| > \|T\|$

$$\begin{aligned} \lambda^{r-1}L(\lambda)^{-1} &= \lambda^{r-1}X(\lambda^{-1}I + \lambda^{-2}T + \lambda^{-3}T^2 + \cdots)Y \\ &= \lambda^{r-1}X(\lambda^{-r}T^{r-1} + \lambda^{-r-1}T^{r} + \cdots)Y \\ &\quad + \lambda^{r-1}X(\lambda^{-1}I + \lambda^{-2}T + \cdots + \lambda^{-(r-1)}T^{r-2})Y. \end{aligned}$$

The last summand is zero for $r = 1, \ldots, l$ and I for $r = l + 1$ (since $XT^{i}Y = 0$ for $i = 0, \ldots, l - 2$ and $XT^{l-1}Y = I$ by the definition of Y), and the first summand is apparently $XT^{r-1}(\lambda I - T)^{-1}Y$, so (2.24) follows.

The same argument shows also that for $r > l + 1$ the equality

$$\lambda^{r-1}L(\lambda)^{-1} = XT^{r-1}(\lambda I - T)^{-1}Y + \sum_{i=l-1}^{r-2} \lambda^{r-2-i}XT^{i}Y$$

holds.

Another consequence of the resolvent form is the following formula (as before, (X, T, Y) stands for a standard triple of a monic matrix polynomial $L(\lambda)$):

$$\frac{1}{2\pi i}\int_{\Gamma} f(\lambda)L^{-1}(\lambda)\,d\lambda = Xf(T)Y, \tag{2.26}$$

where Γ is a contour such that $\sigma(L)$ is inside Γ, and $f(\lambda)$ is a function which is analytic inside and in a neighborhood of Γ. (Taking Γ to be a union of small circles around each eigenvalue of $L(\lambda)$, we can assume that $f(\lambda)$ is analytic in a neighborhood of $\sigma(L)$.) The proof of (2.26) is immediate:

$$\frac{1}{2\pi i}\int_{\Gamma} f(\lambda)L^{-1}(\lambda)\,d\lambda = \frac{1}{2\pi i}\int_{\Gamma} f(\lambda)X(\lambda I - T)^{-1}Y\,d\lambda$$

$$= X\cdot\frac{1}{2\pi i}\int_{\Gamma} f(\lambda)(\lambda I - T)^{-1}\,d\lambda\,Y = Xf(T)Y$$

by the definition of $f(T)$ (see Section S1.8).

We conclude this section with the observation that the resolvent form (2.16) also admits another representation for $L(\lambda)$ itself.

Proposition 2.7. *Let (X, T, Y) be a standard triple for the monic matrix polynomial $L(\lambda)$. Then the functions $E(\lambda)$, $F(\lambda)$ defined by*

$$E(\lambda) = L(\lambda)X(I\lambda - T)^{-1}, \qquad F(\lambda) = (I\lambda - T)^{-1}YL(\lambda)$$

are $n \times ln$, $ln \times n$ matrix polynomials whose degrees do not exceed $l - 1$, and

$$L(\lambda) = E(\lambda)(I\lambda - T)F(\lambda).$$

Proof. It follows from (2.24) that

$$L(\lambda)^{-1}[I \quad \lambda I \quad \cdots \quad \lambda^{l-1}I] = X(I\lambda - T)^{-1}\operatorname{row}(T^jY)_{j=0}^{l-1}.$$

Since the rightmost matrix is nonsingular, it follows that

$$E(\lambda) = L(\lambda)X(I\lambda - T)^{-1} = [I \quad \lambda I \quad \cdots \quad \lambda^{l-1}I][\operatorname{row}(T^jY)_{j=0}^{l-1}]^{-1}$$

and the statement concerning $E(\lambda)$ is proved. The corresponding statement for $F(\lambda)$ is proved similarly.

Now form the product $E(\lambda)(I\lambda - T)F(\lambda)$ and the conclusion follows from (2.16). □

2.3. Resolvent Form and Linear Systems Theory

The notion of a resolvent form for a monic matrix polynomial is closely related to the notion of a transfer function for a time-invariant linear system, which we shall now define.

Consider the system of linear differential equations

$$\frac{dx}{dt} = Ax + Bu, \qquad y = Cx, \quad x(0) = 0 \tag{2.27}$$

where A, B, C are constant matrices and x, y, u are vectors (depending on t), which represent the state variable, the output variable, and the control variable, respectively. Of course, the sizes of all vectors and matrices are such that equalities in (2.27) make sense. We should like to find the behavior of the output y as a function of the control vector u. Applying the Laplace transform

$$\hat{z}(s) = \int_0^\infty e^{-st} z(t)\, dt$$

(we denote by $\hat{z}$ the Laplace image of a vector function z), we obtain from (2.27)

$$s\hat{x}(s) = A\hat{x} + B\hat{u}, \qquad \hat{y} = C\hat{x},$$

or

$$\hat{y} = C(Is - A)^{-1} B\hat{u}.$$

The rational matrix function $F(s) = C(Is - A)^{-1}B$ is called the *transfer function* of the linear system (2.27). Conversely, the linear system (2.27) is called a *realization* of the rational matrix valued function $F(s)$. Clearly, the realization is not unique; for instance, one can always make the size of A bigger: if $F(s) = C(Is - A)^{-1}B$, then also

$$F(s) = [C \quad 0]\left\{Is - \begin{bmatrix} A & 0 \\ 0 & 0 \end{bmatrix}\right\}^{-1} \begin{bmatrix} B \\ 0 \end{bmatrix}.$$

The realization (2.27) is called *minimal* if A has the minimal possible size. All minimal realizations are similar, i.e., if A, B, C and A', B', C' are minimal realizations of the same rational matrix function $F(s)$, then $A = S^{-1}A'S$, $B = S^{-1}B'$, $C = C'S$ for some invertible matrix S. This fact is well known in linear system theory and can be found, as well as other relevant information, in books devoted to linear system theory; see, for instance [80, Chapter 4].

In particular, in view of the resolvent representation, the inverse $L^{-1}(\lambda)$ of a monic matrix polynomial $L(\lambda)$ can be considered as a transfer function

$$L^{-1}(\lambda) = Q(I\lambda - T)^{-1}R,$$

where (Q, T, R) is a standard triple of $L(\lambda)$. The corresponding realization is (2.27) with $A = T$, $B = R$, $C = Q$. Theorem 2.6 ensures that this realization of $L^{-1}(\lambda)$ is unique up to similarity, provided the dimension of the state variable is kept equal to nl, i.e., if (A', B', C') is another triple of matrices such that the size of A' is nl and

$$L^{-1}(\lambda) = C'(I\lambda - A')^{-1}B',$$

then there exists an invertible matrix S such that $C' = QS$, $A' = S^{-1}TS$, $B' = S^{-1}R$. Moreover, the proof of Theorem 2.6 shows (see in particular (2.21)), that the realization of $L^{-1}(\lambda)$ via a standard triple is also minimal.

We now investigate when a transfer function $F(\lambda) = Q(I\lambda - T)^{-1}R$ is the inverse of a monic matrix polynomial. When this is the case, we shall say that the triple (Q, T, R) is a *polynomial triple*.

Theorem 2.8. *A triple (Q, T, R) is polynomial if and only if for some integer l the following conditions hold:*

$$QT^iR = \begin{cases} 0 & \text{for} \quad i = 0, \ldots, l-2 \\ I & \text{for} \quad i = l-1, \end{cases} \tag{2.28}$$

and

$$\operatorname{rank}\begin{bmatrix} QR & QTR & \cdots & QT^{m-1}R \\ QTR & QT^2R & \cdots & QT^mR \\ \vdots & \vdots & \cdots & \vdots \\ QT^{m-1}R & QT^mR & \cdots & QT^{2m-2}R \end{bmatrix} = nl \tag{2.29}$$

for all sufficiently large integers m.

Proof. Assume $F(\lambda) = Q(I\lambda - T)^{-1}R$ is the inverse of a monic matrix polynomial $L(\lambda)$ of degree l. Let (Q_0, T_0, R_0) be a standard triple for $L(\lambda)$; then clearly

$$QT^iR = Q_0T_0^iR_0, \qquad i = 0, 1, \ldots. \tag{2.30}$$

Now equalities (2.28) follow from the conditions (2.12) which are satisfied by the standard triple (Q_0, T_0, R_0). Equality (2.29) holds for every $m \geq l$ in view of (2.30). Indeed,

$$\begin{bmatrix} QR & QTR & \cdots & QT^{m-1}R \\ QTR & QT^2R & \cdots & QT^mR \\ \vdots & \vdots & \cdots & \vdots \\ QT^{m-1}R & QT^mR & \cdots & QT^{2m-2}R \end{bmatrix} = \operatorname{col}(Q_0T_0^i)_{i=0}^{m-1} \cdot \operatorname{row}(T_0^iR_0)_{i=0}^{m-1}, \tag{2.31}$$

and since $\operatorname{rank} \operatorname{col}(Q_0T_0^i)_{i=0}^{m-1} = \operatorname{rank} \operatorname{col}(T_0^iR_0)_{i=0}^{m-1} = nl$ for $m \geq l$, rank $Z_m \leq nl$, where Z_m is the $mn \times mn$ matrix in the left-hand side of (2.31). On the other hand, the matrices $\operatorname{col}(Q_0T_0^i)_{i=0}^{l-1}$ and $\operatorname{row}(T_0^iR_0)_{i=0}^{l-1}$ are nonsingular, so (2.31) implies rank $Z_l = nl$. Clearly, rank $Z_m \geq$ rank Z_l for $m \geq l$, so in fact rank $Z_m = nl$ $(m \geq l)$ as claimed.

Assume now (2.28) and (2.29) hold; then

$$\operatorname{col}(QT^i)_{i=0}^{l-1} \cdot \operatorname{row}(T^jR)_{j=0}^{l-1} = \begin{bmatrix} 0 & & \cdots & 0 & I \\ \vdots & \vdots & & \cdot^{\cdot^{\cdot}} & \\ 0 & I & \cdot^{\cdot^{\cdot}} & & \\ I & & & & * \end{bmatrix}. \tag{2.32}$$

In particular, the size ν of T is greater or equal to nl. If $\nu = nl$, then (2.32) implies nonsingularity of $\operatorname{col}(QT^i)_{i=0}^{l-1}$. Consequently, there exists a monic matrix polynomial of degree l with standard triple (Q, T, R), so (Q, T, R) is a polynomial triple. (Note that in this argument (2.29) was not used.)

Assume $\nu > nl$. We shall reduce the proof to the case when $\nu = nl$. To this end we shall appeal to some well-known notions and results in realization theory for linear systems, which will be stated here without proofs.

Let $W(\lambda)$ be a rational matrix function. For a given $\lambda_0 \in \mathbb{C}$ define the *local degree* $\delta(W; \lambda_0)$ (cf. [3c, Chapter IV]),

$$\delta(W; \lambda_0) = \operatorname{rank} \begin{bmatrix} W_{-q} & W_{-q+1} & \cdots & W_{-1} \\ 0 & W_{-q} & \cdots & W_{-2} \\ \vdots & \vdots & \cdots & \vdots \\ 0 & 0 & \cdots & W_{-q} \end{bmatrix},$$

where $W(\lambda) = \sum_{j=-q}^{\infty} (\lambda - \lambda_0)^j W_j$ is the Laurent expansion of $W(\lambda)$ in a neighborhood of λ_0. Define also $\delta(W; \infty) = \delta(\tilde{W}; 0)$, where $\tilde{W}(\lambda) = W(\lambda^{-1})$. Evidently, $\delta(W; \lambda_0)$ is nonzero only for finitely many complex numbers λ_0. Put

$$\delta(W) = \sum_{\lambda \in \mathbb{C} \cup \{\infty\}} \delta(W; \lambda).$$

The number $\delta(W)$ is called the *McMillan degree* of $W(\lambda)$ (see [3c, 80]). For rational functions $W(\lambda)$ which are analytic at infinity, the following equality holds (see [81] and [3c, Section 4.2]):

$$\delta(W) = \operatorname{rank} \begin{bmatrix} D_1 & D_2 & \cdots & D_m \\ D_2 & D_3 & \cdots & D_{m+1} \\ \vdots & \vdots & \cdots & \vdots \\ D_m & D_{m+1} & \cdots & D_{2m-1} \end{bmatrix}, \tag{2.33}$$

where D_j are the coefficients of the Taylor series for $W(\lambda)$ at infinity, $W(\lambda) = \sum_{j=0}^{\infty} D_j \lambda^{-j}$, and m is a sufficiently large integer. In our case $F(\lambda) = Q(I\lambda - T)^{-1}R$, where (Q, T, R) satisfies (2.28) and (2.29). In particular, equalities (2.29) and (2.33) ensure that $\delta(F) = nl$.

By a well-known result in realization theory (see, for instance, [15a, Theorem 4.4] or [14, Section 4.4]), there exists a triple of matrices (Q_0, T_0, R_0) with sizes $n \times nl$, $nl \times nl$, $nl \times n$, respectively, such that $Q_0(I\lambda - T_0)^{-1}R_0 = Q(I\lambda - T)^{-1}R$. Now use (2.32) for (Q_0, T_0, R_0) in place of (Q, T, R) to prove that (Q_0, T_0, R_0) is in fact a standard triple of some monic matrix polynomial of degree l. $\square$

2.4. Initial Value Problems and Two-Point Boundary Value Problems

In this section we shall obtain explicit formulas for the solutions of the initial value problem and a boundary value problem associated with the differential equation

$$L\left(\frac{d}{dt}\right)x(t) = \frac{d^l x(t)}{dt^l} + \sum_{j=0}^{l-1} A_j \frac{d^j x(t)}{dt^j} = f(t), \qquad t \in [a, b] \tag{2.34}$$

where $L(\lambda) = I\lambda^l + \sum_{j=0}^{l-1} A_j \lambda^j$ is a monic matrix polynomial, $f(t)$ is a known vector function on $t \in [a, b]$ (which will be supposed piecewise continuous), and $x(t)$ is a vector function to be found. The results will be described in terms of a standard triple (X, T, Y) of $L(\lambda)$.

The next result is deduced easily from Theorem 1.5 using the fact that all standard triples of $L(\lambda)$ are similar to each other.

Theorem 2.9. *The general solution of equation* (2.34) *is given by the formula*

$$x(t) = Xe^{tT}c + X\int_a^t e^{(t-s)T} Yf(s)\, ds, \qquad t \in [a, b], \tag{2.35}$$

where (X, T, Y) is a standard triple of $L(\lambda)$, and $c \in \mathbb{C}^{nl}$ is arbitrary.

In particular, the general solution of the homogeneous equation

$$L\left(\frac{d}{dt}\right)x(t) = 0 \tag{2.36}$$

is given by the formula

$$x(t) = Xe^{tT}c, \qquad c \in \mathbb{C}^n.$$

Consider now the initial value problem: find a function $x(t)$ such that (2.34) holds and

$$x^{(i)}(a) = x_i, \qquad i = 0, \ldots, l-1, \tag{2.37}$$

where x_i are known vectors.

Theorem 2.10. *For every given set of vectors $x_0, \ldots, x_{l-1}$ there exists a unique solution $x(t)$ of the initial value problem* (2.34), (2.37). *It is given by the formula* (2.35) *with*

$$c = [Y \quad TY \quad \cdots \quad T^{l-1}Y] \begin{bmatrix} A_1 & A_2 & \cdots & A_{l-1} & I \\ A_2 & & \cdot^{\cdot^{\cdot}} & \cdot^{\cdot^{\cdot}} & \vdots \\ A_{l-1} & I & 0 & \cdots & 0 \\ I & 0 & 0 & \cdots & 0 \end{bmatrix} \begin{bmatrix} x_0 \\ x_1 \\ \vdots \\ x_{l-1} \end{bmatrix}. \tag{2.38}$$

Proof. Let us verify that $x(t)$ given by (2.35) and (2.38) is indeed a solution of the initial value problem. In view of Theorem 2.9 we have only to verify the equalities $x^{(i)}(a) = x_i$, $i = 0, \ldots, l-1$. Differentiating (2.35) it is seen that

$$\begin{bmatrix} x(a) \\ x'(a) \\ \vdots \\ x^{(l-1)}(a) \end{bmatrix} = \begin{bmatrix} X \\ XT \\ \vdots \\ XT^{l-1} \end{bmatrix} c. \tag{2.39}$$

Then substitute for c from (2.38) and the result follows from the biorthogonality condition (2.6).

To prove the uniqueness of the solution of the initial value problem, it is sufficient to check that the vector c from (2.35) is uniquely determined by $x^{(i)}(a)$, $i = 0, \ldots, l-1$. But this follows from (2.39) and the invertibility of $\operatorname{col}(XT^i)_{i=0}^{l-1}$. □

Theorems 2.9 and 2.10 show that the matrix function

$$G(t, s) = \begin{cases} Xe^{T(t-s)}Y & \text{for} \quad t \geq s \\ 0 & \text{for} \quad t < s \end{cases}$$

is the Green's function for the initial value problem. Recall that the *Green's function* is characterized by the following property:

$$x(t) = \int_a^t G(t, s) f(s)\, ds$$

is the solution of (2.34) such that $x^{(i)}(a) = 0$ for $i = 0, \ldots, l-1$.

We pass now to two-point boundary value problems. It will be convenient for us to assume that $T = J$ is a Jordan matrix (although this restriction is not essential).

To facilitate this analysis we begin with the construction of what might be called the pre-Green's function. First, let J_1 and J_2 be Jordan matrices with the property that $J = \operatorname{diag}[J_1, J_2]$ is a Jordan matrix associated with the monic matrix polynomial $L(\lambda) = \sum_{j=0}^{l} A_j \lambda^j$ $(A_l = I)$. Such a direct-sum decomposition of J will be called an *admissible splitting*. Thus, each elementary divisor of $L(\lambda)$ is associated with either a block of J_1 or J_2, but not with both. Let J_1 be $p \times p$ and J_2 be $q \times q$, so that $p + q = ln$. Then there are blocks X_1, X_2 of X and Y_1, Y_2 of Y compatible with the partition of J. With these understandings we have

Lemma 2.11. *The matrix-valued function defined on* $[a, b] \times [a, b]$ *by*

$$G_0(t, \tau) = \begin{cases} -X_1 e^{J_1(t-\tau)} Y_1 & \text{if} \quad a \leq t \leq \tau \\ X_2 e^{J_2(t-\tau)} Y_2 & \text{if} \quad \tau \leq t \leq b \end{cases} \tag{2.40}$$

satisfies the following conditions:

(a) *Differentiating with respect to t,* $L(d/dt)G_0 = 0$, *for* $(t, \tau) \in [a, b] \times [a, b]$, *as long as* $t \neq \tau$.

(b) *Differentiating with respect to t,*

$$G_0^{(r)}|_{t=\tau+} - G_0^{(r)}|_{t=\tau-} = \begin{cases} 0 & \text{for} \quad r = 0, 1, \ldots, l-2 \\ I & \text{for} \quad r = l-1. \end{cases}$$

(c) *The function*

$$u(t) = \int_a^b G_0(t, \tau) f(\tau)\, d\tau$$

is a solution of $L(d/dt)u = f$.

Proof. (a) We have

$$L\left(\frac{d}{dt}\right)G_0 = \begin{cases} -\left(\sum_{r=0}^{l} A_r X_1 J_1^r\right) e^{J_1(t-\tau)} Y_1, & a \leq t < \tau \\ \left(\sum_{r=0}^{l} A_r X_2 J_2^r\right) e^{J_2(t-\tau)} Y_2, & \tau < t \leq b. \end{cases}$$

Proposition 1.10 implies that the summation applied to each elementary divisor associated with J_1 or J_2 is zero, and it follows that $L(d/dt)G_0 = 0$ if $t \neq \tau$.

(b) We have

$$G_0^{(r)}|_{t=\tau+} - G_0^{(r)}|_{t=\tau-} = X_2 J_2^r Y_2 + X_1 J_1^r Y_1 = XJ^r Y,$$

and the conclusion follows from (2.12).

(c) This part is proved by verification. For brevity we treat the case $l = 2$. The same technique applies more generally. Write

$$u(t) = -X_1 e^{J_1 t} \int_t^b e^{-J_1 \tau} Y_1 f(\tau)\, d\tau + X_2 e^{J_2 t} \int_a^t e^{-J_2 \tau} Y_2 f(\tau)\, d\tau.$$

Differentiating and using the fact that $X_1 Y_1 + X_2 Y_2 = XY = 0$, we obtain

$$u^{(1)}(t) = -X_1 J_1 e^{J_1 t} \int_t^b e^{-J_1 \tau} Y_1 f(\tau)\, d\tau + X_2 J_2 e^{J_2 t} \int_a^t e^{-J_2 \tau} Y_2 f(\tau)\, d\tau.$$

Differentiate once more and use $XY = 0$, $XJY = I$ to obtain

$$u^{(2)}(t) = -X_1 J_1^2 e^{J_1 t} \int_t^b e^{-J_1 \tau} Y_1 f(\tau)\, d\tau + X_2 J_2^2 e^{J_2 t} \int_a^t e^{-J_2 \tau} Y_2 f(\tau)\, d\tau + f(t).$$

Then, since $\sum_{r=0}^{2} A_r X_i J_i^r = 0$ for $i = 2$, we obtain

$$L\left(\frac{d}{dt}\right)u = \sum_{r=0}^{2} A_r u^{(r)} = f. \quad \square$$

We now seek to modify G_0 in order to produce a Green's function G which will retain properties (a), (b), and (c) of the lemma but will, in addition, satisfy two-point boundary conditions. To formulate general conditions of this kind, let $\hat{y}_c$ denote the ln-vector determined by $\hat{y}_c = \operatorname{col}(y(c), y^{(1)}(c), \ldots, y^{(l-1)}(c))$ where $y(t)$, $t \in [a, b]$, is an $(l-1)$-times continuously differentiable n-dimensional vector function, and $c \in [a, b]$. Then let M, N be $ln \times ln$ constant matrices, and consider homogeneous boundary conditions of the form

$$M\hat{y}_a + N\hat{y}_b = 0.$$

Our boundary value problem now has the form

$$L\left(\frac{d}{dt}\right)u = f, \qquad M\hat{u}_a + N\hat{u}_b = 0, \tag{2.41}$$

for the case in which the homogeneous problem

$$L\left(\frac{d}{dt}\right)u = 0, \qquad M\hat{u}_a + N\hat{u}_b = 0 \tag{2.42}$$

has only the trivial solution.

The necessary and sufficient condition for the latter property is easy to find.

Lemma 2.12. *The homogeneous problem* (2.42) *has only the trivial solution if and only if*

$$\det(MQe^{Ja} + NQe^{Jb}) \neq 0, \tag{2.43}$$

where $Q = \operatorname{col}(XJ^i)_{i=0}^{l-1}$.

Proof. We have seen that every solution of $L(d/dt)u = 0$ is expressible in the form $u(t) = Xe^{Jt}c$ for some $c \in \mathcal{C}_{ln}$, so the boundary condition implies

$$M\hat{u}_a + N\hat{u}_b = (MQe^{Ja} + NQe^{Jb})c = 0.$$

Since this equation is to have only the trivial solution, the $ln \times ln$ matrix $MQe^{Ja} + NQe^{Jb}$ is nonsingular. $\square$

Theorem 2.13. *If the homogeneous problem* (2.42) *has only the trivial solution, then there is a unique $ln \times n$ matrix $K(\tau)$ independent of t and depending analytically on τ, in $[a, b]$, for which the function*

$$G(t, \tau) = G_0(t, \tau) + Xe^{Jt}K(\tau) \tag{2.44}$$

satisfies conditions (a) *and* (b) *of Lemma* 2.11 (*applied to* G) *as well as*

$$\text{(d)} \qquad V(G) \overset{\text{def}}{=} M\begin{bmatrix} G(a,\tau) \\ G'(a,\tau) \\ \vdots \\ G^{(l-1)}(a,\tau) \end{bmatrix} + N\begin{bmatrix} G(b,\tau) \\ G'(b,\tau) \\ \vdots \\ G^{(l-1)}(b,\tau) \end{bmatrix} = 0.$$

In this case, the unique solution of (2.41) *is given by*

$$u(t) = \int_a^b G(t,\tau)f(\tau)\,d\tau. \tag{2.45}$$

Proof. It is easily seen that G inherits conditions (a) and (b) from G_0 for any $K(\tau)$. For condition (d) observe that $V(G) = 0$ if and only if $V(Xe^{Jt}K(\tau)) = -V(G_0)$. This may be written

$$(MQe^{Ja} + NQe^{Jb})K(\tau) = -V(G_0), \tag{2.46}$$

where $Q = \operatorname{col}(XJ^i)_{i=0}^{l-1}$. We have seen that our assumption on the homogeneous problem implies that the matrix on the left is nonsingular, so that a unique K is obtained. The analytic dependence of K on τ (through $V(G_0)$) is clear.

Finally, with u defined by (2.45) we have the boundary conditions satisfied for u in view of (d), and

$$L\left(\frac{d}{dt}\right)u = L\left(\frac{d}{dt}\right)\left\{\int_a^b G_0(t,\tau)f(\tau)\,d\tau + Xe^{Jt}\int_a^b K(\tau)f(\tau)\,d\tau\right\}.$$

The first integral reduces to f by virtue of condition (c) in Lemma 2.11, and $L(d/dt)Xe^{Jt} = (\sum_{r=0}^{l} A_r XJ^r)e^{Jt}$ is zero because of Theorem 1.23. This completes the proof. □

We remark that there is an admissible splitting of J implicit in the use of G_0 in (2.44). The representation of the solution in the form (2.45) is valid for *any* admissible splitting.

As an important special case of the foregoing analysis we consider second-order problems:

$$L\left(\frac{d}{dt}\right)u = A_0 + A_1u^{(1)} + Iu^{(2)} = f$$

with the boundary conditions $u(a) = u(b) = 0$. These boundary conditions are obtained by defining the $2n \times 2n$ matrices

$$M = \begin{bmatrix} I & 0 \\ 0 & 0 \end{bmatrix}, \qquad N = \begin{bmatrix} 0 & 0 \\ I & 0 \end{bmatrix},$$

where the blocks are all $n \times n$. Insisting that the homogeneous problem have only the trivial solution then implies that the matrix

$$MQe^{Ja} + NQe^{Jb} = \begin{bmatrix} Xe^{Ja} \\ 0 \end{bmatrix} + \begin{bmatrix} 0 \\ Xe^{Jb} \end{bmatrix} = \begin{bmatrix} Xe^{Ja} \\ Xe^{Jb} \end{bmatrix} \tag{2.47}$$

appearing in (2.43) is nonsingular.

In this case we have

$$\begin{aligned} V(G_0) &= -M\begin{bmatrix} X_1 \\ X_1 J_1 \end{bmatrix} e^{J_1(a-\tau)} Y_1 + N\begin{bmatrix} X_2 \\ X_2 J_2 \end{bmatrix} e^{J_2(b-\tau)} Y_2 \\ &= \begin{bmatrix} -X_1 e^{J_1(a-\tau)} Y_1 \\ X_2 e^{J_2(b-\tau)} Y_2 \end{bmatrix} \end{aligned}$$

and the solution of (2.46) is

$$K(\tau) = \begin{bmatrix} Xe^{Ja} \\ Xe^{Jb} \end{bmatrix}^{-1} \begin{bmatrix} X_1 e^{J_1(a-\tau)} Y_1 \\ -X_2 e^{J_2(b-\tau)} Y_2 \end{bmatrix}, \tag{2.48}$$

yielding a more explicit representation of the Green's function in (2.44).

The condition on a, b, and the Jordan chains of $L(\lambda)$, which is necessary and sufficient for the existence of a unique solution to our two-point boundary value problem is, therefore, that the matrix (2.47) be nonsingular.

2.5. Complete Pairs and Second-Order Differential Equations

If $L(\lambda) = I\lambda^2 + A_1\lambda + A_2$ and there exist matrices S, T such that $L(\lambda) = (I\lambda - T)(I\lambda - S)$, then $I\lambda - S$ is said to be a (monic) *right divisor of* L. More general divisibility questions will be considered in subsequent chapters but, for the present discussion, we need the easily verified proposal: *$I\lambda - S$ is a right divisor of L if and only if*

$$L(S) = S^2 + A_1 S + A_2 = 0.$$

In general, $L(\lambda)$ may have many distinct right divisors but pairs of divisors of the following kind are particularly significant: right divisors $I\lambda - S_1$, $I\lambda - S_2$ for $L(\lambda) = I\lambda^2 + A_1\lambda + A_2$ are said to form a *complete pair* if $S_2 - S_1$ is invertible.

Lemma 2.14. *If $I\lambda - S_1$, $I\lambda - S_2$ form a complete pair for $L(\lambda)$, then*

$$L(\lambda) = (S_2 - S_1)(I\lambda - S_2)(S_2 - S_1)^{-1}(I\lambda - S_1)$$

and

$$L^{-1}(\lambda) = \{(I\lambda - S_2)^{-1} - (I\lambda - S_1)\}^{-1}(S_2 - S_1)^{-1}.$$

Proof. By definition of right divisors, there exist matrices T_1, T_2 such that

$$L(\lambda) = (I\lambda - T_1)(I\lambda - S_1) = (I\lambda - T_2)(I\lambda - S_2).$$

Consequently, $T_1 + S_1 = T_2 + S_2$ and $T_1 S_1 = T_2 S_2$. It follows from these two relations, and the invertibility of $S_2 - S_1$ that we may write $T_1 = (S_2 - S_1)S_2(S_2 - S_1)^{-1}$, and the first conclusion follows immediately. Now it is easily verified that

$$(I\lambda - S_1)^{-1} - (I\lambda - S_2)^{-1} = -(I\lambda - S_1)^{-1}(S_2 - S_1)(I\lambda - S_2)^{-1}.$$

If we take advantage of this relation, the second conclusion of the lemma now follows from the first. □

The importance of the lemma is twofold: It shows that a complete pair gives a complete factorization of $L(\lambda)$ and that $\sigma(L) = \sigma(S_1) \cup \sigma(S_2)$. Indeed, the first part of the proof of Theorem 2.16 will show that S_1, S_2 determine a Jordan pair for L.

We shall see that complete pairs can also be useful in the solution of differential equations, but first let us observe a useful sufficient condition ensuring that a pair of right divisors is complete.

Theorem 2.15. *Let $I\lambda - S_1$, $I\lambda - S_2$ be right divisors of $L(\lambda)$ such that $\sigma(S_1) \cap \sigma(S_2) = \varnothing$. Then $I\lambda - S_1$, $I\lambda - S_2$ form a complete pair.*

Proof. Observe first that

$$L(\lambda)[(I\lambda - S_1)^{-1} - (I\lambda - S_2)^{-1}] = S_1 - S_2. \tag{2.49}$$

Indeed, since $L(\lambda)(I\lambda - S_i)^{-1}$ is a monic linear polynomial for $i = 1, 2$, the left-hand side of (2.49) is a constant matrix, i.e., does not depend on λ. Further, for large $|\lambda|$ we have

$$(I\lambda - S_i)^{-1} = I\lambda^{-1} + S_i\lambda^{-2} + S_i^2\lambda^{-3} + \cdots, \qquad i = 1, 2;$$

so

$$\lim_{\lambda\to\infty} \{\lambda^2[(I\lambda - S_1)^{-1} - (I\lambda - S_2)^{-1}]\} = S_1 - S_2.$$

Since $\lim_{\lambda\to\infty}\{\lambda^{-2}L(\lambda)\} = I$, equality (2.49) follows.

Assume now that $(S_1 - S_2)x = 0$ for some $x \in \mathbb{C}^n$. Equality (2.49) implies that for $\lambda \notin \sigma(L)$,

$$(I\lambda - S_1)^{-1}x = (I\lambda - S_2)^{-1}x. \tag{2.50}$$

Since $\sigma(S_1) \cap \sigma(S_2) = \varnothing$, equality (2.50) defines a vector function $E(\lambda)$ which is analytic in the whole complex plane:

$$E(\lambda) = \begin{cases} (I\lambda - S_1)^{-1}x, & \lambda \in \mathbb{C}\backslash\sigma(S_1) \\ (I\lambda - S_2)^{-1}x, & \lambda \in \mathbb{C}\backslash\sigma(S_2). \end{cases}$$

But $\lim_{\lambda\to\infty} E(\lambda) = 0$. By Liouville's theorem, $E(\lambda)$ is identically zero. Then $x = (I\lambda - S_1)E(\lambda) = 0$, i.e., $\operatorname{Ker}(S_1 - S_2) = \{0\}$ and $S_1 - S_2$ is nonsingular. □

Theorem 2.16. *Let S_1, S_2 be a complete pair of solutions of $L(S) = 0$, where $L(\lambda) = I\lambda^2 + A_1\lambda + A_0$. Then*

(a) *Every solution of $L(d/dt)u = 0$ has the form*

$$u(t) = e^{S_1 t}c_1 + e^{S_2 t}c_2$$

for some $c_1, c_2 \in \mathbb{C}^n$.

(b) *If, in addition, $e^{S_2(b-a)} - e^{S_1(b-a)}$ is nonsingular, then the two-point boundary value problem*

$$L\left(\frac{d}{dt}\right)u = f, \qquad u(a) = u(b) = 0, \tag{2.51}$$

has the unique solution

$$u(t) = \int_a^b G_0(t,\tau)f(\tau)\,d\tau - [e^{S_1(t-a)}, \quad e^{S_2(t-a)}]W \int_a^b \begin{bmatrix} e^{S_1(a-\tau)} \\ e^{S_2(b-\tau)} \end{bmatrix} Zf(\tau)\,d\tau \tag{2.52}$$

where

$$Z = (S_2 - S_1)^{-1}, \qquad W = \begin{bmatrix} I & I \\ e^{S_1(b-a)} & e^{S_2(b-a)} \end{bmatrix}^{-1},$$

and

$$G_0(t,\tau) = \begin{cases} e^{S_1(t-\tau)}Z, & a \le t \le \tau \\ e^{S_2(t-\tau)}Z, & \tau \le t \le b. \end{cases} \tag{2.53}$$

The interesting feature of this result is that the solutions of the differential equation are completely described by the solutions S_1, S_2 of the algebraic problem $L(S) = 0$.

Proof. Let S_1, S_2 have Jordan forms J_1, J_2, respectively, and let $S_1 = X_1 J_1 X_1^{-1}$, $S_2 = X_2 J_2 X_2^{-1}$ and write

$$X = [X_1 \quad X_2], \qquad J = \operatorname{diag}[J_1, J_2].$$

We show first that S_1, S_2 a complete pair implies that X, J is a Jordan pair for $L(\lambda)$. We have

$$\begin{bmatrix} X \\ XJ \end{bmatrix} = \begin{bmatrix} X_1 & X_2 \\ X_1 J_1 & X_2 J_2 \end{bmatrix} = \begin{bmatrix} I & I \\ S_1 & S_2 \end{bmatrix} \begin{bmatrix} X_1 & 0 \\ 0 & X_2 \end{bmatrix}.$$

So the nonsingularity of the first matrix follows from that of $S_2 - S_1$. Then $S_i^2 + A_1 S_i + A_0 = 0$, for $i = 1, 2$, implies $X_i J_i^2 + A_1 X_i J_i + A_0 X_i = 0$ for each i and hence that $XJ^2 + A_1 XJ + A_0 X = 0$. Hence (by Theorem 1.23) X, J is a Jordan pair for $L(\lambda)$ and, in the terminology of the preceding section, J_1 and J_2 determine an admissible splitting of J.

To prove part (a) of the theorem observe that

$$u(t) = \begin{bmatrix} e^{S_1 t} & e^{S_2 t} \end{bmatrix} \begin{bmatrix} c_1 \\ c_2 \end{bmatrix} = \begin{bmatrix} X_1 e^{J_1 t} & X_2 e^{J_2 t} \end{bmatrix} \begin{bmatrix} X_1^{-1} c_1 \\ X_2^{-1} c_2 \end{bmatrix} = X e^{Jt} \alpha,$$

where α is, in effect, an arbitrary vector of $\mathbb{C}^{2n}$. Since X, J is a standard pair, the result follows from Theorem 2.9.

For part (b) observe first that, as above,

$$X e^{Jt} = \begin{bmatrix} e^{S_1 t} X_1 & e^{S_2 t} X_2 \end{bmatrix}$$

so that the matrix of (2.47) can be written

$$\begin{bmatrix} X e^{Ja} \\ X e^{Jb} \end{bmatrix} = \begin{bmatrix} e^{S_1 a} X_1 & e^{S_2 a} X_2 \\ e^{S_1 b} X_1 & e^{S_2 b} X_2 \end{bmatrix} = \begin{bmatrix} I & I \\ e^{S_1 (b-a)} & e^{S_2 (b-a)} \end{bmatrix} \begin{bmatrix} e^{S_1 a} X_1 & 0 \\ 0 & e^{S_2 a} X_2 \end{bmatrix}$$

and the hypothesis of part (b) implies the nonsingularity of this matrix and the uniqueness of the solution for part (b).

Thus, according to Theorem 2.13 and remarks thereafter, the unique solution of (2.51) is given by the formula

$$u(t) = \int_a^b G_0(t, \tau) f(\tau)\, d\tau + \int_a^b X e^{Jt} \begin{bmatrix} X e^{Ja} \\ X e^{Jb} \end{bmatrix}^{-1} \begin{bmatrix} X_1 e^{J_1 (a-\tau)} Y_1 \\ -X_2 e^{J_2 (b-\tau)} Y_2 \end{bmatrix} f(\tau)\, d\tau \tag{2.54}$$

where

$$\begin{bmatrix} Y_1 \\ Y_2 \end{bmatrix} = \begin{bmatrix} X \\ XJ \end{bmatrix}^{-1} \begin{bmatrix} 0 \\ I \end{bmatrix}$$

and

$$G_0(t, \tau) = \begin{cases} -X_1 e^{J_1 (t-\tau)} Y_1 & \text{if } a \le t \le \tau, \\ X_2 e^{J_2 (t-\tau)} Y_2 & \text{if } \tau \le t \le b. \end{cases} \tag{2.55}$$

It remains to show that formulas (2.52) and (2.54) are the same.

First note that $Y_1 = -X_1^{-1} Z$, $Y_2 = X_2^{-1} Z$. Indeed,

$$\begin{bmatrix} X \\ XJ \end{bmatrix} \begin{bmatrix} -X_1^{-1} Z \\ X_2^{-1} Z \end{bmatrix} = \begin{bmatrix} X_1 & X_2 \\ X_1 J_1 & X_2 J_2 \end{bmatrix} \begin{bmatrix} -X_1^{-1} \\ X_2^{-1} \end{bmatrix} Z = \begin{bmatrix} 0 \\ S_2 - S_1 \end{bmatrix} Z = \begin{bmatrix} 0 \\ I \end{bmatrix}.$$

Furthermore, the function G_0 of (2.53) can now be identified with the function G_0 of (2.55). To check this use the equality $X_i e^{J_i(t-\tau)} X_i^{-1} = e^{S_i(t-\tau)}$, $i = 1, 2$ (cf. Section S1.8). Further,

$$Xe^{Jt} = [X_1 e^{J_1 t} \quad X_2 e^{J_2 t}] = [e^{S_1 t} \quad e^{S_2 t}] \begin{bmatrix} X_1 & 0 \\ 0 & X_2 \end{bmatrix}$$

$$= [e^{S_1(t-a)} \quad e^{S_2(t-b)}] \begin{bmatrix} e^{aS_1} X_1 & 0 \\ 0 & e^{bS_2} X_2 \end{bmatrix}.$$

So

$$Xe^{Jt} \begin{bmatrix} Xe^{Ja} \\ Xe^{Jb} \end{bmatrix}^{-1} \begin{bmatrix} X_1 e^{J_1(a-\tau)} Y_1 \\ -X_2 e^{J_2(b-\tau)} Y_2 \end{bmatrix}$$

$$= -[e^{S_1(t-a)} \quad e^{S_2(t-b)}] \begin{bmatrix} e^{aS_1} X_1 & 0 \\ 0 & e^{bS_2} X_2 \end{bmatrix} \begin{bmatrix} Xe^{Ja} \\ Xe^{Jb} \end{bmatrix}^{-1} \begin{bmatrix} e^{S_1(a-\tau)} \\ e^{S_2(b-\tau)} \end{bmatrix} Z,$$

and in order to prove that formulas (2.52) and (2.54) are the same, we need only to check that

$$W = \begin{bmatrix} e^{aS_1} X_1 & 0 \\ 0 & e^{bS_2} X_2 \end{bmatrix} \begin{bmatrix} Xe^{Ja} \\ Xe^{Jb} \end{bmatrix}^{-1},$$

or

$$W^{-1} \stackrel{\text{def}}{=} \begin{bmatrix} I & I \\ e^{S_1(b-a)} & e^{S_2(b-a)} \end{bmatrix} = \begin{bmatrix} Xe^{Ja} \\ Xe^{Jb} \end{bmatrix} \begin{bmatrix} X_1^{-1} e^{-aS_1} & 0 \\ 0 & X_2^{-1} e^{-bS_2} \end{bmatrix}.$$

But this follows from the equalities $S_i = X_i J_i X_i^{-1}$, $i = 1, 2$. □

2.6. Initial Value Problem for Difference Equations, and the Generalized Newton Identities

Consider the difference equation

$$A_0 u_r + A_1 u_{r+1} + \cdots + A_{l-1} u_{r+l-1} + u_{r+l} = f_r, \qquad r = 0, 1, \ldots, \tag{2.56}$$

where $L(\lambda) = I\lambda^l + \sum_{j=0}^{l-1} A_j \lambda^j$ is a monic matrix polynomial with a standard triple (X, T, Y).

Using Theorem 1.6 and the similarity between any two standard triples of $L(\lambda)$, we obtain the following result.

Theorem 2.17. *Every solution of* (2.56) *is given by* $u_0 = Xc$ *and for* $i = 1, 2, \ldots,$

$$u_i = XT^i c + X \sum_{k=0}^{i-1} T^{i-k-1} Y f_k, \tag{2.57}$$

where $c \in \mathbb{C}^{nl}$ is arbitrary. In particular, the general solution of the homogeneous problem (with $f_r = 0$, $r = 0, 1, \ldots$) is given by

$$u_i = XT^i c, \qquad i = 0, 1, \ldots, \quad c \in \mathbb{C}^{nl}.$$

Corollary 2.18. *The solution of (2.56) which satisfies the prescribed initial conditions*

$$u_r = a_r, \qquad r = 0, \ldots, l-1,$$

is unique and given by putting

$$c = [Y \quad TY \quad \cdots \quad T^{l-1}Y] \begin{bmatrix} A_1 & A_2 & \cdots & A_{l-1} & I \\ A_2 & & \cdot^{\cdot^{\cdot}} & \cdot^{\cdot^{\cdot}} & \\ \vdots & \cdot^{\cdot^{\cdot}} & \cdot^{\cdot^{\cdot}} & & \vdots \\ A_{l-1} & I & \cdots & & 0 \\ I & 0 & \cdots & & 0 \end{bmatrix} \begin{bmatrix} a_0 \\ a_1 \\ \vdots \\ a_{l-1} \end{bmatrix}$$

in Theorem 2.17.

Proof. Write

$$\begin{bmatrix} u_0 \\ \vdots \\ u_{l-1} \end{bmatrix} = \begin{bmatrix} a_0 \\ \vdots \\ a_{l-1} \end{bmatrix}, \tag{2.58}$$

where u_i is given by (2.57). Since $XT^iY = 0$ for $i = 0, \ldots, l-2$, we obtain $u_i = XT^ic$ for $i = 0, \ldots, l-1$, and (2.58) becomes

$$\operatorname{col}(XT^i)_{i=0}^{l-1} \cdot c = \operatorname{col}(a_i)_{i=0}^{l-1}.$$

Since $\operatorname{col}(XT^i)_{i=0}^{l-1}$ is invertible, $c \in \mathbb{C}^{nl}$ is uniquely defined and, by the biorthogonality condition (2.6),

$$c = [\operatorname{col}(XT^i)_{i=0}^{l-1}]^{-1} \cdot \operatorname{col}(a_i)_{i=0}^{l-1} = [\operatorname{row}(T^iY)_{i=0}^{l-1}]B \cdot \operatorname{col}(a_i)_{i=0}^{l-1}. \qquad \square$$

As an application of Theorem 2.17, we describe here a method of evaluation of the sum σ_r of the rth powers of the eigenvalues of $L(\lambda)$ (including multiplicities).

Consider the matrix difference equation connected with the polynomial $L(\lambda) = I\lambda^l + \sum_{j=0}^{l-1} A_j\lambda^j$:

$$A_0U_r + A_1U_{r+1} + \cdots + A_{l-1}U_{r+l-1} + U_{r+l} = 0, \qquad r = 1, 2, \ldots, \tag{2.59}$$

where $\{U_r\}_{r=1}^{\infty}$ is a sequence of $n \times n$ matrices U_r to be found. In view of Theorem 2.17 the general solution of (2.59) is given by the following formula:

$$U_r = XT^{r-1}Z, \qquad r = 1, 2, \ldots, \tag{2.60}$$

where (X, T) is a part of the standard triple (X, T, Y) of $L(\lambda)$ and Z is an arbitrary $ln \times n$ matrix. Clearly, any particular solution of (2.59) is uniquely defined by the initial values $U_1, \ldots, U_l$.

We now define matrices $S_1, S_2, \ldots, S_l$ by the solution of

$$\begin{bmatrix} A_1 & A_2 & \cdots & I \\ A_2 & \vdots & \cdot^{\cdot^{\cdot}} & 0 \\ \vdots & I & & \vdots \\ I & 0 & \cdots & 0 \end{bmatrix} \begin{bmatrix} S_1 \\ S_2 \\ \vdots \\ S_l \end{bmatrix} = - \begin{bmatrix} lA_0 \\ \vdots \\ 2A_{l-2} \\ A_{l-1} \end{bmatrix} \tag{2.61}$$

and we observe that, in practice, they would most readily be computed recursively beginning with S_1. Note that the matrix on the left is nonsingular. Define $\{S_r\}_{r=1}^{\infty}$ as the unique solution of (2.59) determined by the initial values $U_i = S_i$, $i = 1, \ldots, l$.

Let σ_r be the sum of the rth powers of the eigenvalues of L (including multiplicities), and let $\mathrm{Tr}(A)$ of a matrix A denote the trace of A, i.e., the sum of the elements on the main diagonal of A, or, what is the same, the sum of all eigenvalues of A (counting multiplicities).

Theorem 2.19. *If $\{S_r\}_{r=1}^{\infty}$ is defined as above, then*

$$\sigma_r = \mathrm{Tr}(S_r), \qquad r = 1, 2, \ldots. \tag{2.62}$$

The equalities (2.62) are known as the generalized Newton identities.

Proof. By Corollary 2.18,

$$S_r = -XT^{r-1}[Y \quad TY \quad \cdots \quad T^{l-1}Y] \begin{bmatrix} lA_0 \\ \vdots \\ 2A_{l-2} \\ A_{l-1} \end{bmatrix},$$

or

$$S_r = -(l\Gamma_r A_0 + \cdots + 2\Gamma_{r+l-2}A_{l-2} + \Gamma_{r+l-1}A_{l-1}), \tag{2.63}$$

where $\Gamma_r = XT^{r-1}Y$.

Now we also have (see Proposition 2.1 (iii))

$$\Gamma_r A_0 + \Gamma_{r+1}A_1 + \cdots + \Gamma_{r+l} = 0, \qquad r = 1, 2, \ldots,$$

and this implies that (2.63) can be written

$$S_r = \Gamma_{r+1}A_1 + 2\Gamma_{r+2}A_2 + \cdots + l\Gamma_{r+l}, \qquad r = 1, 2, \ldots. \tag{2.64}$$

To approach the eigenvalues of L we observe that they are also the eigenvalues of the matrix $M = QTQ^{-1}$. Since $Q^{-1} = RB$, where

$$R = [Y \quad TY \quad \cdots \quad T^{l-1}Y]$$

and B is given by (2.5), we have:

$$M^r = QT^rQ^{-1} = QT^rRB$$

$$= \begin{bmatrix} \Gamma_{r+1} & \Gamma_{r+2} & \cdots & \Gamma_{r+l} \\ \Gamma_{r+2} & \Gamma_{r+3} & \cdots & \Gamma_{r+l+1} \\ \vdots & & & \vdots \\ \Gamma_{r+l} & \Gamma_{r+l+1} & \cdots & \Gamma_{r+2l-1} \end{bmatrix} \begin{bmatrix} A_1 & A_2 & \cdots & I \\ A_2 & & ⋰ & \\ \vdots & I & & \\ I & & & 0 \end{bmatrix}.$$

Now observe that the eigenvalues of M^r are just the rth powers of the eigenvalues of M (this fact can be easily observed by passing to the Jordan form of M). Hence

$$\begin{aligned} \sigma_r &= \operatorname{Tr}(M^r) = \operatorname{Tr}(\Gamma_{r+1}A_1 + \cdots + \Gamma_{r+l}A_l) + \cdots \\ &\quad + \operatorname{Tr}(\Gamma_{r+l-1}A_{l-1} + \Gamma_{r+l}A_l) + \operatorname{Tr}(\Gamma_{r+l}A_l) \\ &= \operatorname{Tr}(\Gamma_{r+1}A_1 + 2\Gamma_{r+2}A_2 + \cdots + l\Gamma_{r+l}A_l) \\ &= \operatorname{Tr}(S_r) \end{aligned}$$

by (2.64). □

Comments

The systematization of spectral information in the form of Jordan pairs and triples, leading to the more general standard pairs and triples, originates with the authors in [34a, 34b], as do the representations of Theorem 2.4. The resolvent form, however, has a longer history. More restrictive versions appear in [40, 52a]. As indicated in Section 2.3 it can also be seen as a special case of the well-known realization theorems of system theory (see [80, Chapter 4] and [47]). However, our analysis also gives a complete description of the realization theorem in terms of the spectral data. For resolvent forms of rational matrix functions see [3c, Chapter 2]. Corollary 2.5 was proved in [49a]; see also [38, 55].

Factorizations of matrix polynomials as described in Proposition 2.7 are introduced in [79a].

Our discussion of initial and two-point boundary value problems in Sections 2.4 and 2.5 is based on the papers [34a, 52e]. The concept of "complete pairs" is introduced in [51]. Existence theorems can be found in that paper as well as Lemma 2.14 and Theorem 2.15. Existence theorems and numerical methods are considered in [54]. The applications of Section 2.6 are based on [52f].

Extensions of the main results of this chapter to polynomials whose coefficients are bounded linear operators in a Banach space can be found in [34c].

Some generalizations of this material for analytic matrix functions appear in [37d, 70c].

Chapter 3

Multiplication and Divisibility

In Chapter 2 we used the notion of a standard triple to produce representation theorems having immediate significance in the solution of differential and difference equations. It turns out that these representations of matrix polynomials provide a very natural approach to the study of products and quotients of such polynomials. These questions are to be studied in this chapter.

There are many important problems which require a factorization of matrix- or operator-valued functions in such a way that the factors have certain spectral properties. Using the spectral approach it is possible to characterize right divisors by specific invariant subspaces of a linearization, which are known as supporting subspaces for the right divisors. This is one of the central ideas of the theory to be developed in this chapter, and admits explicit formulas for divisors and quotients. These formulas will subsequently facilitate the study of perturbation and stability of divisors.

If a monic matrix polynomial $L(\lambda) = L_2(\lambda)L_1(\lambda)$ and $L_1(\lambda)$ and $L_2(\lambda)$ are also monic matrix polynomials, then $L_1(\lambda)$, $L_2(\lambda)$ may be referred to as a right divisor and left quotient of $L(\lambda)$, respectively (or, equally well, as a right quotient and left divisor, respectively). It is clear that the spectrum of L must "split" in some way into the spectra of L_1 and L_2. Our first analysis of products and divisors (presented in this chapter), makes no restriction on the nature of this splitting. Subsequently (in the next chapter) we consider in more detail the important special case in which the spectra of L_1 and L_2 are disjoint.

3.1. A Multiplication Theorem

The notion of a standard triple for monic matrix polynomials introduced in the preceding chapter turns out to be a useful and natural tool for the description of multiplication and division of monic matrix polynomials. In this section we deal with the multiplication properties. First we compute the inverse $L^{-1}(\lambda)$ of the product $L(\lambda) = L_2(\lambda)L_1(\lambda)$ of two monic matrix polynomials $L_1(\lambda)$ and $L_2(\lambda)$ in terms of their standard triples.

Theorem 3.1. *Let $L_i(\lambda)$ be a monic operator polynomial with standard triple (Q_i, T_i, R_i) for $i = 1, 2$, and let $L(\lambda) = L_2(\lambda)L_1(\lambda)$. Then*

$$L^{-1}(\lambda) = [Q_1 \quad 0](I\lambda - T)^{-1}\begin{bmatrix} 0 \\ R_2 \end{bmatrix}, \tag{3.1}$$

where

$$T = \begin{bmatrix} T_1 & R_1Q_2 \\ 0 & T_2 \end{bmatrix}.$$

Proof. It is easily verified that

$$(I\lambda - T)^{-1} = \begin{bmatrix} (I\lambda - T_1)^{-1} & (I\lambda - T_1)^{-1}R_1Q_2(I\lambda - T_2)^{-1} \\ 0 & (I\lambda - T_2)^{-1} \end{bmatrix}.$$

The product on the right of (3.1) is then found to be

$$Q_1(I\lambda - T_1)^{-1}R_1Q_2(I\lambda - T_2)^{-1}R_2.$$

But, using Theorem 2.4, this is just $L_1^{-1}(\lambda)L_2^{-1}(\lambda)$, and the theorem follows immediately. □

Theorem 3.2. *If $L_i(\lambda)$ are monic matrix polynomials with standard triples (Q_i, T_i, R_i) for $i = 1, 2$, then $L(\lambda) = L_2(\lambda)L_1(\lambda)$ has a standard triple (Q, T, R) with the representations*

$$Q = [Q_1 \quad 0], \qquad T = \begin{bmatrix} T_1 & R_1Q_2 \\ 0 & T_2 \end{bmatrix}, \qquad R = \begin{bmatrix} 0 \\ R_2 \end{bmatrix}.$$

Proof. Combine Theorem 2.6 with Theorem 3.1. □

Note that if both triples (Q_i, T_i, R_i), $i = 1, 2$, are Jordan, then the triple (Q, T, R) from Theorem 3.2 is not necessarily Jordan (because of the presence of R_1Q_2 in T). But, as we shall see, in some cases it is not hard to write down a Jordan triple for the product, by using Theorem 3.2.

Consider, for instance, the case when $\sigma(T_1) \cap \sigma(T_2) = \emptyset$. In this case one can construct a Jordan triple of the product $L(\lambda) = L_2(\lambda)L_1(\lambda)$, provided the triples (Q_i, T_i, R_i) for $L_i(\lambda)$ ($i = 1, 2$) are also Jordan. Indeed, there exists a unique solution Z of the equation

$$-T_1Z + ZT_2 = R_1Q_2 \tag{3.2}$$

(see Section S2.1); so

$$\begin{bmatrix} T_1 & 0 \\ 0 & T_2 \end{bmatrix} = \begin{bmatrix} I & -Z \\ 0 & I \end{bmatrix}\begin{bmatrix} T_1 & R_1Q_2 \\ 0 & T_2 \end{bmatrix}\begin{bmatrix} I & Z \\ 0 & I \end{bmatrix}.$$

Hence, the triple

$$(Q_1, Q_1Z) = [Q_1 \quad 0]\begin{bmatrix} I & Z \\ 0 & I \end{bmatrix}, \begin{bmatrix} T_1 & 0 \\ 0 & T_2 \end{bmatrix}, \begin{bmatrix} -ZR_2 \\ R_2 \end{bmatrix} = \begin{bmatrix} I & -Z \\ 0 & I \end{bmatrix}\begin{bmatrix} 0 \\ R_2 \end{bmatrix} \tag{3.3}$$

is similar to the standard triple (Q, T, R) of $L(\lambda)$ constructed in Theorem 3.2, and therefore it is also a standard triple of $L(\lambda)$. Now (3.3) is Jordan provided both (Q_1, T_1, R_1) and (Q_2, T_2, R_2) are Jordan.

Recall that equation (3.2) has a solution Z if and only if

$$\begin{bmatrix} T_1 & R_1Q_2 \\ 0 & T_2 \end{bmatrix} \quad \text{is similar to} \quad \begin{bmatrix} T_1 & 0 \\ 0 & T_2 \end{bmatrix}$$

(see Theorem S2.1); then the arguments above still apply, and we obtain the following corollary.

Corollary 3.3. *Let $L_1(\lambda)$, $L_2(\lambda)$ be monic matrix polynomials such that, if $(\lambda - \lambda_0)^{\alpha_{ij}}$, $i = 1, \ldots, k_j$, is the set of the elementary divisors of $L_j(\lambda)$ ($j = 1, 2$) corresponding to any eigenvalue of $L(\lambda) = L_2(\lambda)L_1(\lambda)$, then $(\lambda - \lambda_0)^{\alpha_{ij}}$, $i = 1, \ldots, k_j, j = 1, 2$, is the set of the elementary divisors of $L(\lambda)$ corresponding to λ_0. Let (Q_i, T_i, R_i) be a Jordan triple of $L_i(\lambda)$, $i = 1, 2$. Then Eq. (3.2) has a solution Z, and the triple*

$$[Q_1, Q_1Z], \begin{bmatrix} T_1 & 0 \\ 0 & T_2 \end{bmatrix}, \begin{bmatrix} -ZR_2 \\ R_2 \end{bmatrix}$$

is a Jordan triple for $L(\lambda)$.

If the structure of the elementary divisors of $L(\lambda)$ is more complicated than in Corollary 3.3, it is not known in general how to find the Jordan triples for the product. Another unsolved problem is the following one (see [27, 71a, 76b]): Given the elementary divisors of the factors $L_1(\lambda)$ and $L_2(\lambda)$ (this means, given Jordan matrices T_1, T_2 in the above notation), find the elementary divisors of the product $L(\lambda) = L_2(\lambda)L_1(\lambda)$ (this means, find the Jordan form of the matrix

$$T = \begin{bmatrix} T_1 & R_1 Q_2 \\ 0 & T_2 \end{bmatrix}\Bigg).$$

In some particular cases the solution of this problem is known. For instance, the following result holds:

Lemma 3.4. *Let $J_1, \ldots, J_k$ be Jordan blocks having the same real eigenvalue and sizes $\alpha_1 \geq \cdots \geq \alpha_k$, respectively. Let*

$$T = \begin{bmatrix} \operatorname{diag}[J_i]_{i=1}^k & T_0 \\ 0 & \operatorname{diag}[J_i^{\mathrm{T}}]_{i=1}^k \end{bmatrix}$$

where $T_0 = (t_{ij})_{i,j=1}^{\alpha}$ is a matrix of size $\alpha \times \alpha$ $(\alpha = \alpha_1 + \cdots + \alpha_k)$. Let $\beta_i = \alpha_1 + \cdots + \alpha_i$ $(i = 1, \ldots, k)$, and suppose that the $k \times k$ submatrix $A = (t_{\beta_i \beta_j})_{i,j=1}^k$ of T_0 is invertible and $A = A_1 A_2$, where A_1 is a lower triangular matrix and A_2 is an upper triangular matrix. Then the degrees of the elementary divisors of $I\lambda - T$ are $2\alpha_1, \ldots, 2\alpha_k$.

The proof of Lemma 3.4 is quite technical. Nevertheless, since the lemma will be used later in this book, we give its proof here.

Proof. We shall apply similarity transformations to the matrix T step by step, and eventually obtain a matrix similar to T for which the elementary divisors obviously are $2\alpha_1, \ldots, 2\alpha_k$. For convenience we shall denote block triangular matrices of the form

$$\begin{bmatrix} K & M \\ 0 & N \end{bmatrix}$$

by $\operatorname{triang}[K, M, N]$.

Without loss of generality suppose that $J = \operatorname{diag}[J_i]_{i=1}^k$ is nilpotent, i.e., has only the eigenvalue zero. Let $i \notin \{\beta_1, \beta_2, \ldots, \beta_k\}$ be an integer, $1 \leq i \leq \alpha$. Let U_{i1} be the $\alpha \times \alpha$ matrix such that all its rows (except for the $(i+1)$th row) are zeros, and its $(i+1)$th row is $[t_{i1} \;\; t_{i2} \;\; \cdots \;\; t_{i\alpha}]$. Put $S_{i1} = \operatorname{triang}[I, U_{i1}, I]$; then $S_{i1} T S_{i1}^{-1} = \operatorname{triang}[J, T_{i0}, J^{\mathrm{T}}]$, where the ith row of T_{i0} is zero, and all other rows (except for the $(i+1)$th) of T_{i0} are the same as in T_0. Note also that the submatrix A is the same in T_{i0} and in T_0. Applying this transformation consequently for every $i \in \{1, \ldots, \alpha\} \setminus \{\beta_1, \ldots, \beta_k\}$, we obtain that T is similar to a matrix of the form $T_1 = \operatorname{triang}[J, V, J^{\mathrm{T}}]$, where the ith row of V is zero for $i \notin \{\beta_1, \ldots, \beta_k\}$, and $v_{\beta_p \beta_q} = t_{\beta_p \beta_q}$, $p, q = 1, \ldots, k$, where $V = (v_{ij})_{i,j=1}^{\alpha}$.

We shall show now that, by applying a similarity transformation to T_1, it is possible to make $v_{ij} = 0$ for $i \notin \{\beta_1, \ldots, \beta_k\}$ or $j \notin \{\beta_1, \ldots, \beta_k\}$. Let $V_j = \operatorname{col}(v_{ij})_{i=1}^{\alpha}$ be the jth column of V. For fixed $j \notin \{\beta_1, \ldots, \beta_k\}$ we have $v_{ij} = 0$

for $i \notin \{\beta_1, \ldots, \beta_k\}$, and, since A is invertible, there exist complex numbers $\sigma_1, \ldots, \sigma_k$ such that

$$V_j + \sum_{i=1}^{k} \sigma_{\beta_i} V_{\beta_i} = 0.$$

Let $S_{2j} = \operatorname{diag}[I, I + U_{2j}]$, where all but the jth column of U_{2j} are zeros, and the ith entry in the jth column of U_{2j} is σ_{β_m} if $i = \beta_m$ for some m, and zero otherwise. Then $S_{2j}^{-1} T_1 S_{2j} = \operatorname{triang}[J, V - Z_1, J^{\mathrm{T}} - Z_2]$ with the following structure of the matrices Z_1 and Z_2: $Z_1 = [0 \ \cdots \ 0 \ V_j \ 0 \ \cdots \ 0]$, where V_j is in the jth place; $Z_2 = [0 \ \cdots \ 0 \ Z_{2,j-1} \ 0 \ \cdots \ 0]$, where $Z_{2,j-1} = \operatorname{col}(\operatorname{col}(-\delta_{p,\alpha_q}\sigma_q)_{p=1}^{\alpha_q})_{q=1}^{k}$ is an $\alpha \times 1$ column in the $(j-1)$th place in Z_2 (δ_{p,α_q} denotes the Kronecker symbol). If $j = \beta_m + 1$ for some m, then we put $Z_2 = 0$. It is easy to see that $S_{2j}^{-1} T_1 S_{2j}$ can be reduced by a similarity transformation with a matrix of type $\operatorname{diag}[I, U_{3j}]$ to the form $\operatorname{triang}[J, W, J^{\mathrm{T}}]$, where the mth column of W_1 coincides with the mth column of $V - Z_1$ for $m \geq j$ and for $m \in \{\beta_1, \ldots, \beta_k\}$. Applying this similarity consequently for every $j \notin \{\beta_1, \ldots, \beta_k\}$ (starting with $j = \beta_k - 1$ and finishing with $j = 1$), we obtain that T_1 is similar to the matrix $T_2 = \operatorname{triang}[J, W_2, J^{\mathrm{T}}]$, where the (β_i, β_j) entries of W_2 $(i, j = 1, \ldots, k)$ form the matrix A, and all others are zeros.

The next step is to replace the invertible submatrix A in W_2 by I. We have $A_1^{-1}A = A_2$, where $A_1^{-1} = (b_{ij})_{i,j=1}^{k}$, $b_{ij} = 0$ for $i < j$, is a lower triangular matrix; $A_2 = (c_{ij})_{i,j=1}^{k}$, $c_{ij} = 0$ for $i > j$, is an upper triangular matrix. Define the $2\alpha \times 2\alpha$ invertible matrix $S_3 = (s_{ij}^{(3)})_{i,j=1}^{2\alpha}$ as follows: $s_{\beta_i\beta_j}^{(3)} = b_{ij}$ for $i, j = 1, \ldots, k$; $s_{mm}^{(3)} = 1$ for $m \notin \{\beta_1, \ldots, \beta_k\}$; $s_{ij}^{(3)} = 0$ otherwise. Then $S_3 T_2 S_3^{-1} = \operatorname{triang}[J + Z_3, W_3, J^{\mathrm{T}}]$, where the (β_i, β_j) entries of W_3 $(i, j = 1, \ldots, k)$ form the upper triangular matrix A_2, and all other entries of W_3 are zeros; the $\alpha \times \alpha$ matrix Z_3 can contain nonzero entries only in the places $(\beta_i - 1, \beta_j)$ for $i > j$ and such that $\alpha_i > 1$. It is easy to see that $S_3 T_2 S_3^{-1}$ is similar to $T_3 = \operatorname{triang}[J, W_3, J^{\mathrm{T}}]$. Define the $2\alpha \times 2\alpha$ invertible matrix $S_4 = (s_{ij}^{(4)})_{i,j=1}^{2\alpha}$ by the following equalities: $s_{\beta_i'\beta_j'}^{(4)} = c_{ij}$ for $i, j = 1, \ldots, k$; $s_{mm}^{(4)} = 1$ for $m \notin \{\beta_1', \ldots, \beta_k'\}$, where $\beta_i' = \beta_k + \beta_i$; $s_{ij}^{(4)} = 0$ otherwise. Then $S_4 T_3 S_4^{-1} = \operatorname{triang}[J, W_4, J^{\mathrm{T}} + Z_4]$, where the (β_i, β_j) entries of W_4 form the unit matrix I, and all other entries are zeros; the $\alpha \times \alpha$ matrix Z_4 can contain nonzero entries only in the places $(\beta_i, \beta_j - 1)$ for $i < j$ and such that $\alpha_j > 1$. Again, it is easy to see that $S_4 T_3 S_4^{-1}$ is similar to $T_4 = \operatorname{triang}[J, W_4, J^{\mathrm{T}}]$.

Evidently, the degrees of the elementary divisors of T_4 are $2\alpha_1, \ldots, 2\alpha_k$. So the same is true for T. □

In concrete examples it is possible to find the Jordan form of T by considering the matrix

$$\begin{bmatrix} T_1 & R_1 Q_2 \\ 0 & T_2 \end{bmatrix},$$

as we shall now show by example.

EXAMPLE 3.1. Let

$$L_1(\lambda) = \begin{bmatrix} \lambda^3 & \sqrt{2}\lambda^2 - \lambda \\ \sqrt{2}\lambda^2 + \lambda & \lambda^3 \end{bmatrix}.$$

Using the computations in Example 2.2 and Theorem 3.2, we find a linearization T for the product $L(\lambda) = (L_1(\lambda))^2$:

$$T = \begin{bmatrix} J & YX \\ 0 & J \end{bmatrix},$$

where

$$X = \begin{bmatrix} 1 & 0 & -\sqrt{2}+1 & \sqrt{2}-2 & \sqrt{2}+1 & \sqrt{2}+2 \\ 0 & 1 & 1 & 0 & 1 & 0 \end{bmatrix},$$

$$Y = \begin{bmatrix} 0 & 1 \\ -1 & 0 \\ \frac{1}{4}(\sqrt{2}+2) & 0 \\ \frac{1}{4}(-\sqrt{2}-1) & \frac{1}{4} \\ \frac{1}{4}(-\sqrt{2}+2) & 0 \\ \frac{1}{4}(-\sqrt{2}+1) & -\frac{1}{4} \end{bmatrix}, \quad J = \begin{bmatrix} 0 & & & & & \\ & 0 & & & & \\ & & 1 & 1 & & \\ & & 0 & 1 & & \\ & & & & -1 & 1 \\ & & & & 0 & -1 \end{bmatrix}.$$

By examining the structure of T we find the partial multiplicities of $L(\lambda)$ at $\lambda_0 = 1$. To this end we shall compute the 4×4 submatrix T_0 of T formed by the 3rd, 4th, 9th, and 10th rows and columns (the partial multiplicities of $L(\lambda)$ at $\lambda_0 = 1$ coincide with the degrees of the elementary divisors of $I\lambda - Y_0$). We have

$$T_0 = \begin{bmatrix} 1 & 1 & -\frac{1}{4}\sqrt{2} & -\frac{1}{2} \\ 0 & 1 & \frac{1}{2} & \frac{1}{2}(2\sqrt{2}-3) \\ 0 & 0 & 1 & 1 \\ 0 & 0 & 0 & 1 \end{bmatrix}.$$

Since $\operatorname{rank}(T_0 - I) = 3$, it is clear that the single partial multiplicity of $L(\lambda)$ at $\lambda_0 = 1$ is 4. □

3.2. Division Process

In this section we shall describe the division of monic matrix polynomials in terms of their standard triples.

We start with some remarks concerning the division of matrix polynomials in general.

Let $M(\lambda) = \sum_{j=0}^{l} M_j\lambda^j$ and $N(\lambda) = \sum_{j=0}^{k} N_j\lambda^j$ be matrix polynomials of size $n \times n$ (not necessarily monic). By the division of these matrix polynomials we understand a representation in the form

$$M(\lambda) = Q(\lambda)N(\lambda) + R(\lambda), \tag{3.4}$$

where $Q(\lambda)$ (the *quotient*) and $R(\lambda)$ (the *remainder*) are $n \times n$ matrix polynomials and either the degree of $R(\lambda)$ is less than the degree of $N(\lambda)$ or $R(\lambda)$ is zero.

This definition generalizes the division of scalar polynomials, which is always possible and for which the quotient and remainder are uniquely defined. However, in the matricial case it is very important to point out two factors which do not appear in the scalar case:

(1) Because of the noncommutativity, we have to distinguish between representation (3.4) (which will be referred to as *right* division) and the representation

$$M(\lambda) = N(\lambda)Q_1(\lambda) + R_1(\lambda) \tag{3.5}$$

for some matrix polynomials $Q_1(\lambda)$ and $R_1(\lambda)$, where the degree of $R(\lambda)$ is less than the degree of $N(\lambda)$, or is the zero polynomial. Representation (3.5) will be referred to as *left* division. In general, $Q(\lambda) \neq Q_1(\lambda)$ and $R(\lambda) \neq R_1(\lambda)$, so we distinguish between *right* ($Q(\lambda)$) and *left* ($Q_1(\lambda)$) quotients and between *right* ($R(\lambda)$) and *left* ($R_1(\lambda)$) remainders.

(2) The division is not always possible. The simplest example of this situation appears if we take $M(\lambda) = M_0$ as a constant nonsingular matrix and $N(\lambda) = N_0$ as a constant nonzero singular matrix. If the division is possible, then (since the degree of $N(\lambda)$ is 0) the remainders $R(\lambda)$ and $R_1(\lambda)$ must be zeros, and then (3.4) and (3.5) take the forms

$$M_0 = Q(\lambda)N_0, \qquad M_0 = N_0 Q_1(\lambda), \tag{3.6}$$

respectively. But in view of the invertibility of M_0, neither of (3.6) can be satisfied for any $Q(\lambda)$ or $Q_1(\lambda)$, so the division is impossible.

However, in the important case when $N(\lambda)$ (the divisor) is monic, the situation resembles more the familiar case of scalar polynomials, as the following proposition shows.

Proposition 3.5. *If the divisor $N(\lambda)$ is monic, then the right (or left) division is always possible, and the right (or left) quotient and remainder are uniquely determined.*

Proof. We shall prove Proposition 3.5 for right division; for left division the proof is similar.

Let us prove the possibility of division, i.e., that there exist $Q(\lambda)$ and $R(\lambda)$ ($\deg R(\lambda) < \deg N(\lambda)$) such that (3.4) holds. We can suppose $l \geq k$; otherwise take $Q(\lambda) = 0$ and $R(\lambda) = M(\lambda)$. Let $N(\lambda) = I\lambda^k + \sum_{j=0}^{k-1} N_j\lambda^j$, and write

$$\sum_{j=0}^{l} M_j\lambda^j = \left(\sum_{j=0}^{l-k} Q_j\lambda^j\right) \cdot \left(I\lambda^k + \sum_{j=0}^{k-1} N_j\lambda^j\right) + \sum_{j=0}^{k-1} R_j\lambda^j \tag{3.7}$$

with indefinite coefficients Q_j and R_j. Now compare the coefficients of $\lambda^l, \lambda^{l-1}, \dots, \lambda^k$ in both sides:

$$\begin{aligned} M_l &= Q_{l-k}, \\ M_{l-1} &= Q_{l-k-1} + Q_{l-k}N_{k-1}, \\ &\vdots \\ M_k &= Q_0 + \sum_{i=0}^{k-1} Q_{k-i}N_i. \end{aligned} \tag{3.8}$$

From these equalities we find in succession $Q_{l-k}, Q_{l-k-1}, \dots, Q_0$, i.e., we have found $Q(\lambda) = \sum_{j=0}^{l-k} \lambda^j Q_j$. To satisfy (3.7), define also

$$R_j = M_j - \sum_{q=0}^{j} Q_q N_{j-q} \qquad \text{for} \quad j = 0, \dots, k-1. \tag{3.9}$$

The uniqueness of the quotient and remainder is easy to prove. Suppose

$$M(\lambda) = Q(\lambda)N(\lambda) + R(\lambda) = \tilde{Q}(\lambda)N(\lambda) + \tilde{R}(\lambda),$$

where the degrees of both $R(\lambda)$ and $\tilde{R}(\lambda)$ do not exceed $k - 1$. Then

$$(Q(\lambda) - \tilde{Q}(\lambda))N(\lambda) + R(\lambda) - \tilde{R}(\lambda) = 0.$$

Since $N(\lambda)$ is monic of degree k, it follows that $Q(\lambda) - \tilde{Q}(\lambda) = 0$, and then also $R(\lambda) = \tilde{R}(\lambda) = 0$. □

An interesting particular case of Proposition 3.5 occurs when the divisor $N(\lambda)$ is linear: $N(\lambda) = I\lambda - X$. In this case we have extensions of the remainder theorem:

Corollary 3.6. *Let*

$$\sum_{j=0}^{l} M_j \lambda^j = Q(\lambda)(I\lambda - X) + R = (I\lambda - X)Q_1(\lambda) + R_1.$$

Then $R = \sum_{j=0}^{l} M_j X^j$ *and* $R_1 = \sum_{j=0}^{l} X^j M_j$.

Proof. Use formulas (3.8) and (3.9). It turns out that

$$Q_{l-i} = M_{l-i+1} + M_{l-i+2}X + \cdots + M_l X^{i-1}, \qquad i = 1, \dots, l.$$

According to (3.9),

$$R = M_0 + Q_0 X = \sum_{j=0}^{l} M_j X^j.$$

For R_1 the proof is analogous. □

We go back now to the division of monic matrix polynomials. It follows from Proposition 3.5 that the division of monic matrix polynomials is always possible and uniquely determined. We now express the left and right quotients and remainders in terms of standard triples. This is the first important step toward the application of the spectral methods of this book to factorization problems.

Theorem 3.7. *Let $L(\lambda) = I\lambda^l + \sum_{j=0}^{l-1} A_j\lambda^j$ be a monic matrix polynomial of degree l, and let*

$$L_1(\lambda) = I\lambda^k - X_1T_1^k(V_1 + \cdots + V_k\lambda^{k-1})$$

be a monic matrix polynomial of degree $k \leq l$ in the right canonical form, where (X_1, T_1) is a standard pair of $L_1(\lambda)$, and $[V_1 \ \cdots \ V_k] = [\operatorname{col}(X_1T_1^i)_{i=0}^{k-1}]^{-1}$. Then

$$L(\lambda) = Q(\lambda)L_1(\lambda) + R(\lambda), \tag{3.10}$$

where

$$Q(\lambda) = \sum_{j=0}^{l-k} \lambda^j\left(\sum_{i=j+1}^{l} A_iX_1T_1^{i-j-1}\right)V_k \tag{3.11}$$

and

$$R(\lambda) = \sum_{j=1}^{k} \lambda^{j-1}\left(\sum_{i=0}^{l} A_iX_1T_1^i\right)V_j, \tag{3.12}$$

where we put $A_l = I$.

Before we start to prove Theorem 3.7, let us write down the following important corollary. We say that $L_1(\lambda)$ is a *right divisor* of $L(\lambda)$ if

$$L(\lambda) = Q(\lambda)L_1(\lambda),$$

i.e., the right remainder $R(\lambda)$ is zero.

Corollary 3.8. *$L_1(\lambda)$ is a right divisor of $L(\lambda)$ if and only if, for a standard pair (X_1, T_1) of $L_1(\lambda)$, the equality*

$$A_0X_1 + A_1X_1T_1 + \cdots + A_{l-1}X_1T_1^{l-1} + X_1T_1^l = 0 \tag{3.13}$$

holds.

Proof. If (3.13) holds, then according to (3.12), $R(\lambda) = 0$ and $L_1(\lambda)$ is a right divisor of $L(\lambda)$. Conversely, if $R(\lambda) = 0$, then by (3.12)

$$\left(\sum_{i=0}^{l} A_iX_1T_1^i\right)V_j = 0 \qquad \text{for} \quad j = 1, \ldots, k.$$

Since $[V_1 \ \cdots \ V_k]$ is nonsingular, this means that (3.13) holds. $\square$

Proof of Theorem 3.7. We shall establish first some additional properties of standard triples for the monic polynomials $L(\lambda)$ and $L_1(\lambda)$.

Define

$$G_{\alpha\beta} = X_1 T_1^{\alpha} V_{\beta}, \qquad 1 \le \beta \le k, \qquad \alpha = 0, 1, 2, \ldots. \tag{3.14}$$

Then for each i, $1 \le i \le k$, we have

$$G_{p+1,i} = G_{pk} G_{ki} + G_{p,i-1}, \qquad p = 0, 1, 2, \ldots \tag{3.15}$$

(and we set $G_{p0} = 0$, $p = 0, 1, \ldots$). Indeed, from the definition of V_i we deduce

$$I = [V_1 \quad V_2 \quad \cdots \quad V_k] \begin{bmatrix} X_1 \\ X_1 T_1 \\ \vdots \\ X_1 T_1^{k-1} \end{bmatrix}. \tag{3.16}$$

Consider the composition with $X_1 T_1^p$ on the left and $T_1[V_1 \quad \cdots \quad V_k]$ on the right to obtain

$$[G_{p+1,1} \quad \cdots \quad G_{p+1,k}] = [G_{p,1} \quad \cdots \quad G_{p,k}] \begin{bmatrix} G_{11} & \cdots & G_{1,k} \\ \vdots & & \vdots \\ G_{k,1} & \cdots & G_{k,k} \end{bmatrix}. \tag{3.17}$$

Reversing the order of the factors on the right of (3.16), we see that $G_{ij} = \delta_{i,j-1} I$ for $i = 1, 2, \ldots, k-1$ and $j = 1, 2, \ldots, k$. The conclusion (3.15) then follows immediately.

Now we shall check the following equalities:

$$V_j = \sum_{m=0}^{k-j} T_1^m V_k B_{j+m}, \qquad j = 1, \ldots, k, \tag{3.18}$$

where B_p are the coefficients of $L_1(\lambda)$: $L_1(\lambda) = \sum_{j=0}^{k} B_j \lambda^j$ with $B_k = I$. Since the matrix $\operatorname{col}(X_1 T_1^i)_{i=0}^{k-1}$ is nonsingular, it is sufficient to check that

$$X_1 T_1^i V_j = X_1 T_1^i \sum_{m=0}^{k-j} T_1^m V_k B_{j+m}, \qquad j = 1, \ldots, k, \tag{3.19}$$

for $i = 0, 1, \ldots$. Indeed, using the matrices G_{ij} and the right canonical form for $L_1(\lambda)$ rewrite the right-hand side of (3.19) as follows (where the first equality follows from (3.15)):

$$\begin{aligned} G_{i+k-j,k} &- \sum_{m=0}^{k-j-1} G_{i+m,k} G_{k,j+m+1} \\ &= G_{i+k-j,k} - \sum_{m=0}^{k-j-1} (G_{i+m+1,j+m+1} - G_{i+m,j+m}) \\ &= G_{i+k-j,k} - (G_{i+(k-j-1)+1,k} - G_{ij}) = G_{ij}, \end{aligned}$$

and (3.18) is proved.

Define also the following matrix polynomials: $L_{1,j}(\lambda) = B_j + B_{j+1}\lambda + \cdots + B_k\lambda^{k-j}$, $j = 0, 1, \ldots, k$. In particular, $L_{1,0}(\lambda) = L_1(\lambda)$ and $L_{1,k}(\lambda) = I$. We shall need the following property of the polynomials $L_{1,j}(\lambda)$:

$$V_k L_1(\lambda) = (I\lambda - T_1)\left(\sum_{j=0}^{k-1} T_1^j V_k L_{1,j+1}(\lambda)\right). \tag{3.20}$$

Indeed, since $\lambda L_{1,j+1}(\lambda) = L_{1,j}(\lambda) - B_j$ ($j = 0, \ldots, k-1$), the right-hand side of (3.20) equals

$$\sum_{j=0}^{k-1} T_1^j V_k(L_{1j}(\lambda) - B_j) - \sum_{j=0}^{k-1} T_1^{j+1} V_k L_{1,j+1}(\lambda) = -\sum_{j=0}^{k} T_1^j V_k B_j + V_k L_{10}(\lambda).$$

But since (X_1, T_1, V_k) is a standard triple of $L_1(\lambda)$, in view of (2.10), $\sum_{j=0}^{k} T_1^j V_k B_j = 0$, and (3.20) follows.

We are now ready to prove that the difference $L(\lambda) - R(\lambda)$ is divisible by $L_1(\lambda)$. Indeed, using (3.18) and then (3.20)

$$\begin{aligned} R(\lambda)(L_1(\lambda))^{-1} &= \left(\sum_{i=0}^{l} A_i X_1 T_1^i\right) \cdot \left(\sum_{j=0}^{k-1} T_1^j V_k L_{1,j+1}(\lambda)\right) \cdot L_1^{-1}(\lambda) \\ &= \sum_{i=0}^{l} A_i X_1 T_1^i (I\lambda - T_1)^{-1} V_k. \end{aligned}$$

Thus,

$$L(\lambda)(L_1(\lambda))^{-1} = \sum_{i=0}^{l} A_j \lambda^j \cdot X_1 (I\lambda - T_1)^{-1} V_k$$

(and here we use the resolvent form of $L_1(\lambda)$). So

$$L(\lambda)(L_1(\lambda))^{-1} - R(\lambda)(L_1(\lambda))^{-1} = \sum_{i=0}^{l} A_i X_1 (I\lambda^i - T_1^i)(I\lambda - T_1)^{-1} V_k.$$

Using the equality $I\lambda^i - T_1^i = (\sum_{p=0}^{i-1} \lambda^p T^{i-1-p}) \cdot (\lambda I - T_1)$ $(i > 0)$, we obtain

$$\begin{aligned} L(\lambda)(L_1(\lambda))^{-1} - R(\lambda)(L_1(\lambda))^{-1} &= \sum_{i=1}^{l} A_i X_1 \sum_{p=0}^{i-1} \lambda^p T^{i-1-p} V_k \\ &= \sum_{p=0}^{l-1} \lambda^p \sum_{i=p+1}^{l} A_i X_1 T_1^{i-1-p} V_k. \end{aligned} \tag{3.21}$$

Because of the biorthogonality relations $X_1 T_1^j V_k = 0$ for $j = 0, \ldots, k-1$, all the terms in (3.21) with $p = l - k + 1, \ldots, p = l - 1$ vanish, and formula (3.11) follows. □

In terms of the matrices $G_{\alpha\beta}$ introduced in the proof of Theorem 3.7, the condition that the coefficients of the remainder polynomial (3.12) are zeros (which means that $L_1(\lambda)$ is a right divisor of $L(\lambda)$) can be conveniently expressed in matrix form as follows.

Corollary 3.9. *$L_1(\lambda)$ is a right divisor of $L(\lambda)$ if and only if*

$$[G_{l1} \quad G_{l2} \quad \cdots \quad G_{lk}] = -[A_0 \quad \cdots \quad A_{l-1}] \times \begin{bmatrix} G_{01} & G_{02} & \cdots & G_{0k} \\ G_{11} & G_{12} & \cdots & G_{1k} \\ \vdots & \vdots & & \vdots \\ G_{l-1,1} & G_{l-1,2} & \cdots & G_{l-1,k} \end{bmatrix},$$

where $G_{\alpha\beta}$ are defined by (3.14).

The dual results for left division are as follows.

Theorem 3.10. *Let $L(\lambda)$ be as in Theorem 3.7, and let*

$$L_1(\lambda) = \lambda^k I - (W_1 + \cdots + W_k\lambda^{k-1})T_1^k Y_1$$

be a monic matrix polynomial of degree k in its left canonical form (2.15) (*so (T_1, Y_1) is a left standard pair of $L_1(\lambda)$ and*

$$\operatorname{col}(W_i)_{i=1}^k = [Y_1 \quad T_1 Y_1 \quad \cdots \quad T_1^{k-1}Y_1]^{-1}).$$

Then

$$L(\lambda) = L_1(\lambda)Q_1(\lambda) + R_1(\lambda),$$

where

$$Q_1(\lambda) = \sum_{j=0}^{l-k} \lambda^j \left(\sum_{i=j+1}^{l} W_k \cdot T_1^{i-j-1} Y_1 A_i \right)$$

and

$$R_1(\lambda) = \sum_{j=1}^{k} \lambda^{j-1} \left(W_j \sum_{i=0}^{l} T_1^i Y_1 A_i \right),$$

where $A_l = I$.

Theorem 3.10 can be established either by a parallel line of argument from the proof of Theorem 3.7, or by applying Theorem 3.7 to the transposed matrix polynomials $L^{\mathrm{T}}(\lambda)$ and $L_1^{\mathrm{T}}(\lambda)$ (evidently the left division

$$L(\lambda) = L_1(\lambda)Q_1(\lambda) + R_1(\lambda)$$

of $L(\lambda)$ by $L_1(\lambda)$ gives rise to the right division

$$L^{\mathrm{T}}(\lambda) = Q_1^{\mathrm{T}}(\lambda)L_1^{\mathrm{T}}(\lambda) + R_1^{\mathrm{T}}(\lambda)$$

of the transposed matrix polynomials).

The following definition is now to be expected: a monic matrix polynomial $L_1(\lambda)$ is a *left divisor* of a monic matrix polynomial $L(\lambda)$ if $L(\lambda) = L_1(\lambda)Q_1(\lambda)$ for some matrix polynomial $Q_1(\lambda)$ (which is necessarily monic).

Corollary 3.11. *$L_1(\lambda)$ is a left divisor of $L(\lambda)$ if and only if for a left standard pair (T_1, Y_1) of $L_1(\lambda)$ the equality*

$$\sum_{j=0}^{l-1} Y_1 T_1^j A_j + Y_1 T_1^l = 0$$

holds.

This corollary follows from Theorem 3.10 in the same way as Corollary 3.8 follows from Theorem 3.7.

3.3. Characterization of Divisors and Supporting Subspaces

In this section we begin to study the monic divisors (right and left) for a monic matrix polynomial $L(\lambda) = I\lambda^l + \sum_{j=0}^{l-1} A_j\lambda^j$. The starting points for our study are Corollaries 3.8 and 3.11. As usual, we shall formulate and prove our results for right divisors, while the dual results for the left divisors will be stated without proof. The main result here (Theorem 3.12) provides a geometric characterization of monic right divisors.

A hint for such a characterization is already contained in the multiplication theorem (Theorem 3.2). Let $L(\lambda) = L_2(\lambda)L_1(\lambda)$ be a product of two monic matrix polynomials $L_1(\lambda)$ and $L_2(\lambda)$, and let (X, T, Y) be a standard triple of $L(\lambda)$ such that

$$X = [X_1 \quad 0], \qquad T = \begin{bmatrix} T_1 & X_1 Y_2 \\ 0 & T_2 \end{bmatrix}, \qquad Y = \begin{bmatrix} 0 \\ Y_2 \end{bmatrix}, \tag{3.22}$$

and (X_i, T_i, Y_i) $(i = 1, 2)$ is a standard triple for $L_i(\lambda)$ (cf. Theorem 3.2). The form of T in (3.22) suggests the existence of a T-invariant subspace: namely, the subspace $\mathcal{M}$ spanned by the first nk unit coordinate vectors in $\mathbb{C}^{nl}$ (here l (resp. k) is the degree of $L(\lambda)$ (resp. $L_1(\lambda)$)). This subspace $\mathcal{M}$ can be attached to the polynomial $L_1(\lambda)$, which is considered as a right divisor of $L(\lambda)$, and this correspondence between subspaces and divisors turns out to be one-to-one. It is described in the following theorem.

Theorem 3.12. *Let $L(\lambda)$ be a monic matrix polynomial of degree l with standard pair (X, T). Then for every nk-dimensional T-invariant subspace $\mathcal{M} \subset \mathbb{C}^{nl}$, such that the restriction $\operatorname{col}(X|_{\mathcal{M}}(T|_{\mathcal{M}})^i)_{i=0}^{k-1}$ is invertible, there exists a unique monic right divisor $L_1(\lambda)$ of $L(\lambda)$ of degree k such that its standard pair is similar to $(X|_{\mathcal{M}}, T|_{\mathcal{M}})$.*

Conversely, for every monic right divisior $L_1(\lambda)$ *of* $L(\lambda)$ *of degree* k *with standard pair* (X_1, T_1), *the subspace*

$$\mathcal{M} = \operatorname{Im}([\operatorname{col}(XT^i)_{i=0}^{l-1}]^{-1}[\operatorname{col}(X_1T_1^i)_{i=0}^{l-1}]) \tag{3.23}$$

is T*-invariant,* $\dim \mathcal{M} = nk$, $\operatorname{col}(X|_{\mathcal{M}}(T|_{\mathcal{M}})^i)_{i=0}^{k-1}$ *is invertible and* $(X|_{\mathcal{M}}, T|_{\mathcal{M}})$ *is similar to* (X_1, T_1).

Here $X|_{\mathcal{M}}$ (resp. $T|_{\mathcal{M}}$) is considered as a linear transformation $\mathcal{M} \to \mathbb{C}^n$ (resp. $\mathcal{M} \to \mathcal{M}$). Alternatively, one can think about $X|_{\mathcal{M}}$ (resp. $T|_{\mathcal{M}}$) as an $n \times nk$ (resp. $nk \times nk$) matrix, by choosing a basis in $\mathcal{M}$ (and the standard orthonormal basis in $\mathbb{C}^n$ for representation of $X|_{\mathcal{M}}$).

Proof. Let $\mathcal{M} \subset \mathbb{C}^{nl}$ be an nk-dimensional T-invariant subspace such that $\operatorname{col}(X|_{\mathcal{M}}(T|_{\mathcal{M}})^i)_{i=0}^{k-1}$ is invertible. Construct the monic matrix polynomial $L_1(\lambda)$ with standard pair $(X|_{\mathcal{M}}, T|_{\mathcal{M}})$ (cf. (2.14)):

$$L_1(\lambda) = I\lambda^k - X|_{\mathcal{M}}(T|_{\mathcal{M}})^k(V_1 + V_2\lambda + \cdots + V_k\lambda^{k-1}),$$

where $[V_1 \ \cdots \ V_k] = [\operatorname{col}(X|_{\mathcal{M}}(T|_{\mathcal{M}})^i)_{i=0}^{k-1}]^{-1}$. Appeal to Corollary 3.8 (bearing in mind the equality

$$A_0X + A_1XT + \cdots + A_{l-1}XT^{l-1} + XT^l = 0,$$

where A_j are the coefficients of $L(\lambda)$) to deduce that $L_1(\lambda)$ is a right divisor of $L(\lambda)$.

Conversely, let $L_1(\lambda)$ be a monic right divisor of $L(\lambda)$ of degree k with standard pair (X_1, Y_1). Then Corollary 3.8 implies

$$C_1 \operatorname{col}(X_1T_1^i)_{i=0}^{l-1} = \operatorname{col}(X_1T_1^i)_{i=0}^{l-1}T_1, \tag{3.24}$$

where C_1 is the first companion matrix for $L(\lambda)$. Also

$$C_1 \operatorname{col}(XT^i)_{i=0}^{l-1} = \operatorname{col}(XT^i)_{i=0}^{l-1}T. \tag{3.25}$$

Eliminating C_1 from (3.24) and (3.25), we obtain

$$T[\operatorname{col}(XT^i)_{i=0}^{l-1}]^{-1}[\operatorname{col}(X_1T_1^i)_{i=0}^{l-1}] = [\operatorname{col}(XT^i)_{i=0}^{l-1}]^{-1}[\operatorname{col}(X_1T_1^i)_{i=0}^{l-1}]T_1. \tag{3.26}$$

This equality readily implies that the subspace $\mathcal{M}$ given by (3.23) is T-invariant. Moreover, it is easily seen that the columns of $[\operatorname{col}(XT^i)_{i=0}^{l-1}]^{-1}$ $[\operatorname{col}(X_1T_1^i)_{i=0}^{l-1}]$ are linearly independent; equality (3.26) implies that in the basis of $\mathcal{M}$ formed by these columns, $T|_{\mathcal{M}}$ is represented by the matrix T_1.

Further,

$$X[\operatorname{col}(XT^i)_{i=0}^{l-1}]^{-1}[\operatorname{col}(X_1T_1^i)_{i=0}^{l-1}] = X_1;$$

so $X|_{\mathscr{M}}$ is represented in the same basis in $\mathscr{M}$ by the matrix X_1. Now it is clear that $(X|_{\mathscr{M}}, T|_{\mathscr{M}})$ is similar to (X_1, T_1). Finally, the invertibility of $\operatorname{col}(X|_{\mathscr{M}}(T|_{\mathscr{M}})^i)_{i=0}^{l-1}$ follows from this similarity and the invertibility of $\operatorname{col}(X_1 T_1^i)_{i=0}^{k-1}$. $\square$

Note that the subspace $\mathscr{M}$ defined by (3.23) does not depend on the choice of the standard pair (X_1, T_1) of $L_1(\lambda)$, because

$$\operatorname{Im} \operatorname{col}(X_1 T_1^i)_{i=0}^{l-1} = \operatorname{Im} \operatorname{col}(X_1 S \cdot (S^{-1} T_1 S)^i)_{i=0}^{l-1}$$

for any invertible matrix S. Thus, for every monic right divisor $L_1(\lambda)$ of $L(\lambda)$ of degree k we have constructed an nk-dimensional subspace $\mathscr{M}$, which will be called the *supporting subspace* of $L_1(\lambda)$. As (3.23) shows, the supporting subspace does depend on the standard pair (X, T); but once the standard pair (X, T) is fixed, the supporting subspace depends only on the divisor $L_1(\lambda)$. If we wish to stress the dependence of $\mathscr{M}$ on (X, T) also (not only on $L_1(\lambda)$), we shall speak in terms of a supporting subspace *relative to* the standard pair (X, T).

Note that (for fixed (X, T)) the supporting subspace $\mathscr{M}$ for a monic right divisor $L_1(\lambda)$ is uniquely defined by the property that $(X|_{\mathscr{M}}, T|_{\mathscr{M}})$ is a standard pair of $L_1(\lambda)$. This follows from (3.23) if one uses $(X|_{\mathscr{M}}, T|_{\mathscr{M}})$ in place of (X_1, T_1).

So Theorem 3.12 gives a one-to-one correspondence between the right monic divisors of $L(\lambda)$ of degree k and T-invariant subspaces $\mathscr{M} \subset \mathbb{C}^{nl}$, such that $\dim \mathscr{M} = nk$ and $\operatorname{col}(X|_{\mathscr{M}}(T|_{\mathscr{M}})^i)_{i=0}^{k-1}$ is invertible, which are in fact the supporting subspaces of the right divisors. Thus, Theorem 3.12 provides a description of the algebraic relation (divisibility of monic polynomials) in a geometric language of supporting subspaces.

By the property of a standard pair (Theorem 2.4) it follows also that for a monic right divisor $L_1(\lambda)$ with supporting subspace $\mathscr{M}$ the formula

$$L_1(\lambda) = I\lambda^k - X|_{\mathscr{M}} \cdot (T|_{\mathscr{M}})^k (V_1 + V_2 \lambda + \cdots + V_k \lambda^{k-1}) \tag{3.27}$$

holds, where $[V_1 \ \cdots \ V_k] = [\operatorname{col}(X|_{\mathscr{M}} \cdot (T|_{\mathscr{M}})^i)_{i=0}^{k-1}]^{-1}$.

If the pair (X, T) coincides with the companion standard pair (P_1, C_1) (see Theorem 1.24) of $L(\lambda)$, Theorem 3.12 can be stated in the following form:

Corollary 3.13. *A subspace $\mathscr{M} \subset \mathbb{C}^{nl}$ is a supporting subspace (relative to the companion standard pair (P_1, C_1)) for some monic divisor $L_1(\lambda)$ of $L(\lambda)$ of degree k if and only if the following conditions hold:*

(i) $\dim \mathscr{M} = nk$;

(ii) *$\mathscr{M}$ is C_1-invariant;*

(iii) *the $n(l-k)$-dimensional subspace of $\mathbb{C}^{nl}$ spanned by all nk-dimensional vectors with the first nk coordinates zeros, is a direct complement to $\mathscr{M}$.*

In this case the divisor $L_1(\lambda)$ is uniquely defined by the subspace $\mathcal{M}$ and is given by (3.27) *with $X = P_1$ and $T = C_1$.*

In order to deduce Corollary 3.13 from Theorem 3.12 observe that $\operatorname{col}(P_1(C_1)^i)_{i=0}^{k-1}$ has the form $[I \quad *]$, where I is the $nk \times nk$ unit matrix. Therefore condition (iii) is equivalent to the invertibility of

$$\operatorname{col}(P_1|_{\mathcal{M}}(C_1|_{\mathcal{M}})^i)_{i=0}^{k-1}.$$

We point out the following important property of the supporting subspace, which follows immediately from Theorem 3.12.

Corollary 3.14. *Let $L_1(\lambda)$ be a monic divisor of $L(\lambda)$ with supporting subspace $\mathcal{M}$. Then $\sigma(L_1) = \sigma(J|_{\mathcal{M}})$; moreover, the elementary divisors of $L_1(\lambda)$ coincide with the elementary divisors of $I\lambda - J|_{\mathcal{M}}$.*

Theorem 3.12 is especially convenient when the standard pair (X, T) involved is in fact a Jordan pair, because then it is possible to obtain a deeper insight into the spectral properties of the divisors (when compared to the spectral properties of the polynomial $L(\lambda)$ itself).

The theorem also shows the importance of the structure of the invariant subspaces of T for the description of monic divisors. In the following section we shall illustrate, in a simple example, how to use the description of T-invariant subspaces to find all right monic divisors of second degree.

In the scalar case ($n = 1$), which is of course familiar, the analysis based on Theorem 3.12 leads to the very well-known statement on divisibility of scalar polynomials, as it should. Indeed, consider the scalar polynomial $L(\lambda) = \prod_{i=1}^{p} (\lambda - \lambda_i)^{\alpha_i}$; $\lambda_1, \ldots, \lambda_p$ are different complex numbers and α_i are positive integers. As we have already seen (Proposition 1.18), a Jordan pair (X, J) of $L(\lambda)$ can be chosen as follows:

$$X = [X_1 \quad \cdots \quad X_p], \qquad J = \operatorname{diag}[J_1, \ldots, J_p],$$

where $X_i = [1 \quad 0 \quad \cdots \quad 0]$ is a $1 \times \alpha_i$ row and J_i is an $\alpha_i \times \alpha_i$ Jordan block with eigenvalue λ_i. Every J-invariant subspace $\mathcal{M}$ has the following structure:

$$\mathcal{M} = \mathcal{M}_1 \oplus \cdots \oplus \mathcal{M}_p,$$

where $\mathcal{M}_i \subset \mathbb{C}^{\alpha_i}$ is spanned by the first β_i coordinate unit vectors, $i = 1, \ldots, p$, and β_i is an arbitrary nonnegative integer not exceeding α_i. It is easily seen that each such J-invariant subspace $\mathcal{M}$ is supporting for some monic right divisor $L_{\mathcal{M}}(\lambda)$ of $L(\lambda)$ of degree $\beta = \beta_1 + \cdots + \beta_p$: the pair $(X|_{\mathcal{M}}, J|_{\mathcal{M}}) = ([X_{1,\mathcal{M}} \quad \cdots \quad X_{p,\mathcal{M}}], J_{1,\mathcal{M}} \oplus \cdots \oplus J_{p,\mathcal{M}})$, where $X_{i,\mathcal{M}} = [1 \quad 0 \quad \cdots \quad 0]$ of size $1 \times \beta_i$ and $J_{i,\mathcal{M}}$ is the $\beta_i \times \beta_i$ Jordan block with eigenvalue λ_i, is a Jordan pair for polynomial $\prod_{i=1}^{p} (\lambda - \lambda_i)^{\beta_i}$, and therefore $\operatorname{col}(X|_{\mathcal{M}}(J|_{\mathcal{M}})^i)_{i=0}^{\beta-1}$

is invertible. In this way we also see that the divisor with the supporting subspace $\mathcal{M}$ is just $\prod_{i=1}^{p}(\lambda-\lambda_i)^{\beta_i}$. So, as expected from the very beginning, all the divisors of $L(\lambda)$ are of the form $\prod_{i=1}^{p}(\lambda-\lambda_i)^{\beta_i}$, where $0\le\beta_i\le\alpha_i$, $i=1,\ldots,p$.

For two divisors of $L(\lambda)$ it may happen that one of them is in turn a divisor of the other. In terms of supporting subspaces such a relationship means nothing more than inclusion, as the following corollary shows.

Corollary 3.15. *Let $L_{11}(\lambda)$ and $L_{12}(\lambda)$ be monic right divisors of $L(\lambda)$ then $L_{11}(\lambda)$ is a right divisor of $L_{12}(\lambda)$ if and only if for the supporting subspaces $\mathcal{M}_1$ and $\mathcal{M}_2$ of $L_{11}(\lambda)$ and $L_{12}(\lambda)$, respectively, the relation $\mathcal{M}_1\subset\mathcal{M}_2$ holds.*

Proof. Let (X, T) be the standard pair of $L(\lambda)$ relative to which the supporting subspaces $\mathcal{M}_1$ and $\mathcal{M}_2$ are defined. Then, by Theorem 3.12, $(X|_{\mathcal{M}_i}, T|_{\mathcal{M}_i})$ $(i=1,2)$ is a standard pair of $L_{1i}(\lambda)$. If $\mathcal{M}_1\subset\mathcal{M}_2$, then, by Theorem 3.12 (when applied to $L_{12}(\lambda)$ in place of $L(\lambda)$), $L_{11}(\lambda)$ is a right divisor of $L_{12}(\lambda)$. Suppose now $L_{11}(\lambda)$ is a right divisor of $L_{12}(\lambda)$. Then, by Theorem 3.12, there exists a supporting subspace $\mathcal{M}_{12}\subset\mathcal{M}_2$ of $L_{11}(\lambda)$ as a right divisor of $L_{12}(\lambda)$, so that $(X|_{\mathcal{M}_{12}}, T|_{\mathcal{M}_{12}})$ is a standard pair of $L_{11}(\lambda)$. But then clearly $\mathcal{M}_{12}$ is a supporting subspace of $L_{11}(\lambda)$ as a divisor of $L(\lambda)$. Since the supporting subspace is unique, it follows that $\mathcal{M}_1=\mathcal{M}_{12}\subset\mathcal{M}_2$. □

It is possible to deduce results analogous to Theorem 3.12 and Corollaries 3.14 and 3.15 for left divisors of $L(\lambda)$ (by using left standard pairs and Corollary 3.11). However, it will be more convenient for us to obtain the description for left divisors in terms of the description of quotients in Section 3.5.

3.4. Example

Let

$$L(\lambda)=\begin{bmatrix}\lambda(\lambda-1)^2 & b\lambda\\ 0 & \lambda^2(\lambda-2)\end{bmatrix},\qquad b\in\mathbb{C}.$$

Clearly, $\sigma(L)=\{0,1,2\}$. As a Jordan pair (X, J) of $L(\lambda)$, we can take

$$X=\begin{bmatrix}-b & 0 & 1 & 1 & 0 & -b\\ 1 & 0 & 0 & 0 & 0 & 1\end{bmatrix},$$

$$J=\begin{bmatrix}0 & 1 & & & & \\ & 0 & & & & \\ & & 0 & & & \\ & & & 1 & 1 & \\ & & & & 1 & \\ & & & & & 2\end{bmatrix}.$$

So, for instance, $\left[\begin{smallmatrix}-b\\1\end{smallmatrix}\right]$, 0 and $\left[\begin{smallmatrix}1\\0\end{smallmatrix}\right]$ form a canonical set of Jordan chains of $L(\lambda)$ corresponding to the eigenvalue $\lambda_0 = 0$.

Let us find all the monic right divisors of $L(\lambda)$ of degree 2. According to Theorem 3.12 this means that we are to find all four-dimensional J-invariant subspaces $\mathcal{M}$ such that $\left[\begin{smallmatrix}X\\XJ\end{smallmatrix}\right]|_{\mathcal{M}}$ is invertible. Computation shows that

$$\begin{bmatrix} X \\ XJ \end{bmatrix} = \begin{bmatrix} -b & 0 & 1 & 1 & 0 & -b \\ 1 & 0 & 0 & 0 & 0 & 1 \\ 0 & -b & 0 & 1 & 1 & -2b \\ 0 & 1 & 0 & 0 & 0 & 2 \end{bmatrix}.$$

We shall find first all four-dimensional J-invariant subspaces $\mathcal{M}$ and then check the invertibility condition. Clearly, $\mathcal{M} = \mathcal{M}_0 + \mathcal{M}_1 + \mathcal{M}_2$, where $\mathcal{M}_i$ is J-invariant and $\sigma(J|_{\mathcal{M}i}) = \{i\}$, $i = 0, 1, 2$. Since $\dim \mathcal{M}_0 \leq 3$, $\dim \mathcal{M}_1 \leq 2$, and $\dim \mathcal{M}_2 \leq 1$, we obtain the following five possible cases:

Case	$\dim \mathcal{M}_0$	$\dim \mathcal{M}_1$	$\dim \mathcal{M}_2$
1	3	1	0
2	3	0	1
3	2	2	0
4	2	1	1
5	1	2	1

Case 1 gives rise to a single J-invariant subspace

$$\mathcal{M}_1 = \{(x_1, x_2, x_3, x_4, 0, 0)^{\mathrm{T}} \mid x_j \in \mathbb{C}, j = 1, 2, 3, 4\}.$$

Choosing a standard basis in $\mathcal{M}_1$, we obtain

$$\begin{bmatrix} X \\ XJ \end{bmatrix}\Big|_{\mathcal{M}_1} = \begin{bmatrix} -b & 0 & 1 & 1 \\ 1 & 0 & 0 & 0 \\ 0 & -b & 0 & 1 \\ 0 & 1 & 0 & 0 \end{bmatrix},$$

which is nonsingular for all $b \in \mathbb{C}$. So $\mathcal{M}_1$ is a supporting subspace for the right monic divisor $L_{\mathcal{M}_1}(\lambda)$ of $L(\lambda)$ given by

$$L_{\mathcal{M}_1}(\lambda) = I\lambda^2 - X|_{\mathcal{M}_1} J^2|_{\mathcal{M}_1}(V_1^{(1)} + V_2^{(1)}\lambda), \tag{3.28}$$

where

$$[V_1^{(1)} V_2^{(1)}] = \left[\begin{bmatrix} X \\ XJ \end{bmatrix}\Big|_{\mathcal{M}_1}\right]^{-1} = \begin{bmatrix} 0 & 1 & 0 & 0 \\ 0 & 0 & 0 & 1 \\ 1 & b & -1 & -b \\ 0 & 0 & 1 & b \end{bmatrix}.$$

Substituting in (3.28) also

$$X|_{\mathscr{M}_1}(J|_{\mathscr{M}_1})^2 = \begin{bmatrix} 0 & 0 & 0 & 1 \\ 0 & 0 & 0 & 0 \end{bmatrix},$$

we obtain

$$L_{\mathscr{M}_1}(\lambda) = \begin{bmatrix} \lambda^2 - \lambda & -b\lambda \\ 0 & \lambda^2 \end{bmatrix}.$$

Case 2 also gives rise to a single J-invariant subspace

$$\mathscr{M}_2 = \{(x_1, x_2, x_3, 0, 0, x_6)^{\mathrm{T}} \,|\, x_j \in \mathbb{C}, j = 1, 2, 3, 6\}.$$

But in this case

$$\begin{bmatrix} X \\ XJ \end{bmatrix}\Big|_{\mathscr{M}_2} = \begin{bmatrix} -b & 0 & 1 & -b \\ 1 & 0 & 0 & 1 \\ 0 & -b & 0 & -2b \\ 0 & 1 & 0 & 2 \end{bmatrix}$$

(written in the standard basis) is singular, so $\mathscr{M}_2$ is not a supporting subspace for any monic divisor of $L(\lambda)$ of degree 2.

Consider now case 3. Here we have a J-invariant subspace

$$\mathscr{M}_3 = \{(x_1, 0, x_3, x_4, x_5, 0)^{\mathrm{T}} \,|\, x_j \in \mathbb{C}, j = 1, 3, 4, 5\}$$

and a family of J-invariant subspaces

$$\begin{aligned} \mathscr{M}_3(a) = \operatorname{Span}\{&(1 \;\; 0 \;\; 0 \;\; 0 \;\; 0 \;\; 0)^{\mathrm{T}}, (0 \;\; 1 \;\; a \;\; 0 \;\; 0 \;\; 0)^{\mathrm{T}}, \\ &(0 \;\; 0 \;\; 0 \;\; 1 \;\; 0 \;\; 0)^{\mathrm{T}}, (0 \;\; 0 \;\; 0 \;\; 0 \;\; 1 \;\; 0)^{\mathrm{T}}\}, \end{aligned} \tag{3.29}$$

where $a \in \mathbb{C}$ is arbitrary. For $\mathscr{M}_3$ we have

$$\begin{bmatrix} X \\ XJ \end{bmatrix}\Big|_{\mathscr{M}_3} = \begin{bmatrix} -b & 1 & 1 & 0 \\ 1 & 0 & 0 & 0 \\ 0 & 0 & 1 & 1 \\ 0 & 0 & 0 & 0 \end{bmatrix}$$

which is singular, so there are no monic divisors with supporting subspace $\mathscr{M}_3$; for the family $\mathscr{M}_3(a)$ in the basis which appears in the right-hand side of (3.29)

$$\begin{bmatrix} X \\ XJ \end{bmatrix}\Big|_{\mathscr{M}_3(a)} = \begin{bmatrix} -b & a & 1 & 0 \\ 1 & 0 & 0 & 0 \\ 0 & -b & 1 & 1 \\ 0 & 1 & 0 & 0 \end{bmatrix},$$

which is nonsingular for every $a \in \mathbb{C}$. The monic divisor $L_{\mathcal{M}_3(a)}(\lambda)$ with the supporting subspace $\mathcal{M}_3(a)$ is

$$L_{\mathcal{M}_3(a)}(\lambda) = I\lambda^2 - X|_{\mathcal{M}_3(a)} \cdot (J|_{\mathcal{M}_3(a)})^2 \cdot (V_1^{(3)} + V_2^{(3)}\lambda),$$

where $[V_1^{(3)} V_2^{(3)}] = [\operatorname{col}(X|_{\mathcal{M}_3(a)}(J|_{\mathcal{M}_3(a)})^i)_{i=0}^1]^{-1}$, and computation shows that

$$L_{\mathcal{M}_3(a)}(\lambda) = I\lambda^2 - \begin{bmatrix} 0 & 0 & 1 & 2 \\ 0 & 0 & 0 & 0 \end{bmatrix} \left\{ \begin{bmatrix} 1 & 1 \\ 0 & 0 \\ 1 & b \\ -1 & -b \end{bmatrix} + \begin{bmatrix} 0 & 0 \\ 0 & 1 \\ 0 & -a \\ 1 & a+b \end{bmatrix} \lambda \right\}$$

$$= \begin{bmatrix} \lambda^2 - 2\lambda + 1 & -(a+2b)\lambda + b \\ 0 & \lambda^2 \end{bmatrix}.$$

Case 4. Again, we have a single subspace

$$\mathcal{M}_4 = \{(x_1, 0, x_3, x_4, 0, x_6)^{\mathrm{T}} \mid x_j \in \mathbb{C}, j = 1, 3, 4, 6\}$$

and a family

$$\mathcal{M}_4(a) = \operatorname{Span}\{(1 \ 0 \ 0 \ 0 \ 0 \ 0)^{\mathrm{T}}, (0 \ 1 \ a \ 0 \ 0 \ 0)^{\mathrm{T}}, (0 \ 0 \ 0 \ 1 \ 0 \ 0)^{\mathrm{T}}, (0 \ 0 \ 0 \ 0 \ 0 \ 1)^{\mathrm{T}}\}.$$

Now

$$\begin{bmatrix} X \\ XJ \end{bmatrix}\Big|_{\mathcal{M}_4} = \begin{bmatrix} -b & 1 & 1 & -b \\ 1 & 0 & 0 & 1 \\ 0 & 0 & 1 & -2b \\ 0 & 0 & 0 & 2 \end{bmatrix},$$

which is nonsingular, and the monic divisor $L_{\mathcal{M}_4}(\lambda)$ with the supporting subspace $\mathcal{M}_4$ is

$$L_{\mathcal{M}_4}(\lambda) = \lambda^2 I - X|_{\mathcal{M}_4} \cdot (J|_{\mathcal{M}_4})^2 (V_1^{(4)} + V_2^{(4)}\lambda),$$

where $[V_1^{(4)} V_2^{(4)}] = [\operatorname{col}(X|_{\mathcal{M}_4} \cdot (J|_{\mathcal{M}_4})^i)_{i=0}^1]^{-1}$, and

$$L_{\mathcal{M}_4}(\lambda) = \begin{bmatrix} \lambda^2 - \lambda & b\lambda \\ 0 & \lambda^2 - 2\lambda \end{bmatrix}.$$

We pass now to the family $\mathcal{M}_4(a)$.

$$\begin{bmatrix} X \\ XJ \end{bmatrix}\Big|_{\mathcal{M}_4(a)} = \begin{bmatrix} -b & a & 1 & -b \\ 1 & 0 & 0 & 1 \\ 0 & -b & 1 & -2b \\ 0 & 1 & 0 & 2 \end{bmatrix}.$$

This matrix is nonsingular if and only if $a \neq 0$. So for every complex value of a, except zero, the subspace $\mathcal{M}_4(a)$ is supporting for the monic right divisor $L_{\mathcal{M}_4(a)}(\lambda)$ of degree 2. Computation shows that

$$L_{\mathcal{M}_4(a)}(\lambda) = \begin{bmatrix} \lambda^2 + \left(-1 + \frac{2b}{a}\right)\lambda - \frac{2b}{a} & \frac{ab + 2b^2}{a}\lambda - \frac{2b^2}{a} \\ -\frac{2}{a}\lambda + \frac{2}{a} & \lambda^2 - \frac{2(a+b)}{a}\lambda + \frac{2b}{a} \end{bmatrix} \qquad (a \neq 0).$$

Case 5.

$$\begin{aligned} \mathcal{M}_5 = \mathcal{M}_5(\alpha, \beta) = \operatorname{Span}\{&(\alpha \quad 0 \quad \beta \quad 0 \quad 0 \quad 0)^{\mathrm{T}}, (0 \quad 0 \quad 0 \quad 1 \quad 0 \quad 0)^{\mathrm{T}}, \\ &(0 \quad 0 \quad 0 \quad 0 \quad 1 \quad 0)^{\mathrm{T}}, (0 \quad 0 \quad 0 \quad 0 \quad 0 \quad 1)^{\mathrm{T}}\}, \end{aligned} \tag{3.30}$$

where α, β are complex numbers and at least one of them is different from zero. Then

$$\begin{bmatrix} X \\ XJ \end{bmatrix}\Big|_{\mathcal{M}_5} = \begin{bmatrix} -\alpha b + \beta & 1 & 0 & -b \\ \alpha & 0 & 0 & 1 \\ 0 & 1 & 1 & -2b \\ 0 & 0 & 0 & 2 \end{bmatrix},$$

which is invertible if and only if $\alpha \neq 0$. So $\mathcal{M}_5$ is a supporting subspace for some monic divisor of degree 2 if and only if $\alpha \neq 0$. In fact, for $\alpha = 0$ we obtain only one subspace $\mathcal{M}_5$ (because then $\beta \neq 0$ and we can use $(0 \quad 0 \quad 1 \quad 0 \quad 0 \quad 0)^{\mathrm{T}}$ in place of $(\alpha \quad 0 \quad \beta \quad 0 \quad 0 \quad 0)^{\mathrm{T}}$ in (3.30)). Suppose now $\alpha \neq 0$; then we can suppose that $\alpha = 1$ (using $(1/\alpha)(\alpha \quad 0 \quad \beta \quad 0 \quad 0 \quad 0)^{\mathrm{T}}$ in place of $(\alpha \quad 0 \quad \beta \quad 0 \quad 0 \quad 0)^{T}$ in (3.30)); then with the supporting subspace $\mathcal{M}_5(\beta)$ is

$$L_{\mathcal{M}_5(\beta)}(\lambda) = \begin{bmatrix} \lambda^2 - 2\lambda + 1 & \frac{1}{2}\beta\lambda + b - \beta \\ 0 & \lambda^2 - 2\lambda \end{bmatrix}.$$

(Note for reference that

$$X\,|_{\mathcal{M}_5(\beta)}(J\,|_{\mathcal{M}_5(\beta)})^2 = \begin{bmatrix} 0 & 1 & 2 & -4b \\ 0 & 0 & 0 & 4 \end{bmatrix}.)$$

According to Theorem 3.12, the monic matrix polynomials $L_{\mathcal{M}_1}$, $L_{\mathcal{M}_3(a)}$, $L_{\mathcal{M}_4}$, $L_{\mathcal{M}_4(a)}(a \neq 0)$, $L_{\mathcal{M}_5(\beta)}$ form the complete set of right monic divisors of $L(\lambda)$ of degree 2.

3.5. Description of the Quotient and Left Divisors

In Section 3.3 we have studied the monic right divisors $L_1(\lambda)$ of a given monic matrix polynomial $L(\lambda)$. Here we shall obtain a formula for the quotient $L_2(\lambda) = L(\lambda)L_1^{-1}(\lambda)$. At the same time we provide a description of the left monic divisors $L_2(\lambda)$ of $L(\lambda)$ (because each such divisor has the form $L(\lambda)L_1^{-1}(\lambda)$ for some right monic divisor $L_1(\lambda)$).

We shall present the description in terms of standard triples, which allow us to obtain more symmetric (with respect to left and right) statements of the main results.

Lemma 3.16. *Let $L(\lambda)$ be a monic matrix polynomial of degree l with standard triple (X, T, Y), and let P be a projector in $\mathbb{C}^{nl}$. Then the linear transformation*

$$\operatorname{col}(XT^{i-1})_{i=1}^{k}|_{\operatorname{Im} P}: \operatorname{Im} P \to \mathbb{C}^{nk} \tag{3.31}$$

(where $k < l$) is invertible if and only if the linear transformation

$$(I - P)[\operatorname{row}(T^{l-k-i}Y)_{i=1}^{l-k}]: \mathbb{C}^{n(l-k)} \to \operatorname{Ker} P \tag{3.32}$$

is invertible.

Proof. Put $A = \operatorname{col}(XT^{i-1})_{i=1}^{l}$ and $B = \operatorname{row}(T^{l-i}Y)_{i=1}^{l}$. With respect to the decompositions $\mathbb{C}^{nl} = \operatorname{Im} P \dotplus \operatorname{Ker} P$ and $\mathbb{C}^{nl} = \mathbb{C}^{nk} \oplus \mathbb{C}^{n(l-k)}$ write

$$A = \begin{bmatrix} A_1 & A_2 \\ A_3 & A_4 \end{bmatrix}, \qquad B = \begin{bmatrix} B_1 & B_2 \\ B_3 & B_4 \end{bmatrix}.$$

Thus, A_i are linear transformations with the following domains and ranges: $A_1: \operatorname{Im} P \to \mathbb{C}^{nk}$; $A_2: \operatorname{Ker} P \to \mathbb{C}^{nk}$; $A_3: \operatorname{Im} P \to \mathbb{C}^{n(l-k)}$; $A_4: \operatorname{Ker} P \to \mathbb{C}^{n(l-k)}$; analogously for the B_i.

Observe that A_1 and B_4 coincide with the operators (3.31) and (3.32), respectively. In view of formula (2.11), the product AB has the form

$$AB = \begin{bmatrix} D_1 & 0 \\ * & D_2 \end{bmatrix}$$

with D_1 and D_2 as invertible linear transformations. Recall that A and B are also invertible (by the properties of a standard triple). But then A_1 is invertible if and only if B_4 is invertible. This may be seen as follows.

Suppose that B_4 is invertible. Then

$$B\begin{bmatrix} I & 0 \\ -B_4^{-1}B_3 & B_4^{-1} \end{bmatrix} = \begin{bmatrix} B_1 & B_2 \\ B_3 & B_4 \end{bmatrix}\begin{bmatrix} I & 0 \\ -B_4^{-1}B_3 & B_4^{-1} \end{bmatrix}$$
$$= \begin{bmatrix} B_1 - B_2B_4^{-1}B_3 & B_2B_4^{-1} \\ 0 & I \end{bmatrix}$$

is invertible in view of the invertibility of B, and then also $B_1 - B_2B_4^{-1}B_3$ is invertible. The special form of AB implies $A_1B_2 + A_2B_4 = 0$. Hence $D_1 = A_1B_1 + A_2B_3 = A_1B_1 - A_1B_2B_4^{-1}B_3 = A_1(B_1 - B_2B_4^{-1}B_3)$ and it follows that A_1 is invertible. A similar argument shows that invertibility of A_1 implies invertibility of B_4. This proves the lemma. □

We say that P is a *supporting projector* for the triple (X, T, Y) if $\operatorname{Im} P$ is a nontrivial invariant subspace for T and the linear transformation (3.31) is invertible for some positive integer k. One checks without difficulty that k is unique and $k < l$. We call k the *degree* of the supporting projector. It follows from Theorem 3.12 that P is a supporting projector of degree k if and only if its image is a supporting subspace of some monic divisor of $L(\lambda)$ of degree k.

Let P be a supporting projector for (X, T, Y) of degree k. Define $T_1: \operatorname{Im} P \to \operatorname{Im} P$ and $X_1: \operatorname{Im} P \to \mathbb{C}^n$ by

$$T_1 y = Ty, \qquad X_1 y = Xy.$$

The invertibility of (3.31) now implies that $\operatorname{col}(X_1 T_1^{i-1})_{i=1}^k: \operatorname{Im} P \to \mathbb{C}^{nk}$ is invertible. Hence there exists a unique linear transformation $Y_1: \mathbb{C}^n \to \operatorname{Im} P$ such that the triple (X_1, T_1, Y_1) is a standard triple of some monic matrix polynomial $L_1(\lambda)$ of degree k. The triple (X_1, T_1, Y_1) will be called the *right projection* of (X, T, Y) associated with P. It follows from Theorem 3.12 that the polynomial $L_1(\lambda)$ defined by (X_1, T_1, Y_1) is a right divisor of $L(\lambda)$, and every monic right divisor of $L(\lambda)$ is generated by the right projection connected with some supporting projector of (X, T, Y).

By Lemma 3.16 the linear transformation (3.32) is invertible. Define $T_2: \operatorname{Ker} P \to \operatorname{Ker} P$ and $Y_2: \mathbb{C}^n \to \operatorname{Ker} P$ by

$$T_2 y = (I - P)Ty, \qquad Y_2 x = (I - P)Yx.$$

Since $\operatorname{Im} P$ is an invariant subspace for T, we have $(I - P)T(I - P) = (I - P)T$. This, together with the invertibility of (3.32) implies that

$$\operatorname{row}(T_2^{i-1} Y_2)_{i=1}^{l-k}: \mathbb{C}^{n(l-k)} \to \operatorname{Ker} P$$

is invertible. Therefore there exists a unique $X_2: \operatorname{Ker} P \to \mathbb{C}^n$ such that the triple (X_2, T_2, Y_2) is a standard triple for some monic matrix polynomial $L_2(\lambda)$ (which is necessarily of degree $l - k$). The triple (X_2, T_2, Y_2) will be called the *left projection* of (X, T, Y) associated with P.

Suppose that P is a supporting projector for (X, T, Y) of degree k, and let $P': \mathbb{C}^{nl} \to \mathbb{C}^{nl}$ be another projector such that $\operatorname{Im} P' = \operatorname{Im} P$. Then it follows immediately from the definition that P' is also a supporting projector for (X, T, Y) of degree k. Also the right projections associated with P and P' coincide. Thus what really matters in the construction of the left projection is the existence of a nontrivial invariant subspace $\mathscr{M}_1$ for T such that for some positive integer k

$$\operatorname{col}(XT^i)_{i=0}^{k-1}|_{\mathscr{M}_1}: \mathscr{M}_1 \to \mathbb{C}^{nk}$$

i.e., the existence of a supporting subspace (cf. Section 3.3). The preference for dealing with supporting projectors is due to the fact that it admits a more symmetric treatment of the division (or rather factorization) problem.

The next theorem shows that the monic polynomial $L_2(\lambda)$ defined by the left projection is just the quotient $L(\lambda)L_1^{-1}(\lambda)$ where $L_1(\lambda)$ is the right divisor of $L(\lambda)$ defined by the right projection of (X, T, Y) associated with P.

Theorem 3.17. *Let $L(\lambda)$ be a monic matrix polynomial with standard triple (X, T, Y). Let P be a supporting projector of (X, T, Y) of degree k and let $L_1(\lambda)$ (resp. $L_2(\lambda)$) be the monic matrix polynomial of degree k (resp. $l - k$) generated by the right (resp. left) projection of (X, T, Y) associated with P. Then*

$$L(\lambda) = L_2(\lambda)L_1(\lambda). \tag{3.33}$$

Conversely, every factorization (3.33) of $L(\lambda)$ into a product of two monic factors $L_2(\lambda)$ and $L_1(\lambda)$ of degrees $l - k$ and k, respectively, is obtained by using some supporting projector of (X, T, Y) as above.

Proof. Let $P': \mathbb{C}^{nl} \to \mathbb{C}^{nl}$ be given by

$$P'y = [\operatorname{col}(X_1 T_1^{i-1})_{i=1}^k]^{-1} \cdot [\operatorname{col}(X T^{i-1})_{i=1}^k]y,$$

where $X_1 = X|_{\operatorname{Im} P}$; $T_1 = T|_{\operatorname{Im} P}$. Then P' is a projector and $\operatorname{Im} P' = \operatorname{Im} P$. Also

$$\operatorname{Ker} P' = \left\{ \sum_{i=1}^{l-k} T^{l-k-i} Y x_{i-1} \,|\, x_0, \ldots, x_{l-k-1} \in \mathbb{C}^n \right\}. \tag{3.34}$$

The proof of this based on formulas (2.1).

Define $S: \mathbb{C}^{nl} \to \operatorname{Im} \mathrm{P} \dotplus \operatorname{Ker} P$ by

$$S = \begin{bmatrix} P' \\ I - P \end{bmatrix},$$

where P' and $I - P$ are considered as linear transformations from C^{nl} into $\operatorname{Im} P'$ and $\operatorname{Ker} P$, respectively. One verifies easily that S is invertible. We shall show that

$$[X_1 \quad 0]S = X, \qquad ST = \begin{bmatrix} T_1 & Y_1 X_2 \\ 0 & T_2 \end{bmatrix} S, \tag{3.35}$$

which in view of Theorem 3.1 means that (X, T) is a standard pair for the product $L_2(\lambda)L_1(\lambda)$, and since a monic polynomial is uniquely defined by its standard pair, (3.33) follows.

Take $y \in \mathbb{C}^{nl}$. Then $P'y \in \operatorname{Im} P$ and $\operatorname{col}(X_1 T_1^{i-1} P'y)_{i=1}^k = \operatorname{col}(XT^{i-1}y)_{i=1}^k$. In particular $X_1 P'y = Xy$. This proves that $[X_1 \quad 0]S = X$. The second equality in (3.35) is equivalent to the equalities

$$P'T = T_1 P' + Y_1 X_2 (I - P) \tag{3.36}$$

and $(I - P)T = T_2(I - P)$. The last equality is immediate from the definition of T_2 and the fact that $\operatorname{Im} P$ is an invariant subspace for T. To prove (3.36), take $y \in \mathbb{C}^{nl}$. The case when $y \in \operatorname{Im} P = \operatorname{Im} P'$ is trivial. Therefore assume that $y \in \operatorname{Ker} P'$. We then have to demonstrate that $P'Ty = Y_1 X_2 (I - P)y$. Since $y \in \operatorname{Ker} P'$, there exist $x_0, \ldots, x_{l-k-1} \in \mathbb{C}^n$ such that $y = \sum_{i=1}^{l-k} T^{l-k-i} Y x_{i-1}$. Hence

$$Ty = T^{l-k} Y x_0 + T^{l-k-1} Y x_1 + \cdots + T Y x_{l-k-1} = T^{l-k} Y x_0 + u$$

with $u \in \operatorname{Ker} P'$ and, as a consequence, $P'Ty = P'T^{l-k} Y x_0$. But then it follows from the definition of P' that

$$P'Ty = [\operatorname{row}(T_1^{k-i} Y_1)_{i=1}^k] \operatorname{col}(0, \ldots, 0, x_0) = Y_1 x_0.$$

On the other hand, putting $x = \operatorname{col}(x_{i-1})_{i=1}^{l-k}$,

$$(I - P)y = (I - P) \operatorname{row}(T^{l-k-i} Y)_{i=1}^{l-k} x = \operatorname{row}(T_2^{l-k-i} Y_2)_{i=1}^{l-k} x$$

and so $Y_1 X_2 (I - P)y$ is also equal to $Y_1 x_0$. This completes the proof. □

Using Theorem 3.17, it is possible to write down the decomposition $L(\lambda) = L_2(\lambda) L_1(\lambda)$, where $L_2(\lambda)$ and $L_1(\lambda)$ are written in one of the possible forms: right canonical, left canonical, or resolvent. We give in the next corollary one such decomposition in which $L_2(\lambda)$ is in the left canonical form and $L_1(\lambda)$ is in the right canonical form.

Corollary 3.18. *Let $L(\lambda)$ and (X, T, Y) be as in Theorem 3.17. Let $L_1(\lambda)$ be a right divisor of degree k of $L(\lambda)$ with the supporting subspace $\mathcal{M}$. Then*

$$\begin{aligned} L(\lambda) = {} & [I\lambda^{l-k} - (Z_1 + \cdots + Z_{l-k}\lambda^{l-k-1}) \tilde{P} T^{l-k} \tilde{P} Y] \\ & \cdot [I\lambda^k - X|_{\mathcal{M}} (T|_{\mathcal{M}})^k (W_1 + \cdots + W_k \lambda^{k-1})], \end{aligned}$$

where

$$[W_1 \quad \cdots \quad W_k] = [\operatorname{col}(X|_{\mathcal{M}} (T|_{\mathcal{M}})^i)_{i=0}^{k-1}]^{-1},$$

$\tilde{P}$ is some projector with $\operatorname{Ker} \tilde{P} = \mathcal{M}$, the linear transformation $[\tilde{P}Y \quad \tilde{P}T\tilde{P}Y \quad \cdots \quad \tilde{P}T^{l-k-1}\tilde{P}Y] \colon \mathbb{C}^{n(l-k)} \to \operatorname{Im} \tilde{P}$ is invertible, and

$$\begin{bmatrix} Z_1 \\ \vdots \\ Z_{l-k} \end{bmatrix} = (\operatorname{row}(\tilde{P}Y \cdot \tilde{P}T^i\tilde{P})_{i=0}^{l-k-1})^{-1} \colon \operatorname{Im} \tilde{P} \to \mathbb{C}^{n(l-k)}.$$

Proof. Let P be the supporting projector for the right divisor $L_1(\lambda)$. Then, by Theorem 3.17, (T_2, Y_2) is a left standard pair for $L_2(\lambda)$, where T_2: Ker $P \to$ Ker P and $Y_2: \mathbb{C}^n \to$ Ker P are defined by $T_2 y = (I - P)Ty$, $Y_2 x = (I - P)Yx$. Choose $P = I - \tilde{P}$; then $Y_2 = \tilde{P}Y$ and $T_2 = \tilde{P}T|_{\operatorname{Im}\tilde{P}}$. So Corollary 3.18 will follow from the left standard representation of $L_2(\lambda)$ if we prove that

$$(\tilde{P}T|_{\operatorname{Im}\tilde{P}})^i = \tilde{P}T^i|_{\operatorname{Im}\tilde{P}}, \qquad i = 0, 1, 2, \ldots. \tag{3.37}$$

But (3.37) follows from the representation of T relative to the decomposition $\mathbb{C}^{nl} = \mathcal{M} \dotplus \operatorname{Im}\tilde{P}$:

$$T = \begin{bmatrix} T_{11} & T_{12} \\ 0 & T_{22} \end{bmatrix}$$

where $T_{22} = \tilde{P}T|_{\operatorname{Im}\tilde{P}}$. So

$$T^i = \begin{bmatrix} T_{11}^i & * \\ 0 & T_{22}^i \end{bmatrix} \qquad i = 0, 1, \ldots,$$

i.e., $(\tilde{P}T|_{\operatorname{Im}\tilde{P}})^i = \tilde{P}T^i|_{\operatorname{Im}\tilde{P}}$, which is exactly (3.37). □

Note that Corollary 3.18 does not depend on the choice of the projector $\tilde{P}$.

Using Corollary 3.18 together with Corollary 3.15, one can deduce results concerning the decomposition of $L(\lambda)$ into a product of more than two factors:

$$L(\lambda) = L_m(\lambda)L_{m-1}(\lambda)\cdots L_1(\lambda),$$

where $L_i(\lambda)$ are monic matrix polynomials.

It turns out that such decompositions are in one-to-one correspondence with sequences of nested supporting subspaces $\mathcal{M}_{m-1} \supset \mathcal{M}_{m-2} \supset \cdots \supset \mathcal{M}_1$, where $\mathcal{M}_i$ is the supporting subspace of $L_i(\lambda)\cdots L_1(\lambda)$ as a right divisor of $L(\lambda)$. We shall not state here results of this type, but refer the reader to [34a, 34b].

For further reference we shall state the following immediate corollary of Theorem 3.17 (or Corollary 3.18).

Corollary 3.19. *Let $L(\lambda)$ be a monic matrix polynomial with linearization T and let $L_1(\lambda)$ be a monic right divisor with supporting subspace $\mathcal{M}$. Then for any complementary subspace $\mathcal{M}'$ to $\mathcal{M}$, the linear transformation $PT|_{\mathcal{M}'}$ is a linearization of $L(\lambda)L_1^{-1}(\lambda)$, where P is a projector on $\mathcal{M}'$ along $\mathcal{M}$. In particular, $\sigma(LL_1^{-1}) = \sigma(PT|_{\mathcal{M}'})$.*

We conclude this section with a computational example.

EXAMPLE 3.2. Let

$$L(\lambda) = \begin{bmatrix} \lambda(\lambda-1)^2 & b\lambda \\ 0 & \lambda^2(\lambda-2) \end{bmatrix}, \qquad b \in \mathbb{C}$$

be the polynomial from Section 3.4. We shall use the same canonical pair (X, J) as in Section 4. Consider the family

$$\mathcal{M}_4(\alpha) = \operatorname{Span}\{(1\ \ 0\ \ 0\ \ 0\ \ 0\ \ 0)^{\mathrm{T}}, (0\ \ 1\ \ a\ \ 0\ \ 0\ \ 0)^{\mathrm{T}}, (0\ \ 0\ \ 0\ \ 1\ \ 0\ \ 0)^{\mathrm{T}}, (0\ \ 0\ \ 0\ \ 0\ \ 0\ \ 1)^{\mathrm{T}}\}, \qquad a \neq 0,$$

of supporting subspaces and let

$$L_{\mathcal{M}_4(a)}(\lambda) = \begin{bmatrix} \lambda^2 + \left(-1 + \dfrac{2b}{a}\right)\lambda - \dfrac{2b}{a} & \dfrac{ab + 2b^2}{a}\lambda - \dfrac{2b^2}{a} \\ -\dfrac{2}{a}\lambda + \dfrac{2}{a} & \lambda^2 - \dfrac{2(a+b)}{a}\lambda + \dfrac{2b}{a} \end{bmatrix}$$

be the family of corresponding right divisors (cf. Section 3.4). We compute the quotients $M_a(\lambda) = L(\lambda)\,(L_{\mathcal{M}_4(a)}(\lambda))^{-1} = I\lambda - U_a$. According to Corollary 3.18,

$$U_a = Z_a \cdot \tilde{P}_a J \tilde{P}_a \cdot Y,$$

where $\tilde{P}_a$ is a projector such that $\operatorname{Ker} \tilde{P}_a = \mathcal{M}_4(a)$,

$$Y = \begin{bmatrix} X \\ XJ \\ XJ^2 \end{bmatrix}^{-1} \begin{bmatrix} 0 \\ 0 \\ I \end{bmatrix},$$

and $Z_a = (\tilde{P}_a Y)^{-1} \colon \operatorname{Im} \tilde{P}_a \to \mathbb{C}^2$.

Computation shows that

$$Y = \begin{bmatrix} 0 & -\frac{1}{4} \\ 0 & -\frac{1}{2} \\ 1 & b \\ -1 & -b \\ 1 & b \\ 0 & \frac{1}{4} \end{bmatrix}.$$

Further, let us take $\tilde{P}_a$ the projector with $\operatorname{Ker} \tilde{P}_a = \mathcal{M}_4(a)$ and $\operatorname{Im} \tilde{P}_a = \{(0, x_2, 0, 0, x_5, 0)^{\mathrm{T}} \in \mathbb{C}^6 \mid x_2, x_5 \in \mathbb{C}\}$. In the matrix representation

$$\tilde{P}_a = \begin{bmatrix} 0 & 0 & 0 & 0 & 0 & 0 \\ 0 & 1 & -1/a & 0 & 0 & 0 \\ 0 & 0 & 0 & 0 & 0 & 0 \\ 0 & 0 & 0 & 0 & 0 & 0 \\ 0 & 0 & 0 & 0 & 1 & 0 \\ 0 & 0 & 0 & 0 & 0 & 0 \end{bmatrix}.$$

So $\tilde{P}_a Y: \mathbb{C}^2 \to \operatorname{Im} \tilde{P}_a$ is given by the matrix

$$\begin{bmatrix} -1/a & -\frac{1}{2} - b/a \\ 1 & b \end{bmatrix}$$

and

$$Z_a = \begin{bmatrix} -1/a & -\frac{1}{2} - b/a \\ 1 & b \end{bmatrix}^{-1} = 2\begin{bmatrix} b & \frac{1}{2} + b/a \\ -1 & -1/a \end{bmatrix}.$$

Now let us compute $\tilde{P}_a J \tilde{P}_a$: $\operatorname{Im} \tilde{P}_a \to \operatorname{Im} \tilde{P}_a$. To this end we compute $\tilde{P}_a J e_1$ and $\tilde{P}_a J e_2$, where $e_1 = (0 \quad 1 \quad 0 \quad 0 \quad 0 \quad 0)^{\mathrm{T}}$ and $e_2 = (0 \quad 0 \quad 0 \quad 0 \quad 1 \quad 0)^{\mathrm{T}}$ are basis vectors in $\operatorname{Im} \tilde{P}_a$:

$$\begin{aligned} \tilde{P}_a J e_1 &= \tilde{P}_a (1 \quad 0 \quad 0 \quad 0 \quad 0 \quad 0)^{\mathrm{T}} = 0, \\ \tilde{P}_a J e_2 &= \tilde{P}_a (0 \quad 0 \quad 0 \quad 1 \quad 1 \quad 0)^{\mathrm{T}} = e_2, \end{aligned}$$

so

$$\tilde{P}_a J \tilde{P}_a = \begin{bmatrix} 0 & 0 \\ 0 & 1 \end{bmatrix}.$$

Finally,

$$U_a = 2\begin{bmatrix} b & \frac{1}{2} + b/a \\ -1 & -1/a \end{bmatrix}\begin{bmatrix} 0 & 0 \\ 0 & 1 \end{bmatrix}\begin{bmatrix} -1/a & -\frac{1}{2} - b/a \\ 1 & b \end{bmatrix} = \begin{bmatrix} 1 + 2b/a & b + 2b^2/a \\ -2/a & -2b/a \end{bmatrix}.$$

By direct multiplication one checks that indeed

$$(I\lambda - U_a) \cdot L_{\mathcal{M}_4(a)}(\lambda) = \begin{bmatrix} \lambda(\lambda - 1)^2 & b\lambda \\ 0 & \lambda^2(\lambda - 2) \end{bmatrix} \quad (= L(\lambda)). \quad \square$$

3.6. Divisors and Supporting Subspaces for the Adjoint Polynomial

Let $L(\lambda) = I\lambda^l + \sum_{j=0}^{l-1} A_j \lambda^j$ be a monic matrix polynomial of degree l with a standard triple (X, T, Y). Let

$$L^*(\lambda) = I\lambda^l + \sum_{j=0}^{l-1} A_j^* \lambda^j$$

be the *adjoint* matrix polynomial. As we already know (see Theorem 2.2) the triple (Y^*, T^*, X^*) is a standard triple for $L^*(\lambda)$.

On the other hand, if $L_1(\lambda)$ is a monic right divisor of $L(\lambda)$, we have $L(\lambda) = L_2(\lambda)L_1(\lambda)$ for some monic matrix polynomial $L_2(\lambda)$. So $L^*(\lambda) = L_1^*(\lambda)L_2^*(\lambda)$, and $L_2^*(\lambda)$ is a right divisor of $L^*(\lambda)$. There exists a simple connection between the supporting subspace for $L_1(\lambda)$ (with respect to the standard pair (X, T) of $L(\lambda)$) and the supporting subspace of $L_2^*(\lambda)$ (with respect to the standard pair (Y^*, T^*)).

Theorem 3.20. *If $L_1(\lambda)$ (as a right divisor of $L(\lambda)$) has a supporting subspace $\mathcal{M}$ (with respect to the standard pair (X, T)), then $L_2^*(\lambda)$ (as a right divisor of $L^*(\lambda)$) has supporting subspace $\mathcal{M}^\perp$ (with respect to the standard pair (Y^*, T^*)).*

Proof. We use the notions and results of Section 3.5. Let P be the supporting projector for the triple (X, T, Y) corresponding to the right divisor L_1, so $\mathcal{M} = \operatorname{Im} P$. We can suppose that P is orthogonal, $P = P^*$; then $\operatorname{Im}(I - P) = \mathcal{M}^\perp$. By Theorem 3.17 the quotient $L_2(\lambda) = L(\lambda)L_1^{-1}(\lambda)$ is determined by the left projection (X_2, T_2, Y_2) of (X, T, Y) associated with P:

$$T_2 = (I - P)T: \operatorname{Ker} P \to \operatorname{Ker} P, \qquad Y_2 = (I - P)Y: \mathbb{C}^n \to \operatorname{Ker} P.$$

Since $W \stackrel{\text{def}}{=} \operatorname{row}(T_2^{i-1}Y)_{i=1}^{l-k}: \mathbb{C}^{n(l-k)} \to \operatorname{Ker} P$ is invertible (where k is the degree of $L_1(\lambda)$), so is

$$W^* = \operatorname{col}(Y_2^* T_2^*)|_{\operatorname{Ker} P}: \operatorname{Ker} P \to \mathbb{C}^{n(l-k)}. \tag{3.38}$$

Clearly, Ker P is T^*-invariant (because of the general property which can be checked easily: if a subspace N is A-invariant, then $N^\perp$ is A^*-invariant), and in view of (3.38) and Theorem 3.12 $\mathcal{M}^\perp = \operatorname{Ker} P$ is a supporting subspace such that (Y_2^*, T_2^*) is a standard pair of the divisor $M(\lambda)$ of $L^*(\lambda)$ corresponding to $\mathcal{M}^\perp$ (with respect to the standard triple (Y^*, T^*, X^*) of $L^*(\lambda)$). But (Y_2^*, T_2^*) is also a standard pair for $L_2^*(\lambda)$, as follows from the definition of a left projection associated with P. So in fact $M(\lambda) = L_2^*(\lambda)$, and Theorem 3.20 is obtained. $\square$

3.7. Decomposition into a Product of Linear Factors

It is well known that a scalar polynomial over the complex numbers can be decomposed into a product of linear factors in the same field. Is it true also for monic matrix polynomials? In general, the answer is no, as the following example shows.

EXAMPLE 3.3. The polynomial

$$L(\lambda) = \begin{bmatrix} \lambda^2 & -1 \\ 0 & \lambda^2 \end{bmatrix}$$

has a canonical triple determined by

$$X = \begin{bmatrix} 1 & 0 & 0 & 0 \\ 0 & 0 & 1 & 0 \end{bmatrix}, \qquad J = \begin{bmatrix} 0 & 1 & 0 & 0 \\ 0 & 0 & 1 & 0 \\ 0 & 0 & 0 & 1 \\ 0 & 0 & 0 & 0 \end{bmatrix}.$$

In order that L have a monic right divisor of degree one it is necessary and sufficient that J has a two-dimensional invariant subspace on which X is invertible. However, the only such subspace is $A = \operatorname{Span}(e_1, e_2)$ and X is not invertible on A. Hence L has no (nontrivial) divisors. □

However, in the important case when all the elementary divisors of the polynomial are linear or, in other words, for which there is a *diagonal* canonical matrix J, the answer to the above question is yes, as our next result shows.

Theorem 3.21. *Let $L(\lambda)$ be a monic matrix polynomial having only linear elementary divisors. Then $L(\lambda)$ can be decomposed into a product of linear factors:*

$$L(\lambda) = \prod_i (I\lambda - B_i).$$

Proof. Let (X, J) be a Jordan pair for $L(\lambda)$. Then, by hypothesis, J is diagonal. Since the $ln \times ln$ matrix

$$\begin{bmatrix} X \\ XJ \\ \vdots \\ XJ^{l-1} \end{bmatrix}$$

is nonsingular, the $(l - 1)n \times ln$ matrix A, obtained by omitting the last term, has full rank. Thus, there exist $(l - 1)n$ linearly independent columns of A, say columns $i_1, i_2, \ldots, i_{(l-1)n}$. Since J is diagonal, the unit vectors $e_{i_1}, \ldots, e_{i_{(l-1)n}}$ in $\mathbb{C}^{ln}$ generate an $(l - 1)n$-dimensional invariant subspace $\mathscr{M}$ of J and, obviously, A will be invertible on $\mathscr{M}$. Therefore, by Theorem 3.12 $L(\lambda)$ has a monic right divisor $L_1(\lambda)$ of degree $l - 1$ and we may write $L(\lambda) = (I\lambda - B_1)L_1(\lambda)$.

But we know (again by Theorem 3.12) that the Jordan matrix J_1 associated with $L_1(\lambda)$ is just a submatrix of J and so $L_1(\lambda)$ must also have all linear elementary divisors. We may now apply the argument repeatedly to obtain the theorem. □

In the case $n = 1$ there are precisely $l!$ distinct decompositions of the polynomial into linear factors, provided all the eigenvalues of the polynomial are distinct. This can be generalized in the following way:

Corollary 3.22. *If $L(\lambda)$ (of Theorem 3.20) has all its eigenvalues distinct, then the number of distinct factorizations into products of linear factors is not less than $\prod_{j=0}^{l-1} (nj + 1)$.*

Proof. It is sufficient to show that $L(\lambda)$ has no less than $n(l - 1) + 1$ monic right divisors of first degree since the result then follows by repeated application to the quotient polynomial.

If (X, J) is a Jordan pair for $L(\lambda)$, then each right divisor $I\lambda - B$ of $L(\lambda)$ has the property that the eigenvectors of B are columns of X and the corresponding submatrix of J is a Jordan form for B. To prove the theorem it is sufficient to find $n(l - 1) + 1$ nonsingular $n \times n$ submatrices of X for, since all eigenvalues of L are distinct, distinct matrices B will be derived on combining these matrices of eigenvectors with the (necessarily distinct) corresponding submatrices of J.

By Theorem 3.21 there exists one right divisor $I\lambda - B$. Assume, without loss of generality, that the first n columns of X are linearly independent and correspond to this divisor. Since every column of X is nonzero (as an eigenvector) we can associate with the ith column $(i = n + 1, \ldots, nl)$ $n - 1$ columns $i_1, \ldots, i_{n-1}$, from among the first n columns of X, in such a way that columns $i_1, \ldots, i_{n-1}, i$ form a nonsingular matrix. For each i from $n + 1$ to ln we obtain such a matrix and, together with the submatrix of the first n columns of X we obtain $n(l - 1) + 1$ nonsingular $n \times n$ submatrices of X. □

Comments

The presentation in this chapter is based on the authors' papers [34a, 34b]. The results on divisibility and multiplication of matrix polynomials are essentially of algebraic character, and can be obtained in the same way for matrix polynomials over an algebraically closed field (not necessarily $\mathbb{C}$), and even for any field if one confines attention to standard triples (excluding Jordan triples). See also [20]. These results can also be extended for operator polynomials (see [34c]). Other approaches to divisibility of matrix (and operator) polynomials via supporting subspaces are found in [46, 56d]. A theorem showing (in the infinite-dimensional case) the connection between monic divisors and special invariant subspaces of the companion matrix was obtained earlier in [56d]. Some topological properties of the set of divisors of a monic matrix polynomial are considered in [70e].

Additional information concerning the problem of computing partial multiplicities of the product of two matrix polynomials (this problem was mentioned in Section 3.1) can be found in [27, 71a, 75, 76b]. Our Section 3.3 incorporates some simplifications of the arguments used in [34a] which are based on [73]. The notion of a supporting projector (Section 3.5) originated in [3a]. This idea was further developed in the framework of rational matrix and operator functions, giving rise to a factorization theory for such functions (see [4, 3c]).

Theorem 3.20 (which holds also in the case of operator polynomials acting in infinite dimensional space), as well as Lemma 3.4, is proved in [34f]. Theorem 3.21 is proved by other means in [62b].

Another approach to the theory of monic matrix polynomials, which is similar to the theory of characteristic operator functions, is developed in [3b]. The main results on representations and divisibility are obtained there.

Chapter 4

Spectral Divisors and Canonical Factorization

We consider here the important special case of factorization of a monic matrix polynomial $L(\lambda) = L_2(\lambda)L_1(\lambda)$, in which $L_1(\lambda)$ and $L_2(\lambda)$ are monic polynomials with disjoint spectra. Divisors with this property will be called *spectral*.

Criteria for the existence of spectral divisors as well as explicit formulas for them are given in terms of contour integrals. These results are then applied to the matrix form of Bernoulli's algorithm for the solution of polynomial equations, and to a problem of the stability of solutions of differential equations.

One of the important applications of the results on spectral divisors is to canonical factorization, which plays a decisive role in inversion of block Toeplitz matrices. These applications are included in Sections 4.5 and 4.6. Section 4.7 is devoted to the more general notion of Wiener–Hopf factorization for monic matrix polynomials.

4.1. Spectral Divisors

Much of the difficulty in the study of factorization of matrix-valued functions arises from the need to consider a "splitting" of the spectrum, i.e., the situation in which $L(\lambda) = L_2(\lambda)L_1(\lambda)$ and L_1, L_2, and L all have common eigenvalues. A relatively simple, but important class of right divisors of matrix polynomials $L(\lambda)$ consists of those which have *no* points of spectrum in common with the quotient they generate. Such divisors are described as "spectral" and are the subject of this section.

The point $\lambda_0 \in \mathbb{C}$ is a *regular point* of a monic matrix polynomial L if $L(\lambda_0)$ is nonsingular. Let Γ be a contour in $\mathbb{C}$, consisting of regular points of L. A monic right divisor L_1 of L is a Γ*-spectral right divisor* if $L = L_2 L_1$ and $\sigma(L_1)$, $\sigma(L_2)$ are inside and outside Γ, respectively.

Theorem 4.1. *If L is a monic matrix polynomial with linearization T and L_1 is a Γ-spectral right divisor of L, then the support subspace of L_1 with respect to T is the image of the Riesz projector R_Γ corresponding to T and Γ:*

$$R_\Gamma = \frac{1}{2\pi i} \oint_\Gamma (I\lambda - T)^{-1}\, d\lambda. \tag{4.1}$$

Proof. Let $\mathcal{M}_\Gamma$ be the T-invariant supporting subspace of L_1, and let $\mathcal{M}'_\Gamma$ be some direct complement to $\mathcal{M}_\Gamma$ in $\mathbb{C}^{nl}$ (l is the degree of $L(\lambda)$). By Corollary 3.19 and the definition of a Γ-spectral divisor, $\sigma(T|_{\mathcal{M}_\Gamma})$ is inside Γ and $\sigma(\tilde{P}T|_{\mathcal{M}'_\Gamma})$ is outside Γ, where $\tilde{P}$ is the projector on $\mathcal{M}'_\Gamma$ along $\mathcal{M}_\Gamma$. Write the matrix T with respect to the decomposition $\mathbb{C}^{nl} = \mathcal{M}_\Gamma + \mathcal{M}'_\Gamma$:

$$T = \begin{bmatrix} T_{11} & T_{12} \\ 0 & T_{22} \end{bmatrix}$$

(so that $T_{11} = T|_{\mathcal{M}'_\Gamma}$, $T_{12} = (I - \tilde{P})T|_{\mathcal{M}'_\Gamma}$, $T_{22} = \tilde{P}T|_{\mathcal{M}'_\Gamma}$, where $\tilde{P}$ and $I - \tilde{P}$ are considered as linear transformations on $\mathcal{M}'_\Gamma$ and $\mathcal{M}_\Gamma$ respectively). Then

$$(I\lambda - T)^{-1} = \begin{bmatrix} (I\lambda - T_{11})^{-1} & (I\lambda - T_{11})^{-1}T_{12}(I\lambda - T_{22})^{-1} \\ 0 & (I\lambda - T_{22})^{-1} \end{bmatrix}, \tag{4.2}$$

and

$$R_\Gamma = \frac{1}{2\pi i} \oint (I\lambda - T)^{-1}\, d\lambda = \begin{bmatrix} I & * \\ 0 & 0 \end{bmatrix},$$

and so $\operatorname{Im} R_\Gamma = \mathcal{M}_\Gamma$. $\square$

Note that Theorem 4.1 ensures the uniqueness of a Γ-spectral right divisor, if one exists.

We now give necessary and sufficient conditions for the existence of monic Γ-spectral divisors of degree k of a given monic matrix polynomial L of degree l. One necessary condition is evident: that $\det L(\lambda)$ has exactly nk zeros inside Γ (counting multiplicities). Indeed, the equality $L = L_2 L_1$, where

$L_1(\lambda)$ is a monic matrix polynomial of degree k such that $\sigma(L_1)$ is inside Γ and $\sigma(L_1)$ is outside Γ, leads to the equality

$$\det L = \det L_2 \cdot \det L_1,$$

and therefore $\det L(\lambda)$ has exactly nk zeros inside Γ (counting multiplicities), which coincide with the zeros of $\det L_1(\lambda)$. For the scalar case ($n = 1$) this necessary condition is also sufficient. The situation is completely different for the matrix case:

EXAMPLE 4.1. Let

$$L(\lambda) = \begin{bmatrix} \lambda^2 & 0 \\ 0 & (\lambda - 1)^2 \end{bmatrix},$$

and let Γ be a contour such that 0 is inside Γ and 1 is outside Γ. Then $\det L(\lambda)$ has two zeros (both equal to 0) inside Γ. Nevertheless, $L(\lambda)$ has no Γ-spectral monic right divisor of degree 1. Indeed,

$$X = \begin{bmatrix} 1 & 0 & 0 & 0 \\ 0 & 0 & 1 & 0 \end{bmatrix}, \qquad J = \begin{bmatrix} 0 & 1 & & \\ & 0 & & \\ & & 1 & 1 \\ & & & 1 \end{bmatrix}$$

is a Jordan pair of $L(\lambda)$. The Riesz projector is

$$R_\Gamma = \frac{1}{2\pi i} \int_\Gamma (I\lambda - J)^{-1} d\lambda = \begin{bmatrix} 1 & & & \\ & 1 & & \\ & & 0 & \\ & & & 0 \end{bmatrix},$$

and $\operatorname{Im} R_\Gamma = \operatorname{Span}\{(1 \quad 0 \quad 0 \quad 0)^T, (0 \quad 1 \quad 0 \quad 0)^T\}$. But $X|_{\operatorname{Im} R_\Gamma}$ is not invertible, so R_Γ is not a supporting subspace. By Theorem 4.1 there is no Γ-spectral monic right divisor of $L(\lambda)$. □

So it is necessary to impose extra conditions in order to obtain a criterion for the existence of Γ-spectral divisors. This is done in the next theorem.

Theorem 4.2. *Let L be a monic matrix polynomial and Γ a contour consisting of regular points of L having exactly nk eigenvalues of L (counted according to multiplicities) inside Γ. Then L has a Γ-spectral right divisor if and only if the $nk \times nl$ matrix*

$$M_{k,l} = \frac{1}{2\pi i} \oint_\Gamma \begin{bmatrix} L^{-1}(\lambda) & \cdots & \lambda^{l-1}L^{-1}(\lambda) \\ \vdots & & \vdots \\ \lambda^{k-1}L^{-1}(\lambda) & \cdots & \lambda^{k+l-2}L^{-1}(\lambda) \end{bmatrix} d\lambda$$

has rank kn. If this condition is satisfied, then the Γ*-spectral right divisor* $L_1(\lambda) = I\lambda^k + \sum_{j=0}^{k-1} L_{1j}\lambda^j$ *is given by the formula*

$$[L_{10} \quad \cdots \quad L_{1,k-1}] = -\frac{1}{2\pi i}\int_\Gamma [\lambda^k L^{-1}(\lambda) \quad \cdots \quad \lambda^{k+l-1}L^{-1}(\lambda)]\, d\lambda \cdot M_{k,l}^{\mathrm{I}}, \tag{4.3}$$

where $M_{k,l}^{\mathrm{I}}$ *is any right inverse of* $M_{k,l}$.

Proof. Let X, T, Y be a standard triple for L. Define R_Γ as in (4.1), and let $\mathcal{M}_1 = \operatorname{Im} R_\Gamma$, $\mathcal{M}_2 = \operatorname{Im}(I - R_\Gamma)$. Then $T = T_1 \oplus T_2$, where T_1 and T_2 are the restrictions of T to $\mathcal{M}_1$ and $\mathcal{M}_2$ respectively. In addition, $\sigma(T_1)$ and $\sigma(T_2)$ are inside and outside Γ, respectively.

Define $X_1 = X|_{\mathcal{M}_1}$, $X_2 = X|_{\mathcal{M}_2}$, and $Y_1 = R_\Gamma Y$, $Y_2 = (I - R_\Gamma)Y$. Then, using the resolvent form of L (Theorem 2.4) write

$$L^{-1}(\lambda) = X(I\lambda - T)^{-1}Y = X_1(I\lambda - T_1)^{-1}Y_1 + X_2(I\lambda - T_2)^{-1}Y_2.$$

Since $(I\lambda - T_2)^{-1}$ is analytic inside Γ, then for $i = 0, 1, 2, \ldots$ we have

$$\begin{aligned}\frac{1}{2\pi i}\int_\Gamma \lambda^i L^{-1}(\lambda)\, d\lambda &= \frac{1}{2\pi i}\int_\Gamma \lambda^i X_1(I\lambda - T_1)^{-1} | Y_1\, d\lambda \\ &= X_1 \frac{1}{2\pi i}\int_\Gamma \lambda^i (I\lambda - T_1)^{-1}\, d\lambda Y_1 = X_1 T_1^i Y_1,\end{aligned} \tag{4.4}$$

where the last equality follows from the fact that $\sigma(T_1)$ is inside Γ (see Section S1.8). Thus,

$$\begin{aligned} M_{k,l} &= \begin{bmatrix} X_1 Y_1 & X_1 T_1 Y_1 & \cdots & \cdots & X_1 T_1^{l-1} Y_1 \\ X_1 T_1 Y_1 & & & & \vdots \\ \vdots & & & & \\ X_1 T_1^{k-1} Y_1 & \cdots & & \cdots & X_1 T_1^{k+l-2} Y_1 \end{bmatrix} \\ &= \begin{bmatrix} X_1 \\ X_1 T_1 \\ \vdots \\ X_1 T_1^{k-1} \end{bmatrix} [Y_1 \quad T_1 Y_1 \quad \cdots \quad T_1^{l-1} Y_1], \end{aligned} \tag{4.5}$$

and since the rows of the matrix, $\operatorname{row}(T_1^i Y_1)_{i=0}^{l-1}$, are also rows of the nonsingular matrix, $\operatorname{row}(T^i Y)_{i=0}^{l-1}$, the former matrix is right invertible, i.e., $\operatorname{row}(T_1^i Y_1)_{i=0}^{l-1} \cdot S = I$ for some linear transformation $S: \mathcal{M} \to \mathbb{C}^{nl}$. Now it is clear that $M_{k,l}$ has rank kn only if the same is true of the left factor in (4.5), i.e., the left factor is nonsingular. This implies that $\mathcal{M}$ is a supporting subspace for L, and by Theorem 3.12 L has a right divisor of degree k. Furthermore, X_1, T_1 determine a standard triple for L_1, and by the construction $\sigma(L_1)$ coincides with $\sigma(T_1)$ and the part of $\sigma(L)$ inside Γ.

Finally, we write $L = L_2 L_1$, and by comparing the degrees of det L, det L_1 and det L_2, we deduce that $\sigma(L_2)$ is outside Γ. Thus, L_1 is a Γ-spectral right divisor.

For the converse, if L_1 is a Γ-spectral right divisor, then by Theorem 4.1 the subspace $\mathcal{M} = \operatorname{Im} R_\Gamma \subset \mathbb{C}^{ln}$ is a supporting subspace of L associated with L_1, and the left factor in (4.5) is nonsingular. Since the right factor is right invertible, it follows that $M_{k,l}$ has rank kn.

It remains to prove the formula (4.3). Since the supporting subspace for $L_1(\lambda)$ is Im R_Γ, we have to prove only that

$$X_1 T_1^k \left\{ \left. \begin{bmatrix} X \\ XT \\ \vdots \\ XT^{k-1} \end{bmatrix} \right|_{\mathcal{M}} \right\}^{-1} = \frac{1}{2\pi i} \int_\Gamma [\lambda^k L^{-1}(\lambda) \quad \cdots \quad \lambda^{k+l-1} L^{-1}(\lambda)] \, d\lambda \cdot M_{k,l}^{\mathrm{I}} \tag{4.6}$$

(see Theorem 3.12). Note that

$$M_{k,l}^{\mathrm{I}} = [\operatorname{row}(T_1^i Y_1)_{i=0}^{l-1}]^{\mathrm{I}} \cdot [\operatorname{col}(X_1 T_1^i)_{i=0}^{k-1}]^{-1},$$

where $[\operatorname{row}(T_1^i Y_1)_{i=0}^{l-1}]^{\mathrm{I}}$ is some right inverse of $\operatorname{row}(T_1^i Y_1)_{i=0}^{l-1}$. Using this remark, together with (4.4), the equality (4.6) becomes

$$\begin{aligned} X_1 T_1^k \cdot (\operatorname{col}(X T^i)_{i=0}^{k-1})^{-1} &= [X_1 T_1^k Y_1 \quad \cdots \quad X_1 T_1^{k+l-1} Y_1] \\ &\quad \cdot [\operatorname{row}(T_1^i Y_1)_{i=0}^{l-1}]^{\mathrm{I}} \cdot ((\operatorname{col}(X_1 T_1^i)_{i=0}^{k-1})^{-1}, \end{aligned}$$

which is evidently true. □

The result of Theorem 4.2 can also be written in the following form, which is sometimes more convenient.

Theorem 4.3. *Let L be a monic matrix polynomial, and let Γ be a contour consisting of regular points of L. Then L has a Γ-spectral right divisor of degree k if and only if*

$$\operatorname{rank} M_{k,l} = \operatorname{rank} M_{k+1,l} = nk. \tag{4.7}$$

Proof. We shall use the notation introduced in the proof of Theorem 4.2. Suppose that L has a Γ-spectral right divisor of degree k. Then $\operatorname{col}(X_1 T_1^j)_{j=0}^{k-1}$ is invertible, and T_1, X_1 are $nk \times nk$ and $n \times nk$ matrices, respectively. Since $\operatorname{row}(T_1^i Y_1)_{i=0}^{l-1}$ is right invertible, equality (4.5) ensures that rank $M_{k,l} = nk$. The same equality (4.5) (applied to $M_{k+1,l}$) shows that rank $M_{k+1,l} \leq \{\text{size of } T_1\} = nk$. But in any case rank $M_{k+1,l} \geq$ rank $M_{k,l}$; so in fact rank $M_{k+1,l} = nk$.

Suppose now (4.7) holds. By (4.5) and right invertibility of $\operatorname{row}(T_1^i Y_1)_{i=0}^{l-1}$ we easily obtain

$$\operatorname{rank} \operatorname{col}(X_1 T_1^i)_{i=0}^{k-1} = \operatorname{rank} \operatorname{col}(X_1 T_1^i)_{i=0}^{k} = nk. \tag{4.8}$$

Thus, $\operatorname{Ker} \operatorname{col}(X_1 T_1^i)_{i=0}^{k-1} = \operatorname{Ker} \operatorname{col}(X_1 T_1^i)_{i=0}^{k}$. It follows then that

$$\operatorname{Ker} \operatorname{col}(X_1 T_1^i)_{i=0}^{k-1} = \operatorname{Ker} \operatorname{col}(X_1 T_1^i)_{i=0}^{p} \tag{4.9}$$

for every $p \geq k$. Indeed, let us prove this assertion by induction on p. For $p = k$ it has been established already; assume it holds for $p = p_0 - 1$, $p_0 > k + 1$ and let $x \in \operatorname{Ker} \operatorname{col}(X_1 T_1^i)_{i=0}^{k-1}$. By the induction hypothesis,

$$x \in \operatorname{Ker} \operatorname{col}(X_1 T_1^i)_{i=1}^{p_0-1}.$$

But also

$$x \in \operatorname{Ker} \operatorname{col}(X_1 T_1^i)_{i=0}^{k} \subset \operatorname{Ker}[\operatorname{col}(X_1 T_1^{i-1})_{i=1}^{k} \cdot T_1];$$

so

$$T_1 x \in \operatorname{Ker} \operatorname{col}(X_1 T_1^l)_{i=0}^{k-1},$$

and again by the induction hypothesis,

$$T_1 x \in \operatorname{Ker} \operatorname{col}(X_1 T_1^i)_{i=0}^{p_0-1} \qquad \text{or} \qquad x \in \operatorname{Ker} \operatorname{col}(X_1 T_1^i)_{i=1}^{p_0}.$$

It follows that

$$x \in \operatorname{Ker} \operatorname{col}(X_1 T_1^i)_{i=0}^{p_0},$$

and (4.9) follows from $p = p_0$.

Applying (4.9) for $p = l - 1$ and using (4.8), we obtain that $\operatorname{rank} \operatorname{col}(X_1 T_1^i)_{i=0}^{l-1} = nk$. But $\operatorname{col}(X_1 T_1^i)_{i=0}^{l-1}$ is left invertible (as all its columns are also columns of the nonsingular matrix $\operatorname{col}(XT^i)_{i=0}^{l-1}$), so its rank coincides with the number of its columns. Thus, T_1 is of size $nk \times nk$; in other words, $\det L(\lambda)$ has exactly nk roots inside Γ (counting multiplicities). Now we can apply Theorem 4.2 to deduce the existence of a Γ-spectral right monic divisor of degree k. $\square$

If, as above, Γ is a contour consisting of regular points of L, then a monic left divisor L_2 of L is a *Γ-spectral left divisor* if $L = L_2 L_1$ and the spectra of L_2 and L_1 are inside and outside Γ, respectively. It is apparent that, in this case, L_1 is a Γ_1-spectral right divisor, where the contour Γ_1 contains in its interior exactly those points of $\sigma(L)$ which are outside Γ. Similarly, if L_1 is a Γ-spectral right divisor and $L = L_2 L_1$, then L_2 is a Γ_1-spectral left divisor. Thus, in principle, one may characterize the existence of Γ-spectral left divisors by using the last theorem and a complementary contour Γ_1. We shall show by example that a Γ-spectral divisor may exist from one side but not the other.

The next theorem is the dual of Theorem 4.2. It is proved for instance by applying Theorem 4.2 to the transposed matrix polynomial $L^T(\lambda)$ and its right Γ-spectral divisor.

Theorem 4.4. *Under the hypotheses of Theorem 4.2 L has a Γ-spectral left divisor if and only if the matrix*

$$M_{l,k} = \frac{1}{2\pi i}\int_\Gamma \begin{bmatrix} L^{-1}(\lambda) & \cdots & \lambda^{k-1}L^{-1}(\lambda) \\ \vdots & & \vdots \\ \lambda^{l-1}L^{-1}(\lambda) & \cdots & \lambda^{k+l-2}L^{-1}(\lambda) \end{bmatrix} d\lambda$$

has rank kn. In this case the Γ-spectral left divisor $L_2(\lambda) = I\lambda^k + \sum_{j=0}^{k-1} L_{2j}\lambda^j$ is given by the formula

$$\begin{bmatrix} L_{20} \\ \vdots \\ L_{2,k-1} \end{bmatrix} = -M_{l,k}^{\mathrm{I}} \cdot \frac{1}{2\pi i}\int_\Gamma \begin{bmatrix} \lambda^k L^{-1}(\lambda) \\ \vdots \\ \lambda^{k+l-1}L^{-1}(\lambda) \end{bmatrix} d\lambda, \tag{4.10}$$

where $M_{l,k}^{\mathrm{I}}$ is any left inverse of $M_{l,k}$.

The dual result for Theorem 4.3 is the following:

Theorem 4.5. *Let $L(\lambda)$ be a monic matrix polynomial and Γ be a contour consisting of regular points of L. Then L has a Γ-spectral left divisor of degree k if and only if*

$$\operatorname{rank} M_{l,k} = \operatorname{rank} M_{l,k+1} = nk.$$

Note that the left-hand sides in formulas (4.3) and (4.10) do not depend on the choice of $M_{k,l}^{\mathrm{I}}$ and $M_{l,k}^{\mathrm{I}}$, respectively (which are not unique in general). We could anticipate this property bearing in mind that a (right) monic divisor is uniquely determined by its supporting subspace, and for a Γ-spectral divisor such a supporting subspace is the image of a Riesz projector, which is fixed by Γ and the choice of standard triple for $L(\lambda)$.

Another way to write the conditions of Theorems 4.2 and 4.4 is by using finite sections of block Toeplitz matrices. For a continuous $n \times n$ matrix-valued function $H(\lambda)$ on Γ, such that $\det H(\lambda) \neq 0$ for every $\lambda \in \Gamma$, define $D_j = (2\pi i)^{-1}\int_\Gamma \lambda^{-j-1}H(\lambda)\,d\lambda$. It is clear that if Γ is the unit circle then the D_j are the Fourier coefficients of $H(\lambda)$:

$$H(\lambda) = \sum_{j=-\infty}^{\infty} \lambda^j D_j \tag{4.11}$$

(the series in the right-hand side converges uniformly and absolutely to $H(\lambda)$ under additional requirements; for instance, when $H(\lambda)$ is a rational matrix function). For a triple of integers $\alpha \geq \beta \geq \gamma$ we shall define $T(H_\Gamma; \alpha, \beta, \gamma)$ as the following block Toeplitz matrix:

$$T(H_\Gamma; \alpha, \beta, \gamma) = \begin{bmatrix} D_\beta & D_{\beta-1} & \cdots & D_\gamma \\ D_{\beta+1} & D_\beta & \cdots & D_{\gamma+1} \\ \vdots & \vdots & & \vdots \\ D_\alpha & D_{\alpha-1} & \cdots & D_{\gamma-\beta+\alpha} \end{bmatrix}.$$

We mention that $T(H_\Gamma; \alpha, \alpha, \gamma) = [D_\alpha \quad D_{\alpha-1} \quad \cdots \quad D_\gamma]$ is a block row, and

$$T(H_\Gamma; \alpha, \gamma, \gamma) = \begin{bmatrix} D_\gamma \\ D_{\gamma+1} \\ \vdots \\ D_\alpha \end{bmatrix}$$

is a block column.

Theorem 4.2′. *Let $L(\lambda)$ and Γ be as in Theorem 4.2. Then $L(\lambda)$ has a Γ-spectral right divisor $L_1(\lambda) = I\lambda^k + \sum_{j=0}^{k-1} L_{1j}\lambda^j$ if and only if* rank $T(L_\Gamma^{-1}; -1, -k, -k-l+1) = kn$. *In this case the coefficients of L_1 are given by the formula*

$$[L_{1,k-1} \quad L_{1,k-2} \quad \cdots \quad L_{10}] = -T(L_\Gamma^{-1}; -k-1, -k-1, -k-l) \cdot (T(L_\Gamma^{-1}; -1, -k, -k-l+1))^{\mathrm{I}}, \tag{4.12}$$

where $(T(L_\Gamma^{-1}; -1, -k. -k-l+1)^{\mathrm{I}}$ is any right inverse of

$$T(L_\Gamma^{-1}; -1, -k, -k-l+1).$$

Note that the order of the coefficients L_{1j} in formula (4.12) is opposite to their order in formula (4.3).

Theorem 4.4′. *Let $L(\lambda)$ and Γ be as in Theorem 4.4. Then $L(\lambda)$ has a Γ-spectral left divisor $L_2(\lambda) = I\lambda^k + \sum_{j=0}^{k-1} L_{2j}\lambda^j$ if and only if* rank $T(L_\Gamma^{-1}; -1, -l, -k-l+1) = kn$. *In this case the coefficients of L_2 are given by the formula*

$$\begin{bmatrix} L_{20} \\ L_{21} \\ \vdots \\ L_{2,k-1} \end{bmatrix} = -(T(L_\Gamma^{-1}; -1, -l, -k-l+1))^{\mathrm{I}} \cdot T(L_\Gamma^{-1}; -k-1, -k-l, -k-l), \tag{4.13}$$

where $(T(L_\Gamma^{-1}; -1, -l, -k-l+1))^{\mathrm{I}}$ is any left inverse of

$$T(L_\Gamma^{-1}; -1, -l, -k-l+1).$$

Of course, Theorems 4.3 and 4.5 can also be reformulated in terms of the matrices $T(L_\Gamma^{-1}; \alpha, \beta, \gamma)$.

We conclude this section with two examples:

EXAMPLE 4.2. Let

$$L(\lambda) = \begin{bmatrix} \lambda^3 - 42\lambda^2 & 300\lambda^2 - 301\lambda + 42 \\ -6\lambda^2 & \lambda^3 + 42\lambda^2 - 43\lambda + 6 \end{bmatrix}.$$

Then L has eigenvalues 0 (with multiplicity three), 1, 2, and -3. Let Γ be the circle in the complex plane with centre at the origin and radius $\frac{1}{2}$. We use a Jordan triple X, J, Y. It is found that we may choose

$$X_1 = \begin{bmatrix} 1 & 0 & 0 \\ 0 & 0 & 1 \end{bmatrix}, \qquad J_1 = \begin{bmatrix} 0 & 1 & 0 \\ 0 & 0 & 1 \\ 0 & 0 & 0 \end{bmatrix}, \qquad Y_1 = \begin{bmatrix} -7 & 48 \\ -6 & 42 \\ 1 & -7 \end{bmatrix},$$

and we have

$$\frac{1}{2\pi i} \oint L^{-1}(\lambda)\, d\lambda = X_1 Y_1 = \begin{bmatrix} -7 & 48 \\ 1 & -7 \end{bmatrix},$$

which is invertible. In spite of this, there is *no* Γ-spectral right divisor. This is because there are, in effect, three eigenvalues of L inside Γ, and though $M_{1,3}$ of Theorem 4.2 has full rank, the first hypothesis of the theorem is not satisfied. Note that the columns of X_1 are defined by a Jordan chain, and by Theorem 3.12 the linear dependence of the first two vectors shows that there is no right divisor of degree one. Similar remarks apply to the search for a Γ-spectral left divisor. □

EXAMPLE 4.3. Let a_1, a_2, a_3, a_4 be distinct complex numbers and

$$L(\lambda) = \begin{bmatrix} (\lambda - a_1)(\lambda - a_2) & \lambda - a_1 \\ 0 & (\lambda - a_3)(\lambda - a_4) \end{bmatrix}.$$

Let Γ be a contour with a_1 and a_2 inside Γ and a_3 and a_4 outside Γ. If we choose

$$X_1 = \begin{bmatrix} 1 & 1 \\ 0 & 0 \end{bmatrix},$$

then it is found that

$$Y_1 = \frac{1}{a_1 - a_2} \begin{bmatrix} 1 & 0 \\ -1 & \beta \end{bmatrix}$$

where $\beta = (a_2 - a_3)(a_2 - a_4)^{-1}(a_2 - a_1)$ and

$$X_1 Y_1 = \begin{bmatrix} 0 & \beta(a_1 - a_2)^{-1} \\ 0 & 0 \end{bmatrix}, \qquad X_1 J_1 Y_1 = \begin{bmatrix} 1 & a_2 \beta(a_1 - a_2)^{-1} \\ 0 & 0 \end{bmatrix}.$$

Then, since $M_{1,2} = [X_1 Y_1 \quad X_1 J_1 Y_1]$ and

$$M_{2,1} = \begin{bmatrix} X_1 Y_1 \\ X_1 J_1 Y_1 \end{bmatrix},$$

it is seen that $M_{1,2}$ has rank 1 and $M_{2,1}$ has rank 2. Thus there exists a Γ-spectral left divisor but no Γ-spectral right divisor. □

4.2. Linear Divisors and Matrix Equations

Consider the unilateral matrix equation

$$Y^l + \sum_{j=0}^{l-1} A_j Y^j = 0, \tag{4.14}$$

where $A_j (j = 0, \ldots, l-1)$ are given $n \times n$ matrices and Y is an $n \times n$ matrix to be found. Solutions of (4.14) are closely related to the right linear divisors of the monic matrix polynomial $L(\lambda) = I\lambda^l + \sum_{j=0}^{l-1} A_j \lambda^j$. Namely, it is easy to verify (by a straightforward computation, or using Corollary 3.6, for instance) that Y is a solution of (4.14) if and only if $I\lambda - Y$ is a right divisor of $L(\lambda)$. Therefore, we can apply our results on divisibility of matrix polynomials to obtain some information on the solutions of (4.14).

First we give the following criterion for existence of linear divisors (we have observed in Section 3.7 that, in contrast to the scalar case, not every monic matrix polynomial has a linear divisor).

Theorem 4.6. *The monic matrix polynomial $L(\lambda)$ has a right linear divisor $I\lambda - X$ if and only if there exists an invariant subspace $\mathcal{M}$ of the first companion matrix C_1 of $L(\lambda)$ of the form*

$$\mathcal{M} = \operatorname{Im} \begin{bmatrix} I \\ X \\ X^2 \\ \vdots \\ X^{l-1} \end{bmatrix}.$$

Proof. Let $\mathcal{M}$ be the supporting subspace of $I\lambda - X$ relative to the companion standard pair (P_1, C_1). By Corollary 3.13(iii) we can write

$$\mathcal{M} = \operatorname{Im}[\operatorname{col}(X_i)_{i=0}^{l-1}]$$

for some $n \times n$ matrices $X_0 = I$ and $X_1, \ldots, X_{l-1}$. But C_1-invariance of $\mathcal{M}$ implies that $X_i = X_1^i, i = 1, \ldots, l-1$. Finally, formula (3.27) shows that the right divisor of $L(\lambda)$ corresponding to $\mathcal{M}$ is just $I\lambda - X$, so in fact $X_1 = X$. □

In terms of the solutions of Eq. (4.14) and relative to any standard pair of $L(\lambda)$ (not necessarily the companion pair as in Theorem 4.6) the criterion of Theorem 4.6 can be formulated as follows.

Corollary 4.7. *A matrix Y is a solution of* (4.14) *if and only if Y has the form*

$$Y = X|_{\mathcal{M}} \cdot T|_{\mathcal{M}} \cdot (X|_{\mathcal{M}})^{-1},$$

where (X, T) is a standard pair of the monic matrix polynomial $L(\lambda)$ and $\mathcal{M}$ is a T-invariant subspace such that $X|_{\mathcal{M}}$ is invertible.

To prove Corollary 4.7 recall that equation (4.14) is satisfied if and only if $I\lambda - Y$ is a right divisor of $L(\lambda)$, and then apply Theorem 3.12.

A matrix Y such that equality (4.14) holds, will be called a (*right*) *solvent* of $L(\lambda)$. A solvent S is said to be a *dominant solvent* if every eigenvalue of S exceeds in absolute value every eigenvalue of the quotient $L(\lambda)(I\lambda - S)^{-1}$. Clearly, in this case, $I\lambda - S$ is a spectral divisor of $L(\lambda)$.

The following result provides an algorithm for computation of a dominant solvent, which can be regarded as a generalization of Bernoulli's method for computation of the zero of a scalar polynomial with largest absolute value (see [42]).

Theorem 4.8. *Let $L(\lambda) = I\lambda^l + \sum_{i=0}^{l-1} \lambda^i A_i$ be a monic matrix polynomial of degree l. Assume that $L(\lambda)$ has a dominant solvent S, and the transposed matrix polynomial $L^T(\lambda)$ also has a dominant solvent. Let $\{U_r\}_{r=1}^{\infty}$ be the solution of the system*

$$A_0 U_r + A_1 U_{r+1} + \cdots + A_{l-1} U_{r+l-1} + U_{r+l} = 0, \qquad r = 1, 2, \ldots \tag{4.15}$$

where $\{U_r\}_{r=1}^{\infty}$ is a sequence of $n \times n$ matrices to be found, and is determined by the initial conditions $U_0 = \cdots = U_{l-1} = 0$, $U_l = I$. Then $U_{r+1}U_r^{-1}$ exists for large r and $U_{r+1}U_r^{-1} \to S$ as $r \to \infty$.

We shall need the following lemma for the proof of Theorem 4.8.

Lemma 4.9. *Let W_1 and W_2 be square matrices (not necessarily of the same size) such that*

$$\min\{|\lambda| \mid \lambda \in \sigma(W_1)\} > \max\{|\lambda| \mid \lambda \in \sigma(W_2)\}. \tag{4.16}$$

Then W_1 is nonsingular and

$$\lim_{m\to\infty} \|W_2^m\| \cdot \|W_1^{-m}\| = 0. \tag{4.17}$$

Proof. Without loss of generality we may assume that W_1 and W_2 are in the Jordan form; write

$$W_i = K_i + N_i, \qquad i = 1, 2$$

where K_i is a diagonal matrix, N_i is a nilpotent matrix (i.e., such that $N_i^{r_i} = 0$ for some positive integer r_i), and $K_i N_i = N_i K_i$. Observe that $\sigma(K_i) = \sigma(W_i)$ and condition (4.16) ensures that W_1 (and therefore also K_1) are nonsingular. Further,

$$\|K_1^{-1}\| = [\inf\{|\lambda| \mid \lambda \in \sigma(W_1)\}]^{-1}, \qquad \|K_2\| = \sup\{|\lambda| \mid \lambda \in \sigma(W_2)\}.$$

Thus (again by (4.16))

$$\|K_2^m\| \cdot \|K_1^{-m}\| \le \gamma^m, \qquad 0 < \gamma < 1, \quad m = 1, 2, \ldots. \tag{4.18}$$

Let r_1 be so large that $N_1^{r_1} = 0$, and put $\delta_1 = \max_{0 \le k \le r_1 - 1} \|K_1^{-k} N_1^k\|$. Using the combinatorial identities

$$\sum_{k=0}^{p} (-1)^k \binom{n}{p-k} \binom{n+k-1}{k} = \begin{cases} 0 & \text{for} \quad p > 0 \\ 1 & \text{for} \quad p = 0, \end{cases}$$

it is easy to check that

$$\begin{aligned} (I + K_1^{-1} N_1)^{-m} &= \sum_{k=0}^{\infty} (-1)^k \binom{m+k-1}{k} K_1^{-k} N_1^k \\ &= \sum_{k=0}^{r_1 - 1} (-1)^k \binom{m+k-1}{k} K_1^{-k} N_1^k. \end{aligned}$$

So

$$\begin{aligned} \|W_1^{-m}\| &= \|(K_1 + N_1)^{-m}\| \le \|K_1^{-m}\| \cdot \|(I + K_1^{-1} N_1)^{-m}\| \\ &\le \|K_1^{-m}\| \cdot \left[\sum_{k=0}^{r_1 - 1} \binom{m+k-1}{k} \delta_1 \right] \le \|K_1^{-m}\| \delta_1 r_1 (m + r_1 - 2)^{r_1}. \end{aligned} \tag{4.19}$$

The same procedure applied to W_2^m yields

$$\|W_2^m\| \le \|K_2^m\| \cdot \delta_2 r_2 m^{r_2}, \tag{4.20}$$

where r_2 is such that $N_2^{r_2} = 0$ and $\delta_2 = \max_{0 \le k \le r_2 - 1} \|K_2^k N_2^k\|$. Using (4.18)–(4.20), we obtain

$$\lim_{m \to \infty} \|W_2^m\| \cdot \|W_1^{-m}\| \le \delta_1 \delta_2 r_1 r_2 \cdot \lim_{m \to \infty} [(m + r_1 - 2)^{r_1} m^{r_2} \gamma^m] = 0. \quad \square$$

Proof of Theorem 4.8. Let (X, J, Y) be a Jordan triple for $L(\lambda)$. The existence of a dominant solvent allows us to choose (X, J, Y) in such a way that the following partition holds:

$$X = [X_s \quad X_t], \qquad J = \begin{bmatrix} J_s & 0 \\ 0 & J_t \end{bmatrix}, \qquad Y = \begin{bmatrix} Y_s \\ Y_t \end{bmatrix},$$

where $\sup\{|\lambda| \,|\, \lambda \in \sigma(J_t)\} < \inf\{|\lambda| \,|\, \lambda \in \sigma(J_s)\}$, and the partitions of X and Y are consistent with the partition of J. Here J_s is the part of J corresponding to the dominant solvent S, so the size of J_s is $n \times n$. Moreover, X_s is a nonsingular $n \times n$ matrix (because $I\lambda - S$ is a spectral divisor of $L(\lambda)$), and X_s is the restriction of X to the supporting subspace corresponding to this divisor (cf. Theorems 4.1 and 3.12). Since $L^T(\lambda)$ has also a dominant solvent, and (Y^T, J^T, X^T) is a standard triple of $L^T(\lambda)$, by an analogous argument we obtain that Y_s is also nonsingular.

Let $\{U_r\}_{r=1}^{\infty}$ be the solution of (4.15) determined by the initial conditions $U_1 = \cdots = U_{l-1} = 0$; $U_l = I$. Then it is easy to see that

$$U_r = XJ^{r-1}Y, \qquad r = 1, 2, \ldots.$$

Indeed, we know from the general form (formula (2.57)) of the solution of (4.15) that $U_r = XJ^{r-1}Z$, $r = 1, 2, \ldots$ for some $nl \times n$ matrix Z. The initial conditions together with the invertibility of $\mathrm{col}(XJ^i)_{i=0}^{l-1}$ ensures that in fact $Z = Y$.

Now

$$\begin{aligned} U_r = XJ^{r-1}Y &= X_s J_s^{r-1} Y_s + X_t J_t^{r-1} Y_t \\ &= [X_s J_s^{r-1} X_s^{-1} + (X_t J_t^{r-1} Y_t) Y_s^{-1} X_s^{-1}] X_s Y_s. \end{aligned}$$

Write $M = X_s Y_s$ and $E_r = X_t J_t^r Y_t$, $r = 1, 2, \ldots$, so that (recall that $S = X_s J_s^{r-1} X_s^{-1}$)

$$U_r = (S^{r-1} + E_{r-1}M^{-1})M = (I + E_{r-1}M^{-1}S^{-r+1})S^{r-1}M.$$

Now the fact that S is a *dominant* solvent, together with Lemma 4.9, implies that $E_r M^{-1} S^{-r} \to 0$ in norm as $r \to \infty$ (indeed,

$$\|E_r M^{-1} S^{-r}\| \le \|X_t\|\,\|J_t^r\|\,\|Y_t\|\,\|M^{-1}\|\,\|S^{-r}\| \quad \text{and} \quad \lim_{r\to\infty} \|J_t^r\|\,\|S^{-r}\| = 0$$

by Lemma 4.9). So for large enough r, U_r will be nonsingular. Furthermore, when this is the case,

$$U_{r+1}U_r^{-1} = (I + E_r M^{-1} S^{-r}) S (I + E_{r-1} M^{-1} S^{-r+1})^{-1},$$

and it is clear, by use of the same lemma, that $U_{r+1}U_r^{-1} \to S$ as $r \to \infty$. □

The hypothesis in Theorem 4.8 that $L^T(\lambda)$ should also have a dominant solvent may look unnatural, but the following example shows that it is generally necessary.

EXAMPLE 4.4. Let

$$L(\lambda) = \begin{bmatrix} \lambda^2 - 1 & 0 \\ \lambda - 1 & \lambda^2 \end{bmatrix},$$

with eigenvalues $1, -1, 0, 0$. We construct a right spectral divisor with spectrum $\{1, -1\}$. The following matrices form a canonical triple for L:

$$X = \begin{bmatrix} 1 & 1 & 0 & 0 \\ 0 & 2 & 1 & 0 \end{bmatrix}, \qquad J = \mathrm{diag}\left[1, -1, \begin{bmatrix} 0 & 1 \\ 0 & 0 \end{bmatrix}\right], \qquad Y = \begin{bmatrix} \frac{1}{2} & 0 \\ -\frac{1}{2} & 0 \\ 1 & 0 \\ -1 & 1 \end{bmatrix},$$

and a dominant solvent is defined by

$$S = X_1 J_1 X_1^{-1} = \begin{bmatrix} 1 & -1 \\ 0 & -1 \end{bmatrix}.$$

Indeed, we have

$$L(\lambda) = \begin{bmatrix} \lambda + 1 & -1 \\ 1 & \lambda - 1 \end{bmatrix} \begin{bmatrix} \lambda - 1 & 1 \\ 0 & \lambda + 1 \end{bmatrix}.$$

However, since the corresponding matrix Y_1 is singular, Bernoulli's method breaks down. With initial values $U_0 = U_1 = 0$, $U_2 = I$ it is found that

$$U_3 = \begin{bmatrix} 0 & 0 \\ -1 & 0 \end{bmatrix}, \qquad U_4 = \begin{bmatrix} 1 & 0 \\ 1 & 0 \end{bmatrix}, \qquad U_5 = U_3, \qquad U_6 = U_4,$$

so that the sequence $\{U_r\}_{r=0}^{\infty}$ oscillates and, (for $r > 2$) takes only singular values. □

An important case where the hypothesis concerning $L^T(\lambda)$ is redundant is that in which L is *self-adjoint* (refer to Chapter 10). In this case $L(\lambda) = L^*(\lambda)$ and, if $L(\lambda) = Q(\lambda)(I\lambda - S)$, it follows that $L^T(\lambda) = \bar{Q}(\lambda)(I\lambda - \bar{S})$ and, if S is dominant for L, so is $\bar{S}$ for L^T.

4.3. Stable and Exponentially Growing Solutions of Differential Equations

Let $L(\lambda) = I\lambda^l + \sum_{j=0}^{l-1} A_j \lambda^j$ be a monic matrix polynomial, and let

$$\frac{d^l x(t)}{dt^l} + \sum_{j=0}^{l-1} A_j \frac{d^j x(t)}{dt^j} = 0, \qquad t \in [a, \infty) \tag{4.21}$$

be the corresponding homogeneous differential equation. According to Theorem 2.9, the general solution of (4.21) is given by the formula

$$x(t) = X e^{tJ} c,$$

where (X, J) is a Jordan pair of $L(\lambda)$, and $c \in \mathbb{C}^{nl}$. Assume now that $L(\lambda)$ has no eigenvalues on the imaginary axis. In this case we can write

$$X = [X_1 \quad X_2], \qquad J = \begin{bmatrix} J_1 & 0 \\ 0 & J_2 \end{bmatrix},$$

where $\sigma(J_1)$ (resp. $\sigma(J_2)$) lies in the open left (resp. right) half-plane. Accordingly,

$$x(t) = x_1(t) + x_2(t), \tag{4.22}$$

where $x_1(t) = X_1 e^{tJ_1} c_1$, $x_2(t) = X_2 e^{tJ_2} c_2$, and $c = \left[\begin{smallmatrix} c_1 \\ c_2 \end{smallmatrix}\right]$. Observe that $\lim_{t\to\infty} \|x_1(t)\| = 0$ and, if $c_2 \neq 0$, $\lim_{t\to\infty} \|x_2(t)\| = \infty$. Moreover, $x_2(t)$ is exponentially growing, i.e., for some real positive numbers μ and ν (in fact, μ (resp. ν) is the real part of an eigenvalue of $L(\lambda)$ with the largest (resp. smallest) real part among all eigenvalues with positive real parts) such that

$$\lim_{t\to\infty} \|e^{-\mu t}t^{-p}x_2(t)\| = 0$$

for some positive integer p, but $\lim_{t\to\infty} \|e^{-(\nu-\varepsilon)t}x_2(t)\| = \infty$ for every $\varepsilon > 0$ (unless $c_2 = 0$). Equality (4.22) shows that every solution is a sum of a stable solution (i.e., such that it tends to 0 as $t \to \infty$) and an exponentially growing solution, and such a sum is uniquely determined by $x(t)$.

The following question is of interest: when do initial conditions (i.e., vectors $x^{(j)}(a)$, $j = 0, \ldots, k-1$ for some k) determine a unique stable solution $x(t)$ of (4.21), i.e., such that in (4.22), $x_2(t) \equiv 0$? It turns out that the answer is closely related to the existence of a right spectral divisor of $L(\lambda)$.

Theorem 4.10. *Let $L(\lambda)$ be a monic matrix polynomial such that $\sigma(L)$ does not intersect the imaginary axis. Then for every set of k vectors $x_0(a), \ldots, x_0^{(k-1)}(a)$ there exists a unique stable solution $x(t)$ of (4.21) such that $x^{(j)}(a) = x_0^{(j)}(a)$, $j = 0, \ldots, k-1$, if and only if the matrix polynomial $L(\lambda)$ has a monic right divisor $L_1(\lambda)$ of degree k such that $\sigma(L_1)$ lies in the open left half-plane.*

Proof. Assume first that $L_1(\lambda)$ is a monic right divisor of $L(\lambda)$ of degree k such that $\sigma(L_1)$ lies in the open left half-plane. Then a fundamental result of the theory of differential equations states that, for every set of k vectors $x_0(a), \ldots, x_0^{(k-1)}(a)$ there exists a unique solution $x_1(t)$ of the differential equation $L_1(d/dt)x_1(t) = 0$, such that

$$x_1^{(j)}(a) = x_0^{(j)}(a), \qquad j = 0, \ldots, k-1. \tag{4.23}$$

Of course, $x_1(t)$ is also a solution of (4.21) and, because $\sigma(L_1)$ is in the open left half-plane,

$$\lim_{t\to\infty} \|x_1(t)\| = 0. \tag{4.24}$$

We show now that $x_1(t)$ is the unique solution of $L(\lambda)$ such that (4.23) and (4.24) are satisfied. Indeed, let $\tilde{x}_1(t)$ be another such solution of $L(\lambda)$. Using decomposition (4.22) we obtain that $x_1(t) = X_1 e^{tJ_1}c_1$, $\tilde{x}_1(t) = X_1 e^{tJ_1}\tilde{c}_1$ for some $c_1, \tilde{c}_1$. Now $[x_1(t) - \tilde{x}_1(t)]^{(j)}(a) = 0$ for $j = 0, \ldots, k-1$ implies that

$$\begin{bmatrix} X_1 \\ X_1 J_1 \\ \vdots \\ X_1 J_1^{k-1} \end{bmatrix} e^{aJ_1}(c_1 - \tilde{c}_1) = 0. \tag{4.25}$$

Since (X_1, J_1) is a Jordan pair of $L_1(\lambda)$ (see Theorems 3.12 and 4.1), the matrix $\operatorname{col}(X_1 J_1^i)_{i=0}^{k-1}$ is nonsingular, and by (4.25), $c_1 = \tilde{c}_1$, i.e., $x_1(t) = \tilde{x}_1(t)$.

Suppose now that for every set of k vectors $x_0(a), \ldots, x_0^{(k-1)}(a)$ there exists a unique stable solution $x(t)$ such that $x^{(j)}(a) = x_0^{(j)}(a)$, $j = 0, \ldots, k-1$. Write $x(t) = X_1 e^{tJ_1} c_1$ for some c_1; then

$$\operatorname{col}(X_1 J_1^i)_{i=0}^{k-1} e^{aJ_1} c_1 = \operatorname{col}(x_0^{(i)}(a))_{i=0}^{k-1}. \tag{4.26}$$

For every right-hand side of (4.26) there exists a unique c_1 such that (4.26) holds. This means that $\operatorname{col}(X_1 J_1^i)_{i=0}^{k-1}$ is square and nonsingular and the existence of a right monic divisor $L_1(\lambda)$ of degree k with $\sigma(L_1)$ in the open left half-plane follows from Theorem 3.12. □

4.4. Left and Right Spectral Divisors

Example 4.3 indicates that there remains the interesting question of when a set of kn eigenvalues—and a surrounding contour Γ—determine both a Γ-spectral right divisor and a Γ-spectral left divisor. Two results in this direction are presented.

Theorem 4.11. *Let L be a monic matrix polynomial and Γ a contour consisting of regular points of L having exactly kn eigenvalues of L (counted according to multiplicities) inside Γ. Then L has both a Γ-spectral right divisor and a Γ-spectral left divisor if and only if the $nk \times nk$ matrix $M_{k,k}$ defined by*

$$M_{k,k} = \frac{1}{2\pi i}\int_\Gamma \begin{bmatrix} L^{-1}(\lambda) & \cdots & \lambda^{k-1}L^{-1}(\lambda) \\ \vdots & & \vdots \\ \lambda^{k-1}L^{-1}(\lambda) & \cdots & \lambda^{2k-2}L^{-1}(\lambda) \end{bmatrix} d\lambda \tag{4.27}$$

is nonsingular. If this condition is satisfied, then the Γ-spectral right (resp. left) divisor $L_1(\lambda) = I\lambda^k + \sum_{j=0}^{k-1} L_{1j}\lambda^j$ (resp. $L_2(\lambda) = I\lambda^k + \sum_{j=0}^{k-1} L_{2j}\lambda^j$) is given by the formula:

$$[L_{10} \ \cdots \ L_{1,k-1}] = -\frac{1}{2\pi i}\int_\Gamma [\lambda^k L^{-1}(\lambda) \ \cdots \ \lambda^{2k-1}L^{-1}(\lambda)]\, d\lambda \cdot M_{k,k}^{-1}$$

$$\left(\text{resp.} \begin{bmatrix} L_{20} \\ \vdots \\ L_{2,k-1} \end{bmatrix} = -M_{k,k}^{-1} \cdot \frac{1}{2\pi i}\int_\Gamma \begin{bmatrix} \lambda^k L^{-1}(\lambda) \\ \vdots \\ \lambda^{2k-1}L^{-1}(\lambda) \end{bmatrix} d\lambda\right).$$

Proof. Let X, T, Y be a standard triple for L, and define X_1, T_1, Y_1 and X_2, T_2, Y_2 as in the proof of Theorem 4.2. Then, as in the proof of that theorem, we obtain

$$M_{k,k} = \begin{bmatrix} X_1 \\ X_1 T_1 \\ \vdots \\ X_1 T_1^{k-1} \end{bmatrix} [Y_1 \quad T_1 Y_1 \quad \cdots \quad T_1^{k-1} Y_1]. \tag{4.28}$$

Then nonsingularity of $M_{k,k}$ implies the nonsingularity of both factors and, as in the proof of Theorem 4.2 (Theorem 4.4), we deduce the existence of a Γ-spectral right (left) divisor.

Conversely, the existence of both divisors implies the nonsingularity of both factors on the right of (4.28) and hence the nonsingularity of $M_{k,k}$. The formulas for the Γ-spectral divisors are verified in the same way as in the proof of Theorem 4.2. □

In terms of the matrices $T(L_\Gamma^{-1}; \alpha, \beta, \gamma)$ introduced in Section 4.1, Theorem 4.11 can be restated as follows.

Theorem 4.11′. *Let L and Γ be as in Theorem 4.11. Then L has right and left Γ-spectral divisors $L_1(\lambda) = I\lambda^k + \sum_{j=0}^{k-1} L_{1j}\lambda^j$ and $L_2(\lambda) = I\lambda^k + \sum_{j=0}^{k-1} L_{2j}\lambda^j$, respectively, if and only if* $\det T(L^{-1}; -1, -k, -2k+1) \neq 0$. *In this case the coefficients of $L_1(\lambda)$ and $L_2(\lambda)$ are given by the formulas*

$$[L_{1,k-1} \quad L_{1,k-2} \quad \cdots \quad L_{10}] = -T(L_\Gamma^{-1}; -k-1, -k-1, -2k) \cdot (T(L_\Gamma^{-1}; -1, -k, -2k+1))^{-1},$$

$$\begin{bmatrix} L_{20} \\ L_{21} \\ \vdots \\ L_{2,k-1} \end{bmatrix} = -(T(L_\Gamma^{-1}; -1, -k, -2k+1))^{-1} \cdot T(L_\Gamma^{-1}; -k-1, -2k, -2k).$$

In the next theorem we relax the explicit assumption that Γ contains exactly the right number of eigenvalues. It turns out that this is implicit in the assumptions concerning the choice of l and k.

Theorem 4.12. *Let L be a monic matrix polynomial of degree $l \leq 2k$, and let Γ be a contour of regular points of L. If* $\det M_{k,k} \neq 0$ *($M_{k,k}$ defined by* (4.27)) *then there exist a Γ-spectral right divisor and a Γ-spectral left divisor.*

Proof. We prove Theorem 4.12 only for the case $l = 2k$; in the general case $l < 2k$ we refer to [36b]. Suppose that Γ contains exactly p eigenvalues of L in its interior. As in the preceding proofs, we may then obtain a factorization of $M_{k,k}$ in the form (4.28) where $X_1 = X|_{\mathcal{M}}$, $T_1 = T|_{\mathcal{M}}$ and $\mathcal{M}$ is the p-dimensional invariant subspace of L associated with the eigenvalues inside Γ. The right and left factors in (4.28) are then linear transformations from $\mathbb{C}^{nk}$ to $\mathcal{M}$ and $\mathcal{M}$ to $\mathbb{C}^{nk}$ respectively. The nonsingularity of $M_{k,k}$ therefore implies $p = \dim \mathcal{M} \geq \dim \mathbb{C}^{nk} = kn$.

With Y_1, X_2, T_2, Y_2 defined as in Theorem 4.2 and using the biorthogonality condition (2.14) we have

$$0 = \begin{bmatrix} XY & XTY & \cdots & XT^{k-1}Y \\ XTY & & & \vdots \\ \vdots & & & \\ XT^{k-1}Y & \cdots & \cdots & XT^{2k-2}Y \end{bmatrix}$$

$$= M_{k,k} + \begin{bmatrix} X_2 Y_2 & \cdots & X_2 T_2^{k-1} Y_2 \\ \vdots & & \vdots \\ X_2 T_2^{k-1} Y_2 & \cdots & X_2 T_2^{2k-2} Y_2 \end{bmatrix}.$$

Thus, the last matrix is invertible, and we can apply the argument used in the first paragraph to deduce that $ln - p \geq kn$. Hence $p = kn$, and the theorem follows from Theorem 4.11. □

A closed rectifiable contour Ξ, lying outside Γ together with its interior, is said to be *complementary* to Γ (relative to L) if Ξ contains in its interior exactly those spectral points of $L(\lambda)$ which are outside Γ. Observe that L has right and left Γ-spectral monic divisors of degree k if and only if L has left and right Ξ-spectral monic divisors of degree $l - k$. This observation allows us to use Theorem 4.12 for Ξ as well as for Γ. As a consequence, we obtain, for example, the following interesting fact.

Corollary 4.13. *Let L and Γ be as in Theorem* 4.12, *and suppose $l = 2k$. Then $M_{k,k}$ is nonsingular if and only if $M_{k,k}^{\Xi}$ is nonsingular, where Ξ is a contour complementary to Γ and $M_{k,k}^{\Xi}$ is defined by* (4.27) *replacing Γ by Ξ.*

For $n = 1$ the results of this section can be stated as follows.

Corollary 4.14. *Let $p(\lambda) = \sum_{j=0}^{l} a_j \lambda^j$ be a scalar polynomial with $a_l \neq 0$ and $p(\lambda) \neq 0$ for $\lambda \in \Gamma$. Then an integer k is the number of zeros of $p(\lambda)$ (counting multiplicities) inside Γ if and only if*

$$\operatorname{rank} M_{k,k} = k \qquad \textit{for} \quad k \geq l/2$$

or

$$\operatorname{rank} M_{l-k,l-k}^{\Xi} = l - k \qquad \textit{for} \quad k \leq l/2,$$

where Ξ is a complementary contour to Γ (relative to $p(\lambda)$).

4.5. Canonical Factorization

Let Γ be a rectifiable simple closed contour in the complex plane $\mathbb{C}$ bounding the domain F^+. Denote by F^- the complement of $F^+ \cup \Gamma$ in

$\mathbb{C} \cup \{\infty\}$. We shall always assume that $0 \in F^+$ (and $\infty \in F^-$). The most important example of such a contour is the unit circle $\Gamma_0 = \{\lambda \mid |\lambda| = 1\}$. Denote by $C^+(\Gamma)$ (resp. $C^-(\Gamma)$) the class of all $n \times n$ matrix-valued functions $G(\lambda)$, $\lambda \in \Gamma \cup F^+$ (resp. $\lambda \in \Gamma \cup F^-$) such that $G(\lambda)$ is analytic in F^+ (resp. F^-) and continuous in $F^+ \cup \Gamma$ (resp. $F^- \cup \Gamma$).

A continuous and invertible $n \times n$ matrix function $H(\lambda)$ ($\lambda \in \Gamma$) is said to admit a *right canonical factorization* (relative to Γ) if

$$H(\lambda) = H_+(\lambda)H_-(\lambda), \qquad (\lambda \in \Gamma) \tag{4.29}$$

where $H_+^{\pm 1}(\lambda) \in C^+(\Gamma)$, $H_-^{\pm 1}(\lambda) \in C^-(\Gamma)$. Interchanging the places of $H_+(\lambda)$ and $H_-(\lambda)$ in (4.29), we obtain *left canonical factorization* of H.

Observe that, by taking transposes in (4.29), a right canonical factorization of $H(\lambda)$ determines a left canonical factorization of $(H(\lambda))^T$. This simple fact allows us to deal in the sequel mostly with right canonical factorization, bearing in mind that analogous results hold also for left factorization.

In contrast, the following simple example shows that the existence of a canonical factorization from one side does not generally imply the existence of a canonical factorization from the other side.

EXAMPLE 4.5. Let

$$H(\lambda) = \begin{bmatrix} \lambda^{-1} & 1 \\ 0 & \lambda \end{bmatrix},$$

and let $\Gamma = \Gamma_0$ be the unit circle. Then

$$H(\lambda) = \begin{bmatrix} 0 & 1 \\ -1 & \lambda \end{bmatrix} \begin{bmatrix} 1 & 0 \\ \lambda^{-1} & 1 \end{bmatrix}$$

is a right canonical factorization of $H(\lambda)$, with

$$H_+(\lambda) = \begin{bmatrix} 0 & 1 \\ -1 & \lambda \end{bmatrix} \quad \text{and} \quad H_-(\lambda) = \begin{bmatrix} 1 & 0 \\ \lambda^{-1} & 1 \end{bmatrix}.$$

On the other hand, $H(\lambda)$ does not admit a left canonical factorization. □

In this section we shall consider canonical factorizations with respect to a rectifiable simple contour Γ of rational matrix functions of the type $R(\lambda) = \sum_{j=-r}^{s} A_j\lambda^j$.

We restrict our presentation to the case that $R(\lambda)$ is *monic*, i.e., $A_s = I$. This assumption is made in order to use the results of Sections 4.1 and 4.4, where only monic matrix polynomials are considered; but there is nothing essential in this assumption and most of the results presented below remain valid without the requirement that $A_s = I$ (cf. [36c, 36d, 73]).

We introduce some notation. Denote by $GL(\mathbb{C}^n)$ the class of all nonsingular $n \times n$ matrices (with complex entries). For a continuous $n \times n$ matrix-valued function $H(\lambda)\colon \Gamma \to GL(\mathbb{C}^n)$ define

$$D_j = (2\pi i)^{-1} \int_\Gamma \lambda^{-j-1} H(\lambda)\, d\lambda,$$

and define $T(H_\Gamma; \alpha, \beta, \gamma)$ to be the following block Toeplitz matrix:

$$T(H_\Gamma; \alpha, \beta, \gamma) = \begin{bmatrix} D_\beta & D_{\beta-1} & \cdots & D_\gamma \\ D_{\beta+1} & D_\beta & \cdots & D_{\gamma+1} \\ \vdots & \vdots & & \vdots \\ D_\alpha & D_{\alpha-1} & \cdots & D_{\gamma-\beta+\alpha} \end{bmatrix}. \tag{4.30}$$

We have already used the matrices $T(H_\Gamma; \alpha, \beta, \gamma)$ in Sections 4.1 and 4.4. For convenience, we shall omit the subscript Γ whenever the integration is along Γ and there is no danger of misunderstanding.

The next theorem provides the characterization of right factorization in terms of some finite sections of the infinite Toeplitz matrix $T(\cdot; \infty, 0, -\infty)$.

Theorem 4.15. *A monic rational matrix function* $R(\lambda) = \sum_{j=-r}^{s-1} A_j \lambda^j + I\lambda^s\colon \Gamma \to GL(\mathbb{C}^n)$ *admits a right canonical factorization if and only if*

$$\operatorname{rank} T(R_\Gamma^{-1}; r-1, 0, -r-s+1) = \operatorname{rank} T(R_\Gamma^{-1}; r-1, -1, -r-s) = rn. \tag{4.31}$$

Proof. We first note that the canonical factorization problem can be reduced to the problem of existence of Γ-spectral divisors for monic polynomials. Indeed, let $R(\lambda) = \sum_{j=-r}^{s-1} A_j \lambda^j + I\lambda^s$ be a monic rational function and introduce a monic matrix polynomial $M(\lambda) = \lambda^r R(\lambda)$. Then a right canonical factorization $R(\lambda) = W_+(\lambda) W_-(\lambda)$ for $R(\lambda)$ implies readily the existence of right monic Γ-spectral divisor $\lambda^r W_-(\lambda)$ of degree r for the polynomial $M(\lambda)$. Conversely, if the polynomial $M(\lambda)$ has a right monic Γ-spectral divisor $N_1(\lambda)$ of degree r and $M(\lambda) = N_2(\lambda) N_1(\lambda)$, then the equality $R(\lambda) = N_2(\lambda) \cdot \lambda^{-r} N_1(\lambda)$ yields a right canonical factorization of $R(\lambda)$.

Now we can apply the results of Section 4.1 in the investigation of the canonical factorization. Indeed, it follows from Theorem 4.3 that the polynomial $M(\lambda)$ has a Γ-spectral right divisor if and only if the conditions of Theorem 4.15 are satisfied (here we use the following observation:

$$T(R^{-1}; \alpha, \beta, \gamma) = T(M^{-1}; \alpha + r, \beta + r, \gamma + r)). \quad \square$$

The proof of the following theorem is similar to the proof of Theorem 4.15 (or one can obtain Theorem 4.16 by applying Theorem 4.15 to the transposed rational function).

Theorem 4.16. *A monic rational function* $R(\lambda) = \sum_{j=-r}^{s-1} A_j\lambda^j + I\lambda^s$: $\Gamma \to GL(\mathbb{C}^n)$ *admits a left canonical factorization if and only if*

$$\begin{aligned} &\operatorname{rank} T(R^{-1}; r-1, -s, -r-s+1) \\ &\qquad = \operatorname{rank} T(R^{-1}; r-1, -s, -r-s) = rn. \end{aligned} \tag{4.32}$$

In the particular case $n = 1$ Theorem 4.15 implies the following characterization of the winding number of a scalar polynomial. The winding number (with respect to Γ) of a scalar polynomial $p(\lambda)$ which does not vanish on Γ is defined as the increment of $(1/2\pi)$ $[\arg p(\lambda)]$ when λ runs through Γ in a positive direction. It follows from the argument principle ([63], Vol. 2, Chapter 2) that the winding number of $p(\lambda)$ coincides with the number of zeros of $p(\lambda)$ inside Γ (with multiplicities).

Let us denote by Ξ a closed simple rectifiable contour such that Ξ lies in F^- together with its interior, and $p(\lambda) \neq 0$ for all points $\lambda \in F^- \backslash \{\infty\}$ lying outside Ξ.

Corollary 4.17. *Let* $p(\lambda) = \sum_{j=0}^{l} a_j\lambda^j$ *be a scalar polynomial with* $a_l \neq 0$ *and* $p(\lambda) \neq 0$ *for* $\lambda \in \Gamma$. *An integer* r *is the winding number of* $p(\lambda)$ *if and only if the following condition holds*:

$$\begin{aligned} &\operatorname{rank} T(p^{-1}; -1, -r, -l-r+1) \\ &\qquad = \operatorname{rank} T(p^{-1}; -1, -r-1, -l-r) = r. \end{aligned}$$

This condition is equivalent to the following one:

$$\begin{aligned} &\operatorname{rank} T(p_\Xi^{-1}; -1, -s, -l-s+1) \\ &\qquad = \operatorname{rank} T(p_\Xi^{-1}; -1, -s-1, -l-s) = s, \end{aligned}$$

where $s = l - r$.

The proof of Corollary 4.17 is based on the easily verified fact that r is the winding number of $p(\lambda)$ if and only if the rational function $\lambda^{-r}p(\lambda)$ admits one-sided canonical factorization (because of commutativity, both canonical factorizations coincide).

The last assertion of Corollary 4.17 reflects the fact that the contour Ξ contains in its interior all the zeros of $p(\lambda)$ which are outside Γ. So $p(\lambda)$ has exactly s zeros (counting multiplicities) inside Ξ if and only if it has exactly r zeros (counting multiplicities) inside Γ.

Now we shall write down explicitly the factors of the canonical factorization. The formulas will involve one-sided inverses for operators of the form $T(H; \alpha, \beta, \gamma)$. The superscript I will indicate the appropriate one-sided inverse of such an operator (refer to Chapter S3). We shall also use the notation introduced above.

The formulas for the spectral divisors and corresponding quotients obviously imply formulas for the factorization factors. Indeed, let

$R(\lambda) = I\lambda^s + \sum_{j=-r}^{s-1} A_j\lambda^j: \Gamma \to GL(\mathbb{C}^n)$ be a monic rational function, and let $W_+(\lambda) = I\lambda^s + \sum_{j=0}^{s-1} W_j^+\lambda^j$ and $W_-(\lambda) = I + \sum_{j=1}^{r} W_{r-j}^-\lambda^{-j}$ be the factors of the right canonical factorization of $R(\lambda)$: $R(\lambda) = W_+(\lambda)W_-(\lambda)$. Then the polynomial $\lambda^r W_-(\lambda)$ is a right monic Γ-spectral divisor of the polynomial $M(\lambda) = \lambda^r R(\lambda)$, and formulas (4.12) and (4.13) apply. It follows that

$$[W_{r-1}^- \quad W_{r-2}^- \quad \cdots \quad W_0^-] = -T(R_\Gamma^{-1}; -1, -1, -r-s) \cdot [T(R_\Gamma^{-1}; r-1, 0, -r-s+1)]^{\mathrm{I}}, \tag{4.33}$$

$$\operatorname{col}(W_j^+)_{j=0}^{s-1} = -[T(R_\Xi^{-1}; r-1, -s, -2s+1)]^{\mathrm{I}} \cdot T(R_\Xi^{-1}; r-s-1, -2s, -2s). \tag{4.34}$$

Let us write down also the formulas for left canonical factorization: let $\Lambda_+(\lambda) = I\lambda^s + \sum_{j=0}^{s-1} \Lambda_j^+\lambda^j$ and $\Lambda_-(\lambda) = I + \sum_{j=1}^{r} \Lambda_{r-j}^-\lambda^{-j}$ be the factors of a left canonical factorization of $R(\lambda) = I\lambda^s + \sum_{j=-r}^{s-1} A_j\lambda^j: \Gamma \to GL(\mathbb{C}^n)$, i.e., $R(\lambda) = \Lambda_-(\lambda)\Lambda_+(\lambda)$. Then

$$\operatorname{col}(\Lambda_i^-)_{i=0}^{r-1} = -(T(R_\Gamma^{-1}; r-1, -s, -r-s+1))^{\mathrm{I}} \cdot T(R_\Gamma^{-1}; -1, -r-s, -r-s), \tag{4.35}$$

$$[\Lambda_{s-1}^+ \quad \Lambda_{s-2}^+ \quad \cdots \quad \Lambda_0^+] = -T(R_\Xi^{-1}; -1, -1, -r-s) \cdot (T(R_\Xi^{-1}; r-1, 0, -r-s+1))^{\mathrm{I}}. \tag{4.36}$$

Alternative formulas for identifying factors in canonical factorization can be provided, which do not use formulas (4.12) and (4.13). To this end, we shall establish the following formula for coefficients of the right monic Γ-spectral divisor $N(\lambda)$ of degree r of a monic matrix polynomial $M(\lambda)$ with degree l:

$$[\hat{N}_r \quad \hat{N}_{r-1} \quad \cdots \quad \hat{N}_0] = [I \quad 0 \quad \cdots \quad 0] \cdot [T(M^{-1}; 0, -r, -r-l)]^{\mathrm{I}}, \tag{4.37}$$

where $\hat{N}_j = Y_0 N_j$ $(j = 0, 1, \ldots, r-1)$ and $\hat{N}_r = Y_0$. Here N_j are the coefficients of $N(\lambda)$: $N(\lambda) = I\lambda^r + \sum_{j=0}^{r-1} N_j\lambda^j$; and Y_0 is the lower coefficient of the quotient $M(\lambda)N^{-1}(\lambda)$: $Y_0 = [M(\lambda)N^{-1}(\lambda)]|_{\lambda=0}$.

It is clear that the matrix function $\hat{Y}(\lambda) = Y_0 N(\lambda)M^{-1}(\lambda)$ is analytic in F^+ and $\hat{Y}(0) = I$. Then

$$(2\pi i)^{-1} \int_\Gamma \lambda^{-1} Y_0 N(\lambda)M^{-1}(\lambda)\, d\lambda = I \tag{4.38}$$

and

$$(2\pi i)^{-1} \int_\Gamma \lambda^j Y_0 N(\lambda)M^{-1}(\lambda)\, d\lambda = 0 \qquad \text{for} \quad j = 0, 1, \ldots \tag{4.39}$$

Indeed, (4.39) is a consequence of analyticity of the functions $\lambda^j Y_0 N(\lambda) M^{-1}(\lambda)$ in F^+ for $j = 0, 1, \ldots$. Formula (4.38) follows from the residue theorem, since $\lambda_0 = 0$ is the unique pole of the function $\lambda^{-1} Y_0 N(\lambda) M^{-1}(\lambda)$ in F^+. Combining (4.38) with the first l equalities from (4.39), we easily obtain the following relationship:

$$[Y_0 \quad Y_0 N_{r-1} \quad \cdots \quad Y_0 N_0] \cdot T(M_\Gamma^{-1}; 0, -r, -r-l) = [I \quad 0 \quad \cdots \quad 0],$$

which implies (4.37) immediately.

A similar formula can be deduced for the left Γ-spectral divisor $\Phi(\lambda) = I\lambda^r + \sum_{j=0}^{r-1} \Phi_j \lambda^j$ of $M(\lambda)$:

$$\operatorname{col}(\hat{\Phi}_j)_{j=0}^r = [T(M_\Gamma^{-1}; 0, -l, -r-l)]^{-1} \cdot \operatorname{col}(\delta_{jl} I)_{j=0}^l, \tag{4.40}$$

where $\hat{\Phi}_j = \Phi_j \Psi_0$ $(j = 1, 2, \ldots, r-1)$ and $\hat{\Phi}_r = \Psi_0$. Here $\Psi_0 = [\Phi^{-1}(\lambda) M(\lambda)]_{\lambda=0}$.

Formula (4.40) may be helpful if one is looking for the factorization $R(\lambda) = \hat{W}_+(\lambda)\hat{W}_-(\lambda)$ with factors

$$\hat{W}_+(\lambda) = I + \textstyle\sum_{j=1}^s \hat{W}_j^+ \lambda^j, \quad \hat{W}_-(\lambda) = \textstyle\sum_{j=0}^r \hat{W}_{r-j}^- \lambda^{-j}.$$

Then (4.40) implies the following identification of $\hat{W}_-(\lambda)$:

$$[\hat{W}_r \quad \hat{W}_{r-1} \quad \cdots \quad \hat{W}_0] = [I \quad 0 \quad \cdots \quad 0] \cdot [T(R_\Gamma^{-1}; r, 0, -r-s)]^{\mathrm{I}}.$$

One can use formula (4.39) in an analogous way for left canonical factorization.

We conclude this section with a simple example.

EXAMPLE 4.6. Let

$$R(\lambda) = I\lambda + \begin{bmatrix} -\frac{3}{2} & 1 \\ 4 & 2 \end{bmatrix} + \begin{bmatrix} -1 & -2 \\ 2 & 4 \end{bmatrix} \lambda^{-1},$$

and let Γ be the unit circle. In the notation used above, in our case $r = s = 1$. Let us compute ranks of the matrices $T(R_\Gamma^{-1}; 0, 0, -1)$ and $T(R_\Gamma^{-1}; 0, -1, -2)$. A computation shows that $\det R(\lambda) = (\lambda^2 - 4)(1 + \frac{1}{2}\lambda^{-1})$, so

$$R^{-1}(\lambda) = \frac{1}{(\lambda^2 - 4)(1 + \frac{1}{2}\lambda^{-1})} \begin{bmatrix} \lambda + 2 + 4\lambda^{-1} & -1 + 2\lambda^{-1} \\ -4 - 2\lambda^{-1} & \lambda - \frac{3}{2} - \lambda^{-1} \end{bmatrix}$$

and, multiplying numerator and denominator by λ,

$$R^{-1}(\lambda) = \frac{1}{(\lambda^2 - 4)(\lambda + \frac{1}{2})} \begin{bmatrix} \lambda^2 + 2\lambda + 4 & -\lambda + 2 \\ -4\lambda - 2 & \lambda^2 - \frac{3}{2}\lambda - 1 \end{bmatrix}.$$

Since $\lambda_0 = -\frac{1}{2}$ is the only pole of $R^{-1}(\lambda)$ inside Γ, we can compute the integrals

$$\frac{1}{2\pi i} \int_\Gamma \lambda^j R^{-1}(\lambda)\, d\lambda, \qquad j = 0, 1, -1$$

by the residue theorem. It turns out that

$$\frac{1}{2\pi i}\int_\Gamma R^{-1}(\lambda)\,d\lambda = \begin{bmatrix} -\frac{13}{15} & -\frac{2}{3} \\ 0 & 0 \end{bmatrix}, \qquad \frac{1}{2\pi i}\int_\Gamma \lambda R^{-1}(\lambda)\,d\lambda = \begin{bmatrix} \frac{13}{30} & \frac{1}{3} \\ 0 & 0 \end{bmatrix},$$

$$\frac{1}{2\pi i}\int_\Gamma \lambda^{-1}R^{-1}(\lambda)\,d\lambda = \begin{bmatrix} -\frac{4}{15} & \frac{1}{3} \\ 1 & \frac{1}{2} \end{bmatrix}.$$

So

$$T(R_\Gamma^{-1}; 0, 0, -1) = \begin{bmatrix} -\frac{4}{15} & \frac{1}{3} & -\frac{13}{15} & -\frac{2}{3} \\ 1 & \frac{1}{2} & 0 & 0 \end{bmatrix},$$

$$T(R_\Gamma^{-1}; 0, -1, -2) = \begin{bmatrix} -\frac{13}{15} & -\frac{2}{3} & \frac{13}{30} & \frac{1}{3} \\ 0 & 0 & 0 & 0 \\ -\frac{4}{15} & \frac{1}{3} & -\frac{13}{15} & -\frac{2}{3} \\ 1 & \frac{1}{2} & 0 & 0 \end{bmatrix}.$$

It is easily seen that

$$\operatorname{rank} T(R_\Gamma^{-1}; 0, 0, -1) = \operatorname{rank} T(R_\Gamma^{-1}; 0, -1, -2) = 2,$$

so according to Theorem 4.15, $R(\lambda)$ admits right canonical factorization. Also,

$$\operatorname{rank} T(R^{-1}; 0, -1, -1) = \operatorname{rank} T(R^{-1}; 0, -1, -2) = 2,$$

so $R(\lambda)$ admits a left canonical factorization as well. (In fact, we shall see later in Theorem 4.18 that for a trinomial $R(\lambda) = I\lambda + R_0 + R_{-1}\lambda^{-1}$ the necessary and sufficient condition to admit canonical factorizations from both sides is the invertibility of

$$\frac{1}{2\pi i}\int_\Gamma \lambda^{-1}R^{-1}(\lambda)\,d\lambda,$$

which holds in our example.) It is not difficult to write down canonical factorizations for $R(\lambda)$ using formulas (4.33) and (4.35):

$$R(\lambda) = \begin{bmatrix} \lambda - 2 & 0 \\ 4 & \lambda + 2 \end{bmatrix}\begin{bmatrix} 1 + \frac{1}{2}\lambda^{-1} & \lambda^{-1} \\ 0 & 1 \end{bmatrix}$$

is a right canonical factorization of $R(\lambda)$, and

$$R(\lambda) = \begin{bmatrix} 1 - \frac{13}{14}\lambda^{-1} & -\frac{5}{7}\lambda^{-1} \\ \frac{13}{7}\lambda^{-1} & 1 + \frac{10}{7}\lambda^{-1} \end{bmatrix}\begin{bmatrix} \lambda - \frac{4}{7} & \frac{12}{7} \\ \frac{15}{7} & \lambda + \frac{4}{7} \end{bmatrix}$$

is a left canonical factorization of $R(\lambda)$. □

4.6. Theorems on Two-Sided Canonical Factorization

We apply the results of Section 4.4 to obtain criteria for the simultaneous existence of left and right canonical factorization of a monic rational function.

The main criterion is the following:

Theorem 4.18. *Let $R(\lambda) = \sum_{j=-r}^{s-1} A_j\lambda^j + I\lambda^s: \Gamma \to GL(\mathbb{C}^n)$ be a monic rational function, and let $q_0 = \max(r, s)$. Then $R(\lambda)$ admits a right and a left canonical factorization if and only if the matrix $T(R_\Gamma^{-1}; q-1, 0, -q+1)$ is nonsingular for some $q \geq q_0$. If this condition is satisfied, then $T(R_\Gamma^{-1}; q-1, 0, -q+1)$ is nonsingular for every $q \geq q_0$.*

The proof of Theorem 4.18 is based on the following two lemmas, which are of independent interest and provide other criteria for two-sided canonical factorization.

Lemma 4.19. *A monic rational function $R(\lambda) = \sum_{j=-r}^{s-1} A_j\lambda^j + I\lambda^s: \Gamma \to GL(\mathbb{C}^n)$ admits a right and a left canonical factorization if and only if the matrix $T(R^{-1}; r-1, 0, -r+1)$ is nonsingular and one of the two following conditions holds:*

$$\begin{aligned}\operatorname{rank} T(R_\Gamma^{-1}; r-1, 0, -r-s-1) \\ = \operatorname{rank} T(R_\Gamma^{-1}; r-1, -1, -r-s) = rn\end{aligned} \tag{4.41}$$

or

$$\begin{aligned}\operatorname{rank} T(R_\Gamma^{-1}; r-1, -s, -r-s+1) \\ = \operatorname{rank} T(R_\Gamma^{-1}; r-1, -s, -r-s) = rn.\end{aligned} \tag{4.42}$$

Proof. As usual, we reduce the proof to the case of the monic matrix polynomial $M(\lambda) = \lambda^r R(\lambda)$. Let (X, T, Z) be a standard triple of $M(\lambda)$, and let (X_+, T_+, Z_+) be its part corresponding to the part of $\sigma(M)$ lying inside Γ. As it follows from Theorem 4.1, the right (resp. left) monic Γ-spectral divisor $N(\lambda)$ (resp. $\Phi(\lambda)$) of degree r exists if and only if the matrix $\operatorname{col}(X_+ T_+^j)_{j=0}^{r-1}$ (resp. the matrix $[Z_+ \quad T_+ Z_+ \quad \cdots \quad T_+^{r-1} Z_+]$) is nonsingular. We therefore have to establish the nonsingularity of these two matrices in order to prove sufficiency of the conditions of the lemma. Suppose $T(M^{-1}; -1, -r, -2r+1)$ is nonsingular and, for example, (4.41) holds. The decomposition

$$\begin{aligned}T(M_\Gamma^{-1}; -1, -r, -2r+1) = \operatorname{col}(X_+ T_+^{r-1-j})_{j=0}^{r-1} \\ \cdot [Z_+ \quad T_+ Z_+ \quad \cdots \quad T_+^{r-1} Z_+]\end{aligned} \tag{4.43}$$

(cf. (4.5)) allows us to claim the right invertibility of $\operatorname{col}(X_+ T_+^j)_{j=0}^{r-1}$. On the other hand, equality (4.41) implies that

$$\operatorname{Ker} \operatorname{col}(X_+ T_+^j)_{j=0}^{p-1} = \operatorname{Ker} \operatorname{col}(X_+ T_+^j)_{j=0}^{r-1}$$

holds for every $p \geq r$ (this can be shown in the same way as in the proof of Theorem 4.3). But then

$$\operatorname{Ker}[\operatorname{col}(X_+ T_+^j)_{j=0}^{r-1}] = \operatorname{Ker}[\operatorname{col}(X_+ T_+^j)_{j=0}^{l-1}] = \{0\},$$

which means invertibility of $\operatorname{col}(X_+ T_+^j)_{j=0}^{r-1}$. The invertibility of

$$[Z_+ \quad T_+ Z_+ \quad \cdots \quad T_+^{r-1} Z_+]$$

follows from (4.43). Conversely, suppose that the linear transformations $\operatorname{col}(X_+ T_+^j)_{j=0}^{r-1}$ and $[Z_+ \quad Z_+T_+ \quad \cdots \quad T_+^{r-1}Z_+]$ are invertible. Then by (4.43) $T(M_\Gamma^{-1}; -1, -r, -2r+1)$ is also invertible, and by Theorems 4.15 and 4.16 conditions (4.41) and (4.42) hold. □

Given a contour $\Gamma \subset \mathbb{C}$ and a monic matrix polynomial $M(\lambda): \Gamma \to GL(\mathbb{C}^n)$, a closed rectifiable simple contour Ξ will be called *complementary* to Γ (relative to $M(\lambda)$) if Ξ lies in F^- together with its interior, and $M(\lambda) \in GL(\mathbb{C}^n)$ for all points $\lambda \in F^- \backslash \{\infty\}$ lying outside Ξ. In the sequel we shall use this notion for rational matrix functions as well.

In the next lemma Ξ denotes a contour complementary to Γ (relative to $R(\lambda)$).

Lemma 4.20. *For a monic rational function* $R(\lambda) = I\lambda^s + \sum_{j=-r}^{s-1} A_j\lambda^j$: $\Gamma \to GL(\mathbb{C}^n)$ *to admit simultaneously a right and a left canonical factorization it is necessary and sufficient that the following condition holds*:

(a) *in the case* $r \geq s$, *the matrix* $T_\Gamma = T(R_\Gamma^{-1}; r-1, 0, -r+1)$ *is nonsingular*;

(b) *in the case* $r \leq s$, *the matrix* $T_\Xi = T(R_\Xi^{-1}; r-1, r-s, r-2s+1)$ *is nonsingular.*

Moreover, in case (a), *the nonsingularity of* T_Γ *implies the nonsingularity of* T_Ξ, *while, in case* (b), *the nonsingularity of* T_Ξ *leads to the nonsingularity of* T_Γ.

Proof. Passing, as usual, to the monic polynomial $M(\lambda) = \lambda^r R(\lambda)$ of degree $l = r + s$, we have to prove the following: $M(\lambda)$ has a right and a left monic Γ-spectral divisor of degree r if and only if, in the case $l \geq 2r$, the matrix $T(M_\Gamma^{-1}; -1, -r, -2r+1)$ is nonsingular, or, in the case $l \geq 2r$, the matrix $T(M_\Xi^{-1}; -1, -(l-r), -2(l-r)+1)$ is nonsingular.

This is nothing but a reformulation of Theorem 4.12. The last assertion of Lemma 4.20 can be obtained, for instance, by applying Lemma 4.19. □

Note that in fact we can always realize case (a). Indeed, if $r < s$, regard $R(\lambda)$ as $\sum_{j=-q}^{s-1} A_j\lambda^j + I\lambda^s$ with $A_{-q} = \cdots = A_{-r-1} = 0$, where $q \geq s$ is an arbitrary integer.

Proof of Theorem 4.18. First suppose $s \leq r$, and let q be an arbitrary integer such that $q \geq r$. Consider $R(\lambda)$ as $\sum_{j=-q}^{s-1} A_j\lambda^j + I\lambda^s$ with $A_{-q} = \cdots = A_{-r-1} = 0$. In view of Lemma 4.20 (in case (a)) existence of a two-sided canonical factorization of $R(\lambda)$ is equivalent to the nonsingularity of $T(R_\Gamma^{-1}; q-1, 0, -q+1)$.

Suppose now $s > r$. Considering $R(\lambda)$ as $\sum_{j=-s}^{s-1} A_j\lambda^j + I\lambda^s$ with $A_{-s} = \cdots = A_{-r-1} = 0$, and applying the part of the theorem already proved (the case $s = r$), we obtain that $R(\lambda)$ admits a two-sided canonical factorization iff the matrix $T(R_\Gamma^{-1}; q-1, 0, -q+1)$ is invertible for some $q \geq s$. □

The following corollary from Theorem 4.18 is also of interest.

Corollary 4.21. *Let* $M(\lambda) = I\lambda^l + \sum_{j=0}^{l-1} M_j\lambda^j \colon \Gamma \to GL(\mathbb{C}^n)$ *be a monic matrix polynomial. Then* $\sigma(M) \cap F^+ = \emptyset$ *if and only if the matrix* $T(M^{-1}; q-1, 0, -q+1)$ *is nonsingular for some integer* $q \geq l$.

Proof. Regard $M(\lambda)$ as a monic rational function with $r = 0$, and note that $M(\lambda)$ admits a two-sided canonical factorization if and only if $\sigma(M) \cap F^+ = \emptyset$. $\square$

4.7. Wiener–Hopf Factorization for Matrix Polynomials

We continue to use notations introduced in Section 4.5. A continuous and invertible $n \times n$ matrix function $H(\lambda)$ $(\lambda \in \Gamma)$ is said to admit a *right Wiener–Hopf factorization* (relative to Γ) if the following representation holds:

$$H(\lambda) = H_+(\lambda)\begin{bmatrix} \lambda^{\kappa_1} & & & 0 \\ & \lambda^{\kappa_2} & & \\ & & \ddots & \\ 0 & & & \lambda^{\kappa_n} \end{bmatrix} H_-(\lambda) \qquad (\lambda \in \Gamma), \tag{4.44}$$

where $H_+(\lambda), (H_+(\lambda))^{-1} \in C^+(\Gamma); H_-(\lambda), (H_-(\lambda))^{-1} \in C^-(\Gamma)$, and $\kappa_1 \leq \cdots \leq \kappa_n$ are integers (positive, negative, or zeros). These integers are called *right partial indices* of $H(\lambda)$. Interchanging the places of $H_+(\lambda)$ and $H_-(\lambda)$ in (4.44), we obtain left factorization of $H(\lambda)$. So the *left Wiener–Hopf factorization* is defined by

$$H(\lambda) = H'_-(\lambda)\begin{bmatrix} \lambda^{\kappa_1'} & & & 0 \\ & \lambda^{\kappa_2} & & \\ & & \ddots & \\ 0 & & & \lambda^{\kappa_n'} \end{bmatrix} H'_+(\lambda) \qquad (\lambda \in \Gamma), \tag{4.45}$$

where $H'_+(\lambda)$, $(H'_+(\lambda))^{-1} \in C^+(\Gamma)$; $H'_-(\lambda)$, $(H'_-(\lambda))^{-1} \in C^-(\Gamma)$, and $\kappa_1' \leq \cdots \leq \kappa_n'$ are integers (in general they are different from $\kappa_1, \ldots, \kappa_n$), which naturally are called *left partial indices* of $H(\lambda)$. For brevity, the words "Wiener–Hopf" will be suppressed in this section (so "right factorization" is used instead of "right Wiener–Hopf factorization").

In the scalar case ($n = 1$) there exists only one right partial index and only one left partial index; they coincide (because of the commutativity) and will be referred to as the *index* (which coincides with the winding number introduced in Section 4.5) of a scalar function $H(\lambda)$. Note also that the canonical factorizations considered in Sections 4.5–4.6 are particular cases of (right or left) factorization; namely, when all the partial indices are zero. As in the case of canonical factorization, right factorization of $H(\lambda)$ is equivalent to left factorization of $H^T(\lambda)$. So we shall consider mostly right factorization, bearing in mind that dual results hold for left factorization.

First we make some general remarks concerning the notion of right factorization (relative to Γ). We shall not prove these remarks here but refer to [25, Chapter VIII], (where Wiener–Hopf factorization is called canonical) for proofs and more detail. The factors $H_+(\lambda)$ and $H_-(\lambda)$ in (4.44) are not defined uniquely (for example, replacing in (4.44) $H_+(\lambda)$ by $\alpha H_+(\lambda)$ and $H_-(\lambda)$ by $\alpha^{-1}H_-(\lambda)$, $\alpha \in \mathbb{C}\backslash\{0\}$, we again obtain a right factorization of $H(\lambda)$). But the right factorization indices are uniquely determined by $H(\lambda)$, i.e., they do not depend on the choice of $H_\pm(\lambda)$ in (4.44). Not every continuous and invertible matrix-function on Γ admits a right factorization; description of classes of matrix-functions whose members admit a right factorization, can be found in [25]. We shall consider here only the case when $H(\lambda)$ is a matrix polynomial; in such cases $H(\lambda)$ always admits right factorization provided $\det H(\lambda) \neq 0$ for all $\lambda \in \Gamma$.

We illustrate the notion of factorization by an example.

EXAMPLE 4.7. Let

$$H(\lambda) = \begin{bmatrix} 1 & \lambda \\ 0 & \lambda^2 \end{bmatrix}$$

and Γ be the unit circle. Then

$$H(\lambda) = \begin{bmatrix} 0 & 1 \\ -1 & \lambda \end{bmatrix} \begin{bmatrix} \lambda & 0 \\ 0 & \lambda \end{bmatrix} \begin{bmatrix} 1 & 0 \\ \lambda^{-1} & 1 \end{bmatrix}$$

is a right factorization of $H(\lambda)$, with

$$H_+(\lambda) = \begin{bmatrix} 0 & 1 \\ -1 & \lambda \end{bmatrix} \quad \text{and} \quad H_-(\lambda) = \begin{bmatrix} 1 & 0 \\ \lambda^{-1} & 1 \end{bmatrix}.$$

The right partial indices are $\kappa_1 = \kappa_2 = 1$. A left factorization of $H(\lambda)$ is given by

$$H(\lambda) = \begin{bmatrix} 1 & \lambda^{-1} \\ 0 & 1 \end{bmatrix} \begin{bmatrix} 1 & 0 \\ 0 & \lambda^2 \end{bmatrix},$$

where

$$H_-(\lambda) = \begin{bmatrix} 1 & \lambda^{-1} \\ 0 & 1 \end{bmatrix}, \qquad H_+(\lambda) = I.$$

The left partial indices are $\kappa_1' = 0$, $\kappa_2' = 2$. □

This example shows that in general left and right partial indices are different. But they are not completely independent. For instance, the sums $\sum_{i=1}^n \kappa_i$ and $\sum_{i=1}^n \kappa_i'$ of all right partial indices $\kappa_1 \leq \cdots \leq \kappa_n$ and all left partial indices $\kappa'_1 \le \cdots \le \kappa'_n$, respectively, coincide. This fact is easily verified by taking determinants in (4.44) and (4.45) and observing that $\sum_{i=1}^{n} \kappa_i$, as well as $\sum_{i=1}^{n} \kappa'_i$, coincides with the index of det $H(\lambda)$.

Now we provide formulas for partial indices of matrix polynomials assuming, for simplicity, that the polynomials are monic.

To begin with, consider the linear case $L(\lambda) = I\lambda - A$. Let $A = \operatorname{diag}[A_1, A_2]$ be a block representation where the eigenvalues of A_1 are in F^+ and the eigenvalues of A_2 are in F^-. Then the equality

$$I\lambda - A = \begin{bmatrix} I & 0 \\ 0 & I\lambda - A_2 \end{bmatrix} \begin{bmatrix} \lambda I & 0 \\ 0 & I \end{bmatrix} \begin{bmatrix} I - \lambda^{-1}A_1 & 0 \\ 0 & I \end{bmatrix}$$

is a right factorization of $I\lambda - A$ and the partial indices are $(0, 0, \ldots, 0, 1, 1, \ldots, 1)$, where the number of ones is exactly equal to rank $A_1(=$ rank $\operatorname{Im} \int_\Gamma (I\lambda - A)^{-1}\, d\lambda)$. For a matrix polynomial, this result can be generalized as follows.

Theorem 4.22. *Let $L(\lambda)$ be a monic matrix polynomial of degree m with* det $L(\lambda) \neq 0$ *for $\lambda \in \Gamma$. Then for the right partial indices $\kappa_1 \le \kappa_2 \le \cdots \le \kappa_n$ of $L(\lambda)$ the following equalities hold*

$$\kappa_i = |\{j \mid n + r_{j-1} - r_j \le i - 1, j = 1, 2, \ldots, m\}| \qquad (i = 1, 2, \ldots, n),$$

where

$$r_j = \operatorname{rank} \begin{bmatrix} B_{-1} & B_{-2} & \cdots & B_{-m} \\ B_{-2} & B_{-3} & \cdots & B_{-m-1} \\ \vdots & \vdots & & \vdots \\ B_{-j} & B_{-j-1} & \cdots & B_{-j-m-1} \end{bmatrix}, \qquad B_{-j} = \frac{1}{2\pi i} \int_\Gamma \lambda^{j-1} L^{-1}(\lambda)\, d\lambda$$

and $|\Omega|$ denotes the number of elements in the set Ω.

The same formulas for κ_j hold in case Γ is the unit circle and the B_j are the Fourier coefficients of $L^{-1}(\lambda)$:

$$L^{-1}(\lambda) = \sum_{j=-\infty}^{\infty} \lambda^j B_j \qquad (|\lambda| = 1).$$

In some special cases one can do without computing the contour integrals, as the following result shows.

Theorem 4.23. *Suppose that all the eigenvalues of $L(\lambda) = \sum_{j=0}^{m} A_j \lambda^j$ $(A_m = I)$ are inside Γ and $L(0) = I$. Then*

$$\kappa_j = |\{i \mid n + q_{i-1} - q_i \le j - 1, i = 1, 2, \ldots, n\}|$$

where

$$q_i = \operatorname{rank}\begin{bmatrix} PC^{\nu} \\ PC^{\nu+1} \\ \vdots \\ PC^{\nu+i-1} \end{bmatrix}, \qquad C = \begin{bmatrix} 0 & I & \cdots & 0 \\ \vdots & \vdots & & \vdots \\ 0 & 0 & \cdots & I \\ -A_m & -A_{m-1} & \cdots & -A_1 \end{bmatrix},$$

$\nu \geq 0$ is the minimal integer for which rank C^{ν} = rank $C^{\nu+1}$, *and P is the projector on the first n coordinates.*

In fact, the assumption that $L(\lambda)$ be monic is not essential in Theorems 4.22 and 4.23 and they hold for any matrix polynomial, whose spectrum does not intersect Γ. For the proofs of Theorems 4.22 and 4.23 we refer to [36e] (see also [35b]).

Comments

The results of Sections 4.1 and 4.4 originated in [34b, 36b]. Theorem 4.2 (without formulas for the divisor) is due to Lopatinskii [58]; however, his proof is quite different. Formula (4.3) appears for the first time in [36b]. Section 4.2 is based on the paper [52f]. Theorem 4.6 was first proved in [56c], although historically results like Corollary 4.7 began with the search for complete sets of linearly independent generalized eigenvectors, as in [56b]. Sufficient conditions for the existence of a right divisor of degree one found by methods of nonlinear analysis can be found in [54] as well as references to earlier works in this direction. Lemma 4.9 is taken from [10] and Theorem 4.10 was suggested by Clancey [13].

Sections 4.5 and 4.6 are based on [36b], where analogous results are proved in the infinite dimensional case. The problems concerning spectral factorization for analytic matrix and operator functions have attracted much attention. Other results are available in [36b, 61a, 61b, 61c]. A generalization of spectral factorization, when the spectra of divisor and quotient are not quite separated, appears naturally in some applications ([3c, Chapter 6]).

The canonical factorization is used for inversion of block Toeplitz matrices (see [25, Chapter VIII]). It also plays an important role in the theory of systems of differential and integral equations (see [3c, 25, 32a, 68]).

Existence of a two-sided canonical factorization is significant in the analysis of projection methods for the inversion of Toeplitz matrices (see [25]).

The last section is based on [36a]. For other results connected with Wiener–Hopf factorization and partial indices see [36c, 36d, 70c, 73]. More information about partial indices, their connection with Kronecker indices, and with feedback problems in system theory can be found in [21, 35b].

Extensions of the results of this chapter to the more general case of spectral divisors, simply behaved at infinity, and quasi-canonical factorization in the infinite dimensional case are obtained in [36c, 36d].

Chapter 5

Perturbation and Stability of Divisors

Numerical computation of divisors (using explicit formulas) leads in a natural way to the analysis of perturbation and stability problems. It is generally the case (but not always) that, if the existence of a right divisor depends on a "splitting" of a multiple eigenvalue of the parent polynomial, then the divisor will be practically impossible to compute, because random errors in calculation (such as rounding errors) will "blow up" the divisor. Thus, it is particularly important to investigate those divisors which are "stable" under general small perturbations of the coefficient matrices of $L(\lambda)$. This will be done in this chapter (Section 5.3).

More generally, we study here the dependence of divisors on the coefficients of the matrix polynomial and on the supporting subspace. It turns out that small perturbations of the coefficients of the parent polynomial and of the supporting subspace lead to a small change in the divisor. For the special case when the divisor is spectral, the supporting subspace is the image of a Riesz projector (refer to Theorem 4.1) and these projectors are well known to be stable under small perturbations. Thus, small perturbations of the polynomial automatically generate a small change in the supporting subspace, and therefore also in the divisor. Hence spectral divisors can be approximately computed. Generally speaking, nonspectral divisors do not have this property. The stable divisors which are characterized in this chapter, are precisely those which enjoy this property.

We also study here divisors of monic matrix polynomials with coefficients depending continuously or analytically on a parameter. We discuss both local and global properties of these divisors. Special attention is paid to their singularities. The interesting phenomenon of isolated divisors is also analyzed.

5.1. The Continuous Dependence of Supporting Subspaces and Divisors

With a view to exploring the dependence of divisors on supporting subspaces, and *vice versa*, we first introduce natural topologies into the sets of monic matrix polynomials of fixed size and degree.

Let $\mathscr{P}_k$ be the class of all $n \times n$ monic matrix polynomials of degree k. It is easily verified that the function σ_k defined on $\mathscr{P}_k \times \mathscr{P}_k$ by

$$\sigma_k\left(I\lambda^k + \sum_{i=0}^{k-1} B_i\lambda^i, I\lambda^k + \sum_{i=0}^{k-1} B_i'\lambda^i\right) = \sum_{i=0}^{k-1} \|B_i - B_i'\| \tag{5.1}$$

is a metric, and $\mathscr{P}_k$ will be viewed as the corresponding metric space.

Let $\mathscr{A}$ be the class of all subspaces in $\mathbb{C}^{nl}$. We consider $\mathscr{A}$ also as a metric space where the metric is defined by the *gap* $\theta(\mathscr{M}, \mathscr{N})$ between the subspaces $\mathscr{M}, \mathscr{N} \in \mathbb{C}^{nl}$ (see Section S4.3).

For a monic matrix polynomial $L(\lambda) = I\lambda^l + \sum_{j=0}^{l-1} A_j\lambda^j$ (A_j are $n \times n$ matrices) recall the definition of the first companion matrix C_L:

$$C_L = \begin{bmatrix} 0 & I & 0 & \cdots & 0 \\ 0 & 0 & I & \cdots & 0 \\ \vdots & \vdots & \vdots & & \vdots \\ 0 & 0 & 0 & \cdots & I \\ -A_0 & -A_1 & -A_2 & \cdots & -A_{l-1} \end{bmatrix}.$$

Consider the set $\mathscr{W} \subset \mathscr{A} \times \mathscr{P}_l$ consisting of pairs $\{\mathscr{M}, L(\lambda)\}$ for which $\mathscr{M}$ is an invariant subspace in $\mathbb{C}^{nl}$ for the companion matrix C_L of L. This set $\mathscr{W}$ will be called the *supporting set* and is provided with the metric induced from $\mathscr{A} \times \mathscr{P}_1$ (so an ε-neighborhood of $\{\mathscr{M}, L\} \in \mathscr{W}$ consists of all pairs $\{\tilde{\mathscr{M}}, \tilde{L}\} \in \mathscr{W}$ for which $\theta(\tilde{\mathscr{M}}, \mathscr{M}) + \sigma(\tilde{L}, L) < \varepsilon$).

Then define the subset $\mathscr{W}_k \subset \mathscr{W}$ consisting of all pairs $\{\mathscr{M}, L(\lambda)\}$, where $L(\lambda) \in \mathscr{P}_l$ and $\mathscr{M}$ is a supporting subspace (with respect to the standard pair $([I \ \ 0 \ \ \cdots \ \ 0], C_L)$ of $L(\lambda)$; see Section 1.9), associated with a monic right divisor of L of degree k. The set $\mathscr{W}_k$ will be called the *supporting set* of order k.

Theorem 5.1. *The set $\mathscr{W}_k$ is open in $\mathscr{W}$.*

Proof. Define the subspace $\mathscr{Y}_{l-k}$ of $\mathbb{C}^{nl}$ by the condition $x = (x_1, \ldots, x_l) \in \mathscr{Y}_{l-k}$ if and only if $x_1 = \cdots = x_k = 0$. Suppose that $\mathscr{M}$ is a supporting subspace for $L(\lambda)$; then by Corollary 3.13, $\{\mathscr{M}, L(\lambda)\} \in \mathscr{W}_k$ if and only if $\mathscr{M} + \mathscr{Y}_{l-k} = \mathbb{C}^{nl}$.

Now let $(\mathscr{M}, L) \in \mathscr{W}_k$ and let $(\hat{\mathscr{M}}, \hat{L}) \in \mathscr{W}$ be in an ε-neighborhood of $(\mathscr{M}, L)$. Then certainly $\theta(\mathscr{M}, \hat{\mathscr{M}}) < \varepsilon$. By choosing ε small enough it can be guaranteed, using Theorem S4.7, that $\hat{\mathscr{M}} \dotplus \mathscr{Y}_{l-k} = \mathbb{C}^{nl}$. Consequently, $\hat{\mathscr{M}}$ is a supporting subspace for $\hat{L}$ and $(\hat{\mathscr{M}}, \hat{L}) \in \mathscr{W}_k$. □

We pass now to the description of the continuous dependence of the divisor and the quotient connected with a pair $(\mathscr{M}, L) \in \mathscr{W}_k$. Define a map $F_k : \mathscr{W}_k \to \mathscr{P}_{l-k} \times \mathscr{P}_k$ in the following way: the image of $(\mathscr{M}, L) \in \mathscr{W}_k$ is to be the pair of monic matrix polynomials (L_2, L_1) where L_1 is the right divisor of L associated with $\mathscr{M}$ and L_2 is the quotient obtained on division of L on the right by L_1. It is evident that F_k is one-to-one and surjective so that the map F_k^{-1} exists.

For $(L_2, L_1), (\tilde{L}_2, \tilde{L}_1) \in \mathscr{P}_{l-k} \times \mathscr{P}_k$ put

$$\rho((L_2, L_1), (\tilde{L}_2, \tilde{L}_1)) = \sigma_{l-k}(L_2, \tilde{L}_2) + \sigma_k(\tilde{L}_1, \tilde{L}_1),$$

where σ_i is defined by (5.1); so $\mathscr{P}_{l-k} \times \mathscr{P}_k$ is a metric space with the metric ρ. Define the space $\mathscr{P}_0$ to be the disconnected union: $\mathscr{P}_0 = \bigcup_{k=1}^{l-1} (\mathscr{P}_{l-k} \times \mathscr{P}_k)$. In view of Theorem 5.1, the space $\mathscr{W}_0 = \bigcup_{k=1}^{l-1} \mathscr{W}_k$ is also a disconnected union of its subspaces $\mathscr{W}_1, \ldots, \mathscr{W}_{l-1}$. The map F between topological spaces $\mathscr{W}_0$ and $\mathscr{P}_0$ can now be defined by the component maps $F_1, \ldots, F_{l-1}$, and F will be invertible.

If X_1, X_2 are topological spaces with metrics ρ_1, ρ_2, defined on each connected component of X_1 and X_2, respectively, the map $G: X_1 \to X_2$ is called *locally Lipschitz continuous* if, for every $x \in X_1$, there is a deleted neighborhood U_x of x for which

$$\sup_{y \in U_x} (\rho_2(Gx, Gy)/\rho_1(x, y)) < \infty.$$

Theorem 5.2. *The maps F and F^{-1} are locally Lipschitz continuous.*

Proof. It is sufficient to prove that F_k and F_k^{-1} are locally Lipschitz continuous for $k = 1, 2, \ldots, l-1$. Let $(\mathscr{M}, L) \in \mathscr{W}_k$ and

$$F_k(\mathscr{M}, L) = (L_2, L_1).$$

Recall that by Theorem 3.12 the monic matrix polynomial $L_1(\lambda)$ has the following representation:

$$L_1(\lambda) = I\lambda^k - XC_L^k(W_1 + W_2\lambda + \cdots + W_k\lambda^{k-1}), \tag{5.2}$$

where $X = [I \quad 0 \quad \cdots \quad 0]$, $W_1, \ldots, W_k$ are $nk \times n$ matrices defined by the equality

$$[W_1 \quad \cdots \quad W_k] = (Q_k|_{\mathscr{M}})^{-1}, \tag{5.3}$$

and $Q_k = [I_{kn} \quad 0]$ is a $kn \times ln$ matrix. In order to formulate the analogous representation for $L_2(\lambda)$, let us denote $Y = [0 \quad \cdots \quad 0 \quad I]^{\mathrm{T}}$, $R_{l-k} = [Y \quad C_L Y \quad \cdots \quad C_L^{l-k-1} Y]$. Let P be the projector on

$$\operatorname{Im} R_{l-k} = \{x = (x_1, \ldots, x_l) \in \mathbb{C}^{nl} \mid x_1 = \cdots = x_k = 0\}$$

along $\mathscr{M}$, Then (Corollary 3.18)

$$L_2(\lambda) = I\lambda^{l-k} - (Z_1 + Z_2\lambda + \cdots + Z_{l-k}\lambda^{l-k-1}) P C_L^{l-k} P Y, \tag{5.4}$$

where

$$\begin{bmatrix} Z_1 \\ Z_2 \\ \vdots \\ Z_{l-k} \end{bmatrix} = R_{l-k}^{-1}$$

(in the last equality R_{l-k} is considered as a linear transformation from $\mathbb{C}^{n(l-k)}$ into $\operatorname{Im} R_{l-k}$; with this understanding the inverse R_{l-k}^{-1}: $\operatorname{Im} R_{l-k} \to \mathbb{C}^{n(l-k)}$ exists).

To prove the required continuity properties of F_k: $\mathscr{W}_k \to \mathscr{P}_{l-k} \times \mathscr{P}_k$ it is necessary to estimate the distance between pairs (L_2, L_1), $(\hat{L}_2, \hat{L}_1)$ in $\mathscr{P}_{l-k} \times \mathscr{P}_k$ using the metric ρ defined above. Note first that if $P_{\mathscr{M}}$ is the projector on $\mathscr{M}$ along $\operatorname{Im} R_{l-k}$, then $P = I - P_{\mathscr{M}}$ (the projector appearing in (5.4)), and that $(Q_k|_{\mathscr{M}})^{-1} = P_{\mathscr{M}} C_{kl}$ in (5.3), where C_{kl}: $\mathbb{C}^{kn} \to \mathbb{C}^{ln}$ is the imbedding of $\mathbb{C}^{kn}$ onto the first kn components of $\mathbb{C}^{ln}$. Then observe that in the representations (5.2) and (5.4) the coefficients of L_1 and L_2 are uniformly bounded in some neighborhood of $(\mathscr{M}, L)$. It is then easily seen that in order to establish the continuity required of F_k it is sufficient to verify the assertion: for a fixed $(\mathscr{M}, L) \in \mathscr{W}_k$ there exist positive constants δ and C such that, for any subspace $\mathscr{N}$ satisfying $\theta(\mathscr{M}, \mathscr{N}) < \delta$, it follows that $\mathscr{N} \dotplus \operatorname{Im} R_{l-k} = \mathbb{C}^{nl}$ and

$$\|\tilde{\mathscr{P}}_{\mathscr{M}} - \tilde{\mathscr{P}}_{\mathscr{N}}\| \le C\,\theta(\mathscr{M}, \mathscr{N}),$$

where $\tilde{P}_{\mathscr{N}}$ is the projector of $\mathbb{C}^{nl}$ on $\mathscr{N}$ along $\operatorname{Im} R_{l-k}$. But this conclusion can be drawn from Theorem S4.7, and so the Lipschitz continuity of F_k follows.

To establish the local Lipschitz continuity of F_k^{-1} we consider a fixed $(L_2, L_1) \in \mathscr{P}_{l-k} \times \mathscr{P}_k$. It is apparent that the polynomial $L = L_2 L_1$ will be a Lipschitz continuous function of L_2 and L_1 in a neighborhood of the fixed pair. To examine the behavior of the gap between supporting subspaces associated with neighboring pairs we observe an explicit construction for $P_{\mathscr{M}}$, the projection on $\mathcal{M}$ along Im R_{l-k} (associated with the pair L_2, L_1). In fact, $P_{\mathcal{M}}$ has the representation

$$P_{\mathcal{M}} = \begin{bmatrix} I & 0 \\ F & 0 \end{bmatrix}, \qquad F = \begin{bmatrix} P_1 C_{L_1}^k \\ \vdots \\ P_1 C_{L_1}^{l-1} \end{bmatrix}, \tag{5.5}$$

with respect to the decomposition $\mathbb{C}^{ln} = \mathbb{C}^{kn} \oplus \mathbb{C}^{(l-k)n}$, where $P_1 = [I \;\; 0 \;\; \cdots \;\; 0]$. Indeed, $P_{\mathcal{M}}$ given by (5.5) is obviously a projector along Im R_{l-k}. Let us check that Im $P_{\mathcal{M}} = \mathcal{M}$. The subspace $\mathcal{M}$ is the supporting subspace corresponding to the right divisor $L_1(\lambda)$ of $L(\lambda)$; by formula (3.23), $\mathcal{M} = \operatorname{Im} \operatorname{col}(P_1 C_{L_1}^i)_{i=0}^{l-1} = \operatorname{Im} P_{\mathcal{M}}$.

The local Lipschitz continuity of $P_{\mathcal{M}}$ as a function of L_1 is apparent from formula (5.5).

Given the appropriate continuous dependence of L and $P_{\mathcal{M}}$ the conclusion now follows from the left-hand inequality of (S4.18) in Theorem S4.7. □

Corollary 5.3. *Let $L(\lambda, \mu) = \sum_{i=0}^{l-1} A_i(\mu)\lambda^i + I\lambda^l$ be a monic matrix polynomial whose coefficients $A_i(\mu)$ depend continuously on a parameter μ in an open set $\mathcal{D}$ of the complex plane. Assume that there exists for each $\mu \in \mathcal{D}$ a monic right divisor $L_1(\lambda, \mu) = \sum_{i=0}^{k} B_i(\mu)\lambda^i$ of L with quotient $L_2(\lambda, \mu) = \sum_{i=0}^{l-k} C_i(\mu)\lambda^i$. Let $\mathcal{M}(\mu)$ be the supporting subspace of L with respect to C_L and corresponding to L_1.*

(i) *If $\mathcal{M}(\mu)$ is continuous on $\mathcal{D}$ (in the metric θ), then both L_1 and L_2 are continuous on $\mathcal{D}$.*

(ii) *If one of L_1, L_2 is a continuous function of μ on $\mathcal{D}$, then the second is also continuous on $\mathcal{D}$, as is $\mathcal{M}(\mu)$.*

Proof. Part (i) follows immediately from the continuity of $(\mathcal{M}, L)$ and of F since (L_2, L_1) is the image of the composition of two continuous functions.

For part (ii) suppose first that L_1 is continuous in μ on $\mathcal{D}$. Then, using (5.5) it is clear that $\mathcal{M}(\mu) = \operatorname{Im} P_{\mathcal{M}(\mu)}$ depends continuously on μ. The continuity of L_2 follows as in part (i). □

5.2. Spectral Divisors: Continuous and Analytic Dependence

We maintain here the notation introduced in the preceding section.

We now wish to show that, if $\mathcal{M}$ corresponds to a Γ-spectral divisor of L (see Section 4.1), then there is a neighborhood of $(\mathcal{M}, L)$ in $\mathcal{W}_k$ consisting of pairs $(\mathcal{M}, L)$ for which the spectral property is retained. A pair $(\mathcal{M}, L)$ for which $\mathcal{M}$ corresponds to a Γ-spectral divisor of L will be said to be Γ-*spectral*.

Recall (Theorem 4.1) that for a Γ-spectral pair $(\mathscr{M}, L) \in \mathscr{W}$ we have

$$\mathscr{M} = \operatorname{Im}\left(\frac{1}{2\pi i}\int_\Gamma (I\lambda - C_L)^{-1}\, d\lambda\right).$$

Lemma 5.4. *Let $(\mathscr{M}, L) \in \mathscr{W}$ be Γ-spectral. Then there is a neighborhood of $(\mathscr{M}, L)$ in $\mathscr{W}$ consisting of Γ-spectral pairs.*

Proof. Let $L_0 \in \mathscr{P}_l$ and be close to L in the σ metric. Then C_{L_0} will be close to C_L and, therefore, L_0 can be chosen so close to L that $I\lambda - C_{L_0}$ is nonsingular for all $\lambda \in \Gamma$. (This can be easily verified by assuming the contrary and using compactness of Γ and continuity of $\det(I\lambda - C_{L_0})$ as a function of λ and C_{L_0}.) Let Δ be a closed contour which does not intersect with Γ and contains in its interior exactly those parts of the spectrum of L which are outside Γ. Then we may assume L_0 chosen in such a way that the spectrum of L_0 also consists of two parts; one part inside Γ and the other part inside Δ.

Define $\mathscr{N}, \mathscr{N}_0, \mathscr{M}_0$ to be the images of the Riesz projectors determined by (Δ, L), (Δ, L_0), and (Γ, L_0), respectively, i.e.,

$$\mathscr{N} = \operatorname{Im}\left(\frac{1}{2\pi i}\int_\Delta (I\lambda - C_L)^{-1}\, d\lambda\right),$$

$$\mathscr{N}_0 = \operatorname{Im}\left(\frac{1}{2\pi i}\int_\Delta (I\lambda - C_{L_0})^{-1}\, d\lambda\right), \qquad \mathscr{M}_0 = \operatorname{Im}\left(\frac{1}{2\pi i}\int_\Gamma (I\lambda - C_{L_0})^{-1}\, d\lambda\right).$$

Then $\mathscr{M} \dotplus \mathscr{N} = \mathscr{M}_0 \dotplus \mathscr{N}_0 = \mathbb{C}^{nl}$, $(\mathscr{M}_0, L_0)$ is Γ-spectral, and if $L_0 \to L$ then $\mathscr{M}_0 \to \mathscr{M}$ (the latter assertion follows from Theorem S4.7).

Now let $(\mathscr{M}_1, L_0) \in \mathscr{W}$ be in a neighborhood $\mathscr{U}$ of $(\mathscr{M}, L)$. We are finished if we can show that for $\mathscr{U}$ small enough, $\mathscr{M}_1 = \mathscr{M}_0$. First, $\mathscr{U}$ can be chosen small enough for *both* L_0 to be so close to L (whence $\mathscr{M}_0$ to $\mathscr{M}$) *and* $\mathscr{M}_1$ to be so close to $\mathscr{M}$, that $\mathscr{M}_1$ and $\mathscr{M}_0$ are arbitrarily close. Thus the first statement of Theorem S4.7 will apply and, if $\mathscr{U}$ is chosen small enough, then $\mathscr{M}_1 \dotplus \mathscr{N}_0 = \mathbb{C}^{nl}$.

Now $\mathscr{M}_1$ invariant for C_{L_0} implies the invariance of $\mathscr{M}_1$ for the projector

$$P_0 = \frac{1}{2\pi i}\int_\Gamma (I\lambda - C_{L_0})^{-1}\, d\lambda.$$

Since $\operatorname{Ker} P_0 = \mathscr{N}_0$, and $\mathscr{M}_1 \cap \mathscr{N}_0 = \{0\}$, the P_0-invariance of $\mathscr{M}_1$ implies that $\mathscr{M}_1 \subset \operatorname{Im} P_0 = \mathscr{M}_0$. But since $\mathscr{N}_0$ is a common direct complement to $\mathscr{M}_0$ and $\mathscr{M}_1$, $\dim \mathscr{M}_0 = \dim \mathscr{M}_1$ and therefore $\mathscr{M}_1 = \mathscr{M}_0$. $\square$

Theorem 5.5. *Let $(\mathscr{M}, L) \in \mathscr{W}_k$ be Γ-spectral. Then there is a neighborhood of $(\mathscr{M}, L)$ in $\mathscr{W}$ consisting of Γ-spectral pairs from $\mathscr{W}_k$.*

Proof. This follows immediately from Theorem 5.1 and Lemma 5.4. $\square$

The following result is an analytic version of Theorem 5.5. Assume that a monic matrix polynomial $L(\lambda)$ has the following form:

$$L(\lambda, \mu) = I\lambda^l + \sum_{i=0}^{l-1} A_i(\mu)\lambda^i, \tag{5.6}$$

where the coefficients $A_i(\mu)$, $i = 0, \ldots, l-1$ are analytic matrix-valued functions on the parameter μ, defined in a connected domain Ω of the complex plane.

Theorem 5.6. *Let $L(\lambda, \mu)$ be a monic matrix polynomial of the form* (5.6) *with coefficients analytic in Ω. Assume that, for each $\mu \in \Omega$, there is a separated part of the spectrum of L, say σ_μ, and a corresponding Γ_μ-spectral divisor $L_1(\lambda, \mu)$ of $L(\lambda, \mu)$. Assume also that, for each $\mu_0 \in \Omega$, there is a neighborhood $U(\mu_0)$ such that, for each $\mu \in U(\mu_0)$, σ_μ is inside Γ_{μ_0}.*

Then the coefficients of $L_1(\lambda, \mu)$ and of the corresponding left quotient $L_2(\lambda, \mu)$ are analytic matrix functions of μ in Ω.

Proof. The hypotheses imply that, for $\mu \in U(\mu_0)$, the Riesz projector $P(\mu)$ can be written

$$P(\mu) = \frac{1}{2\pi i} \int_{\Gamma_\mu} \{I\lambda - C_{L(\cdot,\mu)}\}^{-1}\, d\lambda = \frac{1}{2\pi i} \int_{\Gamma_{\mu 0}} \{I\lambda - C_{L(\cdot,\mu)}\}^{-1}\, d\lambda$$

from which we deduce the analytic dependence of $P(\mu)$ on μ in $U(\mu_0)$. Let $\mathcal{M}(\mu) = \operatorname{Im} P(\mu)$ be the corresponding supporting subspace of $L(\lambda, \mu)$. By Theorem S6.1, there exists a basis $x_1(\mu), \ldots, x_q(\mu)$ $(q = nk)$ in $\mathcal{M}(\mu)$ for every $\mu \in \Omega$, such that $x_i(\mu)$, $i = 1, \ldots, q$ are analytic vector functions in $\mu \in \Omega$. Now formulas (5.2) and (5.3) (where $Q_k|_{\mathcal{M}}$ is considered as matrix representation in the basis $x_1(\mu), \ldots, x_q(\mu)$) show that $L_1(\lambda, \mu)$ is analytic for $\mu \in \Omega$. The transpose $(L_2(\lambda, \mu))^{\mathrm{T}}$ to the left quotient is the right spectral divisor of $(L(\lambda, \mu))^{\mathrm{T}}$ corresponding to the separate part $\sigma(L)\backslash\sigma_\mu$ of $\sigma(L)$. Applying the above argument, we find that $(L_2(\lambda, \mu))^{\mathrm{T}}$, and therefore also $L_2(\lambda, \mu)$, is an analytic function of μ. □

5.3. Stable Factorizations

In Section 5.2 it was proved that small perturbations of a polynomial lead to small pertubations in any spectral divisor. In this section we study and describe the class of *all* divisors which are stable under small perturbations. We shall measure small perturbations in terms of the metric σ_k defined by (5.1).

Suppose L, L_1, and L_2 are monic $n \times n$ matrix polynomials of (positive) degree l, r, and q, respectively, and assume $L = L_2 L_1$. We say that this factorization is *stable* if, given $\varepsilon > 0$, there exists $\delta > 0$ with the following property: if $L' \in \mathscr{P}_l$ and $\sigma_l(L', L) < \delta$, then L' admits a factorization $L' = L'_2 L'_1$ such that $L'_1 \in \mathscr{P}_r$, $L'_2 \in \mathscr{P}_q$ and

$$\sigma_r(L'_1, L_1) < \varepsilon, \qquad \sigma_q(L'_2, L_2) < \varepsilon.$$

The aim of this section is to characterize stability of a factorization in terms of the supporting subspace corresponding to it. Theorem 5.5 states that any *spectral* divisor is stable. However, the next theorems show that there also exist stable divisors which are not spectral.

First, we shall relate stable factorization to the notion of stable invariant subspaces (see Section S4.5).

Theorem 5.7. *Let L, L_1, and L_2 be monic $n \times n$ matrix polynomials and assume $L = L_2 L_1$. This factorization is stable if and only if the corresponding supporting subspace (for the companion matrix C_L of L) is stable.*

Proof. Let l be the degree of L and r that of L_1, and put $C = C_L$. Denote the supporting subspace corresponding to the factorization $L = L_2 L_1$ by $\mathscr{M}$. If $\mathscr{M}$ is stable for C then, using Theorem 5.2, one shows that $L = L_2 L_1$ is a stable factorization.

Now conversely, suppose the factorization is stable, but $\mathscr{M}$ is not. Then there exists $\varepsilon > 0$ and a sequence of matrices $\{C_m\}$ converging to C such that, for all $\mathscr{V} \in \Omega_m$,

$$\theta(\mathscr{V}, \mathscr{M}) \geq \varepsilon, \qquad m = 1, 2, \ldots. \tag{5.7}$$

Here $\{\Omega_m\}$ denotes the collection of all invariant subspaces for C_m. Put $Q = \operatorname{col}(\delta_{i1} I)_{i=1}^{l}$ and

$$S_m = \operatorname{col}(Q C_m^{i-1})_{i=1}^{l}, \qquad m = 1, 2, \ldots.$$

Then $\{S_m\}$ converges to $\operatorname{col}(QC^{i-1})_{i=1}^{l}$, which is equal to the unit $nl \times nl$ matrix. So without loss of generality we may assume that S_m is nonsingular for all m, say with inverse $S_m^{-1} = \operatorname{row}(U_{mi})_{i=1}^{l}$. Note that

$$U_{mi} \to \operatorname{col}(\delta_{ji} I)_{i=1}^{l}, \qquad i = 1, \ldots, l. \tag{5.8}$$

A straightforward calculation shows that $S_m C_m S_m^{-1}$ is the companion matrix associated with the monic matrix polynomial

$$L_m(\lambda) = \lambda^l I - \sum_{i=0}^{l-1} \lambda^i Q C_m^l U_{m,i+1}.$$

From (5.8) and the fact that $C_m \to C$ it follows that $\sigma_l(L_m, L) \to 0$. But then we may assume that for all m the polynomial L_m admits a factorization $L_m = L_{m2} L_{m1}$ with $L_{m1} \in \mathscr{P}_r$, $L_{m2} \in \mathscr{P}_{l-r}$, and

$$\sigma_r(L_{m1}, L_1) \to 0, \qquad \sigma_{l-r}(L_{m2}, L_2) \to 0.$$

Let $\mathcal{M}_m$ be the supporting subspace corresponding to the factorization $L_m = L_{m2}L_{m1}$. By Theorem 5.2 we have $\theta(\mathcal{M}_m, \mathcal{M}) \to 0$. Put $\mathcal{V}_m = S_m^{-1}\mathcal{M}_m$. Then $\mathcal{V}_m$ is an invariant subspace for C_m. In other words $\mathcal{V}_m \in \Omega_m$. Moreover, it follows from $S_m \to I$ that $\theta(\mathcal{V}_m, \mathcal{M}_m) \to 0$. (This can be verified easily by using, for example, equality (S4.12).) But then $\theta(\mathcal{V}_m, \mathcal{M}) \to 0$. This contradicts (5.7), and the proof is complete. □

We can now formulate the following criterion for stable factorization.

Theorem 5.8. *Let L, L_1, and L_2 be monic $n \times n$ matrix polynomials and assume $L = L_2L_1$. This factorization is stable if and only if for each common eigenvalue λ_0 of L_1 and L_2 we have* dim Ker $L(\lambda_0) = 1$.

In particular, it follows from Theorem 5.9 that for a spectral divisor L_1 of L the factorization $L = L_2L_1$ is stable. This fact can also be deduced from Theorems 5.5 and 5.2.

The proof of Theorem 5.8 is based on the following lemma.

Lemma 5.9. *Let*

$$A = \begin{bmatrix} A_1 & A_0 \\ 0 & A_2 \end{bmatrix}$$

be a linear transformation from $\mathcal{C}^m$ into $\mathcal{C}^m$, written in matrix form with respect to the decomposition $\mathcal{C}^m = \mathcal{C}^{m_1} \oplus \mathcal{C}^{m_2}$ ($m_1 + m_2 = m$). Then $\mathcal{C}^{m_1}$ is a stable invariant subspace for A if and only if for each common eigenvalue λ_0 of A_1 and A_2 the condition dim Ker$(\lambda_0 I - A) = 1$ *is satisfied.*

Proof. It is clear that $\mathcal{C}^{m_1}$ is an invariant subspace for A. We know from Theorem S4.9 that $\mathcal{C}^{m_1}$ is stable if and only if for each Riesz projector P of A corresponding to an eigenvalue λ_0 with dim Ker$(\lambda_0 I - A) \geq 2$, we have $P\mathcal{C}^{m_1} = 0$ or $P\mathcal{C}^{m_1} = \operatorname{Im} P$.

Let P be a Riesz projector of A corresponding to an arbitrary eigenvalue λ_0. Also for $i = 1, 2$ let P_i be the Riesz projector associated with A_i and λ_0. Then, for ε positive and sufficiently small

$$P = \begin{pmatrix} P_1 & \dfrac{1}{2\pi i}\displaystyle\int_{|\lambda-\lambda_0|=\varepsilon} (I\lambda - A_1)^{-1}A_0(I\lambda - A_2)^{-1}\,d\lambda \\ 0 & P_2 \end{pmatrix}.$$

Observe that the Laurent expansion of $(I\lambda - A_i)^{-1}$ $(i = 1, 2)$ at λ_0 has the form

$$(I\lambda - A_i)^{-1} = \sum_{j=-1}^{-q} (\lambda - \lambda_0)^j P_i Q_{ij} P_i + \cdots, \qquad i = 1, 2, \tag{5.9}$$

where Q_{ij} are some linear transformations of $\operatorname{Im} P_i$ into itself and the ellipsis on the right-hand side of (5.9) represents a series in nonnegative powers of $(\lambda - \lambda_0)$. From (5.9) one sees that P has the form

$$P = \begin{pmatrix} P_1 & P_1 Q_1 + Q_2 P_2 \\ 0 & P_2 \end{pmatrix}$$

where Q_1 and Q_2 are certain linear transformations acting from $\mathbb{C}^{m_2}$ into $\mathbb{C}^{m_1}$. It follows that $P\mathbb{C}^{m_1} \neq \{0\} \neq \operatorname{Im} P$ if and only if $\lambda_0 \in \sigma(A_1) \cap \sigma(A_2)$. Now appeal to Theorem S4.9 (see first paragraph of the proof) to finish the proof of Lemma 5.9. □

Proof of Theorem 5.8. Let $\mathcal{M}$ be the supporting subspace corresponding to the factorization $L = L_2 L_1$. From Theorem 5.7 we know that this factorization is stable if and only if $\mathcal{M}$ is a stable invariant subspace for C_L. Let l be the degree of L, let r be the degree of L_1, and let

$$\mathcal{N}_r = \{(x_1, \ldots, x_l) \in \mathbb{C}^{nl} \mid x_1 = \cdots = x_r = 0\}.$$

Then $\mathbb{C}^{nl} = \mathcal{M} \oplus \mathcal{N}_r$.

With respect to this decomposition we write C_L in the form

$$C_L = \begin{pmatrix} C_1 & * \\ 0 & C_2 \end{pmatrix}.$$

From Corollaries 3.14 and 3.19 it is known that $\sigma(L_1) = \sigma(C_1)$ and $\sigma(L_2) = \sigma(C_2)$. The desired result is now obtained by applying Lemma 5.9. □

5.4. Global Analytic Perturbations: Preliminaries

In Sections 5.1 and 5.2 direct advantage was taken of the explicit dependence of the companion matrix C_L for a monic matrix polynomial $L(\lambda)$ on the coefficients of $L(\lambda)$. Thus the appropriate standard pair for that analysis was $([I \quad 0 \quad \cdots \quad 0], C_L)$. However, this standard pair is used at the cost of leaving the relationship of divisors with invariant subspaces obscure and, in particular, giving no direct line of attack on the *continuation* of divisors from a point to neighboring points.

In this section we shall need more detailed information on the behavior of supporting subspaces and, for this purpose, Jordan pairs X_μ, J_μ will be used. Here the linearization J_μ is relatively simple and its invariant subspaces are easy to describe. Let $A_0(\mu), \ldots, A_{l-1}(\mu)$ be analytic functions on a connected domain Ω in the complex plane taking values in the linear space of complex $n \times n$ matrices. Consider the matrix-valued function

$$L(\lambda, \mu) = I\lambda^l + \sum_{i=0}^{l-1} A_i(\mu)\lambda^i. \tag{5.10}$$

Construct for each $\mu \in \Omega$ a Jordan pair (X_μ, J_μ) for $L(\lambda, \mu)$. The matrix J_μ is in Jordan normal form and will be supposed to have r Jordan blocks $J_\mu^{(i)}$ of size q_i, $i = 1, \ldots, r$ with associated eigenvalues $\lambda_i(\mu)$ (not necessarily distinct). In general, r and q_i depend on μ. We write $J_\mu = \operatorname{diag}[J_\mu^{(1)}, \ldots, J_\mu^{(r)}]$. Partition X_μ as

$$X_\mu = [X_\mu^{(1)} \quad \cdots \quad X_\mu^{(r)}],$$

where $X_\mu^{(i)}$ has q_i columns and each submatrix represents one Jordan chain.

Since (X_μ, J_μ) is also a standard pair for $L(\lambda, \mu)$, and in view of the divisibility results (Theorem 3.12), we may couple the study of divisors of $L(\lambda, \mu)$ with the existence of certain invariant subspaces of J_μ. We are now to determine the nature of the dependence of divisors and corresponding subspaces on μ via the fine structure of the Jordan pairs (X_μ, J_μ).

In order to achieve our objective, some important ideas developed in the monograph of Baumgärtel [5] are needed. The first, a "global" theorem concerning invariance of the Jordan structure of J_μ follows from [5, Section V.7, Theorem 1]. It is necessary only to observe that J_μ is also a Jordan form for the companion matrix

$$C_L(\mu) = \begin{bmatrix} 0 & I & 0 & \cdots & 0 \\ 0 & 0 & I & & 0 \\ \vdots & \vdots & \vdots & \vdots & \vdots \\ 0 & 0 & & 0 & I \\ -A_0(\mu) & -A_2(\mu) & & \cdots & -A_{l-1}(\mu) \end{bmatrix} \tag{5.11}$$

to obtain

Proposition 5.10. *The matrix-valued functions X_μ, J_μ can be defined on Ω in such a way that, for some countable set S_1 of isolated points in Ω, the following statements hold*:

(a) *For every $\mu \in \Omega \backslash S_1$ the number r of Jordan blocks in J_μ and their sizes $q_1, \ldots, q_r$ are independent of μ.*

(b) *The eigenvalues $\lambda_i(\mu)$, $i = 1, \ldots, r$, are analytic functions in $\Omega \backslash S_1$ and may have algebraic branch points at some points of S_1.*

(c) *The blocks $X_\mu^{(i)}$, $i = 1, \ldots, r$, of X_μ are analytic functions of μ on $\Omega \backslash S_1$ which may also be branches of analytic functions having algebraic branch points at some points of S_1.*

The set S_1 is associated with Baumgärtel's *Hypospaltpunkte* and consists of points of discontinuity of J_μ in Ω as well as all branch points associated with the eigenvalues.

Let $\hat{\lambda}_j(\mu)$, $j = 1, 2, \ldots, t$, denote all the *distinct* eigenvalue functions defined on $\Omega \backslash S_1$ (which are assumed analytic in view of Proposition 5.10), and let

$$S_2 = \{\mu \in \Omega \backslash S_1 \mid \hat{\lambda}_i(\mu) = \hat{\lambda}_j(\mu) \text{ for some } i \neq j\}.$$

The Jordan matrix J_μ then has the same set of invariant subspaces for every $\mu \in \Omega \backslash (S_1 \cup S_2)$. (To check this, use Proposition S4.4 and the fact that, for $\alpha, \beta \in \mathbb{C}$, matrices $J_\mu - \alpha I$ and $J_\mu - \beta I$ have the same invariant subspaces.) The set S_2 consists of multiple points (*mehrfache Punkte*) in the terminology of Baumgärtel. The set $S_1 \cup S_2$ is described as the *exceptional set* of $L(\lambda, \mu)$ in Ω and is, at most, countable (however, see Theorem 5.11 below), having its limit points (if any) on the boundary of Ω.

EXAMPLE 5.1. Let $L(\lambda, \mu) = \lambda^2 + \mu\lambda + w^2$ for some constant w, and $\Omega = \mathbb{C}$. Then

$$J_{\pm 2w} = \begin{bmatrix} \pm w & 1 \\ 0 & \pm w \end{bmatrix}$$

and, for $\mu \neq \pm 2w$, $J_\mu = \operatorname{diag}[\lambda_1(\mu), \lambda_2(\mu)]$ where $\lambda_{1,2}$ are the zeros of $\lambda^2 + \mu\lambda + w^2$. Here, $S_1 = \{2w, -2w\}$, $S_2 = \phi$. □

EXAMPLE 5.2. Let

$$L(\lambda, \mu) = \begin{bmatrix} (\lambda - 1)(\lambda - \mu) & 0 \\ \mu & (\lambda - 2)(\lambda - \mu^2) \end{bmatrix}$$

and $\Omega = \mathbb{C}$. The canonical Jordan matrix is found for every $\mu \in \mathbb{C}$ and hence the sets S_1 and S_2. For $\mu \notin \{0, 2, \pm 1, \pm\sqrt{2}\}$,

$$J_\mu = \operatorname{diag}\{\mu, \mu^2, 1, 2\}.$$

Then

$$J_0 = \operatorname{diag}[0, 0, 1, 2], \qquad J_2 = \operatorname{diag}\left\{\begin{bmatrix} 2 & 1 \\ 0 & 2 \end{bmatrix}, 1, 4\right\},$$

$$J_{-1} = \operatorname{diag}\left\{\begin{bmatrix} 1 & 1 \\ 0 & 1 \end{bmatrix}, -1, 2\right\}, \quad J_1 = \operatorname{diag}\left\{\begin{bmatrix} 1 & 1 & 0 \\ 0 & 1 & 1 \\ 0 & 0 & 1 \end{bmatrix}, 2\right\},$$

$$J_{\pm\sqrt{2}} = \operatorname{diag}\left\{\begin{bmatrix} 2 & 1 \\ 0 & 2 \end{bmatrix}, 1, \pm\sqrt{2}\right\}.$$

It follows that $S_1 = \{\pm 1, 2, \pm\sqrt{2}\}$, $S_2 = \{0\}$ and, for $\mu \in \Omega \backslash S_1$,

$$X_\mu = \begin{bmatrix} (\mu - 2)(\mu - 1) & 0 & 1 - \mu^2 & 0 \\ 1 & 1 & \mu & 1 \end{bmatrix}. \quad \square$$

5.5. Polynomial Dependence

In the above examples it turns out that the exceptional set $S_1 \cup S_2$ is finite. The following result shows it will always be this case provided the coefficients of $L(\lambda, \mu)$ are *polynomials* in μ.

Theorem 5.11. *Let $L(\lambda, \mu)$ be given by* (5.10) *and suppose that $A_i(\mu)$ are matrix polynomials in μ ($\Omega = \mathbb{C}$), $i = 0, \ldots, l-1$. Then the set of exceptional points of $L(\lambda, \mu)$ is finite.*

Proof. Passing to the linearization (with companion matrix) we can suppose $L(\lambda, \mu)$ is linear: $L(\lambda, \mu) = I\lambda - C(\mu)$. Consider the scalar polynomial

$$f(\lambda, \mu) = \det(I\lambda - C(\mu)) = \lambda^n + a_{n-1}(\mu)\lambda^{n-1} + \cdots + a_0(\mu)$$

where n is the size of $C(\mu)$, and $a_j(\mu)$ are polynomials in μ. The equation

$$f(\lambda, \mu) = 0 \tag{5.12}$$

determines n zeros $\lambda_1(\mu), \ldots, \lambda_n(\mu)$. Then for μ in $\mathbb{C}\backslash S_2$ (S_3 is some finite set) the number of different zeros of (5.12) is constant. Indeed, consider the matrix (of size $(2n-1) \times (2n-1)$ and in which $a_0, a_1, \ldots, a_{n-1}$ depends on μ)

$$\begin{bmatrix}
1 & a_{n-1} & a_{n-2} & \cdots & a_1 & a_0 & 0 & \cdots & 0 \\
0 & 1 & a_{n-1} & \cdots & a_2 & a_1 & a_0 & & \vdots \\
\vdots & \vdots & \ddots & \vdots & \vdots & \vdots & \vdots & \ddots & 0 \\
0 & & & 0 \quad 1 & a_{n-1} & a_{n-2} & a_{n-3} & & a_0 \\
n & (n-1)a_{n-1} & (n-2)a_{n-2} & \cdots & a_1 & 0 & 0 & \cdots & 0 \\
0 & n & (n-1)a_{n-1} & \cdots & 2a_2 & a_1 & 0 & & \\
\vdots & \vdots & \ddots & \vdots & \vdots & \vdots & \vdots & & \vdots \\
 & & & & & & & \ddots & 0 \\
0 & & \cdots & 0 & n & (n-1)a_{n-1} & & \cdots & a_1
\end{bmatrix}. \tag{5.13}$$

The matrix $M(\mu)$ is just the resultant matrix of $f(\lambda, \mu)$ and $\partial f(\lambda, \mu)/\partial\lambda$, considered as polynomials in λ with parameter μ, and det $M(\mu)$ is the resultant determinant of $f(\lambda, \mu)$ and $\partial f(\lambda, \mu)/\partial\lambda$. (See, for instance [78, Chapter XII] for resultants and their basic properties.) We shall need the following property of $M(\mu)$:

$$\operatorname{rank} M(\mu) = 2n + 1 - \pi(\mu),$$

where $\pi(\mu)$ is the number of common zeros (counting multiplicities) of $f(\lambda, \mu)$ and $\partial f(\lambda, \mu)/\partial\lambda$ (see [26]). On the other hand, $n = \pi(\mu) + \sigma(\mu)$, where $\sigma(\mu)$ is the number of different zeros of (5.12); so

$$\operatorname{rank} M(\mu) = n + 1 + \sigma(\mu). \tag{5.14}$$

Since $C(\mu)$ is a polynomial in μ, we have

$$\operatorname{rank} M(\mu) = \max_{\mu \in \mathbb{C}} \operatorname{rank} M(\mu), \qquad \mu \in \mathbb{C} \backslash S_3, \tag{5.15}$$

where S_3 is a finite set. Indeed, let $\mu_0 \in \mathbb{C}$ be such that

$$\operatorname{rank} M(\mu_0) = \max_{\mu \in \mathbb{C}} \operatorname{rank} M(\mu) \stackrel{\text{def}}{=} r,$$

and let $M_0(\mu)$ be an $r \times r$ submatrix of $M(\mu)$ such that $\det M_0(\mu_0) \neq 0$. Now $\det M_0(\mu)$ is a nonzero polynomial in μ, and therefore has only finitely many zeros. Since $\operatorname{rank} M(\mu) = r$ provided $\det M_0(\mu) \neq 0$, equality (5.15) follows. By (5.14) the number of different zeros of (5.12) is constant for $\mu \in \mathbb{C} \backslash S_3$. Moreover, $\sigma(\mu_1) < \sigma(\mu_2)$ for $\mu_1 \in S_3$ and $\mu_2 \in \mathbb{C} \backslash S_3$.

Consider now the decomposition of $f(\lambda, \mu)$ into the product of irreducible polynomials

$$f(\lambda, \mu) = f_1(\lambda, \mu) \cdots f_m(\lambda, \mu), \tag{5.16}$$

where $f_i(\lambda, \mu)$ are polynomials on λ whose coefficients are polynomials in μ and the leading coefficient is equal to 1; moreover, $f_i(\lambda, \mu)$ are irreducible. Consider one of the factors in (5.16), say, $f_1(\lambda, \mu)$. It is easily seen from the choice of S_3 that the number of different zeros of $f_1(\lambda, \mu)$ is constant for $\mu \in \mathbb{C} \backslash S_3$. Let $\lambda_1(\mu), \ldots, \lambda_s(\mu)$, $\mu \in \mathbb{C} \backslash S_3$ be all the different zeros of equation $f_1(\lambda, \mu) = 0$. For given $\lambda_i(\mu)$, $i = 1, \ldots, s$ and for given $j = 1, 2, \ldots$, denote

$$v_{ij}(\mu) = \operatorname{rank}(\lambda_i(\mu)I - C(\mu))^j \qquad \text{for} \quad \mu \in \mathbb{C} \backslash S_3.$$

Let us prove that $v_{ij}(\mu)$ is constant for $\mu \in (\mathbb{C} \backslash S_3) \backslash S_{ij}$, where S_{ij} is a finite set. Indeed, let

$$v_{ij} = \max_{\mu \in \mathbb{C} \backslash S_3} \operatorname{rank}\, (\lambda_i(\mu)I - C(\mu))^j,$$

and let $\mu_0 = \mu_0(i, j) \in \mathbb{C} \setminus S_3$ be such that $v_{ij}(\mu_0) = v_{ij}$. Let $A_{ij}(\lambda, \mu)$ be a square submatrix of $(I\lambda - C(\mu))^j$ (of size $v_{ij} \times v_{ij}$) such that $\det A_{ij}(\lambda_i(\mu_0), \mu_0) \neq 0$. We claim that

$$\det A_{ij}(\lambda_k(\mu), \mu) \not\equiv 0 \tag{5.17}$$

for $k = 1, 2, \ldots, s$ (so in fact v_{ij} does not depend on i). To check (5.17) we need the following properties of the zeros $\lambda_1(\mu), \ldots, \lambda_s(\mu)$: (1) $\lambda_i(\mu)$ are analytic in $\mathbb{C} \setminus S_3$; (2) $\lambda_1(\mu), \ldots, \lambda_s(\mu)$ are the branches of the same multiple-valued analytic function in $\mathbb{C} \setminus S_3$; i.e., for every $\mu \in \mathbb{C} \setminus S_3$ and for every pair $\lambda_i(\mu)$, $\lambda_j(\mu)$ of zeros of $f_1(\lambda, \mu)$ there exists a closed rectifiable contour $\Gamma \subset \mathbb{C} \setminus S_3$ with initial and terminal point μ such that, after one complete turn along Γ, the branch $\lambda_i(\mu)$ becomes $\lambda_j(\mu)$. Property (1) follows from Proposition 5.10(b) taking into account that the number of different zeros of $f_1(\lambda, \mu)$ is constant in $\mathbb{C} \setminus S_3$. Property (2) follows from the irreducibility of $f_1(\lambda, \mu)$ (this is a well-known fact; see, for instance [63, Vol. III, Theorem 8.22]). Now suppose (5.17) does not hold for some k; so $\det A_{ij}(\lambda_k(\mu), \mu) \equiv 0$. Let Γ be a contour in $\mathbb{C} \setminus S_3$ such that $\mu_0 \in \Gamma$ and after one complete turn (with initial and terminal point μ_0) along Γ, the branch $\lambda_k(\mu)$ becomes $\lambda_i(\mu)$. Then by analytic continuation along Γ we obtain $\det A_{ij}(\lambda_i(\mu_0), \mu_0) = 0$, a contradiction with the choice of μ_0. So (5.17) holds, and, in particular, there exists $\mu_0' = \mu_0'(i, j) \in \mathbb{C} \setminus S_3$ such that

$$\det A_{ij}(\lambda_k(\mu_0'), \mu_0') \neq 0, \qquad k = 1, \ldots, s. \tag{5.18}$$

(For instance, μ_0' can be chosen in a neighborhood of μ_0.)

Consider now the system of two scalar polynomial equations

$$\begin{aligned} f_1(\lambda, \mu) &= 0, \\ \det A_{ij}(\lambda, \mu) &= 0. \end{aligned} \tag{5.19}$$

(Here i and j are assumed to be fixed as above.) According to (5.18), for $\mu = \mu_0'$ this system has no solution. Therefore the resultant $R_{ij}(\mu)$ of $f_1(\lambda, \mu)$ and $\det A_{ij}(\lambda, \mu)$ is not identically zero (we use here the following property of $R_{ij}(\mu)$ (see [78, Chapter XII]): system (5.19) has a common solution λ for fixed μ if and only if $R_{ij}(\mu) = 0$). Since $R_{ij}(\mu)$ is a polynomial in μ, it has only a finite set S_{ij} of zeros. By the property of the resultant mentioned above,

$$v_{ij}(\mu) = v_{ij} \qquad \text{for every} \quad \mu \in (\mathbb{C} \setminus S_3) \setminus S_{ij},$$

as claimed.

Now let $T_1 = \bigcup_{i=1}^{s} \bigcup_{i=1}^{n} S_{ij}$, and let $T = \bigcup_{p=1}^{m} T_p$, where $T_2, \ldots, T_m$ are constructed analogously for the irreducible factors $f_2(\lambda, \mu), \ldots, f_m(\lambda, \mu)$, respectively, in (5.16). Clearly T is a finite set. We claim that the exceptional set $S_1 \cup S_2$ is contained in $S_3 \cup T$ (thereby proving Theorem 5.11). Indeed, from the construction of S_3 and T it follows that the number of different eigenvalues of $I\lambda - C(\mu)$ is constant for $\mu \in \mathcal{C}\setminus(S_3 \cup T)$, and for every eigenvalue $\lambda_0(\mu)$ (which is analytic in $\mathcal{C}\setminus(S_3 \cup T)$) of $I\lambda - C(\mu)$,

$$r(\mu; j) \stackrel{\text{def}}{=} \operatorname{rank}(\lambda_0(\mu)I - C(\mu))^j$$

is constant for $\mu \in \mathcal{C}\setminus(S_3 \cup T)$; $j = 0, 1, \ldots$. The sizes of the Jordan blocks in the Jordan normal form of $C(\mu)$ corresponding to $\lambda_0(\mu)$ are completely determined by the numbers $r(\mu; j)$; namely,

$$n - r(\mu; j) = \gamma_1(\mu) + \cdots + \gamma_j(\mu), \qquad j = 1, \ldots, n, \tag{5.20}$$

where $\gamma_j(\mu)$ is the number of the Jordan blocks corresponding to $\lambda_0(\mu)$ whose size is not less than j; $j = 1, \ldots, n$ (so $\gamma_1(\mu)$ is just the number of the Jordan blocks). Equality (5.20) is easily observed for the Jordan normal form of $C(\mu)$; then it clearly holds for $C(\mu)$ itself. Since $r(\mu; j)$ are constant for $\mu \in \mathcal{C}\setminus(S_3 \cup T)$, (5.20) ensures that the Jordan structure of $C(\mu)$ corresponding to $\lambda_0(\mu)$ is also constant for $\mu \in \mathcal{C}\setminus(S_3 \cup T)$. Consequently, the exceptional set of $I\lambda - C(\mu)$ is disjoint with $S_3 \cup T$. □

Let us estimate the number of exceptional points for $L(\lambda, \mu)$ as in Theorem 5.11, assuming for simplicity that $\det L(\lambda, \mu)$ is an irreducible polynomial. (We shall not aim for the best possible estimate.) Let m be the maximal degree of the matrix polynomials $A_0(\mu), \ldots, A_{l-1}(\mu)$. Then the degrees of the coefficients $a_i(\mu)$ of $f(\lambda, \mu) = \det(I\lambda - C(\mu))$, where $C(\mu)$ is the companion matrix of $L(\lambda, \mu)$, do not exceed mn. Consequently, the degree of any minor (not identically zero) of $M(\mu)$ (given by formula (5.13)) does not exceed $(2nl - 1)mn$. So the set S_3 (introduced in the proof of Theorem 5.11) contains not more than $(2nl - 1)mn$ elements. Further, the left-hand sides of Eq. (5.19) are polynomials whose degrees (as polynomials in λ) are nl (for $f_1(\lambda, \mu) = f(\lambda, \mu)$) and not greater than nlj (for $\det A_{ij}(\lambda, \mu)$). Thus, the resultant matrix of $f(\lambda, \mu)$ and $\det A_{ij}(\lambda, \mu)$ has size not greater than $nl + nlj = nl(j + 1)$. Further, the degrees of the coefficients of $f(\lambda, \mu)$ (as polynomials in μ) do not exceed mn, and the degrees of the coefficients of $\det A_{ij}(\lambda, \mu)$ (as polynomials on μ) do not exceed mnj. Thus, the resultant $R_{ij}(\mu)$ of $f(\lambda, \mu)$ and $\det A_{ij}(\lambda, \mu)$ (which is equal to the determinant of the resultant matrix) has degree less than or equal to $lmn^2j(j + 1)$. Hence the number of elements in S_{ij} (in the proof of Theorem 5.11) does not exceed $lmn^2j(j + 1)$. Finally, the set

$$S_3 \bigcup \bigcup_{i=1}^{nl} \bigcup_{j=1}^{nl} S_{ij}$$

consists of not more than $(2nl - 1)mn + n^3l^2m \sum_{j=1}^{nl} (j + 1)j = (2nl - 1)mn + n^3l^2m(\frac{1}{3}n^3l^3 + n^2l^2 + \frac{2}{3}nl) = mn(\frac{1}{3}p^5 + p^4 + \frac{2}{3}p^3 + 2p - 1)$ points where $p = nl$. Thus for $L(\lambda, \mu)$ as in Theorem 5.11, the number of exceptional points does not exceed $mn(\frac{1}{3}p^5 + p^4 + \frac{2}{3}p^3 + 2p - 1)$, where m is the maximal degree of the matrix polynomials $A_i(\mu)$, $i = 0, \ldots, l-1$, and $p = nl$. Of course, this estimate is quite rough.

5.6. Analytic Divisors

As in Eq. (5.10), let $L(\lambda, \mu)$ be defined for all $(\lambda, \mu) \in \mathbb{C} \times \Omega$. Suppose that for some $\mu_0 \in \Omega$ the monic matrix polynomial $L(\lambda, \mu_0)$ has a right divisor $L_1(\lambda)$. The possibility of extending $L_1(\lambda)$ to a continuous (or an analytic) family of right divisors $L_1(\lambda, \mu)$ of $L(\lambda, \mu)$ is to be investigated. It turns out that this may not be possible, in which case $L_1(\lambda)$ will be described as an isolated divisor, and that this can only occur if μ_0 is in the exceptional set of L in Ω. In contrast, we have the following theorem.

Theorem 5.12. *If $\mu_0 \in \Omega\backslash(S_1 \cup S_2)$, then every monic right divisor $L_1(\lambda)$ of $L(\lambda, \mu_0)$ can be extended to an analytic family $L_1(\lambda, \mu)$ of monic right divisors of $L(\lambda, \mu)$ in the domain $\Omega\backslash(S_1 \cup S_3)$ where S_3 is an, at most, countable subset of isolated points of $\Omega\backslash S_1$ and $L_1(\lambda, \mu)$ has poles or removable singularities at the points of S_3.*

Proof. Let $\mathcal{M}$ be the supporting subspace of divisor $L_1(\lambda)$ of $L(\lambda, \mu_0)$ with respect to the Jordan matrix J_{μ_0}. Then, if L_1 has degree k, define

$$Q_k(\mu) = [\operatorname{col}(X_\mu J_\mu^i)_{i=0}^{k-1}]|_{\mathcal{M}}, \qquad \mu \in \Omega\backslash S_1.$$

It follows from the definition of a supporting subspace that $Q_k(\mu_0)$ is an invertible linear transformation. Since (by Proposition 5.10) $Q_k(\mu)$ is analytic on $\Omega\backslash S_1$ it follows that $Q_k(\mu)$ is invertible on a domain $(\Omega\backslash S_1)\backslash S_3$ where $S_3 = \{\mu \in \Omega\backslash S_1 \,|\, \det Q_k(\mu) = 0\}$ is, at most, a countable subset of isolated points of $\Omega\backslash S_1$. Furthermore, the J_{μ_0}-invariant subspace $\mathcal{M}$ is also invariant for every J_μ with $\mu \in \Omega\backslash S_1$. Indeed, we have seen in Section 5.4 that this holds for $\mu \in \Omega\backslash(S_1 \cup S_2)$. When $\mu \in S_2\backslash S_1$, use the continuity of J_μ. Hence by Theorem 3.12 there exists a family of right divisors $L_1(\lambda, \mu)$ of $L(\lambda, \mu)$ for each $\mu \in \Omega\backslash(S_1 \cup S_3)$, each divisor having the same supporting subspace $\mathcal{M}$ with respect to J_μ. By formula (3.27), it follows that this family is analytic in $\Omega\backslash(S_1 \cup S_3)$.

An explicit representation of $L_1(\lambda, \mu)$ is obtained in the following way (cf. formula (3.27)). Take a fixed basis in $\mathcal{M}$ and for each μ in $\Omega\backslash S_1$ represent $Q_k(\mu)$ as a matrix defined with respect to this basis and the standard basis for $\mathbb{C}^{kn}$. Then for $\mu \notin S_3$, define $ln \times n$ matrix-valued functions $W_1(\mu), \ldots, W_k(\mu)$ by

$$[W_1(\mu) \quad \cdots \quad W_k(\mu)] = RQ_k^{-1}(\mu),$$

where R is the matrix (independent of μ) representing the embedding of $\mathscr{M}$ into $\mathbb{C}^{ln}$. The divisor L_1 has the form

$$L_1(\lambda, \mu) = I\lambda^k - X_\mu J_\mu^k\{W_1(\mu) + W_2(\mu)\lambda + \cdots + W_k(\mu)\lambda^{k-1}\}.$$

The nature of the singularities of $L_1(\lambda, \mu)$ is apparent from this representation. □

An important special case of the theorem is

Corollary 5.13. *If* det $L(\lambda, \mu)$ *has* ln *distinct zeros for every* $\mu \in \Omega$, *and if* $\mu_0 \in \Omega$, *then every monic right divisor of* $L(\lambda, \mu_0)$ *can be extended to a family of monic right divisors for* $L(\lambda, \mu)$ *which is analytic on* $\Omega \backslash S_3$.

Proof. Under these hypotheses S_1 and S_2 are empty and the conclusion follows. □

EXAMPLE 5.3. Let

$$L(\lambda, \mu) = \begin{bmatrix} \lambda^2 & 0 \\ -\lambda\mu & (\lambda-1)^2 \end{bmatrix}.$$

Then a Jordan matrix for L does not depend on μ, i.e., for every $\mu \in \mathbb{C}$,

$$J_\mu = J = \text{diag}\left\{\begin{bmatrix} 0 & 1 \\ 0 & 0 \end{bmatrix}, \begin{bmatrix} 1 & 1 \\ 0 & 1 \end{bmatrix}\right\}.$$

We have S_1 and S_2 both empty and

$$X_\mu = \begin{bmatrix} 1 & 0 & 0 & 0 \\ 0 & \mu & 1 & 0 \end{bmatrix}.$$

The subspace $\mathscr{M}$ spanned by the first two component vectors e_1 and e_2 is invariant under J and

$$X_\mu|_{\mathscr{M}} = \begin{bmatrix} 1 & 0 \\ 0 & \mu \end{bmatrix}.$$

Thus, for $\mu \neq 0$, $\mathscr{M}$ is a supporting subspace for $L(\lambda, \mu)$. The corresponding divisor is $L_1(\lambda, \mu) = I\lambda - X_\mu J_\mu (X_\mu|_{\mathscr{M}})^{-1}$ and computation shows that

$$L_1(\lambda, \mu) = I\lambda - \begin{bmatrix} 0 & \mu^{-1} \\ 0 & 0 \end{bmatrix}.$$

So $S_3 = \{0\}$. □

Corollary 5.14. *If the divisor* $L_1(\lambda)$ *of Theorem* 5.12 *is, in addition, a* Γ*-spectral divisor for some closed contour* Γ *then it has a unique analytic extension to a family* $L_1(\lambda, \mu)$ *of monic right divisors defined on* $\Omega \backslash (S_1 \cup S_3)$.

Proof. Let $L_1(\lambda, \mu)$ be an extension of $L_1(\lambda)$; existence is guaranteed by Theorem 5.12. By Lemma 5.4, for $\mu \in \Omega$ close to μ_0, the divisor $L_1(\lambda, \mu)$ is Γ-spectral. In particular, $L_1(\lambda, \mu)$ is uniquely defined in a neighborhood of μ_0. Because of analyticity, $L_1(\lambda, \mu)$ is a unique extension of $L_1(\lambda)$ throughout $\Omega \backslash (S_1 \cup S_3)$. □

5.7. Isolated and Nonisolated Divisors

As before, let $L(\lambda, \mu)$ be a monic matrix polynomial of degree l with coefficients depending analytically on μ in Ω, and let $L_1(\lambda)$ be a monic right divisor of degree k of $L(\lambda, \mu_0)$, $\mu_0 \in \Omega$. The divisor $L_1(\lambda)$ is said to be *isolated* if there is a neighborhood $U(\mu_0)$ of μ_0 such that $L(\lambda, \mu)$ has no family of right monic divisors $L_1(\lambda, \mu)$ of degree k which (a) depends continuously on μ in $U(\mu_0)$ and (b) has the property that $\lim_{\mu \to \mu_0} L_1(\lambda, \mu) = L_1(\lambda)$.

Theorem 5.12 shows that if, in the definition, we have $\mu_0 \in \Omega \backslash (S_1 \cup S_2)$ then monic right divisors cannot be isolated. We demonstrate the existence of isolated divisors by means of an example.

EXAMPLE 5.4. Let $C(\mu)$ be any matrix depending analytically on μ in a domain Ω with the property that for $\mu = \mu_0 \in \Omega$, $C(\mu_0)$ has a square root and for $\mu \neq \mu_0$, μ in a neighborhood of μ_0, $C(\mu)$ has no square root. The prime example here is

$$\mu_0 = 0, \qquad C(\mu) = \begin{bmatrix} 0 & 0 \\ \mu & 0 \end{bmatrix}.$$

Then define $L(\lambda, \mu) = I\lambda^2 - C(\mu)$. It is easily seen that if $L(\lambda, \mu)$ has a right divisor $I\lambda - A(\mu)$ then $L(\lambda, \mu) = I\lambda^2 - A^2(\mu)$ and hence that $L(\lambda, \mu)$ has a monic right divisor if and only if $C(\mu)$ has a square root. Thus, under the hypotheses stated, $L(\lambda, \mu)$ has an isolated divisor at μ_0. □

It will now be shown that there are some cases where divisors which exist at points of the exceptional set can, nevertheless, be extended analytically. An example of this situation is provided by taking $\mu_0 = 0$ in Example 5.2.

Theorem 5.15. *Let $\mu_0 \in S_2$ and $L_1(\lambda)$ be a nonisolated right monic divisor of degree k of $L(\lambda, \mu_0)$. Then $L_1(\lambda)$ can be extended to an analytic family $L_1(\lambda, \mu)$ of right monic divisors of degree k of $L(\lambda, \mu)$.*

In effect, this result provides an extension of the conclusions of Theorem 5.12. The statements concerning the singularities also carry over.

Proof. Since $L_1(\lambda)$ is nonisolated, there is a sequence $\{\mu_m\}$ in $\Omega \backslash (S_1 \cup S_2)$ such that $\mu_m \to \mu_0$ and there are monic right divisors $L_1(\lambda, \mu_m)$ of $L(\lambda, \mu_m)$ such that $\lim_{m \to \infty} L_1(\lambda, \mu_m) = L_1(\lambda)$. Let $\mathcal{N}(\mu_m)$ be the supporting subspace of $L_1(\lambda, \mu_m)$ relative to J_{μ_m}. As the set of subspaces is compact (Theorem

S4.8), there is a subsequence $m_k, k = 1, 2, \ldots,$ such that $\lim_{k\to\infty} \mathcal{N}(\mu_{m_k}) = \mathcal{N}_0$ for some subspace $\mathcal{N}_0$. We can assume that $\lim_{m\to\infty} \mathcal{N}(\mu_m) = \mathcal{N}_0$. Since $\mathcal{N}(\mu_m)$ is invariant for J_{μ_m} and all invariant subspaces of J_μ are independent of μ in $\Omega\backslash(S_1 \cup S_2)$ it follows that $\mathcal{N}(\mu_0)$ is invariant for J_{μ_m} for each m.

From this point, the proof of Theorem 5.12 can be applied to obtain the required conclusion. □

The following special case is important:

Corollary 5.16. *If the elementary divisors of $L(\lambda, \mu)$ are linear for each fixed $\mu \in \Omega$, then every nonisolated right monic divisor $L_1(\lambda)$ of $L(\lambda, \mu_0)$ can be extended to an analytic family of right monic divisors for $L(\lambda, \mu)$ on $\Omega\backslash S_3$.*

Proof. In this case S_1 is empty and so the conclusion follows from Theorems 5.12 and 5.15. □

Comments

The presentation of Sections 5.1, 5.2, 5.4, 5.6, and 5.7 is based on the authors' paper [34e] (some of the results are given in [34e] in the context of infinite-dimensional spaces). The main results of Section 5.3 are taken from [3b]; see also [3c]. Section 5.5 is probably new.

The results concerning stable factorizations (Section 5.3) are closely related to stability properties of solutions of the matrix Riccati equation (see [3b] and [3c, Chapter 7]).

It will be apparent to the reader that Section 5.4 on global analytic perturbations leans very heavily on the work of Baumgärtel [5], who developed a comprehensive analytic perturbation theory for matrix polynomials.

Chapter 6

Extension Problems

Let $L(\lambda)$ be a monic matrix polynomial with Jordan pairs (X_i, J_i), $i = 1, \ldots, r$, corresponding to its different eigenvalues $\lambda_1, \ldots, \lambda_r$, respectively ($\sigma(L) = \{\lambda_1, \lambda_2, \ldots, \lambda_r\}$). We have seen in Chapter 2 that the pairs (X_i, J_i), $i = 1, \ldots, r$ are not completely independent. The only relationships between these pairs are contained in the requirement that the matrix

$$\begin{bmatrix} X_1 & X_2 & \cdots & X_r \\ X_1 J_1 & X_2 J_2 & \cdots & X_r J_r \\ \vdots & \vdots & & \vdots \\ X_1 J_1^{l-1} & X_2 J_2^{l-1} & \ldots & X_r J_r^{l-1} \end{bmatrix}$$

be nonsingular. Thus, for instance, the pairs $(X_1, J_1), \ldots, (X_{r-1}, J_{r-1})$ impose certain restrictions on (X_r, J_r).

In this chapter we study these restrictions more closely. In particular, we shall describe all possible Jordan matrices J_r which may occur in the pair (X_r, J_r).

It is convenient to reformulate this problem in terms of construction of a monic matrix polynomial given only part of its Jordan pair, namely, (X_1, J_1), $\ldots, (X_{r-1}, J_{r-1})$. It is natural to consider a generalization of this extension problem: construct a monic matrix polynomial if only part of its Jordan pair is given (in the above problem this part consists of the Jordan pairs corresponding to all but one eigenvalue of $L(\lambda)$). Clearly, one can expect many solutions (and it is indeed so). Hence we seek a solution with special properties. In this chapter we present a construction of a monic matrix polynomial, given part of its Jordan pair, by using left inverses.

6.1. Statement of the Problems and Examples

Let (X, J) be a pair of matrices where X is $n \times r$ and J is an $r \times r$ Jordan matrix. We would like to regard the pair (X, J) as a Jordan pair or, at least, a part of a Jordan pair of some monic matrix polynomial $L(\lambda)$.

Consider the problem more closely. We already know that (X, J) is a Jordan pair of a monic matrix polynomial of degree l if and only if $\operatorname{col}(XJ^i)_{i=0}^{l-1}$ is nonsingular; in particular, only if $r = nl$. If this condition is not satisfied (in particular, if r is not a multiple of n), we can still hope that (X, J) is a part of a Jordan pair $(\tilde{X}, \tilde{J})$ of some monic matrix polynomial $L(\lambda)$ of degree l. This means that for some $\tilde{J}$-invariant subspace $\mathscr{M} \subset \mathbb{C}^{nl}$ there exists an invertible linear transformation $S: \mathscr{M} \to \mathbb{C}^r$ such that $X = \tilde{X}|_{\mathscr{M}} S^{-1}$, $J = SJ|_{\mathscr{M}} S^{-1}$. In this case we shall say that (X, J) is a restriction of $(\tilde{X}, \tilde{J})$. (The notions of restriction and extension of Jordan pairs will be studied more extensively in Chapter 7.) Clearly, the property that (X, J) be a restriction of $(\tilde{X}, \tilde{J})$ does not depend on the choice of the Jordan pair $(\tilde{X}, \tilde{J})$ of $L(\lambda)$, but depends entirely on the monic matrix polynomial $L(\lambda)$ itself. So we can reformulate our problems as follows: given sets of Jordan chains and eigenvalues (represented by the pair (X, J)), is there a monic matrix polynomial $L(\lambda)$ such that this set is a part of a canonical set of the Jordan chains of $L(\lambda)$ (represented by the Jordan pair $(\tilde{X}, \tilde{J})$ of $L(\lambda)$)? If such an $L(\lambda)$ exists, how can we find it, and how can we find such an $L(\lambda)$ subject to different additional constraints?

This kind of question will be referred to as an *extension problem*, because it means that (X, J) is to be extended to a Jordan pair $(\tilde{X}, \tilde{J})$ for some monic polynomial.

Of course, we could replace (X, J) by a pair (X, T), where T is an $r \times r$ matrix (not necessarily Jordan), and ask for an extension of (X, T) to a standard pair $(\tilde{X}, \tilde{T})$ of some monic polynomial of degree l. For brevity, we shall say that $(\tilde{X}, \tilde{T})$ is a *standard pair of degree* l. It will be convenient for us to deal with the extension problem in this formally more general setting.

A necessary condition can be given easily:

Proposition 6.1. *Let X and T be matrices of sizes $n \times r$ and $r \times r$, respectively. Suppose that there exists an extension of (X, T) to a standard pair $(\tilde{X}, \tilde{T})$ of degree l. Then*

$$\operatorname{rank} \operatorname{col}(XT^i)_{i=0}^{l-1} = r. \tag{6.1}$$

Proof. Let $(\tilde{X}, \tilde{T})$ be a standard pair of the monic matrix polynomial $L(\lambda)$, and let $\mathscr{M}$ be a T-invariant subspace such that (X, T) is similar to

$(\tilde{X}|_{\mathcal{M}}, \tilde{T}|_{\mathcal{M}})$: $X = \tilde{X}|_{\mathcal{M}}S$, $T = S^{-1}\tilde{T}|_{\mathcal{M}}S$ for some invertible linear transformation $S: \mathbb{C}^r \to \mathcal{M}$ (such an $\mathcal{M}$ exists by the definition of extension). Clearly, it is enough to prove (6.1) for $(\tilde{X}_{\mathcal{M}}, \tilde{T}_{\mathcal{M}})$ in place of (X, T). Let $\mathcal{M}'$ be some direct complement to $\mathcal{M}$ in $\mathbb{C}^{nl}$, and let

$$\tilde{X} = [X_1 \quad X_2], \qquad \tilde{T} = \begin{bmatrix} T_{11} & T_{12} \\ T_{21} & T_{22} \end{bmatrix}$$

be the matrix representations of $\tilde{X}$ and $\tilde{T}$ with respect to the decomposition $\mathbb{C}^{nl} = \mathcal{M} \dotplus \mathcal{M}'$. Since $\mathcal{M}$ is $\tilde{T}$-invariant, we have $T_{21} = 0$. So

$$\begin{bmatrix} \tilde{X} \\ \tilde{X}\tilde{T} \\ \vdots \\ \tilde{X}\tilde{T}^{l-1} \end{bmatrix} = \begin{bmatrix} X_1 & \\ X_1 T_{11} & \\ \vdots & * \\ X_1 T_{11}^{l-1} & \end{bmatrix},$$

and since this matrix is nonsingular, the columns of $\operatorname{col}(X_1 T_{11}^i)_{i=0}^{l-1}$ are linearly independent. But $X_1 = \tilde{X}|_{\mathcal{M}}$, $T_{11} = \tilde{T}|_{\mathcal{M}}$, and (6.1) follows. □

For convenience, we introduce the following definitions: a pair of matrices (X, T), where X is $n \times r$ and T is $r \times r$ (the number r may vary from one pair to another; the number n is fixed) will be called an *admissible pair*. An admissible pair (X, T) satisfying condition (6.1) will be called *l-independent*. Thus, there exists an extension of the admissible pair (X, T) to a standard pair of degree l only if (X, T) is l-independent.

We shall see later that the condition (6.1) is also sufficient for the existence of an extension of (X, T) to a standard pair of degree l. First we present some examples.

EXAMPLE 6.1 (The scalar case). Let (X, J) be a pair of matrices with $1 \times r$ row X and $r \times r$ Jordan matrix J. Let $J = \operatorname{diag}[J_1, \ldots, J_p]$ be a decomposition of J into Jordan blocks and $X = [X_1 \ \cdots \ X_p]$ be the consistent decomposition of X. The necessary condition (6.1) holds for some l if and only if $\sigma(J_i) \cap \sigma(J_j) = \varnothing$ for $i \neq j$, and the leftmost entry in every X_i is different from zero. Indeed, suppose for instance that $\sigma(J_1) = \sigma(J_2)$, and let x_{11} and x_{12} be the leftmost entries in X_1 and X_2, respectively. If α_1 and α_2 are complex numbers (not both zero) such that $\alpha_1 x_{11} + \alpha_2 x_{12} = 0$, then the r-dimensional vector x containing α_1 in its first place, α_2 in its $(r_1 + 1)$th place (where r_1 is the size of J_1), and zero elsewhere has the property that

$$x \in \operatorname{Ker}(XJ^i), \qquad i = 0, 1, \ldots.$$

Thus, (6.1) cannot be satisfied. If, for instance, the left entry x_{11} of X_1 is zero, then

$$(1 \quad 0 \quad \cdots \quad 0)^{\mathrm{T}} \in \operatorname{Ker}(XJ^i), \qquad i = 0, 1, \ldots,$$

and (6.1) again does not hold.

On the other hand, if $\sigma(J_i) \cap \sigma(J_j) = \phi$ for $i \neq j$, and the leftmost entry in every X_i is different from zero, then the polynomial $\prod_{j=1}^{p} (\lambda - \lambda_j)^{r_j}$, where $\{\lambda_j\} = \sigma(J_j)$ and r_j is the size of J_j, has the Jordan pair (X, J). We leave to the reader the verification of this fact. It turns out that in the scalar case every admissible pair of matrices (X, T) which satisfies the necessary condition (6.1) is in fact a standard pair of a scalar polynomial. □

EXAMPLE 6.2 (Existence of an extension to a standard pair of a linear monic polynomial). Another case when the sufficiency of condition (6.1) can be seen directly is when (6.1) is satisfied with $l = 1$, i.e., the matrix X has independent columns (in particular, $r \leq n$). In this case it is easy to construct an extension $(\tilde{X}, \tilde{T})$ of (X, T) such that $(\tilde{X}, \tilde{T})$ is a standard pair of some monic matrix polynomial of degree 1. Namely, put for instance

$$\tilde{X} = [X \quad Y], \qquad \tilde{T} = \begin{bmatrix} T & 0 \\ 0 & 0 \end{bmatrix},$$

where Y is some $n \times (n - r)$ matrix such that $\tilde{X}$ is invertible.

Note that here in place of

$$\begin{bmatrix} T & 0 \\ 0 & 0 \end{bmatrix}$$

we could take as $\tilde{T}$ any $n \times n$ matrix provided its left upper corner of size $r \times r$ is T and the left lower corner is zero. So a solution of the extension problem here is far from *unique*. This fact leads to various questions on uniqueness of the solution of the general extension problem. Also, there arises a problem of classification and description of the set of solutions. □

In the next sections we shall give partial answers to these questions, by using the notion of left inverse.

6.2. Extensions via Left Inverses

Let (X, T) be an admissible l-independent pair of matrices and let r be the size of T. Then there exists an $r \times nl$ matrix U such that

$$U \operatorname{col}(XT^i)_{i=0}^{l-1} = I,$$

and, in general, U is not uniquely defined (see Chapter S3). Let

$$U = [U_1 \quad \cdots \quad U_l]$$

be the partition of U into blocks U_i, each of size $r \times n$. We can construct now a monic $n \times n$ matrix polynomial

$$L(\lambda) = I\lambda^l - XT^l(U_1 + U_2\lambda + \cdots + U_l\lambda^{l-1}). \tag{6.2}$$

Note that the polynomial $L(\lambda)$ is defined as if (X, T) were its standard pair replacing the inverse of $\operatorname{col}(XT^i)_{i=0}^{l-1}$ (which generally does not exist) by its left inverse U. The polynomial $L(\lambda)$ is said to be *associated* with the pair (X, T).

Theorem 6.2. *Let (X, T) be an admissible l-independent pair, and let $L(\lambda)$ be some monic matrix polynomial associated with (X, T). Then a standard pair (X_L, T_L) of $L(\lambda)$ is an extension of (X, T), i.e., from some T_L-invariant subspace $\mathcal{M}$ there exists an invertible linear transformation $S: \mathcal{M} \to \mathbb{C}^r$ such that $X = X_L|_{\mathcal{M}} S^{-1}$ and $T = ST_L|_{\mathcal{M}} S^{-1}$.*

Proof. Evidently, it is sufficient to prove Theorem 6.2 for the case when $X_L = [I \quad 0 \quad \cdots \quad 0]$ and T_L is the first companion matrix of $L(\lambda)$:

$$T_L = \begin{bmatrix} 0 & I & 0 & \cdots & 0 \\ 0 & 0 & I & \cdots & 0 \\ \vdots & \vdots & \vdots & & \vdots \\ 0 & 0 & 0 & \cdots & I \\ XT^lU_1 & XT^lU_2 & & \cdots & XT^lU_l \end{bmatrix}.$$

Let us check that $\mathcal{M} = \operatorname{Im} \operatorname{col}(XT^i)_{i=0}^{l-1}$ is T_L-invariant. Indeed, since

$$[U_1 \quad U_2 \quad \cdots \quad U_l]\operatorname{col}(XT^i)_{i=0}^{l-1} = I,$$

the equality

$$T_L \begin{bmatrix} X \\ XT \\ \vdots \\ XT^{l-1} \end{bmatrix} = \begin{bmatrix} X \\ XT \\ \vdots \\ XT^{l-1} \end{bmatrix} T \tag{6.3}$$

holds, so that T_L-invariance of $\mathcal{M}$ is clear. From (6.3) it follows also that for

$$S^{-1} = \operatorname{col}(XT^i)_{i=0}^{l-1}: \mathbb{C}^r \to \mathcal{M}$$

the conditions of Theorem 6.2 are satisfied. □

In connection with Theorem 6.2 the following question arises: is it true that every extension of (X, T) to a standard pair of a fixed degree l (such that (6.1) is satisfied for this l) can be obtained via some associated monic matrix polynomial, as in Theorem 6.2? In general the answer is no, as the following example shows.

EXAMPLE 6.3. Let

$$X = \begin{bmatrix} 1 & 0 & 1 \\ 0 & 0 & 1 \end{bmatrix}, \qquad T = \begin{bmatrix} 0 & 1 & 0 \\ 0 & 0 & 0 \\ 0 & 0 & 0 \end{bmatrix}.$$

Then the matrix

$$\begin{bmatrix} X \\ XT \end{bmatrix} = \begin{bmatrix} 1 & 0 & 0 \\ 0 & 0 & 1 \\ 0 & 1 & 0 \\ 0 & 0 & 0 \end{bmatrix}$$

has linearly independent columns, so the necessary condition (6.1) is satisfied. However, $XT^2 = 0$, and formula (6.2) gives

$$L(\lambda) = I\lambda^2. \tag{6.4}$$

So any extension via left inverses (as in Theorem 6.2) is in fact an extension to a standard pair $(\tilde{X}, \tilde{T})$ of the single polynomial (6.4). For instance,

$$\tilde{X} = \begin{bmatrix} 1 & 0 & 0 & 0 \\ 0 & 0 & 1 & 0 \end{bmatrix}, \qquad \tilde{T} = \begin{bmatrix} 0 & 1 & 0 & 0 \\ 0 & 0 & 0 & 0 \\ 0 & 0 & 0 & 1 \\ 0 & 0 & 0 & 0 \end{bmatrix}.$$

But there exist many other extensions of (X, T) to nonsimilar standard pairs of degree 2. For example, if $a \in \mathbb{C}\setminus\{0\}$, take

$$\tilde{X} = \begin{bmatrix} 1 & 0 & 0 & 0 \\ 0 & 0 & 1 & 1 \end{bmatrix}, \qquad \tilde{T} = \begin{bmatrix} 0 & 1 & 0 & 0 \\ 0 & 0 & 0 & 0 \\ 0 & 0 & 0 & 0 \\ 0 & 0 & 0 & a \end{bmatrix}. \qquad \square$$

On the other hand, it is easy to produce an example where every possible extension to a standard pair (of fixed degree) can be obtained via associated monic polynomials.

EXAMPLE 6.4. Let

$$X = \begin{bmatrix} 1 & 1 & 0 \\ 0 & 0 & 1 \end{bmatrix}, \qquad T = \begin{bmatrix} 0 & 0 & 0 \\ 0 & 1 & 0 \\ 0 & 0 & 1 \end{bmatrix}.$$

Then the columns of

$$\begin{bmatrix} X \\ XT \end{bmatrix} = \begin{bmatrix} 1 & 1 & 0 \\ 0 & 0 & 1 \\ 0 & 1 & 0 \\ 0 & 0 & 1 \end{bmatrix}$$

are linearly independent, i.e., $\left[\begin{smallmatrix} X \\ XT \end{smallmatrix}\right]$ is left invertible. The general form of its left inverse is

$$V = [V_1 \quad V_2] = \begin{bmatrix} 1 & a & -1 & -a \\ 0 & b & 1 & -b \\ 0 & c & 0 & 1-c \end{bmatrix},$$

where a, b, c are independent complex numbers. The associated monic polynomials $L(\lambda)$ are in the form

$$L(\lambda) = I\lambda^2 - XT^2(V_1 + V_2\lambda) = \begin{bmatrix} \lambda^2 - \lambda & b\lambda - b \\ 0 & \lambda^2 + (c-1)\lambda - c \end{bmatrix}.$$

As a Jordan pair $(\tilde{X}, \tilde{J})$ of $L(\lambda)$ for $c \neq -1, 0$ we can take

$$\tilde{X} = \tilde{X}(b, c) = \begin{bmatrix} 1 & 1 & 0 & b/c \\ 0 & 0 & 1 & 1 \end{bmatrix}, \tag{6.5}$$

$$\tilde{J} = \tilde{J}(c) = \operatorname{diag}(0, 1, 1, -c)$$

(we emphasize in the notation the dependence of $\tilde{X}$ on b and c and the dependence of $\tilde{J}$ on c). For $c = 0$ take

$$\tilde{X}(0, 0) = \begin{bmatrix} 1 & 1 & 0 & 0 \\ 0 & 0 & 1 & 1 \end{bmatrix}, \qquad \tilde{J}(0, 0) = \operatorname{diag}(0, 1, 1, 0) \tag{6.6}$$

if $b = 0$, and

$$\tilde{X} = \tilde{X}(b, 0) = \begin{bmatrix} 1 & 1 & 0 & 0 \\ 0 & 0 & 1 & -b^{-1} \end{bmatrix}, \qquad \tilde{J} = \tilde{J}(b, 0) = \begin{bmatrix} 0 & 0 & 0 & 1 \\ 0 & 1 & 0 & 0 \\ 0 & 0 & 1 & 0 \\ 0 & 0 & 0 & 0 \end{bmatrix}, \tag{6.7}$$

if $b \neq 0$. For $c = -1$, take

$$\tilde{X} = \tilde{X}(b, -1) = \begin{bmatrix} 1 & 1 & -b & 0 \\ 0 & 0 & 1 & 0 \end{bmatrix}, \qquad \tilde{J} = \tilde{J}(-1) = \begin{bmatrix} 0 & 0 & 0 & 0 \\ 0 & 1 & 0 & 0 \\ 0 & 0 & 1 & 1 \\ 0 & 0 & 0 & 1 \end{bmatrix}. \tag{6.8}$$

We show now that every standard extension (X', T') of (X, T) of degree 2 is similar to one of the pairs $(\tilde{X}, \tilde{J})$ given in (6.5)–(6.8), i.e., there exists an invertible 4×4 matrix S (depending on (X', T')) such that $X' = \tilde{X}S$, $T' = S^{-1}\tilde{J}S$. Several cases can occur.

(1) $\det(I\lambda - T')$ has a zero different from 0 and 1. In this case T' is similar to $\tilde{J}(c)$ from (6.5), and without loss of generality we can suppose that $\tilde{J}(c) = T'$ for some $c \neq 0$, -1. Further, since (X', T') is an extension of (X, T), the matrix X' can be taken in the following form:

$$X' = \begin{bmatrix} 1 & 1 & 0 & \alpha_1 \\ 0 & 0 & 1 & \alpha_2 \end{bmatrix}, \qquad \alpha_1, \alpha_2 \in \mathbb{C}.$$

Necessarily $\alpha_2 \neq 0$ (otherwise $\left[\begin{smallmatrix} X' \\ X'T' \end{smallmatrix}\right]$ is not invertible), and then clearly (X', T') is similar to $(\tilde{X}(b, c), \tilde{J}(c))$ from (6.5) with $b = \alpha_1\alpha_2^{-1}c$.

(2) $\det(I\lambda - T') = \lambda^2(\lambda - 1)^2$. Then one can check that (X', T') is similar to $(\tilde{X}, \tilde{J})$ from (6.6) or (6.7) according as T' has two or one lineraly independent eigenvector(s) corresponding to the eigenvalue zero.

(3) $\det(I\lambda - T') = \lambda(\lambda - 1)^3$. In this case (X', T') is similar to $(\tilde{X}, \tilde{J})$ from (6.8). □

The following theorem allows us to recognize the cases when every extension to a standard pair can be obtained via associated monic matrix polynomials.

Theorem 6.3. *Let (X, T) be an admissible l-independent pair of matrices. Then every standard pair (X', T') of degree l, which is an extension of (X, T), is also a standard pair of some monic matrix polynomial of degree l associated with (X, T), if and only if the rows of XT^l are linearly independent.*

Proof. Let (X', T') be a standard pair of degree l, and let $L(\lambda) = I\lambda^l + \sum_{j=0}^{l-1} A_j\lambda^j$ be the monic matrix polynomial with standard pair (X', T'). From the definition of a standard pair (Section 1.9) it follows that (X', T') is an extension of (X, T) if and only if

$$[A_0 \quad A_1 \quad \cdots \quad A_{l-1}]\begin{bmatrix} X \\ XT \\ \vdots \\ XT^{l-1} \end{bmatrix} = -XT^l$$

or

$$[\operatorname{col}(XT^i)_{i=0}^{l-1}]^T \cdot \begin{bmatrix} A_0^{\mathrm{T}} \\ A_1^{\mathrm{T}} \\ \vdots \\ A_{l-1}^{\mathrm{T}} \end{bmatrix} = -(XT^l)^{\mathrm{T}}. \tag{6.9}$$

According to a result described in Chapter S3, a general solution of (6.9) is given by

$$\begin{bmatrix} A_0^{\mathrm{T}} \\ A_1^{\mathrm{T}} \\ \vdots \\ A_{l-1}^{\mathrm{T}} \end{bmatrix} = -\{[\operatorname{col}(XT^i)_{i=0}^{l-1}]^{\mathrm{T}}\}^{\mathrm{I}} \cdot (XT^l)^{\mathrm{T}}, \tag{6.10}$$

(where the superscript "I" designates any right inverse of $[\operatorname{col}(XT^i)_{i=0}^{l-1}]^T$) if and only if the columns of $(XT^l)^T$ are linearly independent. Taking the transpose in (6.10), and using the definition of an associated matrix polynomial, we obtain the assertion of Theorem 6.3. □

6.3. Special Extensions

In Section 6.2 we have seen that an extension of an admissible l-independent pair (X, T) to a standard pair of degree l can always be obtained via associated monic polynomials. Thus, if $L(\lambda)$ is such a polynomial and $(\tilde{X}, \tilde{T})$ its standard pair, then there exists a T-invariant subspace $\mathcal{M}$ such that the pair (X, T) is similar to $(\tilde{X}|_{\mathcal{M}}, \tilde{T}|_{\mathcal{M}})$. Let $\mathcal{M}'$ be some direct complement to $\mathcal{M}$ and let P be the projector on $\mathcal{M}'$ along $\mathcal{M}$. Then the part $(\tilde{X}|_{\mathcal{M}'}, P\tilde{T}|_{\mathcal{M}'})$ of the standard pair $(\tilde{X}, \tilde{T})$ does not originate in (X, T); it is the extra part which is "added" to the pair (X, T). In other words, the monic polynomial $L(\lambda)$ has a spectral structure which originates in (X, T), and an additional spectral structure. In this section we shall construct the associated polynomial $L(\lambda)$ for which this additional spectral structure is the simplest possible. This construction (as for every associated polynomial) uses a certain left inverse of the matrix $\operatorname{col}(XT^i)_{i=0}^{l-1}$, which will be called the *special left inverse*.

The following lemma plays an important role.

For given subspaces $\mathscr{U}_1, \ldots, \mathscr{U}_l \subset \mathbb{C}^n$ we denote by $\operatorname{col}(\mathscr{U}_i)_{i=1}^{l}$ the subspace in $\mathbb{C}^{nl}$ consisting of all l-tuples $(x_1, \ldots, x_l)$ of n-dimensional vectors $x_1, \ldots, x_l$ such that $x_1 \in \mathscr{U}_1, x_2 \in \mathscr{U}_2, \ldots, x_l \in \mathscr{U}_l$.

Lemma 6.4. *Let (X, T) be an admissible l-independent pair with an invertible matrix T. Then there exists a sequence of subspaces $\mathbb{C}^n \supseteq \mathscr{W}_1 \supseteq \cdots \supseteq \mathscr{W}_l$ such that*

$$\operatorname{col}(\mathscr{W}_j)_{j=k+1}^{l} \dotplus \operatorname{Im} \operatorname{col}(XT^{j-1})_{j=k+1}^{l} = \mathbb{C}^{n(l-k)} \tag{6.11}$$

for $k = 0, \ldots, l-1$.

Proof. Let $\mathscr{W}_l$ be a direct complement to $\operatorname{Im} XT^{l-1} = \operatorname{Im} X$ (this equality follows from the invertibility of T) in $\mathbb{C}^n$. Then (6.11) holds for $k = l-1$. Suppose that $\mathscr{W}_{i+1} \supseteq \cdots \supseteq \mathscr{W}_l$ are already constructed so that (6.11) holds for $k = i, \ldots, l-1$. It is then easy to check that $\operatorname{col}(\mathscr{Z}_j)_{j=i}^{l} \cap \operatorname{Im} \operatorname{col}(XT^{j-1})_{j=i}^{l} = \{0\}$, where $\mathscr{Z}_k = \mathscr{W}_k$ for $k = i+1, \ldots, l$, and $\mathscr{Z}_i = \mathscr{W}_{i+1}$. Hence, the sum

$$\mathscr{S} = \operatorname{col}(\mathscr{Z}_j)_{j=1}^{l} \dotplus \operatorname{Im} \operatorname{col}(XT^{j-1})_{j=1}^{l}$$

is a direct sum. Let $\mathscr{Q}$ be a direct complement of $\operatorname{Ker} A$ in $\mathbb{C}^{n(l-i)}$, where $A = \operatorname{col}(XT^{j-1})_{j=i+1}^{l}$. Let A^{I} be the generalized inverse of A (see Chapter S3) such that $\operatorname{Im} A^{\mathrm{I}} = \mathscr{Q}$ and $\operatorname{Ker} A^{\mathrm{I}} = \operatorname{col}(\mathscr{W}_j)_{j=i+1}^{l}$. Let P be the projector on $\operatorname{Ker} A$ along $\mathscr{Q}$. One verifies easily that $A^{\mathrm{I}}A = I - P$. Thus

$$\mathscr{S} = \left\{ \begin{bmatrix} y + XT^{i-1}Pz + XT^{i-1}A^{\mathrm{I}}x \\ x \end{bmatrix} \middle| \, y \in \mathscr{W}_{i+1}, x \in \mathbb{C}^{n(l-i)}, z \in \mathbb{C}^r \right\}, \tag{6.12}$$

where r is the size of T.

Indeed, if $x \in \mathbb{C}^{n(l-i)}$ then by the induction hypothesis $x = Az_1 + x_1$, where $z_1 \in \mathscr{Q}$ and $x_1 \in \operatorname{col}(\mathscr{W}_j)_{j=i+1}^{i}$. Hence

$$\begin{bmatrix} XT^{i-1}A^{\mathrm{I}}x \\ x \end{bmatrix} = \begin{bmatrix} XT^{i-1}A^{\mathrm{I}}(Az_1 + x_1) \\ Az_1 + x_1 \end{bmatrix} = \begin{bmatrix} XT^{i-1}z_1 \\ Az_1 \end{bmatrix} + \begin{bmatrix} 0 \\ x_1 \end{bmatrix}.$$

From the definition of $\mathscr{S}$ it follows that

$$\begin{bmatrix} XT^{i-1}A^{\mathrm{I}}x \\ x \end{bmatrix} \in \mathscr{S}.$$

For any $z \in \mathcal{C}^r$, also

$$\begin{bmatrix} XT^{i-1}Pz \\ 0 \end{bmatrix} = \begin{bmatrix} XT^{i-1}Pz \\ APz \end{bmatrix} \in \mathcal{S},$$

and clearly

$$\begin{bmatrix} y \\ 0 \end{bmatrix} \in \mathcal{S}$$

for any $y \in \mathcal{W}_{i+1}$. The inclusion $\supset$ in (6.12) thus follows. To check the converse inclusion, take $y \in \mathcal{W}_{i+1}$, $x_1 \in \operatorname{col}(\mathcal{W}_j)_{j=i+1}^{i}$, and $z \in \mathcal{C}^r$. Then

$$\begin{aligned}
\begin{bmatrix} y \\ x_1 \end{bmatrix} + \begin{bmatrix} XT^{i-1} \\ A \end{bmatrix} z &= \begin{bmatrix} y \\ x_1 \end{bmatrix} + \begin{bmatrix} XT^{i-1} \\ A \end{bmatrix} [Pz + (I - P)z] \\
&= \begin{bmatrix} y \\ x_1 \end{bmatrix} + \begin{bmatrix} XT^{i-1}Pz \\ 0 \end{bmatrix} + \begin{bmatrix} XT^{i-1}(I - P)z \\ A(I - P)z \end{bmatrix} \\
&= \begin{bmatrix} y \\ 0 \end{bmatrix} + \begin{bmatrix} XT^{i-1}Pz \\ 0 \end{bmatrix} + \begin{bmatrix} XT^{i-1}A^{\mathrm{I}}(x_1 + A(I - P)z) \\ x_1 + A(I - P)z \end{bmatrix}
\end{aligned}$$

(the last equality follows from $A^{\mathrm{I}}x_1 = 0$ and $I - P = A^{\mathrm{I}}A$), and (6.12) is proved.

Now, let $\mathcal{Y}$ be a direct complement of $\mathcal{W}_{i+1} + \operatorname{Im} XT^{i-1}P$ in $\mathcal{C}^n$. Then from (6.12) we obtain

$$\mathcal{S} \dotplus \begin{bmatrix} \mathcal{Y} \\ 0 \end{bmatrix} = \mathcal{C}^{n(l-i+1)},$$

and so we can put $\mathcal{W}_i = \mathcal{W}_{i+1} \dotplus \mathcal{Y}$. $\square$

We note some remarks concerning the subspaces $\mathcal{W}_j$ constructed in Lemma 6.4.

(1) The subspaces $\mathcal{W}_j$ are not uniquely determined. The question arises if one can choose $\mathcal{W}_j$ in a "canonical" way. It follows from the proof of Lemma 6.4, for instance, that $\mathcal{W}_j$ can be chosen as the span of some set of coordinate unit vectors:

$$\mathcal{W}_j = \operatorname{Span}\{e_k \mid k \in K_j\},$$

where $K_l \subset K_{l-1} \subset \cdots \subset K = \{1, \ldots, n\}$, and

$$e_j = (0 \quad \cdots \quad 0 \quad 1 \quad 0 \quad \cdots \quad 0)^T \in \mathcal{C}^n$$

with 1 in the jth place.

(2) The dimensions $\nu_j = \dim \mathscr{W}_j, j = 1, \ldots, l$ are determined uniquely by the pair (X, T): namely,

$$\nu_{k+1} = n + \operatorname{rank} \operatorname{col}(XT^{j-1})_{j=k+1}^{l} - \operatorname{rank} \operatorname{col}(XT^{j-1})_{j=k}^{l}, \qquad k = 0, \ldots, l-1. \tag{6.13}$$

(3) Equality (6.11) for $k = 0$ means

$$\operatorname{col}(\mathscr{W}_j)_{j=1}^{l} \dotplus \operatorname{Im} \operatorname{col}(XT^{j-1})_{j=1}^{l} = \mathbb{C}^{nl},$$

and consequently, there exists an $r \times nl$ matrix V such that

$$V \cdot \operatorname{col}(XT^{j-1})_{j=1}^{l} = I, \qquad V(\operatorname{col}(\mathscr{W}_j)_{j=1}^{l}) = \{0\}. \tag{6.14}$$

The left inverse V of $\operatorname{col}(XT^{j-1})_{j=1}^{l}$ with property (6.14) will be called the *special left inverse* (of course, it depends on the choice of $\mathscr{W}_j$).

Let now (X, T) be an admissible l-independent pair with invertible matrix T, and let $V = [V_1 \quad \cdots \quad V_l]$ be a special left inverse of $\operatorname{col}(XT^i)_{i=0}^{l-1}$. Then we can construct the associated monic matrix polynomial

$$L(\lambda) = I\lambda^l - XT^l(V_1 + V_2 + \cdots + V_l\lambda^{l-1}). \tag{6.15}$$

The polynomial $L(\lambda)$ will be called the *special* matrix polynomial associated with the pair (X, T).

Theorem 6.5. *Let $L(\lambda)$ be a special matrix polynomial associated with (X, T), where T is invertible of size $r \times r$. Then zero is an eigenvalue of $L(\lambda)$ with partial multiplicities $\kappa_1 \geq \kappa_2 \geq \cdots$, where*

$$\kappa_i = |\{j \mid 1 \leq j \leq l \text{ and } \nu_j \geq i\}|$$

and ν_j are defined by (6.13).

Here, as usual, $|A|$ denotes the number of different elements in a finite set A.

Proof. Let $\mathscr{W}_1 \supset \cdots \supset \mathscr{W}_l$ be the sequence of subspaces from Lemma 6.4. Let

$$V = [V_1 \quad \cdots \quad V_l]$$

be a special left inverse of $\operatorname{col}(XT^i)_{i=0}^{l-1}$ such that (6.14) and (6.15) hold. From (6.14) it follows that

$$\operatorname{Ker} V_j \supset \mathscr{W}_j, \qquad j = 1, \ldots, l.$$

It is easy to see that $x \in \mathscr{W}_i \backslash \mathscr{W}_{i+1}$ ($i = 1, \ldots, l$; $\mathscr{W}_{l+1} = \{0\}$) is an eigenvector of $L(\lambda)$ generating a Jordan chain $x, 0, \ldots, 0$ ($i - 1$ zeros). Taking a basis in each $\mathscr{W}_i$ modulo $\mathscr{W}_{i+1}$ (i.e., the maximal set of vectors $x_1, \ldots, x_k \in \mathscr{W}_i$ such that $\sum_{i=1}^{k} \alpha_i x_i \in \mathscr{W}_{i+1}, \alpha_i \in \mathbb{C}$ implies that $\alpha_1 = \cdots = \alpha_k = 0$), we obtain in this way a canonical set of Jordan chains of $L(\lambda)$ corresponding to the eigenvalue zero. Now Theorem 6.5 follows from the definition of partial multiplicities. □

Observe that (under the conditions and notations of Theorem 6.5) the multiplicity of zero as a root of det $L(\lambda)$ is just $nl - r$. Then dimensional considerations lead to the following corollary of Theorem 6.5.

Corollary 6.6. *Under the conditions and notations of Theorem 6.5 let $(\tilde{X}, \tilde{T})$ be a standard pair of $L(\lambda)$. Then (X, T) is similar to $(\tilde{X}|_{\mathscr{M}}, \tilde{T}|_{\mathscr{M}})$, where $\mathscr{M}$ is the maximal $\tilde{T}$-invariant subspace such that $\tilde{T}|_{\mathscr{M}}$ is invertible.*

Proof. Let $\mathscr{M}_1$ and $\mathscr{M}_2$ be the maximal $\tilde{T}$-invariant subspaces such that $\tilde{T}|_{\mathscr{M}_1}$ is invertible and $\tilde{T}|_{\mathscr{M}_2}$ is nilpotent (i.e., $\sigma(\tilde{T}|_{\mathscr{M}_2}) = \{0\}$). We know from Theorem 6.2 that for some T-invariant subspace $\mathscr{M}$, (X, T) is similar to $(\tilde{X}|_{\mathscr{M}}, \tilde{T}|_{\mathscr{M}})$. Since T is invertible, $\mathscr{M} \subset \mathscr{M}_1$. In fact, $\mathscr{M} = \mathscr{M}_1$. Indeed, by Theorem 6.5

$$\dim \mathscr{M}_2 = \kappa_1 + \kappa_2 + \cdots;$$

on the other hand, it is easy to see that

$$\kappa_1 + \kappa_2 + \cdots = \nu_1 + \cdots + \nu_l = nl - \operatorname{rank} \operatorname{col}(XT^i)_{i=0}^{l-1} = nl - r,$$

so $\dim \mathscr{M}_1 = r$. But clearly, also $\dim \mathscr{M} = r$, and $\mathscr{M} = \mathscr{M}_1$. □

It is not hard to generalize Theorem 6.5 in such a way as to remove the invertibility condition on T. Namely, let $\alpha \in \mathbb{C}$ be any number outside the spectrum of T, and consider the admissible pair $(X, -\alpha I + T)$ in place of (X, T). Now the pair $(X, T - \alpha I)$ is also l-independent because of the identity

$$\begin{bmatrix} I & 0 & 0 & \cdots & 0 \\ \alpha\binom{1}{0}I & \binom{1}{1}I & 0 & \cdots & 0 \\ \vdots & \vdots & \vdots & & \vdots \\ \alpha^{l-1}\binom{l-1}{0}I & \alpha^{l-2}\binom{l-1}{1}I & & \cdots & \binom{l-1}{l-1}I \end{bmatrix} \begin{bmatrix} X \\ X(T - \alpha I) \\ \vdots \\ X(T - \alpha I)^{l-1} \end{bmatrix} = \begin{bmatrix} X \\ XT \\ \vdots \\ XT^{l-1} \end{bmatrix}.$$

So we can apply Theorem 6.5 for the pair $(X, -\alpha I + T)$ producing a monic matrix polynomial $L_\alpha(\lambda)$ whose standard pair is an extension of $(X, -\alpha I + T)$ (Theorem 6.2) and having at zero the partial multiplicities

$$\nu_{k+1} = n + \operatorname{rank} \operatorname{col}(X(T - \alpha I)^{j-1})_{j=k+1}^{l} - \operatorname{rank} \operatorname{col}(X(T - \alpha I)^{j-1})_{j=k}^{l}, \quad k = 0, \ldots, l-1. \tag{6.16}$$

Let $(X_\alpha, T_\alpha, Y_\alpha)$ be a standard triple of $L_\alpha(\lambda)$; then $(X_\alpha, T_\alpha + \alpha I, Y_\alpha)$ is a standard triple of the monic matrix polynomial $L(\lambda) = L_\alpha(\lambda + \alpha)$. Indeed, by the resolvent form (2.16),

$$\begin{aligned}(L(\lambda))^{-1} &= (L_\alpha(\lambda + \alpha))^{-1} = X_\alpha(I(\lambda + \alpha) - T_\alpha)^{-1}Y_\alpha \\ &= X_\alpha(I\lambda - (T_\alpha + \alpha I))^{-1}Y_\alpha,\end{aligned}$$

and it remains to use Theorem 2.4. It follows that the standard triple of the polynomial $L(\lambda)$ is an extension of (X, T) and, in addition, its partial multiplicities at the eigenvalue $\lambda_0 = \alpha$ are defined by (6.16).

Thus, the construction of a special monic matrix polynomial associated with a given admissible l-independent pair (X, T) (with invertible T) can be regarded as an extension of (X, T) to a standard pair $(\tilde{X}, \tilde{T})$ of degree l such that $\sigma(\tilde{T}) = \{0\} \cup \sigma(T)$; moreover, elementary divisors of $I\lambda - T$ corresponding to the nonzero eigenvalues are the same as those of $I\lambda - T$, and the degrees of the elementary divisors of $I\lambda - \tilde{T}$ corresponding to zero are defined by (6.13). Again, we can drop the requirement that T is invertible, replacing 0 by some $\alpha \notin \sigma(T)$ and replacing (6.13) by (6.16).

Given an admissible l-independent pair (X, T) (with invertible T of size $r \times r$), it is possible to describe all the possible Jordan structures at zero of the matrix $\tilde{T}$ such that zero is an eigenvalue of $\tilde{T}$ with multiplicity $nl - r$, and for some $n \times nl$ matrix $\tilde{X}$, the pair $(\tilde{X}, \tilde{T})$ is a standard extension of degree l of the pair (X, T). (The standard pair $(\tilde{X}, \tilde{T})$ described in the preceding paragraph is an extension of this type.) The description is given by the following theorem.

Theorem 6.7. *Let (X, T) be an admissible l-independent pair. Suppose that T is invertible of size $r \times r$, and let $\tilde{T}$ be an $nl \times nl$ matrix such that $\lambda = 0$ is a zero of $\det(I\lambda - \tilde{T})$ of multiplicity $nl - r$. Then for some matrix $\tilde{X}$, $(\tilde{X}, \tilde{T})$ is a standard extension of degree l of the pair (X, T) if and only if the degrees $p_1 \geq \cdots \geq p_v$ of the elementary divisors of $I\lambda - \tilde{T}$ corresponding to the eigenvalue 0 satisfy the inequalities*

$$\sum_{j=0}^{i} p_j \geq \sum_{j=0}^{i} s_j \qquad \text{for} \quad i = 1, 2, \ldots, v, \tag{6.17}$$

where s_k is the number of the integers $n + q_{l-j-1} - q_{l-j}$ $(j = 0, 1, \ldots, l-1)$ larger than k, $k = 0, 1, \ldots$, and $q_0 = 0$,

$$q_j = \operatorname{rank} \operatorname{col}(XT^i)_{i=0}^{j-1} \qquad j \geq 1.$$

For the proof of this result we refer to [37c].

Note that the partial multiplicities at zero of the special monic matrix polynomial $L(\lambda)$ associated with (X, T) (which coincide with the degrees of the elementary divisors of $I\lambda - \tilde{T}$ corresponding to zero, where $\tilde{T}$ is some linearization of $L(\lambda)$), are extremal in the sense that equalities hold in (6.17) throughout (cf. (6.13)).

Comments

The results of this chapter are based on [37c]. The exposition of Section 6.3 follows [37b].

Part II

Nonmonic Matrix Polynomials

In this part we shall consider regular $n \times n$ matrix polynomials $L(\lambda)$, i.e., such that $\det L(\lambda) \not\equiv 0$ and the leading coefficient of $L(\lambda)$ is not necessarily the identity, or even invertible. As in the monic case, the spectrum $\sigma(L)$, which coincides with the zeros of $\det L(\lambda)$, is still a finite set. This is the crucial property allowing us to investigate the spectral properties of $L(\lambda)$ along the lines developed in Part I. Such investigations will be carried out in Part II.

Most of the applications which stem from the spectral analysis of monic matrix polynomials, and are presented in Part I, can be given also in the context of regular matrix polynomials. We present some applications to differential and difference equations in Chapter 8 and refer the reader to the original papers for others.

On the other hand, problems concerning least common multiples and greatest common divisors of matrix polynomials fit naturally into the framework of regular polynomials rather than monic polynomials. Such problems are considered in Chapter 9.

It is assumed throughout Part II that all matrix polynomials are regular, unless stated otherwise.

Chapter 7

Spectral Properties and Representations

In this chapter we extend the theory of Part I to include description of the spectral data, and basic representation theorems, for regular matrix polynomials, as well as solution of the inverse problem, i.e., construction of a regular polynomial given its spectral data. The latter problem is considered in two aspects: (1) when full spectral data are given, including infinity; (2) when spectral data are given at finite points only.

7.1. The Spectral Data (Finite and Infinite)

Let $L(\lambda) = \sum_{j=0}^{l} A_j \lambda^j$ be an $n \times n$ matrix polynomial. We shall study first the structure of the spectral data of $L(\lambda)$, i.e., eigenvalues and Jordan chains.

Recall the definition of a canonical set of Jordan chains of $L(\lambda)$ corresponding to the eigenvalue λ_0 (see Section 1.6). Let

$$\varphi_0^{(1)}, \ldots, \varphi_{r_1-1}^{(1)}, \qquad \varphi_0^{(2)}, \ldots, \varphi_{r_2-1}^{(2)}, \qquad \ldots, \qquad \varphi_0^{(k)}, \ldots, \varphi_{r_k-1}^{(k)},$$

be such a canonical set, where $r_j = \operatorname{rank} \varphi_0^{(j)}, j = 1, \ldots, k, k = \dim \operatorname{Ker} L(\lambda_0)$, and $\varphi_0^{(i)}, i = 1, \ldots, k$, are eigenvectors of $L(\lambda)$ corresponding to λ_0. We write this canonical set of Jordan chains in matrix form:

$$X(\lambda_0) = [\varphi_0^{(1)} \ \cdots \ \varphi_{r_1-1}^{(1)} \ \ \varphi_0^{(2)} \ \cdots \ \varphi_{r_2-1}^{(2)} \ \cdots \ \varphi_0^{(k)} \ \cdots \ \varphi_{r_k-1}^{(k)}],$$
$$J(\lambda_0) = \operatorname{diag}(J_1, J_2, \ldots, J_k),$$

where J_i is a Jordan block of size r_i with eigenvalue λ_0. Hence $X(\lambda_0)$ is an $n \times r$ matrix and $J(\lambda_0)$ is an $r \times r$ matrix, where $r = \sum_{j=1}^{k} r_j$ is the multiplicity of λ_0 as a zero of det $L(\lambda)$.

The pair of matrices $(X(\lambda_0), J(\lambda_0))$ is called a *Jordan pair* of $L(\lambda)$ corresponding to λ_0. The following theorem provides a characterization for a Jordan pair corresponding to λ_0.

Theorem 7.1. *Let $(\hat{X}, \hat{J})$ be a pair of matrices, where $\hat{X}$ is an $n \times \mu$ matrix and $\hat{J}$ is a $\mu \times \mu$ Jordan matrix with unique eigenvalue λ_0. Then the following conditions are necessary and sufficient in order that $(\hat{X}, \hat{J})$ be a Jordan pair of $L(\lambda) = \sum_{j=0}^{l} \lambda^j A_j$ corresponding to λ_0:*

(i) det $L(\lambda)$ *has a zero λ_0 of multiplicity μ,*
(ii) rank $\operatorname{col}(\hat{X}\hat{J}^j)_{j=0}^{l-1} = \mu$,
(iii) $A_l \hat{X}\hat{J}^l + A_{l-1}\hat{X}\hat{J}^{l-1} + \cdots + A_0 \hat{X} = 0.$

Proof. Suppose that $(\hat{X}, \hat{J})$ is a Jordan pair of $L(\lambda)$ corresponding to λ_0. Then (i) follows from Corollary 1.14 and (iii) from Proposition 1.10. Property (ii) follows immediately from Theorem 7.3 below.

Suppose now that (i)–(iii) hold. By Proposition 1.10, the part of $\hat{X}$ corresponding to each Jordan block in J, forms a Jordan chain of $L(\lambda)$ corresponding to λ_0. Condition (ii) ensures that the eigenvectors of these Jordan chains are linearly independent. Now in view of Proposition 1.15, it is clear that $(\hat{X}, \hat{J})$ is a Jordan pair of $L(\lambda)$ corresponding to λ_0. □

Taking a Jordan pair $(X(\lambda_j), J(\lambda_j))$ for every eigenvalue λ_j of $L(\lambda)$, we define a *finite Jordan pair* (X_F, J_F) of $L(\lambda)$ as

$$X_F = [X(\lambda_1) \quad X(\lambda_2) \quad \cdots \quad X(\lambda_p)], \qquad J_F = \operatorname{diag}[J(\lambda_1), J(\lambda_2), \ldots, J(\lambda_p)],$$

where p is the number of different eigenvalues of $L(\lambda)$.

Note that the sizes of X_F and J_F are $n \times v$ and $v \times v$, respectively, where $v = \deg(\det L(\lambda))$. Note also that the pair (X_F, J_F) is not determined uniquely by the polynomial $L(\lambda)$: the description of all finite Jordan pairs of $L(\lambda)$ (with fixed J_F) is given by the formula $(X_F U, J_F)$, where (X_F, J_F) is any fixed finite Jordan pair of $L(\lambda)$, and U is any invertible matrix which commutes with J_F. The finite Jordan pair (X_F, J_F) does not determine $L(\lambda)$ uniquely either: any matrix polynomial of the form $V(\lambda)$, $L(\lambda)$, where $V(\lambda)$ is a matrix polynomial with det $V(\lambda) \equiv \text{const} \neq 0$, has the same finite Jordan pairs as $L(\lambda)$ (see Theorem 7.13 below).

In order to determine $L(\lambda)$ uniquely we have to consider (together with (X_F, J_F)) an additional Jordan pair (X_∞, J_∞) of $L(\lambda)$ for $\lambda = \infty$, which is defined below.

Let

$$\psi_0^{(i)}, \ldots, \psi_{s_i-1}^{(i)}, \qquad i = 1, \ldots, q, \tag{7.1}$$

be a canonical set of Jordan chains of the analytic (at infinity) matrix function $\lambda^{-l}L(\lambda)$ corresponding to $\lambda = \infty$, if this point is an eigenvalue of $\lambda^{-l}L(\lambda)$ (where l is the degree of $L(\lambda)$, i.e., the maximal integer j such that $L^{(j)}(\lambda) \neq 0$). By definition, a Jordan chain (resp. canonical set of Jordan chains) of an analytic matrix-valued function $M(\lambda)$ at infinity is just a Jordan chain (resp. canonical set of Jordan chains) of the matrix function $M(\lambda^{-1})$ at zero. Thus, the Jordan chains (7.1) form a canonical set of Jordan chains of the matrix polynomial $\tilde{L}(\lambda) = \lambda^l L(\lambda^{-1})$ corresponding to the eigenvalue zero. We shall use the following notation

$$X_\infty = [\psi_0^{(1)} \quad \cdots \quad \psi_{s_1-1}^{(1)}\psi_0^{(2)} \quad \cdots \quad \psi_{s_2-1}^{(2)} \quad \cdots \quad \psi_0^{(q)} \quad \cdots \quad \psi_{s_q-1}^{(q)}],$$
$$J_\infty = \operatorname{diag}[J_{\infty 1}, J_{\infty 2}, \ldots, J_{\infty q}],$$

where $J_{\infty j}$ is the Jordan block of size s_j with eigenvalue zero, and call (X_∞, J_∞) an *infinite Jordan pair* of $L(\lambda)$. Note that (X_∞, J_∞) is a Jordan pair of the matrix polynomial $\tilde{L}(\lambda) = \lambda^l L(\lambda^{-1})$ corresponding to the eigenvalue $\lambda = 0$. This observation (combined with Theorem 7.1) leads to the following characterization of infinite Jordan pairs.

Theorem 7.2. *Let $(\hat{X}, \hat{J})$ be a pair of matrices, where $\hat{X}$ is $n \times \mu$ and $\hat{J}$ is a $\mu \times \mu$ Jordan matrix with unique eigenvalue $\lambda_0 = 0$. Then the following conditions are necessary and sufficient in order that $(\hat{X}, \hat{J})$ be an infinite Jordan pair of $L(\lambda) = \sum_{j=0}^{l} \lambda^j A_j$:*

(i) $\det(\lambda^l L(\lambda^{-1}))$ *has a zero at $\lambda_0 = 0$ of multiplicity μ,*
(ii) $\operatorname{rank} \operatorname{col}(\hat{X}\hat{J}^j)_{j=0}^{l-1} = \mu$,
(iii) $A_0 \hat{X}\hat{J}^l + A_1 \hat{X}\hat{J}^{l-1} + \cdots + A_l = 0.$

EXAMPLE 7.1. Consider the matrix polynomial

$$L(\lambda) = \begin{bmatrix} -(\lambda - 1)^3 & \lambda \\ 0 & \lambda + 1 \end{bmatrix}.$$

In this case we can put down

$$X_F = \begin{bmatrix} 1 & 0 & 0 & 1 \\ 0 & 0 & 0 & 8 \end{bmatrix}, \qquad J_F = \operatorname{diag}\left(\begin{bmatrix} 1 & 1 & 0 \\ 0 & 1 & 1 \\ 0 & 0 & 1 \end{bmatrix}, -1\right),$$

and

$$X_\infty = \begin{bmatrix} 0 & 0 \\ 1 & 0 \end{bmatrix}, \qquad J_\infty = \begin{bmatrix} 0 & 1 \\ 0 & 0 \end{bmatrix}. \quad \square$$

7.2. Linearizations

Let $L(\lambda) = \sum_{i=0}^{l} A^i \lambda_i$ be a regular $n \times n$ matrix polynomial. An $nl \times nl$ linear matrix polynomial $S_0 + S_1\lambda$ is called a *linearization* of $L(\lambda)$ if

$$\begin{bmatrix} L(\lambda) & 0 \\ 0 & I_{n(l-1)} \end{bmatrix} = E(\lambda)(S_0 + S_1\lambda)F(\lambda)$$

for some $nl \times nl$ matrix polynomials $E(\lambda)$ and $F(\lambda)$ with constant nonzero determinants. For monic matrix polynomials the notion of linearization (with $S_1 = I$) was studied in Chapter 1. In this section we construct linearizations for regular matrix polynomials.

For the matrix polynomial $L(\lambda)$ define its *companion polynomial* $C_L(\lambda)$ as follows:

$$C_L(\lambda) = \begin{bmatrix} I & 0 & \cdots & & 0 \\ 0 & I & \cdots & & 0 \\ \vdots & \vdots & & \vdots & \vdots \\ 0 & 0 & \cdots & I & 0 \\ 0 & 0 & \cdots & 0 & A_l \end{bmatrix} \lambda + \begin{bmatrix} 0 & -I & 0 & \cdots & 0 \\ 0 & 0 & -I & \cdots & 0 \\ \vdots & \vdots & \vdots & \vdots & \vdots \\ 0 & 0 & & \cdots & -I \\ A_0 & A_1 & & \cdots & A_{l-1} \end{bmatrix}.$$

So $C_L(\lambda)$ is a linear $nl \times nl$ matrix polynomial; if $L(\lambda)$ is monic, then $C_L(\lambda) = I\lambda - C_1$, where C_1 is the first companion matrix of $L(\lambda)$ (cf. Theorem 1.1).

It turns out that $C_L(\lambda)$ is a linearization of $L(\lambda)$. Indeed, define a sequence of polynomials $E_i(\lambda)$, $i = 1, \ldots, l$, by induction: $E_l(\lambda) = A_l$, $E_{i-1}(\lambda) = A_{i-1} + \lambda E_i(\lambda)$, $i = l, \ldots, 2$. It is easy to see that

$$\begin{bmatrix} E_1(\lambda) & \cdots & & E_{l-1}(\lambda) & I \\ -I & 0 & \cdots & & 0 \\ 0 & -I & \cdots & & 0 \\ \vdots & \vdots & & \vdots & \vdots \\ 0 & 0 & \cdots & -I & 0 \end{bmatrix} C_L(\lambda)$$

$$= \begin{bmatrix} L(\lambda) & 0 \\ 0 & I \end{bmatrix} \begin{bmatrix} I & 0 & 0 & \cdots & & 0 \\ -\lambda I & I & 0 & \cdots & & 0 \\ 0 & -\lambda I & I & \cdots & & 0 \\ \vdots & \vdots & \vdots & & \vdots & \vdots \\ 0 & 0 & 0 & \cdots & -\lambda I & I \end{bmatrix}, \tag{7.2}$$

so indeed $C_L(\lambda)$ is a linearization of $L(\lambda)$. In particular, every regular matrix polynomial has a linearization.

Clearly, a linearization is not unique; so the problem of a canonical choice of a linearization arises. For monic matrix polynomials the linearization $I\lambda - C_1$ with companion matrix C_1 can serve as a canonical choice.

Another class of polynomials for which an analogous canonical linearization is available is the *comonic polynomials* $L(\lambda)$, i.e., such that $L(0) = I$. Thus, if $L(\lambda) = I + \sum_{j=1}^{l} A_j\lambda^j$ is a comonic matrix polynomial of degree l, the the matrix

$$R = \begin{bmatrix} 0 & I & 0 & \cdots & 0 \\ 0 & 0 & I & \cdots & 0 \\ \vdots & \vdots & \vdots & \cdots & \vdots \\ 0 & 0 & 0 & \cdots & I \\ -A_l & -A_{l-1} & -A_{l-2} & \cdots & -A_1 \end{bmatrix}$$

will be called the *comonic companion* matrix of $L(\lambda)$. (Compare with the definition of the companion matrix for monic matrix polynomials.) Using the comonic companion matrix, we have the linearization $I - R\lambda$; namely,

$$I - R\lambda = B(\lambda)\text{diag}[L(\lambda), I, \ldots, I]D(\lambda), \tag{7.3}$$

where

$$B(\lambda) = \begin{bmatrix} 0 & 0 & 0 & \cdots & 0 & I \\ 0 & 0 & 0 & \cdots & I & 0 \\ \vdots & \vdots & \vdots & \cdots & \vdots & \vdots \\ 0 & I & 0 & \cdots & 0 & 0 \\ I & B_1(\lambda) & B_2(\lambda) & \cdots & B_{l-2}(\lambda) & B_{l-1}(\lambda) \end{bmatrix},$$

with $B_i(\lambda) = \sum_{j=0}^{l-i-1} A_{l-j}\lambda^{l-i-j}$, $i = 1, \ldots, l-1$, and

$$D(\lambda) = \begin{bmatrix} 0 & 0 & 0 & \cdots & I \\ & & & & -\lambda I \\ \vdots & \vdots & \vdots & \cdots & \vdots \\ 0 & I & -\lambda I & \cdots & 0 \\ I & -\lambda I & 0 & \cdots & 0 \end{bmatrix}$$

Clearly, $B(\lambda)$ and $D(\lambda)$ are everywhere invertible matrix polynomials. Indeed, it is easily seen that

$$D^{-1}(\lambda) = \begin{bmatrix} \lambda^{l-1}I & \lambda^{l-2}I & \cdots & \lambda I & I \\ \lambda^{l-2}I & \lambda^{l-3}I & \cdots & I & 0 \\ \vdots & \vdots & & \vdots & \vdots \\ I & 0 & \cdots & 0 & 0 \end{bmatrix}$$

and

$$(I - R\lambda)D^{-1}(\lambda) = B(\lambda) \cdot \text{diag}[L(\lambda), I, \ldots, I].$$

Linearizations of type (7.3) are easily extended to the case in which $L(0) \neq I$. This is done by considering the matrix polynomial $\tilde{L}(\lambda) = (L(a))^{-1}L(\lambda + a)$ instead of $L(\lambda)$, where $a \notin \sigma(L)$, and observing that $\tilde{L}(\lambda)$ is comonic.

In the next section we shall encounter a different type of linearization based on the notion of decomposable pair.

7.3. Decomposable Pairs

In Section 1.8 the important concept of a Jordan pair for a monic matrix polynomial was introduced, and in Section 7.1 we began the organization of the spectral data necessary for generalization of this concept to the nonmonic case. Here, we seek the proper generalization of the concept of a standard pair as introduced in Section 1.10.

First, we call a pair of matrices (X, T) *admissible of order* p if X is $n \times p$ and T is $p \times p$. Then, an admissible pair (X, T) of order nl is called a *decomposable pair* (of degree l) if the following properties are satisfied:

(i)

$$X = [X_1 \quad X_2], \qquad T = \begin{bmatrix} T_1 & 0 \\ 0 & T_2 \end{bmatrix},$$

where X_1 is an $n \times m$ matrix, and T_1 is an $m \times m$ matrix, for some m, $0 \leq m \leq nl$ (so that X_2 and T_2 are of sizes $n \times (nl - m)$ and $(nl - m) \times (nl - m)$, respectively),

(ii) the matrix

$$S_{l-1} = \begin{bmatrix} X_1 & X_2 T_2^{l-1} \\ X_1 T_1 & X_2 T_1^{l-2} \\ \vdots & \vdots \\ X_1 T_1^{l-1} & X_2 \end{bmatrix} \tag{7.4}$$

is nonsingular.

The number m will be called the *parameter* of the decomposable pair (X, T). A decomposable pair (X, T) will be called a decomposable pair of the regular $n \times n$ matrix polynomial $L(\lambda) = \sum_{i=0}^{l} A_i \lambda^i$ if, in addition to (i) and (ii), the following conditions hold:

(iii) $\sum_{i=0}^{l} A_i X_1 T_1^i = 0$, $\sum_{i=0}^{l} A_i X_2 T_2^{l-i} = 0$.

For example, if $L(\lambda)$ is monic, then a standard pair for $L(\lambda)$ is also a decomposable pair with the parameter $m = nl$. So the notion of a decomposable pair can be considered as a natural generalization of the notion of standard pair for monic matrix polynomials. Another example of a decomposable pair (with parameter $m = 0$) is the pair (X, T^{-1}), where (X, T) is a standard pair of a monic matrix polynomial $L(\lambda)$ such that $\det L(0) \neq 0$ (so that T is nonsingular). We see that, in contrast to standard pairs for monic polynomials, there is a large nonuniqueness for their decomposable pairs. We shall touch upon the degree of nonuniqueness of a decomposable pair for a matrix polynomial in the next section.

The next theorem shows that a decomposable pair appears, as in the case of monic polynomials, to carry the full spectral information about its matrix polynomial, but now including spectrum at infinity. The existence of a decomposable pair for every regular matrix polynomial will also follow.

Theorem 7.3. *Let $L(\lambda)$ be a regular matrix polynomial, and let (X_F, J_F) and (X_∞, J_∞) be its finite and infinite Jordan pairs, respectively. Then $([X_F \;\; X_\infty], J_F \oplus J_\infty)$ is a decomposable pair for $L(\lambda)$.*

Proof. Let l be the degree of $L(\lambda)$. By Theorems 7.1 and 7.2, properties (i) and (iii) hold for $(X_1, T_1) = (X_F, J_F)$ and $(X_2, T_2) = (X_\infty, J_\infty)$. In order to show that $([X_F \;\; X_\infty], J_F \oplus J_\infty)$ is a decomposable pair for $L(\lambda)$, it is enough to check the nonsingularity of

$$S_{l-1} = \operatorname{col}(X_F J_F^i, X_\infty J_\infty^{l-1-i})_{i=0}^{l-1}. \tag{7.5}$$

This will be done by reduction to the case of a comonic matrix polynomial.

Let $\alpha \in \mathbb{C}$ be such that the matrix polynomial $L_1(\lambda) = L(\lambda - \alpha)$ is invertible for $\lambda = 0$. We claim that the admissible pair $(X_F, J_F + \alpha I)$ is a finite Jordan pair of $L_1(\lambda)$, and the admissible pair $(X_\infty, J_\infty(I + \alpha J_\infty)^{-1})$ is similar to an infinite Jordan pair of $L_1(\lambda)$. Indeed, write $L(\lambda) = \sum_{j=0}^{l} A_j \lambda^j$, $L_1(\lambda) = \sum_{j=0}^{l} A_{j1} \lambda^j$, then

$$A_k = \sum_{j=k}^{l} \binom{j}{j-k} \alpha^{j-k} A_{j1}, \qquad k = 0, \ldots, l. \tag{7.6}$$

Since (X_∞, J_∞) is an infinite Jordan pair of $L(\lambda)$, we have by Theorem 7.2

$$\sum_{j=0}^{l} A_j X_\infty J_\infty^{l-j} = 0. \tag{7.7}$$

Equations (7.5) and (7.6) imply (by a straightforward computation) that

$$\sum_{j=0}^{l} A_{j1} X_\infty \hat{J}_\infty^{l-j} = 0, \tag{7.8}$$

where $\hat{J}_\infty = J_\infty(I + \alpha J_\infty)^{-1}$. Let μ be the size of J_∞; then for some p,

$$\operatorname{rank} \operatorname{col}(X_\infty \hat{J}_\infty^i)_{i=0}^{p-1} = \mu. \tag{7.9}$$

Indeed, since (X_∞, J_∞) represent a canonical set of Jordan chains, we obtain, in particular, that

$$\operatorname{rank} \operatorname{col}(X_\infty J_\infty^i)_{i=0}^{p-1} = \mu \tag{7.10}$$

for some p. Using the property that $\sigma(J_\infty) = \{0\}$, it is easily seen that (7.10) implies (in fact, is equivalent to) $X_\infty y \neq 0$ for every $y \in \operatorname{Ker} J_\infty \backslash \{0\}$. But apparently $\operatorname{Ker} \hat{J}_\infty = \operatorname{Ker} J_\infty$, so $X_\infty y \neq 0$ for every $y \in \operatorname{Ker} \hat{J}_\infty \backslash \{0\}$, and (7.9) follows. Now (7.8) and (7.9) ensure (Theorem 7.2) that $(X_\infty, \hat{J}_\infty)$ is similar to an infinite Jordan pair of $L_1(\lambda)$ (observe that the multiplicity of $\lambda_0 = 0$ as a zero of $\det(\lambda^l L(\lambda^{-1}))$ and of $\det(\lambda^l L_1(\lambda^{-1}))$ is the same). It is an easy exercise to check that $(X_F, J_F + \alpha I)$ is a finite Jordan pair of $L_1(\lambda)$.

Let $(\tilde{X}_F, \tilde{J}_F)$ and $(\tilde{X}_\infty, \tilde{J}_\infty)$ be finite and infinite Jordan pairs of $L_1(\lambda)$. Since $L_1(0)$ is nonsingular, $\tilde{J}_F$ is nonsingular as well. We show now that $(\tilde{X}, \tilde{T}) \stackrel{\text{def}}{=} ([\tilde{X}_F \quad \tilde{X}_\infty], \tilde{J}_F^{-1} \oplus \tilde{J}_\infty)$ is a standard pair for the monic matrix polynomial $\tilde{L}_1(\lambda) = \lambda^l (L_1(0))^{-1} L_1(\lambda^{-1})$.

Let $B = [x_0 \quad x_1 \quad \cdots \quad x_r]$ be a Jordan chain for $L_1(\lambda)$ taken from $\tilde{X}_F$, and corresponding to the eigenvalue $\lambda_0 \neq 0$. Let K be the Jordan block of size $(r+1) \times (r+1)$ with eigenvalue λ_0. By Proposition 1.10 we have

$$\sum_{j=0}^{l} A_{j1} B K^j = 0, \tag{7.11}$$

where A_{j1} are the coefficients of $L_1(\lambda)$. Since $\lambda_0 \neq 0$, K is nonsingular and so

$$\sum_{j=0}^{l} A_{j1} B K^{j-l} = 0. \tag{7.12}$$

Let K_0 be the matrix obtained from K by replacing λ_0 by λ_0^{-1}. Then K_0 and K^{-1} are similar:

$$K^{-1} = M_r(\lambda_0) K_0 (M_r(\lambda_0))^{-1}, \tag{7.13}$$

where

$$M_r(\lambda_0) = \left[(-1)^j \binom{j-1}{i-1} \lambda_0^{j+i} \right]_{i,j=0}^{r}$$

and it is assumed that $\binom{-1}{-1} = 1$ and $\binom{p}{q} = 0$ for $q > p$ or $q = -1$ and $p > -1$. The equality (7.13) or, what is the same, $K M_r(\lambda_0) K_0 = M_r(\lambda_0)$, can be verified by multiplication. Insert in (7.12) the expression (7.13) for K^{-1}; it follows from Proposition 1.10 that the columns of $B(M_r(\lambda_0))^{-1}$ form a Jordan chain of $\tilde{L}_1(\lambda)$ corresponding to λ_0^{-1}.

Let k be the number of Jordan chains in $\tilde{X}_F$ corresponding to λ_0, and let B_i $(i = 1, \ldots, k)$ be the matrix whose columns form the ith Jordan chain corresponding to λ_0. We perform the above transformation for every B_i. It then follows that the columns of $B_1(M_{r_1}(\lambda_0))^{-1}, \ldots, B_k(M_{r_k}(\lambda_0))^{-1}$ form Jordan chains of $\tilde{L}(\lambda)$ corresponding to λ_0^{-1}. Let us check that these Jordan chains of $\tilde{L}(\lambda)$ form a canonical set of Jordan chains corresponding to λ_0^{-1}. By Proposition 1.15, it is enough to check that the eigenvectors $b_1, \ldots, b_k$ in $B_1(M_{r_1}(\lambda_0))^{-1}, \ldots, B_k(M_{r_k}(\lambda_0))^{-1}$, respectively, are linearly independent.

But from the structure of $(M_{r_i}(\lambda_0))^{-1}$ it is clear that $b_1, \ldots, b_k$ are also the eigenvectors in $B_1, \ldots, B_k$, respectively, of $L(\lambda)$ corresponding to λ_0, and therefore they are linearly independent again by Proposition 1.15.

Now it is clear that $(\tilde{X}_{\lambda_0}, \tilde{J}_{\lambda_0}^{-1})$, where $\tilde{J}_{\lambda_0}$ is the part of $\tilde{J}_F$ corresponding to the eigenvalue λ_0 ($\neq 0$) of $\tilde{J}_F$ and $\tilde{X}_{\lambda_0}$ is the corresponding part of $\tilde{X}_F$, is the part of a standard pair of $\tilde{L}_1(\lambda)$ corresponding to λ_0^{-1}. Recall that by definition $(\tilde{X}_\infty, \tilde{J}_\infty)$ is the part of a Jordan pair of $\tilde{L}_1(\lambda)$ corresponding to 0. So indeed $(\tilde{X}, \tilde{T})$ is a standard pair of $\tilde{L}_1(\lambda)$.

In particular, $\operatorname{col}(\tilde{X}\,\tilde{J}^{\,i})_{i=0}^{l-1} = \operatorname{col}(\tilde{X}_F \tilde{J}_F^{-i}, \tilde{X}_\infty \tilde{J}_\infty^i)_{i=0}^{l-1}$ is nonsingular (cf. Section 1.9). Since we can choose $\tilde{X}_F = \tilde{X}_F$, $\tilde{J}_F = J_F + \alpha I$, and the pair $(X_\infty, J_\infty(I + \alpha J_\infty)^{-1})$ is similar to $(\tilde{X}_\infty, \tilde{J}_\infty)$, the matrix $T_{l-1} \overset{\text{def}}{=} \operatorname{col}(X_F(J_F + \alpha I)^i, X_\infty J_\infty^{l-1-i}(I + \alpha J_\infty)^i)_{i=0}^{l-1}$ is also nonsingular. Finally, observe that

$$T_{l-1} = \begin{bmatrix} I & 0 & 0 & \cdots & 0 \\ \alpha I & I & 0 & \cdots & 0 \\ \binom{2}{0}\alpha^2 I & \binom{2}{1}\alpha I & \binom{2}{2} I & \cdots & 0 \\ \vdots & \vdots & \vdots & & \vdots \\ \binom{l}{0}\alpha^l I & \binom{l}{1}\alpha^{l-1} I & \binom{l}{2}\alpha^{l-2} I & \cdots & \binom{l}{l} I \end{bmatrix} S_{l-1}, \tag{7.14}$$

where S_{l-1} is given by (7.5). (This equality is checked by a straightforward computation.) So S_{l-1} is also invertible, and Theorem 7.3 is proved. □

7.4. Properties of Decomposable Pairs

This section is of an auxiliary character. We display here some simple properties of a decomposable pair, some of which will be used to prove main results in the next two sections.

Proposition 7.4. *Let $(X, Y) = ([X_1 \;\; X_2], T_1 \oplus T_2)$ be a decomposable pair of degree l, and let $S_{l-2} = \operatorname{col}(X_1 T_1^i, X_2 T_2^{l-2-i})_{i=0}^{l-2}$. Then*

(a) *S_{l-2} has full rank (equal to $n(l-1)$);*
(b) *the $nl \times nl$ matrix*

$$P = (I \oplus T_2) S_{l-1}^{-1} \begin{bmatrix} I \\ 0 \end{bmatrix} S_{l-2}, \tag{7.15}$$

where S_{l-1} is defined by (7.4), is a projector with $\operatorname{Ker} P = \operatorname{Ker} S_{l-2}$;
(c) *we have*

$$(I - P)(I \oplus T_2) S_{l-1}^{-1} = (I \oplus T_2) S_{l-1}^{-1} \begin{bmatrix} 0 & 0 \\ 0 & I_n \end{bmatrix}. \tag{7.16}$$

Proof. By forming the products indicated it is easily checked that

$$\begin{bmatrix} \lambda I & -I & 0 & \cdots & & 0 \\ 0 & \lambda I & -I & \cdots & & 0 \\ \vdots & \vdots & \vdots & & \vdots & \vdots \\ 0 & 0 & 0 & \cdots & \lambda I & -I \end{bmatrix} S_{l-1} = S_{l-2} T(\lambda), \tag{7.17}$$

where $T(\lambda) = (I\lambda - T_1) \oplus (T_2\lambda - I)$. Then put $\lambda = \lambda_0$ in (7.17), where λ_0 is such that $T(\lambda_0)$ is nonsingular, and use the nonsingularity of S_{l-1} to deduce that S_{l-2} has full rank.

Dividing (7.17) by λ and letting $\lambda \to \infty$, we obtain

$$[I \quad 0]S_{l-1} = S_{l-2}(I \oplus T_2). \tag{7.18}$$

Now, using (7.18)

$$P^2 = (I \oplus T_2)S_{l-1}^{-1}\begin{bmatrix} I \\ 0 \end{bmatrix} S_{l-2}(I \oplus T_2)S_{l-1}^{-1}\begin{bmatrix} I \\ 0 \end{bmatrix} S_{l-2}$$

$$= (I \oplus T_2)S_{l-1}^{-1}\begin{bmatrix} I \\ 0 \end{bmatrix}[I \quad 0]\begin{bmatrix} I \\ 0 \end{bmatrix} S_{l-2} = (I \oplus T_2)S_{l-1}^{-1}\begin{bmatrix} I \\ 0 \end{bmatrix} S_{l-2} = P,$$

so P is a projector.

Formula (7.15) shows that

$$\operatorname{Ker} P \supset \operatorname{Ker} S_{l-2}. \tag{7.19}$$

On the other hand,

$$S_{l-2}P = S_{l-2}(I \oplus T_2)S_{l-1}^{-1}\begin{bmatrix} I \\ 0 \end{bmatrix} S_{l-2} = S_{l-2}$$

in view of (7.18), so rank $P \geq$ rank $S_{l-2} = n(l-1)$. Combining with (7.19), we obtain Ker $P =$ Ker S_{l-2} as required.

Finally, to check (7.16), use the following steps:

$$P(I \oplus T_2)S_{l-1}^{-1} = (I \oplus T_2)S_{l-1}^{-1}\begin{bmatrix} I \\ 0 \end{bmatrix}\{S_{l-2}(I \oplus T_2)S_{l-1}^{-1}\}$$

$$= (I \oplus T_2)S_{l-1}^{-1}\begin{bmatrix} I \\ 0 \end{bmatrix}[I \quad 0] = (I \oplus T_2)S_{l-1}^{-1}\begin{bmatrix} I & 0 \\ 0 & 0_n \end{bmatrix},$$

where the latter equality follows from (7.18). □

The following technical property of decomposable pairs for matrix polynomials will also be useful.

Proposition 7.5. *Let* $([X_1 \quad X_2], T_1 \oplus T_2)$ *be a decomposable pair for the matrix polynomial* $L(\lambda) = \sum_{j=0}^{l} \lambda^j A_j$. *Then*

$$VP = 0, \tag{7.20}$$

where P *is given by* (7.15), *and*

$$V = \left[A_l X_1 T_1^{l-1}, \; -\sum_{i=0}^{l-1} A_i X_2 T_2^{l-1-i}\right].$$

Proof. Since

$$\sum_{i=0}^{l} A_i X_2 T_2^{l-i} = 0 \quad \text{and} \quad [X_1 T_1^{l-1}, X_2] = [0 \quad \cdots \quad 0 \quad I]S_{l-1},$$

we have

$$\begin{aligned} VP &= \left[A_l X_1 T_1^{l-1}, \; -\sum_{i=0}^{l-1} A_i X_2 T_2^{l-i}\right] S_{l-1}^{-1} \begin{bmatrix} I \\ 0 \end{bmatrix} S_{l-2} \\ &= A_l [X_1 T_1^{l-1}, X_2] S_{l-1}^{-1} \begin{bmatrix} I \\ 0 \end{bmatrix} S_{l-2} = A_l [0 \quad \cdots \quad 0 \quad I] \begin{bmatrix} I \\ 0 \end{bmatrix} S_{l-2} = 0. \quad \square \end{aligned}$$

We conclude this section with some remarks concerning similarity of decomposable pairs. First, decomposable pairs

$$(X, T) = ([X_1 \quad X_2], T_1 \oplus T_2), \qquad (\tilde{X}, \tilde{T}) = ([\tilde{X}_1 \quad \tilde{X}_2], \tilde{T}_1 \oplus \tilde{T}_2) \tag{7.21}$$

with the same parameter m are said to be *similar* if for some nonsingular matrices Q_1 and Q_2 of sizes $m \times m$ and $(nl - m) \times (nl - m)$, respectively, the relations

$$X_i = \tilde{X}_i Q_i, \qquad T_i = Q_i^{-1} \tilde{T}_i Q_i, \qquad i = 1, 2,$$

hold. Clearly, if (X, T) is a decomposable pair of a matrix polynomial $L(\lambda)$ (of degree l), then so is every decomposable pair which is similar to (X, T). The converse statement, that any two decomposable pairs of $L(\lambda)$ with the same parameter are similar, is more delicate. We prove that this is indeed the case if we impose certain spectral conditions on T_1 and T_2, as follows.

Let $L(\lambda) = \sum_{i=0}^{l} A_i \lambda^i$ be a matrix polynomial with decomposable pairs (7.21) and the same parameter. Let T_1 and $\tilde{T}_1$ be nonsingular and suppose that the following conditions hold:

$$\sigma(T_i) = \sigma(\tilde{T}_i), \qquad i = 1, 2, \tag{7.22}$$

$$\sigma(T_1^{-1}) \cap \sigma(T_2) = \sigma(\tilde{T}_1^{-1}) \cap \sigma(\tilde{T}_2) = \varnothing. \tag{7.23}$$

Then (X, T) and $(\tilde{X}, \tilde{T})$ are similar.

Let us prove this statement. We have

$$[A_0 \quad A_1 \quad \cdots \quad A_l] \operatorname{col}(X_1 T_1^i, X_2 T_2^{l-i})_{i=0}^l = 0 \tag{7.24}$$

or

$$A_0[X_1 \quad X_2 T_2^l] = -[A_1 \quad \cdots \quad A_l]S_{l-1}(T_1 \oplus I).$$

Since S_{l-1} and T_1 are nonsingular,

$$\operatorname{Im} A_0 \supset \operatorname{Im}[A_1 \quad \cdots \quad A_l],$$

and therefore A_0 must also be nonsingular (otherwise for all $\lambda \in \mathbb{C}$,

$$\operatorname{Im} L(\lambda) \subset \operatorname{Im}[A_0 \quad A_1 \quad \cdots \quad A_l] \neq \mathbb{C}^n,$$

a contradiction with the regularity of $L(\lambda)$). Consider the monic matrix polynomial $\tilde{L}(\lambda) = A_0^{-1}\lambda^l L(\lambda^{-1})$. Equality (7.24) shows that the standard pair

$$\left\{[I \quad 0 \quad \cdots \quad 0], \begin{bmatrix} 0 & I & 0 & \cdots & 0 \\ 0 & 0 & I & \cdots & 0 \\ \vdots & \vdots & \vdots & & \vdots \\ 0 & 0 & 0 & & I \\ -A_0^{-1}A_l & -A_0^{-1}A_{l-1} & -A_0^{-1}A_{l-2} & \cdots & -A_0^{-1}A_1 \end{bmatrix}\right\}$$

of $\tilde{L}(\lambda)$ is similar to $([X_1 \quad X_2], T_1^{-1} \oplus T_2)$, with similarity matrix $\operatorname{col}(X_1 T_1^{-i}, X_2 T_2^i)_{i=0}^{l-1}$ (this matrix is nonsingular in view of the nonsingularity of S_{l-1}). So $([X_1 \quad X_2], T_1^{-1} \oplus T_2)$ is a standard pair of $\tilde{L}(\lambda)$. Analogous arguments show that $([\tilde{X}_1 \quad \tilde{X}_2], \tilde{T}_1^{-1} \oplus \tilde{T}_2)$ is also a standard pair of $\tilde{L}(\lambda)$. But we know that any two standard pairs of a monic matrix polynomial are similar; so $T_1^{-1} \oplus T_2$ and $\tilde{T}_1^{-1} \oplus \tilde{T}_2$ are similar. Together with the spectral conditions (7.22) and (7.23) this implies the similarity of T_i and $\tilde{T}_i$, $i = 1, 2$. Now, using the fact that in a standard triple of $\tilde{L}(\lambda)$, the part corresponding to each eigenvalue is determined uniquely up to similarity, we deduce that (X, T) and $(\tilde{X}, \tilde{T})$ are similar, as claimed.

Note that the nonsingularity condition on T_1 and $\tilde{T}_1$ in the above statement may be dropped if, instead of (7.23), one requires that

$$\begin{aligned}\sigma((T_1 + \alpha I)^{-1} &\cap \sigma(T_2(I + \alpha T_2)^{-1}) \\ &= \sigma((\tilde{T}_1 + \alpha I)^{-1} \cap \sigma(\tilde{T}_2(I + \alpha \tilde{T}_2)^{-1}) = \varnothing, \end{aligned} \tag{7.25}$$

for some $\alpha \in \mathbb{C}$ such that the inverse matrices in (7.25) exist. This case is easily reduced to the case considered above, bearing in mind that

$$([X_1, X_2(I + \alpha T_2)^{l-1}], (T_1 + \alpha I) \oplus T_2(I + \alpha T_2)^{-1})$$

is a decomposable pair for $L(\lambda - \alpha)$, provided $([X_1 \quad X_2], T_1 \oplus T_2)$ is a decomposable pair for $L(\lambda)$. (This can be checked by straightforward calculation using (7.14).) Finally, it is easy to check that similarity of $T_2(I+\alpha T_2)^{-1}$ and $\tilde{T}_2(I + \alpha\tilde{T}_2)^{-1}$ implies (in fact, is equivalent to) the similarity of T_2 and $\tilde{T}_2$.

7.5. Decomposable Linearization and a Resolvent Form

We show now that a decomposable pair for a matrix polynomial $L(\lambda)$, introduced in Section 7.4, determines a linearization of $L(\lambda)$, as follows.

Theorem 7.6. *Let $L(\lambda) = \sum_{i=0}^{l} A_i\lambda^i$ be a regular matrix polynomial, and let $([X_1 \quad X_2], T_1 \oplus T_2)$ be its decomposable pair. Then $T(\lambda) = (I\lambda - T_1) \oplus (T_2\lambda - I)$ is a linearization of $L(\lambda)$. Moreover*

$$C_L(\lambda)S_{l-1} = \begin{bmatrix} S_{l-2} \\ V \end{bmatrix} T(\lambda), \tag{7.26}$$

where $V = [A_l X_1 T_1^{l-1}, -\sum_{i=0}^{l-1} A_i X_2 T_2^{l-1-i}]$ and

$$C_L(\lambda) = \begin{bmatrix} I & 0 & \cdots & 0 & 0 \\ 0 & I & \cdots & 0 & 0 \\ \vdots & \vdots & & \vdots & \vdots \\ 0 & 0 & \cdots & I & 0 \\ 0 & 0 & \cdots & 0 & A_l \end{bmatrix}\lambda + \begin{bmatrix} 0 & -I & 0 & \cdots & 0 \\ 0 & 0 & -I & \cdots & 0 \\ \vdots & \vdots & \vdots & & \vdots \\ 0 & 0 & 0 & \cdots & -I \\ A_0 & A_1 & A_2 & \cdots & A_{l-1} \end{bmatrix}$$

is the companion polynomial of $L(\lambda)$.

Proof. The equality of the first $n(l-1)$ rows of (7.26) coincides with (7.17). The equality in the last n rows of (7.26) follows from the definition of a decomposable pair for a matrix polynomial (property (iii)). The matrix $\left[\begin{smallmatrix} S_{l-2} \\ V \end{smallmatrix}\right]$ is nonsingular, because otherwise (7.26) would imply that $\det C_L(\lambda) \equiv 0$, which is impossible since $C_L(\lambda)$ is a linearization of $L(\lambda)$ (see Section 7.2) and $L(\lambda)$ is regular. Consequently, (7.26) implies that $T(\lambda)$ is a linearization of $L(\lambda)$ together with $C_L(\lambda)$, □

The linearization $T(\lambda)$ introduced in Theorem 7.6 will be called a *decomposable linearization* of $L(\lambda)$ (corresponding to the decomposable pair $([X_1 \quad X_2], T_1 \oplus T_2)$ of $L(\lambda)$).

Using a decomposable linearization, we now construct a resolvent form for a regular matrix polynomial.

Theorem 7.7. *Let $L(\lambda)$ be a matrix polynomial with decomposable pair $([X_1 \quad X_2], T_1 \oplus T_2)$ and corresponding decomposable linearization $T(\lambda)$. Put*

$V = [A_l X_1 T_1^{l-1}, -\sum_{i=0}^{l-1} A_i X_2 T_2^{l-1-i}]$, $S_{l-2} = \operatorname{col}(X_1 T_1^i, \ X_2 T_2^{l-2-i})_{i=0}^{l-2}$,

and

$$Z = [I \oplus T_2^{l-1}] \begin{bmatrix} S_{l-2} \\ V \end{bmatrix}^{-1} [0 \ \cdots \ 0 \ \ I]^{\mathrm{T}}.$$

Then

$$L^{-1}(\lambda) = [X_1 \quad X_2] T(\lambda)^{-1} Z. \tag{7.27}$$

Observe that the matrix $[{S_{l-2} \atop V}]$ in Theorem 7.7 is nonsingular, as we have seen in the proof of Theorem 7.6.

Proof. We shall use the equality (7.2), which can be rewritten in the form

$$\operatorname{diag}[L^{-1}(\lambda), I, \ldots, I] = D(\lambda) C_L^{-1}(\lambda) B^{-1}(\lambda),$$

where

$$B(\lambda) = \begin{bmatrix} B_1(\lambda) & B_2(\lambda) & \cdots & B_{l-1}(\lambda) & I \\ -I & 0 & \cdots & 0 & 0 \\ 0 & -I & \cdots & 0 & 0 \\ \vdots & \vdots & & \vdots & \vdots \\ 0 & 0 & \cdots & -I & 0 \end{bmatrix} \tag{7.28}$$

with some matrix polynomials $B_i(\lambda)$, $i = 1, \ldots, l-1$, and

$$D(\lambda) = \begin{bmatrix} I & 0 & 0 & \cdots & & 0 \\ -\lambda I & I & 0 & \cdots & & 0 \\ 0 & -\lambda I & I & \cdots & & 0 \\ \vdots & \vdots & \vdots & & \vdots & \vdots \\ 0 & 0 & 0 & \cdots & -\lambda I & I \end{bmatrix}. \tag{7.29}$$

Multiplying this relation by $[I \quad 0 \quad \cdots \quad 0]$ from the left and by

$$[I \quad 0 \quad \cdots \quad 0]^{\mathrm{T}}$$

from the right, and taking into account the form of $B(\lambda)$ and $D(\lambda)$ given by (7.28) and (7.29), respectively, we obtain

$$L^{-1}(\lambda) = [I \quad 0 \quad \cdots \quad 0] C_L(\lambda)^{-1} [0 \quad \cdots \quad 0 \quad I]^{\mathrm{T}}.$$

Using (7.26) it follows that

$$L^{-1}(\lambda) = [I \quad 0 \quad \cdots \quad 0] S_{l-1} T(\lambda)^{-1} \begin{bmatrix} S_{l-2} \\ V \end{bmatrix}^{-1} [0 \quad \cdots \quad 0 \quad I]^{\mathrm{T}}.$$

Now

$$[I \quad 0 \quad \cdots \quad 0]S_{l-1}T(\lambda)^{-1}\begin{bmatrix} S_{l-2} \\ V \end{bmatrix}^{-1} = [X_1 \quad X_2T_2^{l-1}]T(\lambda)^{-1}\begin{bmatrix} S_{l-2} \\ V \end{bmatrix}^{-1}$$

$$= [X_1 \quad X_2](I \oplus T_2^{l-1})T(\lambda)^{-1}\begin{bmatrix} S_{l-2} \\ V \end{bmatrix}^{-1}$$

$$= [X_1 \quad X_2]T(\lambda)^{-1}(I \oplus T_2^{l-1})\begin{bmatrix} S_{l-2} \\ V \end{bmatrix}^{-1},$$

and (7.27) follows. □

7.6. Representation and the Inverse Problem

We consider now the inverse problem, i.e., given a decomposable pair (X, T) of degree l, find all regular matrix polynomials (of degree l) which have (X, T) as their decomposable pair. The following theorem provides a description of all such matrix polynomials.

Theorem 7.8. *Let* $(X, T) = ([X_1 \quad X_2], T_1 \oplus T_2)$ *be a decomposable pair of degree* l, *and let* $S_{l-2} = \operatorname{col}(X_1T_1^i, X_2T_2^{l-2-i})_{i=0}^{l-2}$. *Then for every* $n \times nl$ *matrix* V *such that the matrix* $\left[\begin{smallmatrix} S_{l-2} \\ V \end{smallmatrix}\right]$ *is nonsingular, the matrix polynomial*

$$L(\lambda) = V(I - P)[(I\lambda - T_1) \oplus (T_2\lambda - I)](U_0 + U_1\lambda + \cdots + U_{l-1}\lambda^{l-1}), \tag{7.30}$$

where $P = (I \oplus T_2)[\operatorname{col}(X_1T_1^i, X_2T_2^{l-1-i})_{i=0}^{l-1}]^{-1}\left[\begin{smallmatrix} I \\ 0 \end{smallmatrix}\right]S_{l-2}$ *and*

$$[U_0 \quad U_1 \quad \cdots \quad U_{l-1}] = [\operatorname{col}(X_1T_1^i, X_2T_2^{l-1-i})_{i=0}^{l-1}]^{-1},$$

has (X, T) *as its decomposable pair.*

Conversely, if $L(\lambda) = \sum_{i=0}^{l} A_i\lambda^i$ *has* (X, T) *as its decomposable pair, then* $L(\lambda)$ *admits representation* (7.30) *with*

$$V = V(I - P) = \left[A_lX_1T_1^{l-1}, \ -\sum_{i=0}^{l-1} A_iX_2T_2^{l-1-i}\right]. \tag{7.31}$$

For future reference, let us write formula (7.30) explicitly, with V given by (7.31):

$$L(\lambda) = \left[A_lX_1T_1^{l-1}, \ -\sum_{i=0}^{l-1} A_jX_2T_2^{l-1-i}\right][(I\lambda - T_1) \oplus (T_2\lambda - I)]$$
$$\times (U_0 + U_1\lambda + \cdots + U_{l-1}\lambda^{l-1}). \tag{7.32}$$

Proof. Let V be an $n \times nl$ matrix such that $\left[\begin{smallmatrix} S_{l-2} \\ V \end{smallmatrix}\right]$ is nonsingular, and let $L(\lambda) = \sum_{i=0}^{l} A_i \lambda^i$ be defined by (7.30). We show that (X, T) is a decomposable pair of $L(\lambda)$.

Define S_{l-1} and P by (7.4) and (7.15), respectively, and put $W = V(I - P)$. Since P is a projector,

$$WP = 0. \tag{7.33}$$

Besides,

$$\begin{bmatrix} S_{l-2} \\ W \end{bmatrix} = \begin{bmatrix} I & 0 \\ -V(I \oplus T_2)S_{l-1}^{-1}\left[\begin{smallmatrix} I \\ 0 \end{smallmatrix}\right] & I \end{bmatrix} \begin{bmatrix} S_{l-2} \\ V \end{bmatrix};$$

hence $\left[\begin{smallmatrix} S_{l-2} \\ W \end{smallmatrix}\right]$ is nonsingular.

Using (7.33) and Proposition 7.4(c), we obtain

$$\begin{aligned} W(I \oplus T_2)S_{l-1}^{-1} &= W(I - P)(I \oplus T_2)S_{l-1}^{-1} \\ &= W(I \oplus T_2)S_{l-1}^{-1}\begin{bmatrix} 0 & 0 \\ 0 & I_n \end{bmatrix} = [0 \quad \cdots \quad 0 \quad A_l]; \end{aligned} \tag{7.34}$$

and therefore by the definition of $L(\lambda)$

$$W(T_1 \oplus I)S_{l-1}^{-1} = -[A_0 \quad A_1 \quad \cdots \quad A_{l-1}]. \tag{7.35}$$

Combining (7.34) and (7.35), we obtain

$$C_L(\lambda)S_{l-1} = \begin{bmatrix} S_{l-1} \\ W \end{bmatrix} T(\lambda), \tag{7.36}$$

where $C_L(\lambda)$ is the companion polynomial of $L(\lambda)$ and $T(\lambda) = (I\lambda - T_1) \oplus (T_2\lambda - I)$. Comparison with (7.26) shows that in fact

$$W = \left[A_l X_1 T_1^{l-1}, -\sum_{i=0}^{l-1} A_i X_2 T_2^{l-1-i}\right]. \tag{7.37}$$

The bottom row of the equation obtained by substituting (7.37) in (7.36) takes the form

$$\begin{aligned} ([0 \quad \cdots \quad 0 \quad A_l]\lambda + [A_0, \ldots, A_{l-1}]) &\begin{bmatrix} X_1 & X_2 T_2^{l-1} \\ X_1 T_1 & X_2 T_2^{l-2} \\ \vdots & \vdots \\ X_1 T_1^{l-1} & X_2 \end{bmatrix} \\ = \left[A_l X_1 T_1^{l-1}, -\sum_{i=0}^{l=1} A_i X_2 T_2^{l-1-i}\right] &\begin{bmatrix} I\lambda - T_1 & 0 \\ 0 & T_2\lambda - I \end{bmatrix}. \end{aligned}$$

This implies immediately that

$$\sum_{i=0}^{l} A_i X_1 T_1^i = 0, \qquad \sum_{l=0}^{l} A_i X_2 T_2^{l-i} = 0,$$

i.e., (X, T) is a decomposable pair for $L(\lambda)$.

We now prove the converse statement. Let $L(\lambda) = \sum_{i=0}^{l} A_i \lambda^i$ be a regular matrix polynomial with decomposable pair (X, T). First observe that, using (7.17), we may deduce

$$S_{l-2} T(\lambda) S_{l-1}^{-1} \operatorname{col}(I\lambda^i)_{i=0}^{l-1} = 0. \tag{7.38}$$

Then it follows from (7.2) that

$$L(\lambda) \oplus I_{n(l-1)} = \begin{bmatrix} * & I_n \\ -I_{n(l-1)} & 0 \end{bmatrix} C_L(\lambda) \begin{bmatrix} I & 0 & \cdots & 0 \\ I\lambda & I & & \vdots \\ \vdots & \vdots & \ddots & 0 \\ I\lambda^{l-1} & & \cdots & I \end{bmatrix}$$

where the star denotes a matrix entry of no significance. Hence,

$$L(\lambda) = [* \quad \cdots \quad * \quad I_n] C_L(\lambda) \operatorname{col}(I\lambda^i)_{i=0}^{l-1}. \tag{7.39}$$

Now use Theorem 7.6 (Eq. (7.26)) to substitute for $C_L(\lambda)$ and write

$$\begin{aligned} L(\lambda) &= [* \quad \cdots \quad * \quad I_n] \begin{bmatrix} S_{l-2} \\ V \end{bmatrix} T(\lambda) S_{l-1}^{-1} \operatorname{col}(I\lambda^i)_{i=0}^{l-1} \\ &= [* \quad \cdots \quad *] S_{l-2} T(\lambda) S_{l-1}^{-1} \operatorname{col}(I\lambda^i)_{i=0}^{l-1} + V T(\lambda) S_{l-1}^{-1} \operatorname{col}(I\lambda^i)_{i=0}^{l-1}. \end{aligned}$$

But the first term on the right is zero by (7.38), and the conclusion (7.30) follows since $VP = 0$ by Proposition 7.5. □

Note that the right canonical form of a monic matrix polynomial (Theorem 2.4) can be easily obtained as a particular case of representation (7.32). Indeed, let $L(\lambda) = I\lambda^l + \sum_{i=0}^{l-1} A_i \lambda^i$, and let $(X, T) = (X_F, J_F)$ be its finite Jordan pair, which is at the same time its decomposable pair (Theorem 7.3). Then formula (7.32) gives

$$\begin{aligned} L(\lambda) &= XT^{l-1} T(\lambda) \left(\sum_{i=0}^{l-1} U_i \lambda^i \right) = XT^{l-1} (I\lambda - T) \left(\sum_{i=0}^{l-1} U_i \lambda^i \right) \\ &= \lambda X T^{l-1} [\operatorname{col}(XT^i)_{i=0}^{l-1}]^{-1} [\lambda I, \lambda^2 I, \ \ldots, \ \lambda^l I]^{\mathrm{T}} - XT^l \left(\sum_{i=0}^{l-1} U_i \lambda^i \right) \\ &= \lambda [0 \quad \cdots \quad 0 \quad I] [\lambda I, \lambda^2 I, \ldots, \lambda^l I]^{\mathrm{T}} - XT^l \left(\sum_{i=0}^{l-1} U_i \lambda^i \right) \\ &= I\lambda^l - XT^l \left(\sum_{i=0}^{l-1} U_i \lambda^i \right), \end{aligned}$$

which coincides with (2.14).

Theorem 7.8 shows, in particular, that for each decomposable pair (X, T) there are associated matrix polynomials $L(\lambda)$, and all of them are given by formula (7.30). It turns out that such an $L(\lambda)$ is essentially unique: namely, if $L(\lambda)$ and $\tilde{L}(\lambda)$ are regular matrix polynomials of degree l with the same decomposable pair (X, Y), then

$$L(\lambda) = Q\tilde{L}(\lambda) \tag{7.40}$$

for some constant nonsingular matrix Q. (Note that the converse is trivial: if $L(\lambda)$ and $\tilde{L}(\lambda)$ are related as in (7.40), then they have the same decomposable pairs.) More exactly, the following result holds.

Theorem 7.9. *Let (X, T) be a decomposable pair as in Theorem 7.8, and let V_1 and V_2 be $n \times nl$ matrices such that $\left[\begin{smallmatrix} S_{l-2} \\ V_i \end{smallmatrix}\right]$ is nonsingular, $i = 1, 2$. Then*

$$L_{V_1}(\lambda) = QL_{V_2}(\lambda),$$

where $L_{V_i}(\lambda)$ is given by (7.30) *with $V = V_i$, and*

$$Q = (V_{1|\operatorname{Ker} S_{l-2}}) \cdot (V_{2|\operatorname{Ker} S_{l-2}})^{-1}. \tag{7.41}$$

Note that, in view of the nonsingularity of $\left[\begin{smallmatrix} S_{l-2} \\ V_i \end{smallmatrix}\right]$, the restrictions $V_{i|\operatorname{Ker} S_{l-2}}$, $i = 1, 2$ are invertible; consequently, Q is a nonsingular matrix.

Proof. We have to check only that

$$QV_2(I - P) = V_1(I - P). \tag{7.42}$$

But both sides of (7.42) are equal to zero when restricted to Im P. On the other hand, in view of Proposition 7.4(c) and (7.41), both sides of (7.42) are equal when restricted to Ker P. Since $\mathbb{C}^{nl}$ is a direct sum of Ker P and Im P, (7.42) follows. □

We conclude this section by an illustrative example.

EXAMPLE 7.2. Let $X_1 = X_2 = I$, $T_1 = 0$, $T_2 = aI$, $a \neq 0$ (all matrices are 2×2). Then $([X_1 \ \ X_2], T_1 \oplus T_2)$ is a decomposable pair (where $l = 2$). According to Theorem 7.8, all matrix polynomials $L(\lambda)$ for which $[(X_1 \ \ X_2], T_1 \oplus T_2)$ is their decomposable pair are given by (7.30). A computation shows that

$$L(\lambda) = (\lambda - a\lambda^2)(V_1 - V_2), \tag{7.43}$$

where V_1 and V_2 are any 2×2 matrices such that

$$\begin{bmatrix} I & I \\ V_1 & V_2 \end{bmatrix}$$

is nonsingular. Of course, one can rewrite (7.43) in the form $L(\lambda) = V(\lambda - a\lambda^2)$, where V is any nonsingular 2×2 matrix, as it should be by Theorem 7.9. □

7.7. Divisibility of Matrix Polynomials

In this section we shall characterize divisibility of matrix polynomials in terms of their spectral data at the finite points of the spectrum, i.e., the data contained in a finite Jordan pair. This characterization is given by the following theorem.

Theorem 7.10. *Let $B(\lambda)$ and $L(\lambda)$ be regular matrix polynomials. Assume $B(\lambda) = \sum_{j=0}^{m} B_j \lambda^j$, $B_m \neq 0$, and let (X_F, J_F) be the finite Jordan pair of $L(\lambda)$. Then $L(\lambda)$ is a right divisor of $B(\lambda)$, i.e., $B(\lambda)L(\lambda)^{-1}$ is also a polynomial, if and only if*

$$\sum_{j=0}^{m} B_j X_F J_F^j = 0. \tag{7.44}$$

We shall need some preparation for the proof of Theorem 7.10.

First, without loss of generality we can (and will) assume that $L(\lambda)$ is comonic, i.e., $L(0) = I$. Indeed, a matrix polynomial $L(\lambda)$ is a right divisor of a matrix polynomial $B(\lambda)$ if and only if the comonic polynomial $L^{-1}(\alpha)L(\lambda + \alpha)$ is a right divisor of the comonic polynomial $B^{-1}(\alpha)B(\lambda + \alpha)$. Here $\alpha \in \mathbb{C}$ is such that both $L(\alpha)$ and $B(\alpha)$ are nonsingular, existence of such an α is ensured by the regularity conditions $\det B(\lambda) \not\equiv 0$ and $\det L(\lambda) \not\equiv 0$. Observe also that (X_F, J_F) is a finite Jordan pair of $L(\lambda)$ if and only if $(X_F, J_F + \alpha I)$ is a finite Jordan pair of $L^{-1}(\alpha)L(\lambda + \alpha)$. This fact may be verified easily using Theorem 7.1.

Second, we describe the process of division of $B(\lambda)$ by $L(\lambda)$ (assuming $L(\lambda)$ is comonic). Let l be the degree of $L(\lambda)$, and let (X_∞, J_∞) be the infinite Jordan pair of $L(\lambda)$. Put

$$X = [X_F \quad X_\infty], \qquad J = J_F^{-1} \oplus J_\infty. \tag{7.45}$$

We shall use the fact that (X, J) is a standard pair for the monic matrix polynomial $\tilde{L}(\lambda) = \lambda^l L(\lambda^{-1})$ (see Theorem 7.15 below). In particular, $\mathrm{col}(XJ^{i-1})_{i=1}^{l}$ is nonsingular. For $\alpha \geq 0$ and $1 \leq \beta \leq l$ set $F_{\alpha\beta} = XJ^\alpha Z_\beta$, where

$$\mathrm{row}(Z_\beta)_{\beta=1}^{l} = [\mathrm{col}(XJ^{i-1})_{i=1}^{l}]^{-1}. \tag{7.46}$$

Further, for each $\alpha \geq 0$ and $\beta \leq 0$ or $\beta > l$ put $F_{\alpha\beta} = 0$. With this choice of $F_{\alpha\beta}$ the following formulas hold:

$$L(\lambda) = I - \sum_{j=1}^{\infty} \lambda^j F_{l,l+1-j}, \tag{7.47}$$

and

$$F_{\alpha+1,\beta} = F_{\alpha l} F_{l\beta} + F_{\alpha,\beta-1}. \tag{7.48}$$

Indeed, formula (7.47) is an immediate consequence from the right canonical form for $\tilde{L}(\lambda)$ using its standard pair (X, J). Formula (7.48) coincides (up to the notation) with (3.15).

The process of division of $B(\lambda)$ by $L(\lambda)$ can now be described as follows:

$$B(\lambda) = Q_k(\lambda)L(\lambda) + R_k(\lambda), \qquad k = 1, 2, \ldots, \tag{7.49}$$

where

$$Q_k(\lambda) = \sum_{j=0}^{k-1} \lambda^j \left(B_j + \sum_{i=0}^{j-1} B_i F_{l+j-1-i,\,l} \right),$$

$$R_k(\lambda) = \sum_{j=k}^{\infty} \lambda^j \left(B_j + \sum_{i=0}^{k-1} B_i F_{l+k-1-i,\,l+k-j} \right), \tag{7.50}$$

and $B_j = 0$ for $j > m$. We verify (7.49) using induction on k. For $k = 1$ (7.49) is trivial. Assume that (7.49) holds for some $k \geq 1$. Then using the recursion (7.48), one sees that

$$\begin{aligned} B(\lambda) - Q_{k+1}(\lambda)L(\lambda) &= R_k(\lambda) - \lambda^k \left(B_k + \sum_{i=0}^{k-1} B_i F_{l+k-1-i,\,l} \right) L(\lambda) \\ &= \sum_{j=k+1}^{\infty} \lambda^j \left(B_j + \sum_{i=0}^{k-1} B_i F_{l+k-1-i,\,l+k-j} \right) \\ &\quad + \sum_{j=k+1}^{\infty} \lambda^j \left(B_k + \sum_{i=0}^{k-1} B_i F_{l+k-1-i,\,l} \right) F_{l,\,l+k+1-j} \\ &= \sum_{j=k+1}^{\infty} \lambda^j \left(B_j + B_k F_{l,\,l+k+1-j} + \sum_{i=0}^{k-1} B_i F_{l+k-i,\,l+k+1-j} \right) \\ &= R_{k+1}(\lambda), \end{aligned}$$

and the induction is complete.

We say that a matrix polynomial $R(\lambda)$ has *codegree* k if $R^{(i)}(0) = 0$, $i = 0, \ldots, k-1$, and $R^{(k)}(0) \neq 0$ (i.e., the first nonzero coefficient in $R(\lambda) = \sum_{j=0}^{p} R_j \lambda^j$ is R_k). If $B(\lambda) = Q(\lambda)L(\lambda) + R(\lambda)$ where $Q(\lambda)$ and $R(\lambda)$ are $n \times n$ matrix polynomials such that degree $Q(\lambda) \leq k - 1$ and codegree $R(\lambda) \geq k$, then $Q(\lambda) = Q_k(\lambda)$ and $R(\lambda) = R_k(\lambda)$. Indeed, since $L(\lambda)$ is comonic, under the stated conditions the coefficients of $Q(\lambda) = \sum_{i=0}^{k-1} Q_i \lambda^i$ are uniquely determined by those of $B(\lambda)$ and $L(\lambda)$ (first we determine Q_0, then Q_1, Q_2, and so on), and hence the same is true for $R(\lambda)$. It follows that $L(\lambda)$ *is a right divisor of* $B(\lambda)$ *if and only if for some* k *the remainder* $R_k(\lambda) = 0$.

Proof of Theorem 7.10. Using the definition of $F_{\alpha\beta}$ and the property $XJ^{i-1}Z_j = \delta_{ij} I$ $(1 \leq i, j \leq l)$, rewrite (7.50) in the form

$$R_k(\lambda) = \sum_{j=k+l}^{\infty} B_j \lambda^j + \sum_{i=0}^{l+k-1} B_i X J^{l+k-1-i} \left(\sum_{\beta=1}^{l} Z_\beta \lambda^{l+k-\beta} \right).$$

It follows that $R_k(\lambda) = 0$ if and only if

$$l + k > m, \qquad \sum_{i=0}^{m} B_i X J^{l+k-1-i} = 0. \tag{7.51}$$

Now suppose that $L(\lambda)$ is a right divisor of $B(\lambda)$. Then there exists $k \geq 1$ such that (7.51) holds. Because of formula (7.45) this implies

$$\sum_{i=0}^{m} B_i X_{\mathrm{F}} J_{\mathrm{F}}^{-l-k+1+i} = 0.$$

Multiplying the left- and right-hand sides of this identity by J_{F}^{l+k-1} yields the desired formula (7.44).

Conversely, suppose that (7.44) holds. Let v be a positive integer such that $J_\infty^v = 0$. Choose $k \geq 1$ such that $l + k \geq m + v$. Multiply the left- and right-hand sides of (7.44) by J_{F}^{-l-k+1}. This gives

$$\sum_{i=0}^{m} B_i X_{\mathrm{F}} J_{\mathrm{F}}^{-(l+k-1-i)} = 0.$$

As $X_\infty J_\infty^{l+k-1-i} = 0$ for $i = 0, \ldots, m$, we see that with this choice of k, formula (7.51) obtains. Hence $L(\lambda)$ is a right divisor of $B(\lambda)$. □

Corollary 7.11. *Let $B(\lambda)$ and $L(\lambda)$ be regular $n \times n$ matrix polynomials. Then $L(\lambda)$ is a right divisor of $B(\lambda)$ if and only if each Jordan chain of $L(\lambda)$ is a Jordan chain of $B(\lambda)$ corresponding to the same eigenvalue.*

Proof. If $L(\lambda)$ is a right divisor of $B(\lambda)$, then one sees through a direct computation (using root polynomials, for example) that each Jordan chain of $L(\lambda)$ is a Jordan chain for $B(\lambda)$ corresponding to the same eigenvalue as for $L(\lambda)$.

Conversely, if each Jordan chain of $L(\lambda)$ is a Jordan chain for $B(\lambda)$ corresponding to the same eigenvalue, then by Proposition 1.10, (7.44) holds and so $L(\lambda)$ is a right divisor of $B(\lambda)$. □

Note that in the proof of Theorem 7.10 we did not use the comonicity of $B(\lambda)$. It follows that Theorem 7.10 is valid for any pair of $n \times n$ matrix polynomials $B(\lambda)$ and $L(\lambda)$, provided $L(\lambda)$ is regular. A similar remark applies to the previous corollary.

Next we show that a description of divisibility of regular polynomials can also be given in terms of their finite Jordan chains. In order to formulate this criterion we shall need the notions of restriction and extension of admissible pairs.

As before, a pair of matrices (X, T) is called an *admissible* pair of *order p* if X is $n \times p$ and T is $p \times p$. The space

$$\operatorname{Ker}(X, T) = \bigcap_{j=0}^{\infty} \operatorname{Ker} XT^j = \bigcap_{s=0}^{\infty} \operatorname{Ker}[\operatorname{col}(XT^{j-1})_{j=1}^{s}]$$

is called the *kernel* of the pair (X, T). The least positive integer s such that

$$\operatorname{Ker}[\operatorname{col}(XT^{j-1})_{j=1}^{s}] = \operatorname{Ker}[\operatorname{col}(XT^{j-1})_{j=1}^{s+1}]$$

will be called the *index of stabilization* of the pair (X, T); it will be denoted by $\operatorname{ind}(X, T)$. Since the subspaces $\mathscr{M}_s = \operatorname{Ker}[\operatorname{col}(XT^{j-1})_{j=1}^{s}] \subset \mathbb{C}^p$ form a descending chain $\mathscr{M}_1 \supset \mathscr{M}_2 \supset \cdots \supset \mathscr{M}_s \supset \mathscr{M}_{s+1}$, there always exists a minimal s such that $\mathscr{M}_s = \mathscr{M}_{s+1}$, i.e., $\operatorname{ind}(X, T)$ is defined correctly.

Given two admissible pairs (X, T) and (X_1, T_1) of orders p and p_1, respectively, we call (X, T) an *extension* of (X_1, T_1) (or, equivalently, (X_1, T_1) a *restriction* of (X, T)) if there exists an injective linear transformation $S: \mathbb{C}^{p_1} \to \mathbb{C}^p$ such that $XS = X_1$ and $TS = ST_1$. Observe that in this case $\operatorname{Im} S$ is invariant under T. Admissible pairs (X, T) and (X_1, T_1) are called *similar* if each is an extension of the other; in this case their orders are equal and $XU = X_1$, $U^{-1}TU = T_1$ for some invertible linear transformation U. The definition of extension given above is easily seen to be consistent with the discussion of Section 6.1. Thus, the pair $(X|_{\operatorname{Im} S}, T|_{\operatorname{Im} S})$ is similar to (X_1, T_1), with the similarity transformation $S: \mathbb{C}^{p_1} \to \operatorname{Im} S$.

An important example of an admissible pair is a finite Jordan pair (X_F, J_F) of a comonic matrix polynomial $L(\lambda)$. It follows from Theorem 7.3 that the kernel of the admissible pair (X_F, J_F) is zero.

We shall need the following characterization of extension of admissible pairs.

Lemma 7.12. *Let (X, T) and (X_1, T_1) be admissible pairs. If (X, T) is an extension of (X_1, T_1), then*

$$\operatorname{Im}[\operatorname{col}(X_1T_1^{i-1})_{i=1}^{m}] \subset \operatorname{Im}[\operatorname{col}(XT^{i-1})_{i=1}^{m}], \qquad m \geq 1.$$

Conversely, if (X_1, T_1) has a zero kernel and if

$$\operatorname{Im}[\operatorname{col}(X_1T_1^{i-1})_{i=1}^{k+1}] \subset \operatorname{Im}[\operatorname{col}(XT^{i-1})_{i=1}^{k+1}], \tag{7.52}$$

where k is some fixed positive integer such that $k \geq \operatorname{ind}(X, T)$ and $k \geq \operatorname{ind}(X_1, T_1)$, then (X, T) is an extension of (X_1, T_1).

Proof. The first part of the lemma is trivial. To prove the second part: first suppose that both (X_1, T_1) and (X, T) have zero kernel. Our choice of k implies

$$\operatorname{Ker}[\operatorname{col}(XT^{i-1})_{i=1}^{k}] = \{0\}, \qquad \operatorname{Ker}[\operatorname{col}(X_1T_1^{i-1})_{i=1}^{k}] = \{0\}.$$

Put $\Omega = \operatorname{col}(XT^{i-1})_{i=1}^{i}$ and $\Omega_1 = \operatorname{col}(X_1T_1^{i-1})_{i=1}^{k}$, and let p and p_1 be the orders of the pairs (X, T) and (X_1, T_1). From our hypothesis if follows that $\operatorname{Im} \Omega_1 \subset \operatorname{Im} \Omega$. Hence, there exists a linear transformation $S: \mathbb{C}^{p_1} \to \mathbb{C}^p$ such that $\Omega S = \Omega_1$. As $\operatorname{Ker} \Omega_1 = \{0\}$, we see that S is injective. Further, from

$\operatorname{Ker} \Omega = \{0\}$ it follows that S is uniquely determined. Note that $\Omega S = \Omega_1$ implies that $XS = X_1$. To prove that $TS = ST_1$, let us consider

$$\tilde{\Omega} = \operatorname{col}(XT^{i-1})_{i=1}^{k+1}, \qquad \tilde{\Omega}_1 = \operatorname{col}(X_1 T_1^{i-1})_{i=1}^{k+1}.$$

As before, there exists a linear transformation $\tilde{S}: \mathbb{C}^{p_1} \to \mathbb{C}^p$ such that $\tilde{\Omega}\tilde{S} = \tilde{\Omega}_1$. The last equality implies $\Omega\tilde{S} = \Omega_1$, and thus $S = \tilde{S}$. But then $\Omega TS = \Omega T\tilde{S} = \Omega_1 T_1 = \Omega S T_1$. It follows that $\Omega(TS - ST_1) = 0$. As $\operatorname{Ker} \Omega = \{0\}$, we have $TS = ST_1$. This completes the proof of the lemma for the case that both (X, T) and (X_1, T_1) have zero kernel.

Next, suppose that $\operatorname{Ker}(X, T) \neq \{0\}$. Let p be the order of the pair (X, T), and let Q be a projector of $\mathbb{C}^p$ along $\operatorname{Ker}(X, T)$. Put

$$X_0 = X|_{\operatorname{Im} Q}, \qquad T_0 = QT|_{\operatorname{Im} Q}.$$

Then $\operatorname{Ker}(X_0, T_0) = \{0\}$ and for all $m \geq 1$

$$\operatorname{Im}[\operatorname{col}(X_0 T_0^{i-1})_{i=1}^m] = \operatorname{Im}[\operatorname{col}(XT^{i-1})_{i=1}^m].$$

As, moreover, $\operatorname{ind}(X_0, T_0) = \operatorname{ind}(X, T)$, we see from the result proved in the previous paragraph that (7.52) implies that (X_0, T_0) is an extension of (X_1, T_1). But, trivially, (X, T) is an extension of (X_0, T_0). So it follows that (X, T) is an extension of (X_1, T_1), and the proof is complete. □

Now we can state and prove the promised characterization of divisibility of regular matrix polynomials in terms of their finite Jordan pairs.

Theorem 7.13. *Let $B(\lambda)$ and $L(\lambda)$ be regular $n \times n$ matrix polynomials and let (X_B, J_B) and (X_L, J_L) be finite Jordan pairs of $B(\lambda)$ and $L(\lambda)$, respectively. Then $L(\lambda)$ is a right divisor of $B(\lambda)$ if and only if (X_B, J_B) is an extension of (X_L, J_L).*

Proof. Let q be the degree of $\det B(\lambda)$, and let p be the degree of $\det L(\lambda)$. Then (X_B, J_B) and (X_L, J_L) are admissible pairs of orders q and p, respectively. Suppose that there exists an injective linear transformation $S: \mathbb{C}^p \to \mathbb{C}^q$ such that $X_B S = X_L$, $J_B S = SJ_L$. Then $X_B J_B^j S = X_L J_L^j$ for each $j \geq 0$. Now assume that $B(\lambda) = \sum_{j=0}^m B_j \lambda^j$. Then

$$\sum_{j=0}^m B_j X_L J_L^j = \left(\sum_{j=0}^m B_j X_B J_B^j\right) S = 0.$$

So we can apply Theorem 7.10 to show that $L(\lambda)$ is a right divisor of $B(\lambda)$.

Conversely, suppose that $L(\lambda)$ is a right divisor of $B(\lambda)$. Then each chain of $L(\lambda)$ is a Jordan chain of $B(\lambda)$ corresponding to the same eigenvalue as for $L(\lambda)$ (see Corollary 7.11). Hence

$$\operatorname{Im}[\operatorname{col}(X_L J_L^{i-1})_{i=1}^m] \subset \operatorname{Im}[\operatorname{col}(X_B J_B^{i-1})_{i=1}^m]$$

for each $m \geq 1$. As $\mathrm{Ker}(X_L, J_L)$ and $\mathrm{Ker}(X_B, J_B)$ consist of the zero element only, it follows that we can apply Lemma 7.12 to show that (X_B, J_B) is an extension of (X_L, J_L). □

7.8. Representation Theorems for Comonic Matrix Polynomials

In this section the results of Section 7.5 and 7.6 will be considered in the special case of comonic matrix polynomials, in which case simpler representations are available. Recall that $L(\lambda)$ is called *comonic* if $L(0) = I$. Clearly, such a polynomial is regular.

Let (X_F, J_F) and (X_∞, J_∞) be finite and infinite Jordan pairs, respectively, for the comonic polynomial $L(\lambda) = I + \sum_{j=1}^{l} A_j \lambda^j$. By Theorem 7.3, $([X_F \quad X_\infty], J_F \oplus J_\infty)$ is a decomposable pair of $L(\lambda)$. Moreover, since $L(0) = I$, the matrix J_F is nonsingular. Let $J = J_F^{-1} \oplus J_\infty$, and call the pair $(X, J) = ([X_F \quad X_\infty], J_F^{-1} \oplus J_\infty)$ a *comonic Jordan pair* of $L(\lambda)$. Representations of a comonic matrix polynomial are conveniently expressed in terms of a comonic Jordan pair, as we shall see shortly.

We start with some simple properties of the comonic Jordan pair (X, J). It follows from Theorem 7.3 and property (iii) in the definition of a decomposable pair of $L(\lambda)$ that the $nl \times nl$ matrix $\mathrm{col}(XJ^i)_{i=0}^{l-1}$ is nonsingular. Further,

$$\mathrm{col}(XJ^i)_{i=0}^{l-1} J = R\ \mathrm{col}(XJ^i)_{i=0}^{l-1}, \tag{7.53}$$

where R is the comonic companion matrix of $L(\lambda)$ (see Section 7.2). Indeed, (7.53) is equivalent to

$$XJ^l + A_l X + A_{l-1} XJ + \cdots + A_1 XJ^{l-1} = 0,$$

which follows again from Theorem 7.3. In particular, (7.53) proves that J and R are similar.

The following theorem is a more explicit version of Theorem 7.7 applicable in the comonic case.

Theorem 7.14. *Let $L(\lambda)$ be a comonic matrix polynomial of degree l, and let (X, J) be its comonic Jordan pair. Then the $nl \times nl$ matrix $Y = \mathrm{col}(XJ^i)_{i=0}^{l-1}$ is nonsingular and, if $\lambda^{-1} \notin \sigma(J)$,*

$$L^{-1}(\lambda) = XJ^{l-1}(I - J\lambda)^{-1} Y^{-1} [0 \quad 0 \quad \cdots \quad I]^{\mathrm{T}}. \tag{7.54}$$

Proof. Nonsingularity of Y has been observed already.

Apply formula (7.27) with $X_1 = X_F, X_2 = X_\infty, T_1 = J_F, T_2 = J_\infty$;

$$L^{-1}(\lambda) = X\begin{bmatrix} I\lambda - J_F & 0 \\ 0 & J_\infty\lambda - I \end{bmatrix}^{-1} Z = X(I - J\lambda)^{-1}\begin{bmatrix} -J_F^{-1} & 0 \\ 0 & -I \end{bmatrix} Z,$$

where

$$Z = [I \oplus J_\infty^{l-1}]\begin{bmatrix} S_{l-2} \\ W \end{bmatrix}^{-1} [0 \quad \cdots \quad 0 \quad I]^T$$

in the notation of Theorem 7.7. To prove (7.54), it remains to check the equality

$$\begin{bmatrix} -J_F^{-1} & 0 \\ 0 & -I \end{bmatrix}\begin{bmatrix} I & 0 \\ 0 & J_\infty^{l-1} \end{bmatrix}\begin{bmatrix} S_{l-2} \\ W \end{bmatrix}^{-1}\begin{bmatrix} 0 \\ \vdots \\ 0 \\ I \end{bmatrix} = \begin{bmatrix} J_F^{-(l-1)} & 0 \\ 0 & J_\infty^{l-1} \end{bmatrix} Y^{-1}\begin{bmatrix} 0 \\ 0 \\ \vdots \\ I \end{bmatrix}. \tag{7.55}$$

From (7.26) we have

$$\begin{bmatrix} -J_F^{-1} & 0 \\ 0 & -I \end{bmatrix}\begin{bmatrix} S_{l-2} \\ W \end{bmatrix}^{-1} [0 \quad \cdots \quad 0 \quad I]^T$$

$$= \begin{bmatrix} -J_F^{-1}\lambda + I & 0 \\ 0 & -J_\infty\lambda + I \end{bmatrix} S_{l-1}^{-1} C_L(\lambda)^{-1}\begin{bmatrix} 0 \\ \vdots \\ 0 \\ I \end{bmatrix}.$$

Put $\lambda = 0$ in the right-hand side to obtain

$$\begin{bmatrix} -J_F^{-1} & 0 \\ 0 & -I \end{bmatrix}\begin{bmatrix} S_{l-2} \\ W \end{bmatrix}^{-1} [0 \quad \cdots \quad 0 \quad I]^T = S_{l-1}^{-1}[I \quad 0 \quad \cdots \quad 0]^T. \tag{7.56}$$

(Here we use the assumption of comonicity, so that $C_L(0)$ is nonsingular.) But

$$S_{l-1} = \begin{bmatrix} 0 & & \cdots & 0 & I \\ \vdots & & & & \vdots \\ 0 & I & \cdots & & 0 \\ I & 0 & \cdots & & 0 \end{bmatrix} Y \begin{bmatrix} J_F^{-(l-1)} & 0 \\ 0 & I \end{bmatrix}.$$

Inverting this formula and substituting in (7.56) yields (7.55). □

Note that for any pair of matrices similar to (X, J) a representation similar to (7.54) holds. In particular, (7.54) holds, replacing X by $[I \quad 0 \quad \cdots \quad 0]$ and J by the comonic companion matrix R of $L(\lambda)$ (cf. (7.53)).

An alternative way to derive (7.54) would be by reduction to monic polynomials. Let us outline this approach. The following observation is crucial:

Theorem 7.15. *Let $L(\lambda)$ be a comonic matrix polynomial of degree l, and let (X, J) be its comonic Jordan pair. Then (X, J) is a (right) standard pair for the monic matrix polynomial $\tilde{L}(\lambda) = \lambda^l L(\lambda^{-1})$.*

The proof of Theorem 7.15 can be obtained by repeating arguments from the proof of Theorem 7.3.

We already know a resolvent representation for $\tilde{L}(\lambda) = \lambda^l L(\lambda^{-1})$ (Theorem 2.4):

$$\tilde{L}(\lambda) = X(I\lambda - J)^{-1}Z, \qquad \lambda \notin \sigma(J),$$

where $Z = [\operatorname{col}(XJ^i)_{i=0}^{l-1}]^{-1}[0 \quad \cdots \quad 0 \quad I]^{\mathrm{T}}$.

The last relation yields the biorthogonality conditions $XJ^jZ = 0$ for $j = 0, \ldots, l-2$, $XJ^{l-1}Z = I$. Using these we see that for λ close enough to zero,

$$\begin{aligned} L^{-1}(\lambda) &= \lambda^{-l}[\tilde{L}(\lambda^{-1})]^{-1} = \lambda^{-l+1}X(I - \lambda J)^{-1}Z \\ &= \lambda^{-l+1}X(I + \lambda J + \lambda^2J^2 + \cdots)Z \\ &= \lambda^{-l+1}X(\lambda^{l-1}J^{l-1} + \lambda^l J^l + \cdots)Z = XJ^{l-1}(I - \lambda J)^{-1}Z. \end{aligned}$$

Consequently, (7.54) holds for λ close enough to zero, and, since both sides of (7.54) are rational matrix functions, (7.54) holds for all λ such that $\lambda^{-1} \notin \sigma(J)$.

For a comonic matrix polynomial $L(\lambda)$ of degree l representation (7.32) reduces to

$$L(\lambda) = I - XJ^l(V_1\lambda^l + V_2\lambda^{l-1} + \cdots + V_l\lambda), \tag{7.57}$$

where (X, J) is a comonic Jordan pair of $L(\lambda)$, and $[V_1 \quad \cdots \quad V_l] = [\operatorname{col}(XJ^i)_{i=0}^{l-1}]^{-1}$. The easiest way to verify (7.57) is by reduction to the monic polynomial $\lambda^l L(\lambda^{-1})$ and the use of Theorem 7.3. Again, (7.57) holds for any pair similar to (X, J). In particular, (7.57) holds for $([I \quad 0 \quad \cdots \quad 0], R)$ with the comonic companion matrix R of $L(\lambda)$. It also follows that a comonic polynomial can be uniquely reconstructed from its comonic Jordan pair.

7.9. Comonic Polynomials from Finite Spectral Data

We have seen already that a comonic matrix polynomial of given degree is uniquely defined by its comonic Jordan pair (formula (7.57)). On the other hand, the finite Jordan pair alone is not sufficient to determine uniquely the comonic polynomial. We describe here how to determine a comonic matrix polynomial by its finite Jordan pair in a way which may be described as canonical. To this end we use the construction of a special left inverse introduced in Section 6.3.

We shall need the following definition of s-indices and k-indices of an admissible pair (X, T) with nonsingular matrix T and $\operatorname{ind}(X, T) = l$. Let $k_i = n + q_{l-i-1} - q_{l-i}$ for $i = 0, \ldots, l-1$, where $q_i = \operatorname{rank} \operatorname{col}(XT^j)_{j=0}^{i-1}$ for $i \geq 1$ and $q_0 = 0$; define s_i for $i = 0, \ldots$ as the number of integers $k_0, k_1, \ldots, k_{l-1}$ which are larger than i. The numbers $s_0, s_1, \ldots$ will be called *s-indices* of (X, T), and the numbers $k_0, k_1, \ldots, k_{l-1}$ will be called *k-indices* of (X, T). Observe the following easily verified properties of the k-indices and s-indices:

(1) $n \geq k_0 \geq k_1 \geq \cdots \geq k_{l-1}$;
(2) $s_0 \geq s_1 \geq \cdots$;
(3) $s_j \leq l$ for $i = 0, 1, \ldots$, and $s_i = 0$ for $i > n$;
(4) $\sum_{i=0}^{l-1} k_i = \sum_{j=0}^{n} s_j$.

Now we can formulate the theorem which describes a canonical construction of the comonic polynomial via its finite spectral data.

Theorem 7.16. *Let (X_F, J_F) be an admissible pair with zero kernel and nonsingular Jordan matrix J_F. The minimal degree of a comonic polynomial $L(\lambda)$ such that (X_F, J_F) is its finite Jordan pair, is equal to the index of stabilization* $\operatorname{ind}(X_F, J_F)$. *One such polynomial is given by the formula*

$$L_0(\lambda) = I - X_F J_F^{-l}(V_1 \lambda^l + \cdots + V_l \lambda),$$

where $V = [V_1, \ldots, V_l]$ is a special left inverse of $\operatorname{col}(X_F J_F^{-j})_{j=0}^{l-1}$. *The infinite part (X_∞, J_∞) of a comonic Jordan pair for $L_0(\lambda)$ has the form*

$$X_\infty = [x_0 \quad 0 \quad \cdots \quad 0 \quad x_1 \quad 0 \quad \cdots \quad 0 \quad \cdots \quad x_v \quad 0 \quad \cdots \quad 0],$$

$$J_\infty = \operatorname{diag}[J_{\infty 0}, J_{\infty 1}, \ldots, J_{\infty v}],$$

where $J_{\infty j}$ is a nilpotent Jordan block of size s_j, and $s_0 \geq \cdots \geq s_v$ are the nonzero s-indices of (X_F, J_F^{-1}).

Proof. Let (X_F, J_F) be a finite Jordan pair of some comonic polynomial $L(\lambda)$. Then (see Theorem 7.3) (X_F, J_F) is a restriction of a decomposable pair of $L(\lambda)$. In view of property (iii) in the definition of a decomposable pair (Section 7.3), $\operatorname{ind}(X_F, J_F) \leq m$, where m is the degree of $L(\lambda)$. So the first assertion of the theorem follows.

We now prove that $L_0(\lambda)$ has (X_F, J_F) as its finite Jordan pair. Let $\tilde{L}(\lambda) = \lambda^l L_0(\lambda^{-1}) = I\lambda^l - X_F J_F^{-l}(V_1 + \cdots + V_l \lambda^{l-1})$. By Theorem 6.5, (X_F, J_F^{-1}) is a part of a standard pair $(\tilde{X}, \tilde{T})$ for $\tilde{L}(\lambda)$. Further, Corollary 6.6 ensures that (X_F, J_F^{-1}) is similar to $(\tilde{X}|_{\mathcal{M}}, \tilde{T}|_{\mathcal{M}})$, where $\mathcal{M}$ is the maximal $\tilde{T}$-invariant subspace such that $\tilde{T}|_{\mathcal{M}}$ is invertible. By Theorem 7.15, $(\tilde{X}|_{\mathcal{M}}, \tilde{T}|_{\mathcal{M}})$ is similar to the part $(\hat{X}_F, \hat{J}_F^{-1})$ of a comonic Jordan pair $((\hat{X}_F, \hat{X}_\infty), \hat{J}_F^{-1} \oplus \hat{J}_\infty)$ of $L_0(\lambda)$. So in fact (X_F, J_F) is a finite Jordan pair for $L_0(\lambda)$.

In order to study the structure of (X_∞, J_∞), it is convenient to consider the companion matrix

$$C = \begin{bmatrix} 0 & I & 0 & \cdots & 0 \\ 0 & 0 & I & \cdots & 0 \\ \vdots & \vdots & \vdots & & \vdots \\ 0 & 0 & 0 & \cdots & I \\ X_F J_F^{-1} V_1 & X_F J_F^{-1} V_2 & & \cdots & X_F J_F^{-1} V_l \end{bmatrix}$$

of $L(\lambda)$. Let $\mathscr{W}_1 \supset \cdots \supset \mathscr{W}_l$ be the subspaces taken from the definition of the special left inverse. The subspace $\mathscr{W} = \text{col}(\mathscr{W}_j)_{j=1}^l \subset \mathbb{C}^{nl}$ is C-invariant, because $C\mathscr{W} = \text{col}(\mathscr{W}_{j+1})_{j=1}^l \subset \mathscr{W}$ (by definition $\mathscr{W}_{l+1} = \{0\}$). The relation $C[\text{col}(X_F J_F^{-j})_{j=0}^{l-1}] = [\text{col}(X_F J_F^{-j})_{j=0}^{l-1}] \cdot J_F^{-1}$ shows that the subspace $\mathscr{L} = \text{Im col}(X_F J_F^{-j})_{j=0}^{l-1}$ is also C-invariant. It is clear that $C|_{\mathscr{W}}$ is nilpotent, and $C|_{\mathscr{L}}$ is invertible. Let $k_0 \geq k_1 \geq \cdots \geq k_{l-1}$ be the *k-indices* of (X_F, J_F^{-1}). Let $z_1^{(i)}, \ldots, z_{k_{i-1}-k_i}^{(i)}$ be a basis in $\mathscr{W}_i$ modulo $\mathscr{W}_{i+1}$, for $i = 1, \ldots, l$ (by definition $k_l = 0$). Then it is easy to see that for $j = 1, \ldots, k_{i-1} - k_i$, $\text{col}(\delta_{1p} z_j^{(i)})_{p=1}^l, \text{col}(\delta_{2p} z_j^{(i)})_{p=1}^l, \ldots, \text{col}(\delta_{ip} z_j^{(i)})_{p=1}^l$ is a Jordan chain of C corresponding to the eigenvalue 0, of length i. Taking into account the definition of the s-indices and Theorem 6.5, we see that (X_∞, J_∞) has the structure described in the theorem. □

By the way, Theorem 7.16 implies the following important fact.

Corollary 7.17. *Let (X, T) be an admissible pair with zero kernel and nonsingular Jordan matrix T. Then there exists a comonic matrix polynomial such that (X, T) is its finite Jordan pair.*

In other words, every admissible pair (subject to natural restrictions) can play the role of a finite Jordan pair for some comonic matrix polynomial.

The following corollary gives a description of all the comonic matrix polynomials for which (X_F, J_F) is a finite Jordan pair.

Corollary 7.18. *A comonic matrix polynomial $L(\lambda)$ has (X_F, J_F) as its finite Jordan pair if and only if $L(\lambda)$ admits the representation*

$$L(\lambda) = U(\lambda) L_0(\lambda),$$

where $L_0(\lambda) = I - X_F J_F^{-l}(V_1 \lambda^l + \cdots + V_l \lambda)$, $[V_1 \quad \cdots \quad V_l]$ is a special left inverse of $\text{col}(X_F J_F^{-j})_{j=1}^l$, *and $U(\lambda)$ is a comonic matrix polynomial with constant determinant.*

This is an immediate corollary of Theorems 7.13 and 7.16.

We have now solved the problem of reconstructing a comonic polynomial, given its finite Jordan pair (X_F, J_F). Clearly, there are many comonic polynomials of minimal degree with the same (X_F, J_F), which differ by their spectral data at infinity. Now we shall give a description of all possible Jordan structures at infinity for such polynomials.

Theorem 7.19. *Let (X_F, J_F) be an admissible pair of order r and zero kernel with nonsingular Jordan matrix J_F. Let $l = \operatorname{ind}(X_F, J_F)$. If (X_F, J_F) is a finite Jordan pair of a comonic matrix polynomial $L(\lambda)$ of degree l, and (X_∞, J_∞) is its infinite Jordan pair, then the sizes $p_1 \geq p_2 \geq \cdots \geq p_v$ of the Jordan blocks in J_∞ satisfy the conditions*

$$\sum_{i=1}^{j} p_i \geq \sum_{i=1}^{j} s_{i-1} \qquad \text{for} \quad j = 1,2, \ldots, v, \tag{7.58}$$

$$\sum_{i=1}^{v} p_i = nl - r, \tag{7.59}$$

where $s_0, s_1, \ldots,$ are the s-indices of (X_F, J_F^{-1}).

Conversely, if for some positive integers $p_1 \geq p_2 \geq \cdots \geq p_v$, the conditions (7.58), (7.59) are satisfied, then there exists a comonic matrix polynomial $L(\lambda)$ of degree l such that $L(\lambda)$ has a comonic Jordan pair of the form $((X_F X_\infty), J_F^{-1} \oplus J_\infty)$, where J_∞ is a nilpotent Jordan matrix with Jordan blocks of sizes $p_1, \ldots, p_v$.

The proof of Theorem 7.19 follows easily from Theorem 6.7.

7.10. Description of Divisors via Invariant Subspaces

The results of the preceding section allow us to give the following description of (right) comonic divisors of a given comonic matrix polynomial $L(\lambda)$ in terms of its finite Jordan pair (X_F, J_F).

Theorem 7.20. *Let $\mathcal{M}$ be a J_F-invariant subspace, and let $m = \operatorname{ind}(X_{F|\mathcal{M}}, J_{F|\mathcal{M}}^{-1})$. Then a comonic polynomial*

$$L_{\mathcal{M}}(\lambda) = I - (X_{F|\mathcal{M}}) \cdot (J_{F|\mathcal{M}}^{-m}) \cdot (V_1 \lambda^m + \cdots + V_m \lambda), \tag{7.60}$$

where $[V_1 \ \cdots \ V_m]$ is a special left inverse of $\operatorname{col}((X_{F|\mathcal{M}})(J_{F|\mathcal{M}})^{-j+1})_{j=1}^m$, is a right divisor of $L(\lambda)$. Conversely, for every comonic right divisor $L_1(\lambda)$ of $L(\lambda)$ there exists a unique J_F-invariant subspace $\mathcal{M}$ such that $L_1(\lambda) = U(\lambda) L_{\mathcal{M}}(\lambda)$, where $U(\lambda)$ is a matrix polynomial with $\det U(\lambda) \equiv 1$ and $U(0) = I$.

Proof. From Theorems 7.13 and 7.16 it follows that $L_{\mathcal{M}}(\lambda)$ is a right divisor of $L(\lambda)$ for any J_F-invariant subspace $\mathcal{M}$.

Conversely, let $L_1(\lambda)$ be a comonic right divisor of $L(\lambda)$ with finite Jordan pair $(X_{1,\mathrm{F}}, J_{1,\mathrm{F}})$. By Theorem 7.13, $(X_{1,\mathrm{F}}, J_{1,\mathrm{F}})$ is a restriction of $(X_\mathrm{F}, J_\mathrm{F})$; so $(X_{1,\mathrm{F}}, J_{1,\mathrm{F}})$ is a finite Jordan pair of $L_{\mathscr{M}}(\lambda)$, where $L_{\mathscr{M}}(\lambda)$ is given by (7.60), and $\mathscr{M}$ is the unique J_F-invariant subspace such that $(X_{1,\mathrm{F}}, J_{1,\mathrm{F}})$ is similar to $(X_{\mathrm{F}|\mathscr{M}}, J_{\mathrm{F}|\mathscr{M}})$. By the same Theorem 7.13, $L_1(\lambda) = U(\lambda)L_{\mathscr{M}}(\lambda)$, where $U(\lambda)$ is a matrix polynomial without eigenvalues. Moreover, since $L_1(0) = L_{\mathscr{M}}(0) = I$, also $U(0) = I$, and $\det U(\lambda) \equiv 1$. □

If $\mathscr{M}$ is the J_F-invariant subspace, which corresponds to the right divisor $L_1(\lambda)$ of $L(\lambda)$ as in Theorem 7.20, we shall say that $L_1(\lambda)$ is *generated* by $\mathscr{M}$, and $\mathscr{M}$ is the *supporting subspace* of $L_1(\lambda)$ (with respect to the fixed finite Jordan pair $(X_\mathrm{F}, J_\mathrm{F})$ of $L(\lambda)$). Theorem 7.20 can be regarded as a generalization of Theorem 3.12 for nonmonic matrix polynomials.

EXAMPLE 7.3. To illustrate the theory consider the following example. Let

$$L(\lambda) = \begin{bmatrix} (\lambda+1)^3 & \lambda \\ 0 & (\lambda-1)^2 \end{bmatrix}.$$

Then a finite Jordan pair $(X_\mathrm{F}, J_\mathrm{F})$ of $L(\lambda)$ is given by

$$X_\mathrm{F} = \begin{bmatrix} 1 & 0 & 0 & 1 & 0 \\ 0 & 0 & 0 & -8 & -4 \end{bmatrix}, \qquad J_\mathrm{F} = \begin{bmatrix} -1 & 1 & 0 \\ 0 & -1 & 1 \\ 0 & 0 & -1 \end{bmatrix} \oplus \begin{bmatrix} 1 & 1 \\ 0 & 1 \end{bmatrix}.$$

Then

$$J_\mathrm{F}^{-1} = \begin{bmatrix} -1 & -1 & -1 \\ 0 & -1 & -1 \\ 0 & 0 & -1 \end{bmatrix} \oplus \begin{bmatrix} 1 & -1 \\ 0 & 1 \end{bmatrix}$$

We compute the divisors generated by the J_F^{-1}-invariant subspace $\mathscr{M}$ spanned by the vectors $(1\ \ 0\ \ 0\ \ 0\ \ 0)^\mathrm{T}$, $(0\ \ 1\ \ 0\ \ 0\ \ 0)^\mathrm{T}$, $(0\ \ 0\ \ 0\ \ 1\ \ 0)^\mathrm{T}$. We have

$$X_\mathrm{F}|_{\mathscr{M}} = \begin{bmatrix} 1 & 0 & 1 \\ 0 & 0 & -8 \end{bmatrix}, \qquad J_\mathrm{F}^{-1}|_{\mathscr{M}} = \begin{bmatrix} -1 & -1 & 0 \\ 0 & -1 & 0 \\ 0 & 0 & 1 \end{bmatrix}.$$

The columns of

$$\begin{bmatrix} X_\mathrm{F}|_{\mathscr{M}} \\ X_\mathrm{F}|_{\mathscr{M}} J_\mathrm{F}^{-1}|_{\mathscr{M}} \end{bmatrix} = \begin{bmatrix} 1 & 0 & 1 \\ 0 & 0 & -8 \\ -1 & -1 & 1 \\ 0 & 0 & -8 \end{bmatrix}$$

are independent, and a special left inverse is

$$[V_1 \quad V_2] = \begin{bmatrix} 1 & 0 & 0 & \frac{1}{8} \\ -1 & 0 & -1 & -\frac{2}{8} \\ 0 & 0 & 0 & -\frac{1}{8} \end{bmatrix}.$$

Therefore,

$$I - (X_{\mathrm{F}|\mathscr{M}})(J_{\mathrm{F}|\mathscr{M}}^{-2})(V_1\lambda^2 + V_2\lambda) = \begin{bmatrix} (\lambda+1)^2 & \frac{\lambda}{2} \\ 0 & \lambda - 1 \end{bmatrix},$$

and the general form of a right comonic divisor generated by $\mathscr{M}$, is

$$U(\lambda) \cdot \begin{bmatrix} (\lambda+1) & \frac{\lambda}{2} \\ 0 & \lambda - 1 \end{bmatrix},$$

where $U(\lambda)$ is matrix polynomial with $U(0) = I$ and $\det U(\lambda) \equiv 1$. □

7.11. Construction of a Comonic Matrix Polynomial via a Special Generalized Inverse

In Section 7.9 we have already considered the construction of a comonic matrix polynomial $L(\lambda)$ given its finite Jordan pair $(X_{\mathrm{F}}, J_{\mathrm{F}})$. The spectral data in $(X_{\mathrm{F}}, J_{\mathrm{F}})$ are "well organized" in the sense that for every eigenvalue λ_0 of $L(\lambda)$ the part of $(X_{\mathrm{F}}, J_{\mathrm{F}})$ corresponding to λ_0 consists of a canonical set of Jordan chains of $L(\lambda)$ corresponding to λ_0 and Jordan blocks with eigenvalue λ_0 whose sizes match the lengths of the Jordan chains in the canonical set. However, it is desirable to know how to construct a comonic matrix polynomial $L(\lambda)$ if its finite spectral data are "badly organized," i.e., it is not known in advance if they form a canonical set of Jordan chains for each eigenvalue.

In this section we shall solve this problem. The solution is based on the notion of a special generalized inverse.

We start with some auxiliary considerations. Let (X, T) be an admissible pair of order p, and let $A = \mathrm{col}(XT^{i-1})_{i=1}^{s}$, where $s = \mathrm{ind}(X, T)$. Let A^{I} be a generalized inverse of A (see Chapter S3). It is clear that A^{I} can be written as $A^{\mathrm{I}} = [V_1 \quad \cdots \quad V_s]$, where $V_1, V_2, \ldots, V_s$ are $n \times n$ matrices. Put

$$\tilde{L}(\lambda) = I\lambda^s - XT^s(V_1 + \lambda V_2 + \cdots + \lambda^{s-1}V_s). \tag{7.61}$$

Then for every projector P of $\mathbb{C}^p$ along $\mathrm{Ker}(X, T)$ (recall that $\mathrm{Ker}(X, T) = \{x \in \mathbb{C}^p \mid XT^i x = 0, i = 0, 1, \ldots\}$) there exists a generalized inverse A^{I} of A such that a standard pair of $\tilde{L}(\lambda)$ is an extension of the pair

$$(X|_{\mathrm{Im}\,P}, PT|_{\mathrm{Im}\,P}). \tag{7.62}$$

More exactly, the following lemma holds.

Lemma 7.21. *For a projector P of $\mathbb{C}^p$ along* $\mathrm{Ker}(X, T)$, *let A^{I} be a generalized inverse of $A = \mathrm{col}(XT^{i-1})_{i=1}^{s}$ ($s = \mathrm{ind}(X, T)$) such that $A^{\mathrm{I}}A = P$.*

Let $Q = [I \quad 0 \quad \cdots \quad 0]$ and let C be the first companion matrix of the monic matrix polynomial (7.61) *(so (Q, C) is a standard pair for $\tilde{L}(\lambda)$). Then the space* Im A *is invariant for C and the pair $(Q|_{\operatorname{Im} A}, C|_{\operatorname{Im} A})$ is similar to $(X|_{\operatorname{Im} P}, PT|_{\operatorname{Im} P})$.*

Proof. We have

$$C = \begin{bmatrix} 0 & I & & \cdots & 0 \\ 0 & & I & \cdots & 0 \\ \vdots & \vdots & \vdots & & \vdots \\ 0 & 0 & & \cdots & I \\ XT^sV_1 & XT^sV_2 & & \cdots & XT^sV_s \end{bmatrix}.$$

As $AA^{\mathrm{I}}A = A$, we have $XT^{i-1}A^{\mathrm{I}}A = XT^{i-1}$ for $i = 1, \ldots, s$. So $CA = ATA^{\mathrm{I}}A$. Since $A^{\mathrm{I}}A = P$, the equalities $A^{\mathrm{I}}AP = P$ and $AP = A$ hold. So $CA = APTP$, and hence

$$C(A|_{\operatorname{Im} P}) = (A|_{\operatorname{Im} P})(PT|_{\operatorname{Im} P}).$$

Note that $S = A|_{\operatorname{Im} P}$ has a zero kernel (as a linear transformation from Im P to $\mathbb{C}^{ns}$). From what we proved above we see that $CS = S(PT|_{\operatorname{Im} P})$. Further

$$[I \quad 0 \quad \cdots \quad 0]S = [I \quad 0 \quad \cdots \quad 0]A|_{\operatorname{Im} P} = X|_{\operatorname{Im} P}.$$

It follows that $\mathcal{M} = \operatorname{Im} S = \operatorname{Im} A$ is a C-invariant subspace, and

$$([I \quad 0 \quad \cdots \quad 0]|_{\mathcal{M}}, C|_{\mathcal{M}})$$

is similar to $(X|_{\operatorname{Im} P}, PT|_{\operatorname{Im} P})$. □

For the purpose of further reference let us state the following proposition.

Proposition 7.22. *Let (X, T) be an admissible pair of order p, and suppose that T is nonsingular. Then, in the notation of Lemma* 7.21, *the restriction $C|_{\operatorname{Im} A}$ is nonsingular.*

Proof. In view of Lemma 7.21, it is sufficient to prove that $PT|_{\operatorname{Im} P}$ is invertible, where P is a projector of $\mathbb{C}^p$ along Ker $\operatorname{col}(XT^i)_{i=0}^{s-1}$. Write X and T in matrix form with respect to the decomposition $\mathbb{C}^p = \operatorname{Im} P \dotplus \operatorname{Ker} P$:

$$X = [X_1 \quad 0], \qquad T = \begin{bmatrix} T_{11} & T_{12} \\ T_{21} & T_{22} \end{bmatrix}.$$

For example, $T_{11} = PT|_{\operatorname{Im} P}$. Let us prove that $T_{12} = 0$. Indeed, the matrix $\operatorname{col}(XT^i)_{i=0}^{s-1}$ has the following matrix form with respect to the decomposition $\mathbb{C}^p = \operatorname{Im} P \dotplus \operatorname{Ker} P$:

$$\begin{bmatrix} X \\ XT \\ \vdots \\ XT^{s-1} \end{bmatrix} = \begin{bmatrix} X_1 & 0 \\ X_1T_{11} & 0 \\ \vdots & \vdots \\ X_1T_{11}^{s-1} & 0 \end{bmatrix}. \tag{7.63}$$

Since $s = \operatorname{ind}(X, T)$, Ker $\operatorname{col}(XT^i)_{i=0}^{s} =$ Ker $\operatorname{col}(XT^i)_{i=0}^{s-1}$, and therefore $XT^s = [X_1 T_{11}^s \quad 0]$. Taking into account (7.63) we obtain that $X_1 T_{11}^j \cdot T_{12} = 0$ for $j = 0, \ldots, s-1$, and in view of the left invertibility of $\operatorname{col}(X_1 T_{11}^j)_{j=0}^{s-1}$, it follows that $T_{12} = 0$. Now clearly the nonsingularity of T implies that $T_{11} = PT|_{\operatorname{Im} P}$ is also invertible. □

So if T in the admissbile pair (X, T) is nonsingular, then by Proposition 7.22, $C|_{\operatorname{Im} A}$ is invertible too. But Im A is not necessarily the largest C-invariant subspace $\mathscr{M}$ of $\mathbb{C}^{ns}$ such that $C|_{\mathscr{M}}$ is invertible. To get this maximal subspace we have to make a special choice for A^{I}.

Recall that if $\mathscr{W}_1, \ldots, \mathscr{W}_s$ are subspaces in $\mathbb{C}^n$, then $\operatorname{col}(\mathscr{W}_j)_{j=1}^{s}$ will denote the subspace of $\mathbb{C}^{ns}$ consisting of all vectors $x = \operatorname{col}(x_j)_{j=1}^{s}$ with $x_j \in \mathscr{W}_j$ for $j = 1, 2, \ldots, s$. As before, let (X, T) be an admissible pair of order p, let $s = \operatorname{ind}(X, T)$ and assume that T is nonsingular. Then there exist linear subspaces $\mathscr{W}_s \subset \mathscr{W}_{s-1} \subset \cdots \subset \mathscr{W}_1 \subset \mathbb{C}^n$ such that

$$\operatorname{Im} \operatorname{col}(XT^{i-1})_{i=1}^{s} \oplus \operatorname{col}(\mathscr{W}_j)_{j=1}^{s} = \mathbb{C}^{ns}. \tag{7.64}$$

For the case in which $\operatorname{Ker}(X, T) = \{0\}$, formula (7.64) follows from Lemma 6.4. By replacing (X, T) by the pair (7.62) the general case may be reduced to the case $\operatorname{Ker}(X, T) = \{0\}$, because

$$\operatorname{Im} \operatorname{col}(XT^{i-1})_{i=1}^{s} = \operatorname{Im}[\operatorname{col}((X|_{\operatorname{Im} P})(PT|_{\operatorname{Im} P})^{i-1})_{i=1}^{s}].$$

We call A^{I} a *special generalized inverse* of $A = \operatorname{col}(XT^{i-1})_{i=1}^{s}$ whenever A^{I} is a generalized inverse of A and Ker $A^{\mathrm{I}} = \operatorname{col}(\mathscr{W}_j)_{j=1}^{s}$ for some choice of subspaces $\mathscr{W}_s \subset \cdots \subset \mathscr{W}_1$ such that (7.64) holds. If A is left invertible, and A^{I} is a special left inverse of A, as defined in Section 6.3, then A^{I} is also a special generalized inverse of A.

The following result is a refinement of Lemma 7.21 for the case of special generalized inverse.

Lemma 7.23. *Let (X, T) be an admissible pair of order p, let $s = \operatorname{ind}(X, T)$ and assume that T is nonsingular. Let A^{I} be a special generalized inverse of $A = \operatorname{col}(XT^{i-1})_{i=1}^{s}$, and write $A^{\mathrm{I}} = [V_1 \quad V_2 \quad \cdots \quad V_s]$. Let (Q, C) be the standard pair with the first companion matrix C of*

$$\tilde{L}(\lambda) = I\lambda^s - XT^s(V_1 + V_2\lambda + \cdots + V_s\lambda^{s-1}).$$

Then C is completely reduced by the subspaces Im A *and* Ker A^{I}, *the pairs*

$$(X|_{\operatorname{Im} A^{\mathrm{I}}A}, A^{\mathrm{I}}AT|_{\operatorname{Im} A^{\mathrm{I}}A}), (Q|_{\operatorname{Im} A}, C|_{\operatorname{Im} A}) \tag{7.65}$$

are similar, and

(i) *$C|_{\operatorname{Im} A}$ is invertible,*
(ii) $(C|_{\operatorname{Ker} A^{\mathrm{I}}})^s = 0$.

Proof. The similarity of the pairs (7.65) has been established in the proof of Lemma 7.21. There it has also been shown that Im A is C-invariant. As T is nonsingular, the same is true for

$$A^{\mathrm{I}}AT|_{\operatorname{Im} A^{\mathrm{I}}A}$$

(cf. Proposition 7.22). But then, because of similarity, statement (i) must be true. The fact that Ker A^{I} is C-invariant and statement (ii) follow from the special form of Ker A^{I} (cf. formula (7.64)). □

The next theorem gives the solution of the problem of construction of a comonic matrix polynomial given its "badly organized" data.

Theorem 7.24. *Let (X, T) be an admissible pair of order p, let $s = \operatorname{ind}(X, T)$, and assume that T is nonsingular. Let A^{I} be a special generalized inverse of $A = \operatorname{col}(XT^{1-i})_{i=1}^{s}$.*

Write $A^{\mathrm{I}} = [W_1 \quad W_2 \quad \cdots \quad W_s]$, and put

$$L(\lambda) = I - XT^{-s}(W_s\lambda + \cdots + W_1\lambda^s).$$

Let $P = A^{\mathrm{I}}A$. Then the finite Jordan pair of $L(\lambda)$ is similar to the pair

$$(X|_{\operatorname{Im} P}, PT|_{\operatorname{Im} P}). \tag{7.66}$$

Further, the minimal possible degree of a comonic $n \times n$ matrix polynomial with a finite Jordan pair similar to the pair (7.66) is precisely s.

Proof. Note that $\operatorname{ind}(X, T^{-1}) = \operatorname{ind}(X, T) = s$. Put

$$\tilde{L}(\lambda) = I\lambda^s - XT^{-s}(W_1 + \cdots + W_s\lambda^{s-1}). \tag{7.67}$$

From the previous lemma it is clear that the pair

$$(X|_{\operatorname{Im} P}, PT^{-1}|_{\operatorname{Im} P}) \tag{7.68}$$

is similar to the pair $(Q|_{\operatorname{Im} A}, C|_{\operatorname{Im} A})$, where (Q, C) is the standard pair with the companion matrix C of the polynomial (7.67).

Let (X_L, J_L) be a finite Jordan pair of $L(\lambda)$, and let (X_∞, J_∞) be a Jordan pair of $\tilde{L}(\lambda)$ corresponding to the eigenvalue 0. By Theorem 7.15 the pair $([X_L \quad X_\infty], J_L^{-1} \oplus J_\infty)$ is similar to (Q, C). From the previous lemma we know that

$$C = C|_{\operatorname{Im} A} \oplus C|_{\operatorname{Ker} A^{\mathrm{I}}},$$

where $C|_{\operatorname{Im} A}$ is invertible and $C|_{\operatorname{Ker} A^{\mathrm{I}}}$ is nilpotent. It follows that the pair (X_L, J_L^{-1}) is similar to the pair $(Q|_{\operatorname{Im} A}, C|_{\operatorname{Im} A})$.

But then we have proved that the finite Jordan pair of $L(\lambda)$ is similar to the pair (7.66).

Now, let (X_B, J_B) be a finite Jordan pair of some comonic $n \times n$ matrix polynomial $B(\lambda)$, and assume that (X_B, J_B) is similar to the pair (7.66). Since the index of stabilization of the pair (7.66) is equal to s again, we see that $\operatorname{ind}(X_B, J_B) = s$. Let $\tilde{B}(\lambda) = \lambda^m B(\lambda^{-1})$, where m is the degree of $B(\lambda)$. Then (X_B, J_B^{-1}) is a restriction of a standard pair of $\tilde{B}(\lambda)$ (cf. Theorem 7.15). It follows that $s = \operatorname{ind}(X_B, J_B) = \operatorname{ind}(X_B, J_B^{-1}) \leq m$. This completes the proof of the theorem. □

In conclusion we remark that one can avoid the requirement that T be invertible in Theorem 7.24. To this end replace T by $T - \alpha I$, where $\alpha \notin \sigma(T)$, and shift the argument $\lambda \to \lambda - \alpha$. The details are left to the reader.

Comments

The presentation in Section 7.1 follows the paper [37a]; the contents of Sections 7.2–7.6 are based on [14], and that of Section 7.7 (including Lemma 7.13) on [29a]. The main results of Section 7.7 also appear in [37a]. Another version of division of matrix polynomials appears in [37b]. For results and discussion concerning s-indices and k-indices, see [37c, 35b]. The results of Sections 7.8–7.10 are based on [37a, 37b], and those of Section 7.11 appear in [29a]. Direct connections between a matrix polynomial and its linearization are obtained in [79a, 79b].

Chapter 8

Applications to Differential and Difference Equations

In this chapter we shall consider systems of lth-order linear differential equations with constant coefficients, as well as systems of linear difference equations with constant coefficients. In each case, there will be a characteristic matrix polynomial of degree l with determinant not identically zero. The results presented here can be viewed as extensions of some results in Chapters 2 and 3, in which the characteristic matrix polynomial is monic. It is found that, in general, the system of differential equations corresponding to a nonmonic matrix polynomial cannot be solved for every continuous right-hand part. This peculiarity can be attributed to the nonvoid spectral structure at infinity, which is always present for nonmonic matrix polynomials (see Chapter 7). One way to ensure the existence of a solution of the differential equation for every continuous right-hand side is to consider the equation in the context of distributions. We shall not pursue this approach here, referring the reader to the paper [29a] for a detailed exposition. Instead we shall stick to the "classical" approach, imposing differentiability conditions on the right-hand side (so that the differentiation which occurs in the course of solution can always be performed). In the last section of this chapter we solve the problem of reconstructing a differential or difference equation when the solutions are given for the homogeneous case.

8.1. Differential Equations in the Nonmonic Case

Let $L(\lambda) = \sum_{j=0}^{l} A_j \lambda^j$ be a regular matrix polynomial. We now consider the corresponding system of lth-order differential equations

$$L\left(\frac{d}{dt}\right)u = A_0 u + A_1 \frac{d}{dt} u + \cdots + A_l \frac{d^l}{dt^l} u = f, \tag{8.1}$$

where f is a $\mathbb{C}^n$-valued function of the real variable t. Our aim is to describe the solution of (8.1) in terms of a decomposable pair of $L(\lambda)$.

First, observe that in general equation (8.1) is not solvable for every continuous function f. For example, consider the system of differential equations

$$u_1 + \frac{d^l}{dt^l} u_2 = f_1, \qquad u_2 = f_2, \tag{8.2}$$

which corresponds to the matrix polynomial

$$L(\lambda) = \begin{bmatrix} 1 & 0 \\ 0 & 1 \end{bmatrix} + \begin{bmatrix} 0 & 1 \\ 0 & 0 \end{bmatrix} \lambda^l. \tag{8.3}$$

Clearly, the system (8.3) has a continuous solution if and only if f_2 is l-times continuously differentiable, and in that case the solution $u_1 = f_1 - f_2^{(l)}$, $u_2 = f_2$ is unique.

To formulate the main theorem on solutions of (8.1) it is convenient to introduce the notion of a resolvent triple of $L(\lambda)$, as follows. Let $(X, T) = ([X_1 \quad X_2], T_1 \oplus T_2)$ be a decomposable pair of $L(\lambda)$. Then there exists a matrix Z such that

$$L^{-1}(\lambda) = XT(\lambda)^{-1}Z,$$

where $T(\lambda) = (I\lambda - T_1) \oplus (T_2 \lambda - I)$ (see Theorem 7.7). The triple (X, T, Z) will be called a *resolvent triple* of $L(\lambda)$.

Theorem 8.1. *Let $L(\lambda)$ be a matrix polynomial of degree l, and let (X, T, Z) be a resolvent triple for $L(\lambda)$, where $X = [X_1 \quad X_2]$, $T = T_1 \oplus T_2$, and let $Z = \left[\begin{smallmatrix} Z_1 \\ Z_2 \end{smallmatrix}\right]$ be the corresponding partition of Z. Assume that the size of T_1 is $m \times m$, where $m =$ degree(det $L(\lambda)$), and that $T_2^{\nu} = 0$ for some positive integer ν. Then for every $(\nu + l - 1)$-times continuously differentiable $\mathbb{C}^n$-valued function $f(t)$ the general solution of (8.1) is given by the formula*

$$u(t) = X_1 e^{T_1 t} x + \int_{t_0}^{t} X_1 e^{T_1(t-s)} Z_1 f(s)\, ds - \sum_{i=0}^{\nu-1} X_2 T_2^i Z_2 f^{(i)}(t), \tag{8.4}$$

where $x \in \mathbb{C}^m$ is an arbitrary vector.

Proof. We check first that formula (8.4) does indeed produce a solution of (8.1). We have (for $j = 0, 1, \ldots, l$)

$$u^{(j)}(t) = X_1 T_1^j e^{T_1 t} x + \sum_{k=0}^{j-1} X_1 T_1^{j-1-k} Z_1 f^{(k)}(t) + \int_{t_0}^{t} X_1 T_1^j e^{T_1(t-s)} Z_1 f(s)\, ds - \sum_{i=0}^{\nu-1} X_2 T_2^i Z_2 f^{(i+j)}(t).$$

So

$$\sum_{j=0}^{l} A_j u^{(j)}(t) = \sum_{j=0}^{l} A_j X_1 T_1^j e^{T_1 t} x + \sum_{j=1}^{l} \sum_{k=0}^{j-1} A_j X_1 T_1^{j-1-k} Z_1 f^{(k)}(t) + \int_{t_0}^{t} \left[\sum_{j=0}^{l} A_j X_1 T_1^j \right] e^{T_1(t-s)} Z_1 f(s)\, ds - \sum_{i=0}^{\nu-1} \sum_{j=0}^{l} A_j X_2 T_2^i Z_2 f^{(i+j)}(t).$$

Using the property $\sum_{j=0}^{l} A_j X_1 T_1^j = 0$ and rearranging terms, we have

$$\sum_{j=0}^{l} A_j u^{(j)}(t) = \sum_{k=0}^{l+\nu-1} \left\{ \sum_{j=k+1}^{l} A_j X_1 T_1^{j-1-k} Z_1 - \sum_{j=0}^{k} A_j X_2 T_2^{k-j} Z_2 \right\} f^{(k)}(t). \tag{8.5}$$

(As usual, the sum $\sum_{j=k+1}^{l}$ is assumed to be empty if $k + 1 > l$.) By the definition of a resolvent triple,

$$L^{-1}(\lambda) = [X_1 \quad X_2] \begin{bmatrix} (I\lambda - T_1)^{-1} & 0 \\ 0 & (T_2\lambda - I)^{-1} \end{bmatrix} \begin{bmatrix} Z_1 \\ Z_2 \end{bmatrix},$$

and for $|\lambda|$ large enough,

$$L^{-1}(\lambda) = -X_2 T_2^{\nu-1} Z_2 \lambda^{\nu-1} - \cdots - X_2 T_2 Z_2 \lambda - X_2 Z_2 + X_1 Z_1 \lambda^{-1} + X_1 T_1 Z_1 \lambda^{-2} + \cdots. \tag{8.6}$$

So the coefficient of $f^{(k)}(t)$ is the right-hand side of (8.5) is nothing but the coefficient of λ^k in the product $L(\lambda)L^{-1}(\lambda) = I$. Hence, $\sum_{j=0}^{l} A_j u^{(j)}(t) = f(t)$.

It remains to show that (8.4) gives the general solution of (8.1). As every solution of (8.1) is a sum of a fixed solution of (8.1) and some solution $u_0(t)$ of the homogeneous equation

$$\sum_{i=0}^{l} A_i \frac{d^i u_0(t)}{dt^i} = 0, \tag{8.7}$$

it is sufficient to check that

$$u_0(t) = X_1 e^{T_1 t} x, \qquad x \in \mathbb{C}^m, \tag{8.8}$$

is the general solution of (8.7).

Since the set of all solutions of (8.7) is an m-dimensional linear space (Theorem S1.6), we have only to check that $u_0(t) \equiv 0$ is possible only if $x = 0$. In fact, $u_0(t) \equiv 0$ implies $u_0^{(i)}(t) = X_1 T_1^i e^{T_1 t} x = 0$ for $i = 1, \ldots, l-1$. So $\operatorname{col}(X_1 T_1^i)_{i=0}^{l-1} e^{T_1 t} x = 0$, and since the columns of $\operatorname{col}(X_1 T_1^i)_{i=0}^{l-1}$ are linearly independent (refer to the definition of a decomposable pair in Section 7.3), we obtain $e^{T_1 t} x = 0$ and $x = 0$. □

The following theorem is a more detailed version of Theorem 8.1 for the case of comonic matrix polynomials.

Theorem 8.2. *Let $L(\lambda)$ be a comonic matrix polynomial of degree l, and let (X_F, J_F) and (X_∞, J_∞) be its finite and infinite Jordan pairs, respectively. Define Z to be the $nl \times n$ matrix given by*

$$[\operatorname{col}(XJ^{i-1})_{i=1}^{l}]^{-1} = [* \quad \cdots \quad * \quad Z],$$

where $X = [X_F \quad X_\infty]$, $J = J_F^{-1} \oplus J_\infty$.

Make a partition $Z = \left[\begin{smallmatrix} Z_F \\ Z_\infty \end{smallmatrix}\right]$ corresponding to the partition $X = [X_F \quad X_\infty]$. Assume that the $\mathbb{C}^n$-valued function $f(t)$ is v-times continuously differentiable, where v is the least nonnegative integer such that $J_\infty^v = 0$. Then the general solution of (8.1) is an l-times continuously differentiable function of the form

$$u(t) = X_F e^{J_F t} x + \sum_{k=0}^{v-l} X_\infty J_\infty^{l-1+k} Z_\infty f^{(k)}(t)$$
$$- X_F J_F^{-(l-2)} \int_a^t e^{J_F(t-\tau)} Z_F f(\tau)\, d\tau, \tag{8.9}$$

where x is an arbitrary vector in $\mathbb{C}^m$, $m = \deg\det L(\lambda)$.

Proof. By Theorem 7.3, $([X_F \quad X_\infty], J_F \oplus J_\infty)$ is a decomposable pair of $L(\lambda)$. Using formula (7.55) it is easily seen that

$$\left(X, J, -\begin{bmatrix} J_F^{-(l-2)} & 0 \\ 0 & J_\infty^{l-1} \end{bmatrix} Z\right)$$

is a resolvent triple of $L(\lambda)$. It remains to apply Theorem 8.1 to obtain (8.9). □

Observe that according to (8.9) it is sufficient to require v-times differentiability of $f(\lambda)$ in Theorem 8.2 (instead of $(v + l - 1)$-times differentiability in Theorem 8.1).

If $v = 0$, i.e., if the leading coefficient of $L(\lambda)$ is invertible, then Theorem 2.9 shows that (8.1) is solvable for any continuous $\mathbb{C}^n$-valued function f. Note that $v = 0$ implies that deg det $L(\lambda) = nl$, and the general solution of (8.1) (assuming comonicity of $L(\lambda)$) is given by (8.9)

$$u(t) = X_{\mathrm{F}} e^{J_{\mathrm{F}} t} x - X_{\mathrm{F}} J_{\mathrm{F}}^{-(l-2)} \int_a^t e^{J_{\mathrm{F}}(t-\tau)} Z_{\mathrm{F}} f(\tau)\, d\tau, \tag{8.10}$$

where x is an arbitrary vector in $\mathbb{C}^{nl}$. In the particular case that the leading coefficient of $L(\lambda)$ is equal to the identity, formula (8.10) is the same as formula (2.35) with $X = X_{\mathrm{F}}$, $T = J_{\mathrm{F}}$, because then

$$L(\lambda)^{-1} = X_{\mathrm{F}}(I\lambda - J_{\mathrm{F}})^{-1} Y = X_{\mathrm{F}} J_{\mathrm{F}}^{-(l-1)} (I - J_{\mathrm{F}}^{-1}\lambda)^{-1} Z_{\mathrm{F}}$$

(cf. Theorem 2.6), and thus

$$X_{\mathrm{F}} J_{\mathrm{F}}^{\alpha} Y = -X_{\mathrm{F}} J_{\mathrm{F}}^{-(l-2)+\alpha} Z_{\mathrm{F}}, \qquad \alpha \geq 0. \tag{8.11}$$

To illustrate the usefulness of the above approach, we shall write out the general solution of the matrix differential equation

$$A\dot{u} + Bu = f \tag{8.12}$$

in another closed form in which the coefficient matrices appear explicity. We assume that A and B are $n \times n$ matrices such that $\det(A\lambda + B)$ does not vanish identically. Choose $a \in \mathbb{C}$ such that $aA + B$ is nonsingular, and let Γ_0 and Γ_1 be two positively oriented closed contours around the zeros of $\det(A\lambda + B)$ such that a is inside Γ_0 and outside Γ_1. Then the general solution of $A\dot{u} + Bu = f$ may be written as

$$u(t) = u_1(t) + u_2(t) + u_3(t), \tag{8.13}$$

where

$$u_1(t) = \left[-\frac{1}{2\pi i} \int_{\Gamma_1} \frac{e^{t\lambda}}{\lambda - a} (A\lambda + B)^{-1}\, d\lambda \right] x_0,$$

$$u_2(t) = \sum_{k=0}^{\mu-1} \left[\frac{1}{2\pi i} \int_{\Gamma_0} \left(\frac{1}{\lambda - a} \right)^{k+1} (\lambda A + B)^{-1}\, d\lambda \right] \left(-a + \frac{d}{dt} \right)^k f(t),$$

$$u_3(t) = \frac{1}{2\pi i} \int_{\Gamma_1} (A\lambda + B)^{-1} \left(\int_a^t e^{(t-\tau)\lambda} f(\tau)\, d\tau \right) d\lambda.$$

Here x_0 is an arbitrary vector in $\mathbb{C}^n$ and μ is the least nonnegative integer such that

$$\int_{\Gamma_0} \left(\frac{1}{\lambda - a} \right)^{\mu+1} (A\lambda + B)^{-1}\, d\lambda$$

is the zero matrix.

Let us deduce the formula (8.13). Let $L(\lambda) = A\lambda + B$ and let $\tilde{L}(\lambda) = A\lambda + aA + B = L(\lambda + a)$. Consider the differential equation

$$(aA + B)^{-1}\tilde{L}\left(\frac{d}{dt}\right)v = g. \tag{8.14}$$

We apply Theorem 8.2 to find a general solution of (8.14). A comonic Jordan pair $X = [X_F \quad X_\infty]$, $J = J_F^{-1} \oplus J_\infty$ of $\tilde{L}(\lambda)$ is given by the formula $(X, J) = (I, -(aA + B)^{-1}A)$, and

$$(X_\infty, J_\infty) = (X|_{\mathscr{M}_0}, J|_{\mathscr{M}_0}) \qquad \text{where} \quad \mathscr{M}_0 = \operatorname{Im} \int_{\Delta_0} (I\lambda - J)^{-1}\, d\lambda,$$

$$(X_F, J_F^{-1}) = (X|_{\mathscr{M}_1}, J|_{\mathscr{M}_1}) \qquad \text{where} \quad \mathscr{M}_1 = \operatorname{Im} \int_{\Delta_1} (I\lambda - J)^{-1}\, d\lambda.$$

Here Δ_0 and Δ_1 are closed rectifiable contours such that $(\Delta_0 \cup \Delta_1) \cap \sigma(J) = \varnothing$; zero is the only eigenvalue of J lying inside Δ_0, and Δ_1 contains in its interior all but the zero eigenvalue of J. The matrix Z from Theorem 8.2 is just the identity. Write formula (8.9):

$$v(t) = v_1(t) + v_2(t) + v_3(t),$$

where

$$\begin{aligned} v_1(t) &= X_F e^{J_F t} x = e^{(J|_{\mathscr{M}_1})^{-1} t} x \\ &= \left(\frac{1}{2\pi i}\int_{\Delta_1} e^{\lambda^{-1} t}[I\lambda + (aA + B)^{-1}A]^{-1}\, d\lambda\right)x, \qquad x \in \mathbb{C}^n, \\ v_2(t) &= \sum_{k=0}^{\mu-1} X_\infty J_\infty^k Z_\infty g^{(k)}(t) \\ &= \sum_{k=0}^{\mu-1}\left(\frac{1}{2\pi i}\int_{\Delta_0} \lambda^k [I\lambda + (aA + B)^{-1}A]^{-1}\, d\lambda\right) g^{(k)}(t), \end{aligned}$$

where μ is the least nonnegative integer such that $J_\infty^\mu = 0$;

$$\begin{aligned} v_3(t) &= -X_F J_F \int_a^t e^{J_F(t-\tau)} Z_F g(\tau)\, d\tau \\ &= -\frac{1}{2\pi i}\int_{\Delta_1} \lambda^{-1}[I\lambda + (aA + B)^{-1}A]^{-1} \int_a^t e^{\lambda^{-1}(t-\tau)} g(\tau)\, d\tau\, d\lambda. \end{aligned}$$

Observe now that the contours Δ_0 and Δ_1, while satisfying the above requirements, can be chosen also in such a way that $\Gamma_0 = -\{\lambda^{-1} + a \mid \lambda \in \Delta_0\}$, $\Gamma_1 = \{\lambda^{-1} + a \mid \lambda \in \Delta_1\}$ (the minus in the formula for Γ_0 denotes the opposite orientation). Make the change of variable in the above formulas for $v_1(t)$, $v_2(t)$, $v_3(t)$: $\lambda = (\mu - a)^{-1}$. A calculation gives (note that

$$[I\lambda + (aA + B)^{-1}A]^{-1} = (\mu - a)(\mu A + B)^{-1}(aA + B))$$

$$v_1(t) = \left(-\frac{1}{2\pi i}\int_{\Gamma_1} e^{(\mu - a)t}(A\mu + B)^{-1}\frac{d\mu}{\mu - a}\right)(aA + B)x,$$

$$v_2(t) = \overset{\mu-1}{\underset{k=0}{X}} \frac{1}{2\pi i}\int_{\Gamma_0} (\mu - a)^{-k-1}(A\mu + B)^{-1}\,d\mu)(aA + B)g^{(k)}(t),$$

$$v_3(t) = \frac{1}{2\pi i}\int_{\Gamma_1} (A\mu + B)^{-1}\int_a^t e^{(\mu - a)(t - \tau)}(aA + B)g(\tau)\,d\tau\,d\mu.$$

Finally, the general solution $u(t)$ of (8.12) is given by the formula $u(t) = e^{at}v(t)$, where $v(t)$ is a general solution of (8.14) with $g(t) = (aA + B)^{-1}e^{-at}f(t)$. Using the above formulas for $v(t)$, we obtain the desired formula (8.13) for $u(t)$.

These arguments can also be used to express formulas (8.9) in terms of the coefficients of $L(\lambda) = I + \sum_{j=1}^{l} A_j\lambda^j$. To see this, let

$$R = \begin{bmatrix} 0 & I & \cdots & 0 \\ \vdots & \vdots & & \vdots \\ 0 & 0 & \cdots & I \\ -A_l & -A_{l-1} & \cdots & -A_1 \end{bmatrix}$$

be the comonic companion matrix of $L(\lambda)$. Further, put $Q = \mathrm{row}(\delta_{1j}I)_{j=1}^{l}$ and $Y = \mathrm{col}(\delta_{lj}I)_{j=1}^{l}$. Take P_0 to be the Riesz projector of R corresponding to the eigenvalue 0, i.e.,

$$P_0 = \frac{1}{2\pi i}\int_{\Gamma_0} (I\lambda - R)^{-1}\,d\lambda, \tag{8.15}$$

where Γ_0 is a suitable contour around zero separating 0 from the other eigenvalues of R. Finally, let $P = I - P_0$. Then the general solution of (8.1) (with $A_0 = I$) is equal to

$$u(t) = u_1(t) + u_2(t) + u_3(t),$$

where

$$u_1(t) = (Q|_{\mathrm{Im}\,P})[\exp((R|_{\mathrm{Im}\,P})^{-1}t)]Pz_0,$$

$$u_2(t) = \sum_{k=0}^{\mu-l} QP_0R^{l-1+k}P_0Yf^{(k)}(t),$$

$$u_3(t) = -(Q|_{\mathrm{Im}\,P})(R|_{\mathrm{Im}\,P})^{l-2}\int_a^t [\exp((R|_{\mathrm{Im}\,P})^{-1}(t - \tau))]PYf(\tau)\,d\tau.$$

Here z_0 is an arbitrary vector in $\operatorname{Im} P$ and μ is the least nonnegative integer such that $P_0 R^{\mu} P_0 = 0$.

8.2. Difference Equations in the Nonmonic Case

We move on now to the study of the corresponding finite difference equations.

Theorem 8.3. *Let $L(\lambda)$ and (X, T, Z) be as in Theorem 8.1. Let $\{f_r\}_{r=0}^{\infty}$ be a sequence of vectors in $\mathbb{C}^n$. Then the general solution of a finite difference equation*

$$A_0 u_r + A_1 u_{r+1} + \cdots + A_l u_{r+l} = f_r, \qquad r \geq 0, \tag{8.16}$$

where $L(\lambda) = \sum_{j=0}^{l} A_j \lambda^j$, is given by the formulas

$$u_0 = X_1 x - \sum_{i=0}^{\nu-1} X_2 T_2^i Z_2 f_i,$$

$$u_r = X_1 T_1^r x - \sum_{i=0}^{\nu-1} X_2 T_2^i Z_2 f_{i+r} + \sum_{j=0}^{r-1} X_1 T_1^{r-j-1} Z_1 f_j, \qquad r \geq 1, \tag{8.17}$$

where $x \in \mathbb{C}^m$ is arbitrary and m is the degree of $\det L(\lambda)$.

Proof. It is easy to see (using Theorem S1.8) that the general solution of the corresponding homogeneous finite difference equation is given by

$$u_r = X_1 T_1^r x, \qquad r = 0, 1, \ldots.$$

Hence it suffices to show that the sequence $\{\varphi_r\}_{r=0}^{\infty}$ given by

$$\varphi_0 = -\sum_{i=0}^{\nu-1} X_2 T_2^i Z_2 f_i,$$

$$\varphi_r = -\sum_{i=0}^{\nu-1} X_2 T_2^i Z_2 f_{i+r} + \sum_{j=0}^{r-1} X_1 T_1^{r-j-1} Z_1 f_j, \qquad r \geq 1,$$

is a solution of (8.16). Indeed,

$$\begin{aligned}
&A_0 \varphi_r + A_1 \varphi_{r+1} + \cdots + A_l \varphi_{r+l} \\
&\quad = \sum_{k=0}^{l} \sum_{j=0}^{r+k-1} A_k X_1 T_1^{r+k-j-1} Z_1 f_j - \sum_{k=0}^{l} \sum_{i=0}^{\nu-1} A_k X_2 T_2^i Z_2 f_{r+k+i} \\
&\quad = \sum_{j=0}^{\infty} \left(\sum_{k=0}^{\infty} A_k X_1 T_1^{r+k-j-1} Z_1 - \sum_{k=0}^{\infty} A_k X_2 T_2^{j-k-r} Z_2 \right) f_j,
\end{aligned} \tag{8.18}$$

where it is assumed $T_1^p = 0$, $T_2^p = 0$ for $p < 0$. Using (8.6), it is seen that the coefficient of f_j in the right-hand side of (8.18) is just the coefficient of λ^{j-r} in the product $L(\lambda)L^{-1}(\lambda)$, which is equal to I for $j = r$, and zero otherwise. Hence the right-hand side of (8.18) is just f_r. $\square$

Again, a simplified version of Theorem 8.3 for comonic polynomials can be written down using finite and infinite Jordan pairs for the construction of a decomposable pair of $L(\lambda)$. In this case the general solution of (8.16) (with $A_0 = I$) has the form

$$u_0 = X_F x + \sum_{j=0}^{v-l} X_\infty J_\infty^{l-1+j} Z_\infty f_j,$$

$$u_r = X_F J_F^r x + \sum_{j=0}^{v-l} X_\infty J_\infty^{l-1+j} Z_\infty f_{j+r} - \sum_{j=0}^{r-1} X_F J_F^{-l+r-j} Z_F f_j, \qquad r \geq 1, \tag{8.19}$$

where (X_F, J_F, Z_F), $(X_\infty, J_\infty, Z_\infty)$ are as in Theorem 8.2. As in the case of a differential equation, the general solution (8.19) can be expressed in terms of the comonic companion matrix R of $L(\lambda)$, and the Riesz projector P_0 given by (8.15). We leave it to the reader to obtain these formulas. The following similarity properties may be used for this purpose:

$$[X_F \quad X_\infty] = \operatorname{row}(\delta_{1j} I)_{j=1}^{l} S, \qquad J_F^{-1} \oplus J_\infty = S^{-1} R S,$$

$$\begin{bmatrix} Z_F \\ Z_\infty \end{bmatrix} = S^{-1} \operatorname{col}(\delta_{lj} I)_{j=1}^{l},$$

where R is the comonic companion matrix of $L(\lambda)$, and

$$S = \operatorname{col}(X_F J_F^{-i}, X_\infty J_\infty^{i})_{i=0}^{l-1}.$$

Theorem 8.3 allows us to solve various problems concerning the behavior of the solution $(u_r)_{r=0}^{\infty}$ of (8.16) provided the behavior of the data $(f_r)_{r=0}^{\infty}$ is known. The following corollary presents one result of this type (problems of this kind are important for stability theory of difference schemes for approximate solution of partial differential equations).

We say the sequence $(f_r)_{r=0}^{\infty}$ is bounded if $\sup_{r=0,1,\ldots} \|f_r\| < \infty$.

Corollary 8.4. *If the spectrum of $L(\lambda)$ is inside the unit circle, then all solutions $(u_r)_{r=0}^{\infty}$ of (8.16) are bounded provided $(f_r)_{r=0}^{\infty}$ is bounded.*

Proof. Use the decomposable pair $([X_F \quad X_\infty], J_F \oplus J_\infty)$ in Theorem 8.3, where (X_F, J_F) and (X_∞, J_∞) are finite and infinite Jordan pairs of $L(\lambda)$, respectively. Since the eigenvalues of J_F are all inside the unit circle, we have

$\|J_F^i\| \le K\rho^i$, $i = 0, 1, \ldots$, where $0 < \rho < 1$, and $K > 0$ is a fixed constant. Then (8.17) gives (with $X_1 = X_F$, $T_1 = J_F$, $X_2 = X_\infty$, $T_2 = J_\infty$):

$$\|u_r\| \le \|X_F\|K\rho^r\|x\| + \left(\sum_{i=0}^{v-1} \|X_\infty J_\infty^i Z_2\|\right) \sup_r \|f_r\| + \sum_{j=0}^{r-1} \|X_F\|K\rho^{r-j-1}\|Z_1\| \sup_r \|f_r\|,$$

and the right-hand side is bounded uniformly in r provided $\sup_r \|f_r\| < \infty$. □

8.3. Construction of Differential and Difference Equations with Given Solutions

Consider the $\mathbb{C}^n$-vector functions $\varphi_j(t) = p_j(t)e^{\lambda_j t}$, $j = 1, \ldots, r$, where $p_j(t)$ are polynomials in t whose coefficients are vectors in $\mathbb{C}^n$. In this section we show how one can find an $n \times n$ matrix polynomial $L(\lambda)$ with a finite number of eigenvalues, such that the functions $\varphi_1, \ldots, \varphi_r$ are solutions of the equation

$$L\left(\frac{d}{dt}\right)\varphi = 0. \tag{8.20}$$

Without any additional restriction, this problem is easy to solve. At first consider the scalar case ($n = 1$). Then we may choose

$$L(\lambda) = \prod_{j=1}^{r} (\lambda - \lambda_j)^{s_j + 1},$$

where s_j is the degree of the polynomial $p_j(t)$.

In the general case ($n > 1$), one first chooses scalar polynomials $L_1(\lambda), \ldots, L_n(\lambda)$ such that for each j the kth coordinate function of φ_j is a solution of $L_k(d/dt)\varphi = 0$. Next, put $L(\lambda) = L_1(\lambda) \oplus \cdots \oplus L_n(\lambda)$. Then the functions $\varphi_1, \ldots, \varphi_r$ are solutions of $L(d/dt)\varphi = 0$. But with this choice of $L(\lambda)$, Eq. (8.20) may have solutions which are not linear combinations of the vector functions $\varphi_1, \ldots, \varphi_r$ and their derivatives. So we want to construct the polynomial $L(\lambda)$ in such a way that the solution space of Eq. (8.20) is precisely equal to the space spanned by the functions $\varphi_1, \ldots, \varphi_r$ and their derivatives. This "inverse" problem is solved by Theorem 8.5 below. In the solution of this problem we use the notions and results of Section 7.11. In particular, the special generalized inverse will play an important role.

Theorem 8.5. *Let $\varphi_j(t) = (\sum_{k=0}^{s_j} p_{jk} t^k)e^{\lambda_j t}$ where each p_{jk} is a vector in $\mathbb{C}^n$, $0 \le k \le s_j$, $1 \le j \le r$. Put*

$$X = \text{row}(\text{row}((s_j - k)!\, p_{j, s_j - k})_{k=0}^{s_j})_{j=1}^{r},$$

and let $J = J_1 \oplus \cdots \oplus J_r$, *where* J_i *is the Jordan block of order* $s_j + 1$ *with eigenvalue* λ_i. *Let* α *be some complex number different from* $\lambda_1, \ldots, \lambda_r$, *and define*

$$L(\lambda) = I - X(J - \alpha I)^{-l}\{(\lambda - \alpha)V^l + \cdots + (\lambda - \alpha)^l V_1\}, \qquad (8.21)$$

where $l = \operatorname{ind}(X, J)$ *and* $[V_1 \quad \cdots \quad V_l]$ *is a special generalized inverse of* $\operatorname{col}(X(J - \alpha I)^{1-i})_{i=1}^{l}$. *Then* $\varphi_1, \ldots, \varphi_r$ *are solutions of the equation*

$$L\left(\frac{d}{dt}\right)\varphi = 0, \qquad (8.22)$$

and every solution of this equation is a linear combination of $\varphi_1, \ldots, \varphi_r$ *and their derivatives. Further,* l *is the minimal possible degree of any* $n \times n$ *matrix polynomial with this property.*

Proof. Note that $l = \operatorname{ind}(X, J) = \operatorname{ind}(X, J - \alpha I)$; because of the identity

$$\begin{bmatrix} I & 0 & \cdots & 0 \\ \alpha\binom{1}{0}I & \binom{1}{1}I & \cdots & 0 \\ \vdots & \vdots & & \vdots \\ \alpha^{l-1}\binom{l-1}{0}I & \alpha^{l-2}\binom{l-1}{1}I & \cdots & \binom{l-1}{l-1}I \end{bmatrix} \begin{bmatrix} X \\ X(J - \alpha I) \\ \vdots \\ X(J - \alpha I)^{l-1} \end{bmatrix} = \begin{bmatrix} X \\ XJ \\ \vdots \\ XJ^{l-1} \end{bmatrix}.$$

It follows that without loss of generality we may suppose that $\alpha = 0$, and let (X_L, J_L) be a finite Jordan pair of $L(\alpha)$. Then the general solution of Eq. (8.22) is equal to $X_L e^{tJ_L} x$, where x is an arbitrary vector of $\mathbb{C}^m$, $m = \operatorname{degree}(\det L(\lambda))$ (see Theorem 8.1). By Theorem 7.24 the pair (X_L, J_L) is similar to the pair

$$(X|_{\operatorname{Im} P}, PJ|_{\operatorname{Im} P}),$$

where P is a projector of $\mathbb{C}^\mu$, $\mu = \sum_{j=1}^{r} (s_j + 1)$, with $\operatorname{Ker} P = \operatorname{Ker}(X, J)$. It follows that the general solution of (8.22) is of the form $Xe^{tJ}z$, where z is an arbitrary vector in $\mathbb{C}^\mu$. But this means that the solution space of (8.22) is equal to the space spanned by the function $\varphi_1, \ldots, \varphi_r$ and their derivatives.

Let $B(\lambda)$ be an $n \times n$ matrix polynomial with the property that the solution space of $B(d/dt)\varphi = 0$ is the space spanned by $\varphi_1, \ldots, \varphi_r$ and their derivatives. Then $B(\lambda)$ can have only a finite number of eigenvalues, and the finite Jordan pairs of $B(\lambda)$ and $L(\lambda)$ are similar. But then we can apply Theorem 7.24 to show that degree $B(\lambda) \geq$ degree $L(\lambda) = l$. So l is the minimal possible degree. □

Now consider the analogous problem for difference equations:

Let $L(\lambda)$ be an $n \times n$ regular matrix polynomial and consider the finite difference equation

$$L(\mathscr{E})u_i = 0, \qquad i = 1, 2, \ldots, \qquad (8.23)$$

where $\mathscr{E}u_i = u_{i+1}$ and $u_1, u_2, \ldots,$ is a sequence of vectors in $\mathbb{C}^n$. Let $x_0, \ldots, x_{k-1}$ be a Jordan chain for $L(\lambda)$ corresponding to the eigenvalue λ_0. Put $X_0 = [x_0 \quad x_1 \quad \cdots \quad x_{k-1}]$, and let J_0 be the Jordan block of order k with λ_0 as an eigenvalue. Then for each $z \in \mathbb{C}^k$ the sequence

$$u_{r+1} = X_0 J_0^r z, \qquad r \geq 0$$

is a solution of (8.23) (Theorem 8.3). By taking $z = (0 \quad \cdots \quad 0 \quad 1 \quad 0 \quad \cdots \quad 0)^{\mathrm{T}}$ with 1 in the qth place, we see that

$$u_{r+1} = \sum_{j=0}^{r} \binom{r}{r-q+j} \lambda_0^{r-q+j} x_j, \qquad r \geq 0 \tag{8.24}$$

is a solution of (8.23). Moreover, each solution of (8.23) is a linear combination of solutions of the form (8.24). The next theorem solves the inverse problem; its proof is similar to the proof of Theorem 8.5 and therefore is omitted.

Theorem 8.6. *Suppose we have given a finite number of sequences of the form*

$$u_{r+1}^{(i)} = \sum_{j=0}^{q_i} \binom{r}{r-q_i+j} \lambda_i^{r-q_i+j} x_{ji}, \qquad r \geq 0,$$

where $x_{ji} \in \mathbb{C}^n$, $0 \leq j \leq q_i$, $1 \leq i \leq p$. Put

$$X = \operatorname{row}(\operatorname{row}(x_{ji})_{j=0}^{q_i})_{i=1}^{p}$$

and let $J = J_1 \oplus \cdots \oplus J_p$, where J_i is the Jordan block of order $q_i + 1$ with eigenvalue λ_i. Let α be some complex number different from $\lambda_1, \ldots, \lambda_p$, and define

$$L(\lambda) = I - X(J - \alpha I)^{-l}\{(\lambda - \alpha)V_l + \cdots + (\lambda - \alpha)^l V_1\}, \tag{8.25}$$

where $l = \operatorname{ind}(X, J)$ and $[V_1 \quad \cdots \quad V_l]$ is a special generalized inverse of $\operatorname{col}(X(J - \alpha I)^{1-i})_{i=1}^{l}$. Then the sequences $u^{(i)} = (u_1^{(i)}, u_2^{(i)}, \ldots)$, $1 \leq j \leq p$, are solutions of the equation

$$L(\mathscr{E})u_i = 0, \qquad i = 1, 2, \ldots,$$

and every solution of this equation is a linear combination of $u^{(1)}, \ldots, u^{(p)}$. Further, l is the minimal possible degree of any $n \times n$ matrix polynomial with this property.

Consider a matrix polynomial $M(\lambda)$ such that the set of solutions of the equation $M(\mathscr{E})u_i = 0$ consists of all linear combinations of the sequences $u^{(1)}, \ldots, u^{(p)}$. The general form of $M(\lambda)$ is given by $M(\lambda) = E(\lambda)L(\lambda)$, where

$L(\lambda)$ is given by (8.25) and $E(\lambda)$ is a unimodular $n \times n$ matrix polynomial (i.e., with $\det E(\lambda) \equiv \text{const} \neq 0$).

Comments

Theorems 8.1 and 8.3 are proved in [14]. All other theorems are in [29a]. In a less general form an analysis of the singular linear differential equation (8.12) was made in [11].

Chapter 9

Least Common Multiples and Greatest Common Divisors of Matrix Polynomials

Let $L_1(\lambda), \ldots, L_s(\lambda)$ be a finite set of matrix polynomials. A matrix polynomial $N(\lambda)$ is called a (left) *common multiple* of $L_1(\lambda), \ldots, L_s(\lambda)$ if $L_i(\lambda)$ is a right divisor of $N(\lambda)$, $i = 1, \ldots, s$. A common multiple $N(\lambda)$ is called a *least common multiple* (l.c.m.) of $L_1(\lambda), \ldots, L_s(\lambda)$ if $N(\lambda)$ is a right divisor of every other common multiple. The notions of a common divisor and a greatest common divisor are given in an analogous way: namely, a matrix polynomial $D(\lambda)$ is a (right) *common divisor* of $L_1(\lambda), \ldots, L_s(\lambda)$ if $D(\lambda)$ is a right divisor of every $L_i(\lambda)$; $D(\lambda)$ is a *greatest common divisor* (g.c.d.) if $D(\lambda)$ is a common divisor of $L_1(\lambda), \ldots, L_s(\lambda)$ and every other common divisor is in turn a right divisor of $D(\lambda)$.

It will transpire that the spectral theory of matrix polynomials, as developed in the earlier chapters, provides the appropriate machinery for solving problems concerning l.c.m. and g.c.d. In particular, we give in this chapter an explicit construction of the l.c.m. and g.c.d. of a finite family of matrix polynomials $L_1(\lambda), \ldots, L_s(\lambda)$. The construction will be given in terms of both the spectral data of the family and their coefficients. The discussion in terms of coefficient matrices is based on the use of Vandermonde and resultant matrices, which will be introduced later in the chapter.

Throughout Chapter 9 we assume for simplicity that the matrix polynomials $L_1(\lambda), \ldots, L_s(\lambda)$ are of size $n \times n$ and comonic, i.e., $L_i(0) = I$, $i = 1, \ldots, s$. In fact, the construction of l.c.m. and g.c.d. of a finite family of regular matrix polynomials $L_1(\lambda), \ldots, L_s(\lambda)$ can be reduced to a comonic case by choosing a point a which does not belong to $\bigcup_{i=1}^{s} \sigma(L_i)$, and considering the matrix polynomials $L_i(a)^{-1}L_i(\lambda + a)$ in place of $L_i(\lambda), i = 1, \ldots, s$.

9.1. Common Extensions of Admissible Pairs

In this and the next two sections we shall develop the necessary background for construction of the l.c.m. and g.c.d. of matrix polynomials. This background is based on the notions of restrictions and extensions of admissible pairs. We have already encountered these ideas (see Section 7.7), but let us recall the main definitions and properties. A pair of matrices (X, T) is called an *admissible pair* of *order* p if X is an $n \times p$ matrix and T is a $p \times p$ matrix (the number n is fixed). The space

$$\operatorname{Ker}(X, T) = \bigcap_{j=0}^{\infty} \operatorname{Ker} XT^j = \bigcap_{s=1}^{\infty} \operatorname{Ker}[\operatorname{col}(XT^{j-1})_{j=1}^{s}]$$

is called the *kernel* of the pair (X, T). The least positive integer s such that

$$\operatorname{Ker}[\operatorname{col}(XT^{j-1})_{j=1}^{s}] = \operatorname{Ker}[\operatorname{col}(XT^{j-1})_{j=1}^{s+1}] \tag{9.1}$$

is the *index of stabilization* of (X, T) and is denoted by $\operatorname{ind}(X, T)$. Moreover, if $s = \operatorname{ind}(X, T)$, then

$$\mathcal{M}_{s+r} = \mathcal{M}_s \qquad \text{for} \quad r = 1, 2, \ldots, \tag{9.2}$$

where $\mathcal{M}_i = \operatorname{Ker}[\operatorname{col}(XT^{j-1})_{j=1}^{s}] \subset \mathbb{C}^p$. Indeed, for $r = 1$ the property (9.2) is just the definition of $\operatorname{ind}(X, T)$. To prove (9.2) for $r > 1$, take $x \in \mathcal{M}_s$. Then (by (9.2) for $r = 1$), $x \in \mathcal{M}_{s+1}$. In particular, this means that $\operatorname{col}(XT^i)_{i=1}^{s} x = 0$, i.e., $Tx \in \mathcal{M}_s$. Again, by (9.2) for $r = 1$, $Tx \in \mathcal{M}_{s+1}$ and, consequently, $T^2x \in \mathcal{M}_s$. Continuing this process, we obtain $T^r x \in \mathcal{M}_s$ for $r = 1, 2, \ldots$, and (9.2) follows.

Two admissible pairs (X, T) and (X_1, T_1) are said to be *similar* if there exists a nonsingular matrix S such that $XS = X_1$ and $T = ST_1S^{-1}$. In other words, (X, T) and (X_1, T_1) are similar if (X, T) is an extension of (X_1, T_1) and (X_1, T_1) is an extension of (X, T).

We recall the necessary and sufficient condition for extension of two admissible pairs (X, T) and (X_1, T_1) given by Lemma 7.12 (see Section 7.7 for the definition of extension of admissible pairs): namely, if $\operatorname{Ker}(X, T) = \{0\}$ and $\operatorname{Ker}(X_1, T_1) = \{0\}$, then (X, T) is an extension of (X_1, T_1) if and only if

$$\operatorname{Im}[\operatorname{col}(X_1T_1^{i-1})_{i=1}^{k+1}] \subset \operatorname{Im}[\operatorname{col}(XT^{i-1})_{i=1}^{k+1}] \tag{9.3}$$

for some $k \geq \max\{\operatorname{ind}(X, T), \operatorname{ind}(X_1, T_1)\}$. If this is the case, then (9.3) holds for all integers $k \geq 0$.

Let $(X_1, T_1), \ldots, (X_r, T_r)$ be admissible pairs. The admissible pair (X, T) is said to be a *common extension* of $(X_1, T_1), \ldots, (X_r, T_r)$ if (X, T) is an extension of each (X_j, T_j), $j = 1, \ldots, r$. We call (X_0, T_0) a *least common extension* of $(X_1, T_1), \ldots, (X_r, T_r)$ if (X_0, T_0) is a common extension of $(X_1, T_1), \ldots, (X_r, T_r)$, and any common extension of $(X_1, T_1), \ldots, (X_r, T_r)$ is an extension of (X_0, T_0).

Theorem 9.1. *Let $(X_1, T_1), \ldots, (X_r, T_r)$ be admissible pairs of orders $p_1, \ldots, p_r$, respectively, and suppose that* $\operatorname{Ker}(X_j, T_j) = \{0\}$ *for $1 \leq j \leq r$. Then up to similarity there exists a unique least common extension of $(X_1, T_1), \ldots, (X_r, T_r)$.*

Put $X = [X_1 \ \cdots \ X_r]$, $T = T_1 \oplus \cdots \oplus T_r$, and $p = p_1 + \cdots + p_r$, and let P be a projector of $\mathbb{C}^p$ along $\operatorname{Ker}(X, T)$ *(i.e.,* $\operatorname{Ker} P = \operatorname{Ker}(X, T)$*). Then one such least common extension is given by*

$$(X|_{\operatorname{Im} P}, PT|_{\operatorname{Im} P}). \tag{9.4}$$

Proof. Let X_0 be the first term in (9.4) and T_0 the second. By choosing some basis in $\operatorname{Im} P$ we can interpret (X_0, T_0) as an admissible pair of matrices. As $\operatorname{Ker}(X, T)$ is invariant for T and $Xu = 0$ for $u \in \operatorname{Ker}(X, T)$, one sees that $\operatorname{Ker}(X_0, T_0) = \{0\}$ and

$$\operatorname{Im}[\operatorname{col}(X_0 T_0^{i-1})_{i=1}^m] = \operatorname{Im}[\operatorname{col}(X T^{i-1})_{i=1}^m], \qquad m \geq 1. \tag{9.5}$$

From the definitions of X and T it is clear that

$$\begin{aligned}\operatorname{Im}[\operatorname{col}(X T^{i-1})_{i=1}^m] = \operatorname{Im}[\operatorname{col}(X_1 T_1^{i-1})_{i=1}^m] + \cdots \\ + \operatorname{Im}[\operatorname{col}(X_r T_r^{i-1})_{i=1}^m], \qquad m \geq 0.\end{aligned} \tag{9.6}$$

But then we apply the criterion (9.3) to show that (X_0, T_0) is a common extension of $(X_1, T_1), \ldots, (X_r, T_r)$.

Now assume that (Y, R) is a common extension of $(X_1, T_1), \ldots, (X_r, T_r)$. Then (cf. (9.3)) for $1 \leq j \leq r$ we have

$$\operatorname{Im}[\operatorname{col}(Y R^{i-1})_{i=1}^m] \supset \operatorname{Im}[\operatorname{col}(X_j T_j^{i-1})_{j=1}^m], \qquad m \geq 1.$$

Together with (9.5) and (9.6) this implies that

$$\operatorname{Im}[\operatorname{col}(Y R^{i-1})_{i=1}^m] \supset \operatorname{Im}[\operatorname{col}(X_0 T_0^{i-1})_{i=1}^m], \qquad m \geq 1.$$

But then we can apply (9.3) again to show that (Y, R) is an extension of (X_0, T_0). It follows that (X_0, T_0) is a least common extension of $(X_1, T_1), \ldots, (X_r, T_r)$.

Let $(\tilde{X}_0, \tilde{T}_0)$ be another least common extension of $(X_1, T_1), \ldots, (X_r, T_r)$. Then $(\tilde{X}_0, \tilde{T}_0)$ is an extension of (X_0, T_0) and conversely (X_0, T_0) is an extension of $(\tilde{X}_0, \tilde{T}_0)$. Hence both pairs are similar. □

Corollary 9.2. *Let $(X_1, T_1), \ldots, (X_r, T_r)$ be as in Theorem* 9.1, *and let (X_0, T_0) be a least common extension of $(X_1, T_1), \ldots, (X_r, T_r)$. Then* $\operatorname{Ker}(X_0, T_0) = 0$ *and*

$$\sigma(T_0) \subset \bigcup_{i=1}^{r} \sigma(T_i). \tag{9.7}$$

Proof. The equality $\operatorname{Ker}(X_0, T_0) = \{0\}$ is an immediate consequence of formula (9.4). To verify (9.7) observe that (in the notation of Theorem 9.1) $\operatorname{Ker}(X, T)$ is a T-invariant subspace and T has the following form with respect to the decomposition $\mathbb{C}^p = \operatorname{Ker}(X, T) \dotplus \operatorname{Im} P$:

$$T = \begin{bmatrix} T|_{\operatorname{Ker}(X,T)} & * \\ 0 & PT|_{\operatorname{Im} P} \end{bmatrix}.$$

So

$$\sigma(T_0) = \sigma(PT|_{\operatorname{Im} P}) \subset \sigma(T) = \bigcup_{i=1}^{r} \sigma(T_i),$$

and (9.7) follows. □

The first statement of Theorem 9.1 holds true for any finite set of admissible pairs $(X_1, T_1), \ldots, (X_r, T_r)$ (not necessarily satisfying the condition $\operatorname{Ker}(X_j, T_j) = \{0\}$, $1 \le j \le r$. To this end, one first applies the previous theorem to the pairs

$$(X_j|_{\operatorname{Im} P_j}, P_j T_j|_{\operatorname{Im} P_j}), \qquad 1 \le j \le r, \tag{9.8}$$

where P_j is a projection of $\mathbb{C}^{p_j}$ along $\operatorname{Ker}(X_j, T_j)$. This yields a least common extension (X_0, T_0) of the pairs (9.7). Next we put $X = [X_0 \ \ 0_1]$ and $T = T_0 \oplus 0_2$, where 0_1 is the $n \times m$ zero-matrix and 0_2 is the $m \times m$ zero-matrix, $m = \sum_{j=1}^{r} \dim \operatorname{Ker}(X_j, T_j)$.

The next theorem explains the role of formula (9.6).

Theorem 9.3. *Let $(X_0, T_0), (X_1, T_1), \ldots, (X_r, T_r)$ be admissible, and suppose that* $\operatorname{Ker}(X_j, T_j) = \{0\}$ *for $j = 0, 1, \ldots, r$. Then (X_0, T_0) is a least common extension of $(X_1, T_1), \ldots, (X_r, T_r)$ if and only if for each $m \ge 1$:*

$$\begin{aligned} \operatorname{Im}[\operatorname{col}(X_0 T_0^{i-1})_{i=1}^{m}] = \operatorname{Im}[\operatorname{col}(X_1 T_1^{i-1})_{i=1}^{m}] + \cdots \\ + \operatorname{Im}[\operatorname{col}(X_r T_r^{i-1})_{i=1}^{m}]. \end{aligned} \tag{9.9}$$

Proof. Let $\tilde{X}_0$ be the first term in (9.4) and $\tilde{T}_0$ the second. Then $(\tilde{X}_0, \tilde{T}_0)$ is a least common extension of $(X_1, T_1), \ldots, (X_r, T_r)$ and

$$\begin{aligned} \operatorname{Im}[\operatorname{col}(\tilde{X}_0 \tilde{T}_0^{i-1})_{i=1}^{m}] = \operatorname{Im}[\operatorname{col}(X_1 T_1^{i-1})_{i=1}^{m}] + \cdots \\ + \operatorname{Im}[\operatorname{col}(X_r T_r^{i-1})_{i=1}^{m}], \qquad m \ge 1. \end{aligned}$$

(See Theorem 9.1 and the first part of its proof.) Now the pair (X_0, T_0) is a least common extension of $(X_1, T_1), \ldots, (X_r, T_r)$ if and only if (X_0, T_0) and $(\tilde{X}_0, \tilde{T}_0)$ are similar. By (9.3) the last statement is equivalent to the requirement that $\operatorname{Im}[\operatorname{col}(X_0 T_0^{i-1})_{i=0}^m = \operatorname{Im}[\operatorname{col}(\tilde{X}_0 \tilde{T}_0^{i-1})_{i=1}^m]$ for all $m \geq 1$, and hence the proof is complete. □

9.2. Common Restrictions of Admissible Pairs

Let $(X_1, T_1), \ldots, (X_r, T_r)$ be admissible pairs. The admissible pair (X, T) is said to be a *common restriction* of $(X_1, T_1), \ldots, (X_r, T_r)$ if each (X_j, T_j), $j = 1, \ldots, r$, is an extension of (X, T). We call (X_0, T_0) a *greatest common restriction* of $(X_1, T_1), \ldots, (X_r, T_r)$ if (X_0, T_0) is a common restriction of $(X_1, T_1), \ldots, (X_r, T_r)$ and (X_0, T_0) is an extension of any common restriction of $(X_1, T_1), \ldots, (X_r, T_r)$. To construct a greatest common restriction we need the following lemma.

Lemma 9.4. *Let $(X_1, T_1), \ldots, (X_r, T_r)$ be admissible pairs of orders $p_1, \ldots, p_r$, respectively, and suppose that $\operatorname{Ker}(X_j, T_j) = \{0\}$ for $1 \leq j \leq r$. Let $\mathscr{K}$ be the linear space of all $(\varphi_1, \varphi_2, \ldots, \varphi_r) \in \mathbb{C}^p$ $(p = p_1 + p_2 + \cdots + p_r)$ such that*

$$X_1 T_1^\alpha \varphi_1 = X_2 T_2^\alpha \varphi_2 = \cdots = X_r T_r^\alpha \varphi_r, \qquad \alpha \geq 0.$$

Then $\mathscr{K} = \{(\varphi, S_2\varphi, \ldots, S_r\varphi) \mid \varphi \in \mathscr{M}\}$, where $\mathscr{M}$ is the largest T_1-invariant subspace of $\mathbb{C}^{p_1}$ such that for every $j = 2, \ldots, r$, there exists a linear transformation $S_j \colon \mathscr{M} \to \mathbb{C}^{p_j}$ with the property that

$$X_1|_{\mathscr{M}} = X_j S_j, \qquad S_j T_1|_{\mathscr{M}} = T_j S_j \qquad (j = 2, \ldots, r). \tag{9.10}$$

Proof. Note that $\mathscr{K}$ is a linear space invariant under $T = T_1 \oplus \cdots \oplus T_r$. Put

$$\mathscr{M} = \{\varphi \in \mathbb{C}^{p_1} \mid \exists \varphi_j \in \mathbb{C}^{p_j}, \qquad (2 \leq j \leq r), \qquad (\varphi_1, \varphi_2, \ldots, \varphi_r) \in \mathscr{K}\}.$$

Take $(\varphi_1, \varphi_2, \ldots, \varphi_r)$ and $(\varphi_1, \hat{\varphi}_2, \ldots, \hat{\varphi}_r)$ in $\mathscr{K}$. Then $(0, \varphi_2 - \hat{\varphi}_2, \ldots, \varphi_r - \hat{\varphi}_r) \in \mathscr{K}$, and hence for $2 \leq j \leq r$ we have $X_j T_j^\alpha(\varphi_j - \hat{\varphi}_j) = 0$, $\alpha \geq 0$. As $\operatorname{Ker}(X_j, T_j) = \{0\}$ for each j, we have $\varphi_j = \hat{\varphi}_j$. So each $(\varphi_1, \varphi_2, \ldots, \varphi_r) \in \mathscr{K}$ may be written as

$$(\varphi_1, \varphi_2, \ldots, \varphi_r) = (\varphi_1, S_2\varphi_1, \ldots, S_r\varphi_1),$$

where S_j is a map from $\mathscr{M}$ into $\mathbb{C}^{p_j}$ $(2 \leq j \leq r)$. In other words

$$\mathscr{K} = \{(\varphi, S_2\varphi, \ldots, S_r\varphi) \mid \varphi \in \mathscr{M}\}. \tag{9.11}$$

As $\mathscr{K}$ is a linear space, the maps $S_2, \ldots, S_r$ are linear transformations. Since $\mathscr{K}$ is invariant for $T_1 \oplus \cdots \oplus T_r$, the space $\mathscr{K}$ is invariant for T_1 and

$$S_j T_1|_{\mathscr{M}} = T_j S_j, \qquad j = 2, \ldots, r.$$

Further, from the definition of $\mathscr{K}$ and the identity (9.11) it is clear that $X_jS_j = X_1|_{\mathscr{M}}$.

It remains to prove that $\mathscr{M}$ is the largest subspace of $\mathbb{C}^{p_1}$ with the desired properties. Suppose that $\mathscr{M}_0$ is a T_1-invariant subspace of $\mathbb{C}^{p_1}$ and let S_j^0: $\mathscr{M}_0 \to \mathbb{C}^{p_1}$ $(j = 2, \ldots, r)$ be a linear transformation such that

$$X_1|_{\mathscr{M}_0} = X_jS_j^0, \qquad S_j^0T_1|_{\mathscr{M}_0} = T_jS_j^0. \tag{9.12}$$

Then $S_j^0T_1^\alpha\varphi = T_j^\alpha S_j^0\varphi$ for each $\varphi \in \mathscr{M}_0$ and $\alpha \geq 1$. This together with the first identity in (9.12) shows that

$$(\varphi, S_2^0, \ldots, S_r^0\varphi) \in \mathscr{K} \qquad (\varphi \in \mathscr{M}_0).$$

But then $\mathscr{M}_0 \subset \mathscr{M}$, and the proof is complete. □

The linear transformations S_j $(2 \leq j \leq r)$ in the previous theorem are injective. Indeed, suppose that $S_j\varphi = 0$ for some $j \geq r$. Then (9.10) implies that $\varphi \in \operatorname{Ker}(X_1, T_1)$. But $\operatorname{Ker}(X_1, T_1) = \{0\}$, and hence $\varphi = 0$.

We can now prove an existence theorem for a greatest common restriction under the hypotheses used in Theorem 9.1 to obtain the corresponding result for a least common extension.

Theorem 9.5. *Let $(X_1, T_1), \ldots, (X_r, T_r)$ be admissible pairs of orders $p_1, \ldots, p_r$, respectively, and suppose that* $\operatorname{Ker}(X_j, T_j) = \{0\}$ *for $j = 1, \ldots, r$. Then up to similarity there exists a unique greatest common restriction of $(X_1, T_1), \ldots, (X_r, T_r)$.*

If the subspace $\mathscr{M}$ is defined as in Lemma 9.4, *then one such greatest common restriction is given by*

$$(X_1|_{\mathscr{M}}, T_1|_{\mathscr{M}}).$$

Proof. Put $X_0 = X_1|_{\mathscr{M}}$ and $T_0 = T_1|_{\mathscr{M}}$. By choosing some basis in $\mathscr{M}$ we can interpret (X_0, T_0) as an admissible pair of matrices. By definition (X_0, T_0) is a restriction of (X_1, T_1). As the linear transformations $S_2, \ldots, S_r$ in Lemma 9.4 are injective, we see from formula (9.12) that (X_0, T_0) is a restriction of each (X_j, T_j), $2 \leq j \leq r$. Thus (X_0, T_0) is a common restriction of $(X_1, T_1), \ldots, (X_r, T_r)$.

Next, assume that (Y, R) is a common restriction of $(X_1, T_1), \ldots, (X_r, T_r)$. Let q be the order of the pair (Y, R). Then there exist injective linear transformations G_j: $\mathbb{C}^q \to \mathbb{C}^{p_j}$, $j = 1, \ldots, r$, such that

$$Y = X_jG_j, \qquad G_jR = T_jG_j, \qquad j = 1, \ldots, r.$$

It follows that for each $\varphi \in \mathbb{C}^q$ we have

$$YR^\alpha\varphi = X_1T_1^\alpha G_1\varphi = \cdots = X_rT_r^\alpha G_r\varphi, \qquad \alpha \geq 0.$$

Hence $G_1\varphi \in \mathcal{M}$, and thus $Y = X_0 G_1$ and $G_1 R = T_0 G_1$. But this implies that (X_0, T_0) is an extension of (Y, R). So we have proved that (X_0, T_0) is a greatest common restriction of $(X_1, T_1), \ldots, (X_r, T_r)$.

Finally, let $(\tilde{X}_0, \tilde{T}_0)$ be another greatest common restriction of $(X_1, T_1), \ldots, (X_r, T_r)$. Then $(\tilde{X}_0, \tilde{T}_0)$ is an extension of (X_0, T_0) and conversely (X_0, T_0) is an extension of $(\tilde{X}_0, \tilde{T}_0)$. Hence both pairs are similar. □

As in the case of least common extensions, Theorem 9.5 remains true, with appropriate modification, if one drops the condition that $\operatorname{Ker}(X_j, T_j) = \{0\}$ for $j = 1, \ldots, r$. To see this, one first replaces the pair (X_j, T_j) by

$$(X_j|_{\operatorname{Im} P_j}, P_j T_j|_{\operatorname{Im} P_j}), \qquad j = 1, \ldots, r, \tag{9.13}$$

where P_j is a projector of $\mathbb{C}^{p_j}$ along $\operatorname{Ker}(X_j, T_j)$. Let (X_0, T_0) be a greatest common restriction of the pairs (9.13), and put $X = [X_0 \quad 0_1]$ and $T = T_0 \oplus 0_2$, where 0_1 is the $n \times t$ zero-matrix and 0_2 the $t \times t$ zero-matrix, $t = \min_j \dim \operatorname{Ker}(X_j, T_j)$. Then the pair (X, T) is a greatest common restriction of $(X_1, T_1), \ldots, (X_r, T_r)$.

We conclude this section with another characterization of the greatest common restriction. As before, let $(X_1, T_1), \ldots, (X_r, T_r)$ be admissible pairs of orders $p_1, \ldots, p_r$, respectively. For each positive integer s let $\mathcal{K}_s$ be the linear space of all $(\varphi_1, \varphi_2, \ldots, \varphi_r) \in \mathbb{C}^p$, $(p = p_1 + \cdots + p_r)$ such that

$$X_1 T_1^\alpha \varphi_1 = \cdots = X_r T_r^\alpha \varphi_r, \qquad \alpha = 0, \ldots, s-1.$$

Obviously, $\bigcap_{s=1}^{\infty} \mathcal{K}_s = \mathcal{K}$, where $\mathcal{K}$ is the linear space defined in Lemma 9.4. The subspaces $\mathcal{K}_1, \mathcal{K}_2, \ldots$, form a descending sequence in $\mathbb{C}^p$, and hence there exists a positive integer q such that $\mathcal{K}_q = \mathcal{K}_{q+1} = \cdots$. The least least q with this property will be denoted by $q\{(X_j, T_j)_{j=1}^r\}$.

Theorem 9.6. *Let $(X_0, T_0), (X_1, T_1), \ldots, (X_r, T_r)$ be admissible pairs, and suppose that $\operatorname{Ker}(X_j, T_j) = \{0\}$ for $j = 0, 1, \ldots, r$. Then (X_0, T_0) is a greatest common restriction of $(X_1, T_1), \ldots, (X_r, T_r)$ if and only if*

$$\operatorname{Im}[\operatorname{col}(X_0 T_0^{i-1})_{i=1}^m] = \bigcap_{j=1}^{r} \operatorname{Im}[\operatorname{col}(X_j T_j^{i-1})_{i=1}^m] \tag{9.14}$$

for each $m \geq q\{(X_j, T_j)_{j=1}^r\}$.

Proof. Let (Y_0, R_0) be the greatest common restriction of $(X_1, T_1), \ldots, (X_r, T_r)$ as defined in the second part of Theorem 9.5. Take a fixed $m \geq q\{(X_j, T_j)_{j=1}^r\}$, and let

$$\operatorname{col}(\varphi_i)_{i=1}^m \in \bigcap_{j=1}^{r} \operatorname{Im}[\operatorname{col}(X_j T_j^{i-1})_{i=1}^m].$$

Then there exist $\psi_j \in \mathbb{C}^{p_j}$ ($j = 1, \ldots, r$) such that $\varphi_i = X_j T_j^{i-1} \psi_j$ for $1 \leq i \leq m$ and $1 \leq j \leq r$. It follows that $(\psi_1, \ldots, \psi_r) \in \mathscr{K}_m$. But $\mathscr{K}_m = \mathscr{K}$. So $\psi_1 \in \mathscr{M}$ and $\psi_j = S_j \psi_1, j = 2, \ldots, r$ (cf. Lemma 9.4). But then

$$\varphi_i = X_j T_j^{i-1} S_j \psi_1 = X_1 T_1^{i-1} \psi_1 = Y_0 R_0^{i-1} \psi_1.$$

It follows that

$$\bigcap_{j=1}^{r} \operatorname{Im}[\operatorname{col}(X_j T_j^{i-1})_{i=1}^m] \subset \operatorname{Im}[\operatorname{col}(Y_0 R_0^{i-1})_{i=1}^m].$$

As (Y_0, R_0) is a common restriction of $(X_1, T_1), \ldots, (X_r, T_r)$, the reverse inclusion is trivially true. So we have proved (9.14) for $X_0 = Y_0$ and $T_0 = R_0$. Since all greatest common restrictions of $(X_1, T_1), \ldots, (X_r, T_r)$ are similar, we see that (9.14) has been proved in general.

Conversely, assume that (9.14) holds for each

$$m \geq q = q\{(X_j, T_j)_{j=1}^r\}.$$

Let (Y_0, R_0) be as in the first part of the proof. Then for $m \geq q$ and hence for each m, we have

$$\operatorname{Im}[\operatorname{col}(X_0 T_0^{i-1})_{i=1}^m] = \operatorname{Im}[\operatorname{col}(Y_0 R_0^{i-1})_{i=1}^m].$$

By (9.3) this implies that (X_0, T_0) and (Y_0, R_0) are similar, and hence (X_0, T_0) is a greatest common restriction of $(X_1, T_1), \ldots, (X_r, T_r)$. □

The following remark on greatest common restrictions of admissible pairs will be useful.

Remark 9.7. Let $(X_1, T_1), \ldots, (X_r, T_r)$ be admissible pairs with $\operatorname{Ker}(X_j, T_j) = \{0\}$ for $j = 1, \ldots, r$, and let (X_0, T_0) be a greatest common restriction of $(X_1, T_1), \ldots, (X_r, T_r)$. As the proof of Theorem 9.6 shows, $q\{(X_j, T_j)_{j=1}^r\}$ is the smallest integer $m \geq 1$ such that the equality

$$\operatorname{Im}[\operatorname{col}(X_0 T_0^{i-1})_{i=1}^m] = \bigcap_{j=1}^{r} \operatorname{Im}[\operatorname{col}(X_j T_j^{i-1})_{i=1}^m]$$

holds. Further,

$$\dim\left(\bigcap_{j=1}^{r} \operatorname{Im} \operatorname{col}(X_j T_j^{i-1})_{i=1}^m\right) = d_0 \tag{9.15}$$

for every $m \geq q\{(X_j, T_j)_{j=1}^r\}$, where d_0 is the size of T_0. Indeed, in view of formula (9.14), we have to show that the index of stabilization $\operatorname{ind}(X_0, T_0)$ does not exceed $q\{(X_j, T_j)_{j=1}^r\}$. But this is evident, because

$$\operatorname{ind}(X_0, T_0) \leq \operatorname{ind}(X_1, T_1) \leq q\{(X_j, T_j)_{j=1}^r\},$$

as one checks easily using the definitions.

9.3. Construction of l.c.m. and g.c.d. via Spectral Data

Let $L_1(\lambda), \ldots, L_s(\lambda)$ be comonic matrix polynomials with finite Jordan pairs $(X_{1F}, J_{1F}), \ldots, (X_{sF}, J_{sF})$, respectively. To construct an l.c.m. and a g.c.d. of $L_1(\lambda), \ldots, L_s(\lambda)$ via the pairs $(X_{1F}, J_{1F}), \ldots, (X_{sF}, J_{sF})$, we shall use extensively the notions and results presented in Sections 9.1 and 9.2. Observe that (X_{iF}, J_{iF}) is an admissible pair of order p_i = degree(det $L_i(\lambda)$) and kernel zero:

$$\operatorname{Ker}[\operatorname{col}(X_{iF} J_{iF}^k)_{k=0}^{l_i-1}] = \{0\},$$

where l_i is the degree of $L_i(\lambda)$ (this fact follows, for instance, from Theorem 7.15). Let (X_r, T_r) and (X_e, T_e) be a greatest common restriction and a least common extension, respectively, of the admissible pairs $(X_{1F}, J_{1F}), \ldots, (X_{sF}, J_{sF})$. By Corollary 9.2, the admissible pairs (X_r, T_r) and (X_e, T_e) also have kernel zero, and T_r, T_e are nonsingular matrices. Let

$$l_r = \operatorname{ind}(X_r, T_r), \qquad l_e = \operatorname{ind}(X_e, T_e).$$

Theorem 9.8. *The comonic matrix polynomial*

$$M(\lambda) = I - X_e T_e^{-l_e}(V_1\lambda^{l_e} + V_2\lambda^{l_e-1} + \cdots + V_{l_e}\lambda), \tag{9.16}$$

where $[V_1 \quad V_2 \quad \cdots \quad V_{l_e}]$ *is a special left inverse of* $\operatorname{col}(X_e T_e^{-j})_{j=0}^{l_e-1}$, *is an l.c.m. of minimal possible degree of* $L_1(\lambda), \ldots, L_s(\lambda)$. *Any other l.c.m. of* $L_1(\lambda), \ldots, L_s(\lambda)$ *has the form* $U(\lambda)M(\lambda)$, *where* $U(\lambda)$ *is an arbitrary matrix polynomial with* $\det U(\lambda) \equiv \text{const} \neq 0$.

Proof. Let us prove first that a comonic matrix polynomial $N(\lambda)$ is an l.c.m. of $L_1(\lambda), \ldots, L_s(\lambda)$ if and only if the finite Jordan pair (X_{NF}, J_{NF}) of $N(\lambda)$ is a least common extension of $(X_{1F}, J_{1F}), \ldots, (X_{sF}, J_{sF})$. Indeed, assume that $N(\lambda)$ is an l.c.m. of $L_1(\lambda), \ldots, L_s(\lambda)$. By Theorem 7.13, (X_{NF}, J_{NF}) is a common extension of (X_{iF}, J_{iF}), $i = 1, \ldots, s$. Let (X, T) be an extension of (X_{NF}, J_{NF}), and let $\tilde{N}(\lambda)$ be a matrix polynomial (not necessarily comonic) whose finite Jordan pair is similar to (X, T) ($\tilde{N}(\lambda)$ can be constructed using Theorem 7.16; if T is singular, an appropriate shift of the argument λ is also needed). By Theorem 7.13, $\tilde{N}(\lambda)$ is a common multiple of $L_1(\lambda), \ldots, L_s(\lambda)$, and, therefore, $N(\lambda)$ is a right divisor of $\tilde{N}(\lambda)$. Applying Theorem 7.13 once more we see that (X, T) is an extension of (X_{NF}, J_{NF}). Thus, (X_{NF}, J_{NF}) is a least common extension of $(X_{1F}, J_{1F}), \ldots, (X_{sF}, J_{sF})$. These arguments can be reversed to show that if (X_{NF}, J_{NF}) is a least common extension of (X_{iF}, J_{iF}), $i = 1, \ldots, s$, then $N(\lambda)$ is an l.c.m. of $L_1(\lambda), \ldots, L_s(\lambda)$.

After the assertion of the preceding paragraph is verified, Theorem 9.8 (apart from the statement about the uniqueness of an l.c.m.) follows from Theorem 7.16.

We prove the uniqueness assertion. Observe that, from the definition of an l.c.m., if $N_1(\lambda)$ is an l.c.m. of $L_1(\lambda), \ldots, L_s(\lambda)$, then $N_1(\lambda)$ and $M(\lambda)$ (where $M(\lambda)$ is given by (9.16)) are right divisors of each other. So

$$N_1(\lambda) = U_1(\lambda)M(\lambda), \qquad M(\lambda) = U_2(\lambda)N_1(\lambda)$$

for some matrix polynomials $U_1(\lambda)$ and $U_2(\lambda)$. These equalities imply, in particular, that $U_1(\lambda)U_2(\lambda) = I$ for every $\lambda \in \sigma(N_1)$. By continuity, $U_1(\lambda)U_2(\lambda) = I$ for all $\lambda \in \mathbb{C}$, and therefore both matrix polynomials $U_1(\lambda)$ and $U_2(\lambda)$ have constant nonzero determinant.

So we have proved that any l.c.m. of $L_1(\lambda), \ldots, L_s(\lambda)$ has the form

$$U(\lambda)M(\lambda) \tag{9.17}$$

where $U(\lambda)$ is a matrix polynomial with empty spectrum. Conversely, it is easily seen that every matrix polynomial of the form (9.17) is an l.c.m. of $L_1(\lambda), \ldots, L_s(\lambda)$. □

The proof of the following theorem is analogous to the proof of Theorem 9.8 and therefore is omitted.

Theorem 9.9. *The comonic matrix polynomial*

$$D(\lambda) = I - X_r T^{-l_r}(W_1\lambda^{l_r} + W_2\lambda^{l_r-1} + \cdots + W_{l_r}\lambda),$$

where $[W_1 \quad W_2 \quad \cdots \quad W_{l_r}]$ *is a special left inverse of* $\operatorname{col}(X_r T_r^{-j})_{j=0}^{l_r-1}$, *is a g.c.d. of minimal possible degree of* $L_1(\lambda), \ldots, L_s(\lambda)$. *Any other g.c.d. of* $L_1(\lambda), \ldots, L_s(\lambda)$ *is given by the formula* $U(\lambda)D(\lambda)$, *where* $U(\lambda)$ *is an arbitrary matrix polynomial without eigenvalues.*

9.4. Vandermonde Matrix and Least Common Multiples

In the preceding section we constructed an l.c.m. and a g.c.d. of comonic matrix polynomials $L_1(\lambda), \ldots, L_s(\lambda)$ in terms of their finite Jordan pairs. However, the usage of finite Jordan pairs is not always convenient (in particular, note that the Jordan structure of a matrix polynomial is generally unstable under small perturbation). So it is desirable to construct an l.c.m. and a g.c.d. directly in terms of the coefficients of $L_1(\lambda), \ldots, L_s(\lambda)$. For the greatest common divisor this will be done in the next section, and for the least common multiple we shall do this here.

We need the notion of the Vandermonde matrix for the system $L_1(\lambda), \ldots, L_s(\lambda)$ of comonic matrix polynomials, which will be introduced now. Let p_i be the degree of $L_i(\lambda)$, and let $(X_i, T_i) = ([X_{iF}, X_{i\infty}], J_{iF}^{-1} \oplus J_\infty)$ be a comonic Jordan pair of $L_i(\lambda)$, $i = 1, \ldots, s$. According to Theorem 7.15, (X_i, T_i) is a standard pair for the monic matrix polynomial $\tilde{L}_i(\lambda) = \lambda^{p_i} L_i(\lambda^{-1})$. In particular, $\operatorname{col}(X_i T_i^k)_{k=0}^{p_i-1}$ is nonsingular; denote

$$U_j = [\operatorname{col}(X_j T_j^{i-1})_{i=1}^{p_j}]^{-1}, \qquad j = 1, \ldots, s,$$

and define the following $mn \times pn$ matrix, where $p = \sum_{j=1}^{s} p_j$ and m is an arbitrary positive integer:

$$V_m(L_1, \ldots, L_s) = \begin{bmatrix} X_1 U_1 & X_2 U_2 & \cdots & X_s U_s \\ X_1 T_1 U_1 & X_2 T_2 U_2 & \cdots & X_s T_s U_s \\ \vdots & \vdots & & \vdots \\ X_1 T_1^{m-1} U_1 & X_2 T_2^{m-1} U_2 & \cdots & X_s T_s^{m-1} U_s \end{bmatrix}.$$

The matrix $V_m(L_1, \ldots, L_s)$ will be called the *comonic Vandermonde* matrix of $L_1(\lambda), \ldots, L_s(\lambda)$.

We point out that $V_m(L_1, \ldots, L_s)$ does not depend on the choice of the comonic Jordan pairs (X_i, T_i) and can be expressed directly in terms of the coefficients of $L_1(\lambda), \ldots, L_s(\lambda)$. In fact, let $L_1(\lambda) = I + A_{p_1-1}\lambda + A_{p_1-2}\lambda^2 + \cdots + A_0 \lambda^{p_1}$, and let $U_1 = [U_{11} \quad U_{12} \quad \cdots \quad U_{1p_1}]$, where U_{1j} is an $n \times n$ matrix. Then the expressions $X_1 T_1^m U_{1\beta}$ $(1 \le \beta \le p_1)$ are given by the following formulas (see Proposition 2.3):

$$X_1 T_1^m U_{1\beta} = \begin{cases} 0 & \text{if} \quad m = 0, \ldots, p_1 - 1 \quad \text{and} \quad \beta \ne m + 1 \\ I & \text{if} \quad m = 0, \ldots, p_1 - 1 \quad \text{and} \quad \beta = m + 1 \\ -A_{\beta-1} & \text{if} \quad m = p_1, \end{cases} \tag{9.18}$$

and in general for $m > p_1$,

$$X_1 T_1^m U_{1\beta} = \sum_{k=1}^{m-p_1} \left[\sum_{q=1}^{k} \sum_{i_1 + \cdots + i_q = k} \prod_{j=1}^{q} (-A_{p_1 - i_j}) \right] \cdot (-A_{\beta+k-m+p_1-1}) + (-A_{\beta-m+p_1-1}), \tag{9.19}$$

where, by definition, $A_i = 0$ for $i \le 0$ and

$$\prod_{j=1}^{q} (-A_{p_1-i_j}) = (-A_{p_1-i_1})(-A_{p_1-i_2}) \cdots (-A_{p_1-i_q}).$$

To formulate the next theorem we need some further concepts.

For a given family $L_1, \ldots, L_s$ of comonic matrix polynomials with infinite Jordan pairs $(X_{i\infty}, J_{i\infty})$, $i = 1, \ldots, s$, denote by $\nu_\infty(L_1, \ldots, L_s)$ the minimal integer j such that $J_{i\infty}^j = 0$ for $i = 1, \ldots, s$. For integer $i \le j$, denote by $S_{ij}(L_1, \ldots, L_s)$ the lower $(j - i + 1)$ matrix blocks of the comonic Vandermonde matrix $V_j(L_1, \ldots, L_s)$:

$$S_{ij}(L_1, \ldots, L_s) = \begin{bmatrix} X_1 J_1^{i-1} U_1 & \cdots & X_s J_s^{i-1} U_s \\ \vdots & & \vdots \\ X_1 J_1^{j-1} U_1 & \cdots & X_s J_s^{j-1} U_s \end{bmatrix}.$$

If $i = j$, we shall write $S_i(L_1, \ldots, L_s)$ in place of $S_{ii}(L_1, \ldots, L_s)$.

In Theorem 9.10 below we shall use also the notion of a special generalized inverse (see Section 7.11 for its definition).

Theorem 9.10. *Let $L_1(\lambda), \ldots, L_s(\lambda)$ be a family of comonic matrix polynomials, and let $v \geq v_\infty(L_1, \ldots, L_s)$ be an arbitrary integer. Then the minimal possible degree m of a least common multiple of $L_1(\lambda), \ldots, L_s(\lambda)$ is*

$$m = \min\{j \geq 1 \mid \operatorname{Ker} S_{v+1,v+j}(L_1, \ldots, L_s) = \operatorname{Ker} S_{v+1,v+j+1}(L_1, \ldots, L_s)\} \tag{9.20}$$

and does not depend on the choice of v. One of the least common multiples of $L_1(\lambda), \ldots, L_s(\lambda)$ of the minimal degree m is given by the formula

$$M(\lambda) = I - S_{v+m+1,v+m+1}(L_1, \ldots, L_r) \cdot [W_1\lambda^m + W_2\lambda^{m-1} + \cdots + W_m\lambda], \tag{9.21}$$

where $[W_1 \quad W_2 \quad \cdots \quad W_m]$ is the special generalized inverse of

$$S_{v+1,v+m}(L_1, \ldots, L_s).$$

Proof. Let $(X_i, T_i) = ([X_{iF} \quad X_{i\infty}], J_{iF}^{-1} \oplus J_\infty)$ be a comonic Jordan pair of $L_i(\lambda)$, $i = 1, \ldots, s$; let $U_i = [\operatorname{col}(X_i J_i^j)_{j=0}^{p_i-1}]^{-1}$, where p_i is the degree of $L_i(\lambda)$. Then for $v \geq v_\infty(L_1, \ldots, L_s)$ we have

$$\begin{aligned} S_{v+1,v+j}&(L_1, \ldots, L_s) \\ &= \begin{bmatrix} X_1T_1^v & \cdots & X_sT_s^v \\ \vdots & & \vdots \\ X_1T_1^{v+j-1} & \cdots & X_sT_s^{v+j-1} \end{bmatrix} \operatorname{diag}[U_1, \ldots, U_s] \\ &= \begin{bmatrix} X_{1F} & 0 & X_2 & 0 & \cdots & X_{sF} & 0 \\ \vdots & \vdots & \vdots & \vdots & & \vdots & \vdots \\ X_{1F}J_{1F}^{-(j-1)} & 0 & X_{2F}J_{2F}^{-(j-1)} & 0 & \cdots & X_{sF}J_{sF}^{-(j-1)} & 0 \end{bmatrix} \\ &\quad \cdot \operatorname{diag}[(J_{iF}^{-v} \oplus I)U_i]_{i=1}^s. \end{aligned} \tag{9.22}$$

Since the matrix $\operatorname{diag}[(J_{iF}^{-v} \oplus I)U_i]_{i=1}^s$ is nonsingular, the integer m (defined by (9.20)) indeed does not depend on $v \geq v_\infty(L_1, \ldots, L_s)$.

Let (X_e, T_e) be the least common extension of the pairs $(X_{1F}, J_{1F}), \ldots, (X_{sF}, J_{sF})$. In view of Theorem 9.8 it is sufficient to show that $m = \operatorname{ind}(X_e, T_e)$, and

$$X_eT_e^{-m}[V_1 \quad V_2 \quad \cdots \quad V_m] = S_{v+m+1}(L_1, \ldots, L_r)[W_1 \quad W_2 \quad \cdots \quad W_m], \tag{9.23}$$

where $[V_1 \quad V_2 \quad \cdots \quad V_m]$ is a special left inverse of $\operatorname{col}(X_eT_e^{-j})_{j=0}^{m-1}$.

From the equality (9.22) it follows that m coincides with the least integer j such that

$$\begin{aligned}&\operatorname{Ker}\operatorname{col}[X_{1F}J_{1F}^{-k}, X_{2F}J_{2F}^{-k}, \ldots, X_{sF}J_{sF}^{-k}]_{k=0}^{j-1}\\ &\quad= \operatorname{Ker}\operatorname{col}[X_{1F}J_{1F}^{-k}, X_{2F}J_{2F}^{-k}, \ldots, X_{sF}J_{sF}^{-k}]_{k=0}^{j}.\end{aligned}$$

Using the construction of a least common extension given in Theorem 9.1, we find that indeed $m = \operatorname{ind}(X_e, T_e)$. Further, we have (by analogy with (9.22))

$$\begin{aligned}S_{v+m+1} &= [X_{1F}J_{1F}^{-m} \quad 0 \quad X_{2F}J_{2F}^{-m} \quad 0 \quad \cdots \quad X_{sF}J_{sF}^{-m} \quad 0]\\ &\quad\cdot \operatorname{diag}[(J_{iF}^{-v} \oplus I)U_i]_{i=1}^{s}\end{aligned}$$

and taking into account (9.22) it is easy to see that the right-hand side of (9.23) is equal to

$$\operatorname{row}(X_{jF}J_{jF}^{-m})_{j=1}^{s} \cdot [W_1' \quad W_2' \quad \cdots \quad W_m'], \tag{9.24}$$

where $[W_1' \quad W_2' \quad \cdots \quad W_m']$ is a special generalized inverse of

$$\operatorname{col}[X_{1F}J_{1F}^{-j}, X_{2F}J_{2F}^{-j}, \ldots, X_{sF}J_{sF}^{-j}]_{j=0}^{m-1}.$$

We can assume (according to Theorem 9.1) that $(X_e, T_e) = (X|_{\operatorname{Im}P}, PT|_{\operatorname{Im}P})$, where

$$X = [X_{1F} \quad \cdots \quad X_{sF}] \qquad T = \operatorname{diag}[J_{1F}^{-1}, \ldots, J_{sF}^{-1}],$$

and P is a projector such that

$$\operatorname{Ker} P = \operatorname{Ker}(X, T) \qquad \left(= \bigcap_{i=0}^{\infty} \operatorname{Ker} XT^i\right).$$

The matrices X and T have the following form with respect to the decomposition $\mathbb{C}^p = \operatorname{Ker} P \dotplus \operatorname{Im} P$, where $p = p_1 + p_2 + \cdots + p_s$:

$$X = [0 \quad X_e], \qquad T = \begin{bmatrix} * & * \\ 0 & T_e \end{bmatrix}.$$

It follows that $XT^i = [0 \quad X_e T_e^i]$, $i = 0, 1, \ldots$. Now from the definition of a special generalized inverse we obtain that the left-hand side of (9.23) is equal to (9.24). So the equality (9.23) follows. □

The integer $v_\infty(L_1, \ldots, L_s)$ which appears in Theorem 9.10 can be easily estimated:

$$v_\infty(L_1, \ldots, L_s) \le \min\{j \ge 1 \,|\, \operatorname{Ker} V_j(L_1, \ldots, L_m) = \operatorname{Ker} V_{j+1}(L_1, \ldots, L_m)\}. \tag{9.25}$$

Indeed, for the integer $v = v_\infty(L_1, \ldots, L_r)$ we have $J_{i\infty}^v = 0$ and $J_{i\infty}^{v-1} \neq 0$ for some i. Then also $X_{i\infty} J_{i\infty}^{v-1} \neq 0$, and there exists a vector x in the domain of definition of $J_{i\infty}$ such that $X_{i\infty} J_{i\infty}^{v-1} x \neq 0$, $X_{i\infty} J_{i\infty}^v x = 0$. Therefore,

$$\operatorname{Ker} V_v(L_1, \ldots, L_m) \not\supset \operatorname{Ker} S_{v+1}(L_1, \ldots, L_m),$$

and (9.25) follows.

9.5. Common Multiples for Monic Polynomials

Consider the following problem: given monic matrix polynomials $L_1(\lambda), \ldots, L_s(\lambda)$, construct a monic matrix polynomial $L(\lambda)$ which is a common multiple of $L_1(\lambda), \ldots, L_s(\lambda)$ and is, in a certain sense, the smallest possible. Because of the monicity requirement for $L(\lambda)$, in general one cannot demand that $L(\lambda)$ be an l.c.m. of $L_1(\lambda), \ldots, L_s(\lambda)$ (an l.c.m. of $L_1(\lambda), \ldots, L_s(\lambda)$ may never be monic). Instead we shall require that $L(\lambda)$ has the smallest possible degree.

To solve this problem, we shall use the notion of the Vandermonde matrix. However, in the case of monic polynomials it is more convenient to use standard pairs (instead of comonic Jordan pairs in the case of comonic polynomials). We arrive at the following definition: let $L_1(\lambda), \ldots, L_s(\lambda)$ be monic matrix polynomials with degrees $p_1, \ldots, p_s$ and Jordan pairs $(X_1, T_1), \ldots, (X_s, T_s)$, respectively. For an arbitrary integer $m \geq 1$ put

$$W_m(L_1, \ldots, L_s) = \begin{bmatrix} X_1 U_1 & X_2 U_2 & \cdots & X_s U_s \\ X_1 T_1 U_1 & X_2 T_2 U_2 & \cdots & X_s T_s U_s \\ \vdots & \vdots & & \vdots \\ X_1 T_1^{m-1} U_1 & X_2 T_2^{m-1} U_2 & \cdots & X_s T_s^{m-1} U_s \end{bmatrix},$$

where $U_j = [\operatorname{col}(X_j T_j^{i-1})_{i=1}^{p_j}]^{-1}$. Call $W_m(L_1, \ldots, L_s)$ the *monic Vandermonde matrix* of $L_1(\lambda), \ldots, L_s(\lambda)$. From formulas (9.18) and (9.19) (where $L_1(\lambda) = \sum_{j=0}^{p_1} A_j \lambda^j$) it is clear that $W_m(L_1, \ldots, L_s)$ can be expressed in terms of the coefficients of $L_1(\lambda), \ldots, L_s(\lambda)$ and, in particular, does not depend on the choice of the standard pairs.

For simplicity of notation we write W_m for $W_m(L_1, \ldots, L_s)$ in the next theorem.

Theorem 9.11. *Let $L_1(\lambda), \ldots, L_s(\lambda)$ be monic matrix polynomials, and let*

$$r = \min\{m \geq 1 \mid \operatorname{Ker} W_m = \operatorname{Ker} W_{m+1}\}.$$

Then there exists a monic common multiple $L(\lambda)$ of $L_1(\lambda), \ldots, L_s(\lambda)$ of degree r. One such monic common multiple is given by the formula

$$L_1(\lambda) = I\lambda^r - S_{r+1}|_{\operatorname{Im} P} \cdot (V_1 + V_2 \lambda + \cdots + V_r \lambda^{r-1}) \qquad (9.26)$$

where P is a projector along Ker W_r, $[V_1, V_2, \ldots, V_r]$ *is some left inverse for* $W_r|_{\operatorname{Im} P}$, *and* S_{r+1} *is formed by the lower n rows of* W_{r+1}.

Conversely, if $L(\lambda)$ *is a monic common multiple of* $L_1(\lambda), \ldots, L_s(\lambda)$, *then* $\deg L \geq r$.

Observe that a monic common multiple of minimal degree is not unique in general.

Proof. Since $U_1, \ldots, U_s$ are nonsingular,

$$r = \operatorname{ind}([X_1 \quad \cdots \quad X_s], \operatorname{diag}[T_1, \ldots, T_s]).$$

Let (X_0, T_0) be the least common extension of $(X_1, T_1), \ldots, (X_s, T_s)$ (Section 9.1). Then $\operatorname{Ker}(X_0, T_0) = \{0\}$ and $\operatorname{ind}(X_0, T_0) = r$ (see Theorem 9.1). Let $L(\lambda)$ be a monic matrix polynomial associated with the r-independent admissible pair (X_0, T_0), i.e.,

$$L(\lambda) = I\lambda^r - X_0 T_0^r(\tilde{V}_1 + \tilde{V}_2\lambda + \cdots + \tilde{V}_r\lambda^{r-1}), \tag{9.27}$$

where $[V_1, V_2, \ldots, V_r]$ is a left inverse of $\operatorname{col}(X_0 T_0^i)_{i=0}^{r-1}$. By Theorem 6.2, a standard pair of $L(\lambda)$ is an extension of (X_0, T_0) and, consequently, it is also an extension of each (X_i, T_i), $i = 1, \ldots, s$. By Theorem 7.13, $L(\lambda)$ is a common multiple of $L_1(\lambda), \ldots, L_s(\lambda)$. In fact, formula (9.27) coincides with (9.26). This can be checked easily using the description of a least common extension given in Theorem 9.1.

Conversely, let $L(\lambda)$ be a monic common multiple of $L_1(\lambda), \ldots, L_s(\lambda)$ of degree l. Then (Theorem 7.13) a standard pair (X, T) of $L(\lambda)$ is a common extension of $(X_1, T_1), \ldots, (X_s, T_s)$. In particular,

$$\operatorname{ind}(X, T) \geq \operatorname{ind}(X_0, T_0) = r,$$

(As before, (X_0, T_0) stands for a least common extension of $(X_1, T_1), \ldots, (X_s, T_s)$.) On the other hand, nonsingularity of $\operatorname{col}(XT^i)_{i=0}^{l-1}$ implies $\operatorname{ind}(X, T) = l$; so $l \geq r$ as claimed. $\square$

Theorem 9.11 becomes especially simple for the case of linear polynomials $L_i(\lambda) = I\lambda - X_i$, $i = 1, \ldots, s$. In this case

$$W_m(L_1, \ldots, L_s) = \operatorname{col}([X_1^i, X_2^i, \ldots, X_s^i])_{i=0}^{m-1}.$$

Thus:

Corollary 9.12. *Let* $X_1, \ldots, X_s$ *be* $n \times n$ *matrices. Then there exist* $n \times n$ *matrices* $A_0, \ldots, A_{l-1}$ *with the property*

$$X_i^l + \sum_{j=0}^{l-1} A_j X_i^j = 0, \qquad i = 1, \ldots, s, \tag{9.28}$$

if and only if the integer l satisfies the inequality

$$\begin{aligned} l \geq \min\{m \geq 1 \mid &\operatorname{Ker} \operatorname{col}([X_1^k, X_2^k, \ldots, X_s^k])_{k=0}^{m-1} \\ &= \operatorname{Ker} \operatorname{col}([X_1^k, X_2^k, \ldots, X_s^k])_{k=0}^{m}\}. \end{aligned} \tag{9.29}$$

If (9.29) *is satisfied, one of the possible choices for* $A_0, \ldots, A_{l-1}$ *such that* (9.28) *holds is given by the formula*

$$[A_0 \quad A_1 \quad \cdots \quad A_{l-1}] = -[X_1^l, X_2^l, \ldots, X_s^l]|_{\operatorname{Im} P}[V_1 \quad V_2 \quad \cdots \quad V_l],$$

where P is a projector along $\operatorname{Ker} \operatorname{col}([X_1^k, X_2^k, \ldots, X_s^k])_{k=0}^{l-1}$, *and*

$$[V_1 \quad V_2 \quad \cdots \quad V_l]$$

is some left inverse for $\operatorname{col}([X_1^k, X_2^k, \ldots, X_s^k])_{k=0}^{l-1}|_{\operatorname{Im} P}$.

9.6. Resultant Matrices and Greatest Common Divisors

In order to describe the construction of a g.c.d. for a finite family of matrix polynomials, we shall use resultant matrices, which are described below.

The notion of a *resultant matrix* $R(a, b)$ for a pair of scalar polynomials $a(\lambda) = \sum_{i=0}^{r} a_i \lambda^i$ and $b(\lambda) = \sum_{i=0}^{s} b_i \lambda^i$ $(a_i, b_j \in \mathbb{C})$ is well known:

$$R(a,b) = \begin{bmatrix} a_0 & a_1 & & \cdots & & a_r & 0 & & \cdots & 0 \\ 0 & a_0 & a_1 & \cdots & & & a_r & 0 & \cdots & 0 \\ \vdots & \vdots & \vdots & & \vdots & \vdots & \vdots & \vdots & & \vdots \\ 0 & 0 & & \cdots & 0 & a_0 & a_1 & & \cdots & a_r \\ b_0 & b_1 & & \cdots & b_s & 0 & 0 & & \cdots & 0 \\ 0 & b_0 & b_1 & \cdots & & b_s & 0 & & \cdots & 0 \\ \vdots & \vdots & \vdots & & \vdots & \vdots & \vdots & \vdots & & \vdots \\ 0 & 0 & & \cdots & & 0 & b_0 & b_1 & \cdots & b_s \end{bmatrix} \begin{matrix} \left.\vphantom{\begin{matrix}a\\a\\a\\a\end{matrix}}\right\} s \text{ rows} \\ \left.\vphantom{\begin{matrix}a\\a\\a\\a\end{matrix}}\right\} r \text{ rows} \end{matrix}$$

Its determinant $\det R(a, b)$ is called the *resultant* of $a(\lambda)$ and $b(\lambda)$ and has the property that $\det R(a, b) \neq 0$ if and only if $a(\lambda)$ and $b(\lambda)$ have no common zeros (it is assumed here that $a_r b_s \neq 0$). A more general result is known (see, for instance, [26]): the number of common zeros (counting multiplicities) of $a(\lambda)$ and $b(\lambda)$ is equal to $r + s - \operatorname{rank} R(a, b)$. Thus, the resultant matrix is closely linked to the greatest common divisor of $a(\lambda)$ and $b(\lambda)$. We shall establish analogous connections between the resultant matrix (properly generalized to the matrix case) of several matrix polynomials and their greatest common divisor.

Let $L(\lambda) = \sum_{j=0}^{l} A_j \lambda^j$ be a regular $n \times n$ matrix polynomial. The following $n(q-l) \times nq$ block matrices $(q > l)$

$$R_q(L) = \begin{bmatrix} A_0 & A_1 & \cdots & & A_l & & 0 \\ & A_0 & A_1 & & \cdots & A_l & \\ & & \ddots & \ddots & & & \ddots \\ 0 & & & A_0 & A_1 & \cdots & A_l \end{bmatrix} \tag{9.30}$$

are called *resultant matrices* of the polynomial $L(\lambda)$. For a family of matrix polynomials $L_j(\lambda) = \sum_{k=0}^{p_j} A_{kj}\lambda^k$ (A_{kj} is an $n \times n$ matrix, $k = 0, 1, \ldots, p_j$, $j = 1, 2, \ldots, s$) we define the resultant matrices as

$$R_q(L_1, L_2, \ldots, L_r) = \begin{bmatrix} R_q(L_1) \\ R_q(L_2) \\ \vdots \\ R_q(L_r) \end{bmatrix} \qquad \left(q > \max_{1 \le j \le r} p_j\right). \tag{9.31}$$

This definition is justified by the fact that the matrices R_q play a role which is analogous to that of the resultant matrix for two scalar polynomials, as we shall see in the next theorem.

Let $L_1, \ldots, L_s$ be comonic matrix polynomials with comonic Jordan pairs $(X_1, T_1), \ldots, (X_s, T_s)$, respectively. Let m_j be the degree of $\det L_j$, $j = 1, \ldots, s$, and let $m = m_1 + \cdots + m_s$. For every positive integer q define

$$\mathscr{K}_q = \{(\varphi_1, \ldots, \varphi_s) \in \mathbb{C}^m \mid \varphi_j \in \mathbb{C}^{m_j} \quad \text{and}$$

$$X_1 T_1^\alpha \varphi_1 = X_2 T_2^\alpha \varphi_2 = \cdots = X_s T_s^\alpha \varphi_s, \qquad \alpha = 0, \ldots, q-1\}.$$

As the subspaces $\mathscr{K}_1, \mathscr{K}_2, \ldots$ form a descending sequence in $\mathbb{C}^m$, there exists a positive integer q_0 such that $\mathscr{K}_{q_0} = \mathscr{K}_{q_0+1} = \cdots$. The least integer q_0 with this property will be denoted by $q(L_1, \ldots, L_s)$. It is easily seen that this definition of $q(L_1, \ldots, L_s)$ does not depend on the choice of $(X_1, T_1), \ldots, (X_s, T_s)$.

The following result provides a description for the kernels of resultant matrices. This description will serve as a basis for construction of a g.c.d. for a family of matrix polynomials (see Theorem 9.15).

Theorem 9.13. *Let $L_1, \ldots, L_s$ be comonic matrix polynomials with comonic Jordan pairs $[X_i, T_i] = ([X_{iF} \; X_{i\infty}], J_{iF}^{-1} \oplus J_{i\infty})$, $i = 1, \ldots, s$, respectively. Let d_0 be the maximal degree of the polynomials $L_1, \ldots, L_s$ (so $R_q(L_1, \ldots, L_s)$ is defined for $q > d_0$). Then $q_0 \overset{\text{def}}{=} \max\{q(L_1, \ldots, L_s), d_0 + 1\}$ is the minimal integer $q > d_0$ such that*

$$\dim \operatorname{Ker} R_q(L_1, \ldots, L_s) = \dim \operatorname{Ker} R_{q+1}(L_1, \ldots, L_s), \tag{9.32}$$

and for every integer $q \geq q_0$ *the following formula holds:*

$$\operatorname{Ker} R_q(L_1, \ldots, L_s) = \operatorname{Im} \operatorname{col}(X_{\mathrm{F}} J_{\mathrm{F}}^{i-1})_{i=1}^q \dotplus \operatorname{Im} \operatorname{col}(X_\infty J_\infty^{q-i})_{i=1}^q, \quad (9.33)$$

where $(X_{\mathrm{F}}, J_{\mathrm{F}})$ *(resp.* (X_∞, J_∞)*) is the greatest common restriction of the pairs* $(X_{1\mathrm{F}}, J_{1\mathrm{F}}), \ldots, (X_{s\mathrm{F}}, J_{s\mathrm{F}})$ *(resp.* $(X_{1\infty}, J_{1\infty}), \ldots, (X_{s\infty}, J_{s\infty})$*).*

The number $q(L_1, \ldots, L_s)$ can be estimated in terms of the degrees $p_1, \ldots, p_s$ of $L_1(\lambda), \ldots, L_s(\lambda)$, respectively:

$$q(L_1, \ldots, L_s) \leq n \cdot \min_{1 \leq j \leq s} p_j + \max_{1 \leq j \leq s} p_j. \quad (9.34)$$

For the proof of this estimation see [29b]. So (9.33) holds, in particular, for every $q \geq n \cdot \min_{1 \leq j \leq s} p_j + \max_{1 \leq j \leq s} p_j$.

Theorem 9.13 holds also in the case that $L_1(\lambda), \ldots, L_s(\lambda)$ are merely regular matrix polynomials. For a regular matrix polynomial $L_i(\lambda)$, let $(X_{i\infty}, J_{i\infty})$ be an infinite Jordan pair for the comonic matrix polynomial $L_i^{-1}(a)L_i(\lambda + a)$ $(a \notin \sigma(L_i))$. The pair $(X_{i\infty}, J_{i\infty}) = (X_{i\infty}(a), X_{i\infty}(a))$ may depend on the choice of a, i.e., it can happen that the pairs $(X_{i\infty}(a), J_{i\infty}(a))$ and $(X_{i\infty}(b), J_{i\infty}(b))$ are not similar for $a \neq b$, $a, b \notin \sigma(L_i)$. However, the subspace

$$\operatorname{Im} \operatorname{col}(X_{i\infty}(a)(J_{i\infty}(a))^{q-j})_{j=1}^q$$

does not depend on the choice of a. In view of Theorem 9.6, the subspace $\operatorname{Im} \operatorname{col}(X_\infty J_\infty^{q-j})_{j=1}^q$ also does not depend on the choices of $a_i \notin \sigma(L_i)$, $i = 1, \ldots, s$, where (X_∞, J_∞) is the greatest common restriction of $(X_{i\infty}(a_i), J_{i\infty}(a_i))$, $i = 1, \ldots, s$. So formula (9.33) makes sense also in the case when $L_1(\lambda), \ldots, L_s(\lambda)$ are regular matrix polynomials. Moreover, the formula (9.33) is true in this case also. For additional information and proofs of the facts presented in this paragraph, see [29b].

For the proof of Theorem 9.13 we need the following lemma.

Lemma 9.14. *Let* $L(\lambda) = I + \sum_{j=0}^{l} A_j \lambda^j$ *be a comonic* $n \times n$ *matrix polynomial, and let* (X, T) *be a comonic Jordan pair for L. Then for* $q > l$ *we have*

$$\operatorname{Ker} R_q(L) = \operatorname{Im} \operatorname{col}(XT^{q-\alpha})_{\alpha=1}^q.$$

Proof. Put $F_{\alpha\beta} = XT^\alpha Z_\beta$ $(\alpha \geq 0, 1 \leq \beta \leq l)$, where

$$[Z_1 \quad Z_2 \quad \cdots \quad Z_l] = [\operatorname{col}(XT^{i-1})_{i=1}^l]^{-1}.$$

Then (see Theorem 7.15 and (2.14))

$$R_q(L) = \begin{bmatrix} I & -F_{ll} & \cdots & & -F_{l1} & 0 \\ & \ddots & \ddots & & & \ddots \\ 0 & & I & -F_{ll} & \cdots & & -F_{l1} \end{bmatrix}.$$

Introduce

$$S = \begin{bmatrix} 0 & 0 & \cdots & -I \\ 0 & & \iddots & -F_{ll} \\ \vdots & \iddots & & \vdots \\ -I & -F_{ll} & \cdots & -F_{q-2l} \end{bmatrix}.$$

Using Theorem 7.15 and equalities (3.15) we deduce that

$$SR_q(L) = [U \quad V],$$

where

$$U = \begin{bmatrix} 0 & & -I \\ & \iddots & \\ -I & & 0 \end{bmatrix}, \qquad V = \begin{bmatrix} F_{ll} & \cdots & F_{l1} \\ \vdots & & \vdots \\ F_{q-1,l} & \cdots & F_{q-1,1} \end{bmatrix}.$$

As S is nonsingular, we see that

$$\begin{aligned} \operatorname{Ker} R_q(L) &= \left\{ \begin{bmatrix} \varphi_1 \\ \varphi_2 \end{bmatrix} \middle| U\varphi_1 + V\varphi_2 = 0 \right\} \\ &= \left\{ \begin{bmatrix} \varphi_1 \\ \varphi_2 \end{bmatrix} \middle| \varphi_1 = -U^{-1}V\varphi_2 \right\} = \operatorname{Im} \begin{bmatrix} -U^{-1}V \\ I \end{bmatrix}. \end{aligned}$$

Now

$$-U^{-1}V = \begin{bmatrix} F_{q-1,l} & \cdots & F_{q-1,1} \\ \vdots & & \vdots \\ F_{ll} & \cdots & F_{l1} \end{bmatrix}, \qquad I = \begin{bmatrix} F_{l-1,l} & \cdots & F_{l-1,1} \\ \vdots & & \vdots \\ F_{0l} & \cdots & F_{0,1} \end{bmatrix}.$$

Since $F_{\alpha\beta} = XT^{\alpha}Z_{\beta}$, it follows that

$$\operatorname{Ker} R_q(L) = \operatorname{Im} \left\{ \begin{bmatrix} XT^{q-1} \\ \vdots \\ XT \\ X \end{bmatrix} \cdot [Z_l \quad Z_{l-1} \quad \cdots \quad Z_1] \right\}.$$

But $\operatorname{row}(Z_{l-j})_{j=0}^{l-1}$ is nonsingular, and hence the lemma is proved. □

Proof of Theorem 9.13. Let $(X_1, T_1), \ldots, (X_s, T_s)$ be comonic Jordan pairs for $L_1, \ldots, L_s$, respectively. From the previous lemma we know that for $q > d_0 = \max_{1 \le j \le s} p_j$, where p_j is the degree of $L_j(\lambda)$, we have

$$\operatorname{Ker} R_q(L_1, \ldots, L_s) = \bigcap_{j=1}^{s} \operatorname{Im} \operatorname{col}(X_j T_j^{q-\alpha})_{\alpha=1}^{q}.$$

The assertion about minimality of q_0 follows now from this formula and Remark 9.7.

Assume now $q \geq q_0$. From Theorem 9.6 we infer that

$$\operatorname{Ker} R_q(L_1, \ldots, L_s) = \operatorname{Im} \operatorname{col}(X_0 T_0^{q-\alpha})_{\alpha=1}^q,$$

where (X_0, T_0) is any greatest common restriction of the pairs $(X_1, T_1), \ldots, (X_s, T_s)$.

Let (X_{iF}, J_{iF}) be a finite Jordan pair for $L_i(\lambda)$, and let $(X_{i\infty}, J_{i\infty})$ be the infinite Jordan pair of $L_i(\lambda)$. So $X_i = [X_{iF} \quad X_{i\infty}]$, $T_i = J_{iF}^{-1} \oplus J_{i\infty}$, $1 \leq i \leq s$. Let (X_F, J_F) be a greatest common restriction of the pairs $(X_{1F}, J_{1F}), \ldots, (X_{sF}, J_{sF})$, and similarly let (X_∞, J_∞) be a greatest common restriction of the pairs $(X_{1\infty}, J_{1\infty}), \ldots, (X_{s\infty}, J_{s\infty})$. Then it is not difficult to see that the pair

$$([X_F \quad X_\infty], J_F^{-1} \oplus J_\infty)$$

is a greatest common restriction of the pairs $(X_1, T_1), \ldots, (X_s, T_s)$. It follows that for $q \geq q(L_1, L_2, \ldots, L_s)$,

$$\operatorname{Ker} R_q(L_1, \ldots, L_s) = \operatorname{Im} \operatorname{col}(X_F J_F^{\alpha-q}, X_\infty J_\infty^{q-\alpha})_{\alpha=1}^q.$$

Multiplying on the right by $J_F^{q-1} \oplus I$ one sees that

$$\operatorname{Ker} R_q(L_1, \ldots, L_r) = \operatorname{Im} \operatorname{col}(X_F J_F^{\alpha-1})_{\alpha=1}^q + \operatorname{Im}(X_\infty J_\infty^{q-\alpha})_{\alpha=1}^q$$

for $q \geq q_0$. As $q > \max_{1 \leq j \leq s} p_j$, the sum on the right-hand side is a direct sum, and formula (9.33) follows. □

We give now a rule for construction of a greatest common divisor which is based on (9.33). This is the main result of this section.

Theorem 9.15. *Let $L_1, \ldots, L_s$ be comonic matrix polynomials. Assume (X_0, T_0) is an admissible pair with* $\operatorname{Ker}(X_0, T_0) = \{0\}$ *such that*

$$Q(\operatorname{Ker} R_q(L_1, \ldots, L_s)) = \operatorname{Im} \operatorname{col}(X_0 T_0^i)_{i=0}^{l-1},$$

where $Q = [I \quad 0]$ is a matrix of size $nl \times n(f + l)$ with integers f and l large enough. Then the matrix T_0 is nonsingular, the matrix $\operatorname{col}(X_0 T_0^{-i})_{i=0}^{l-1}$ *is left invertible, and the comonic matrix polynomial*

$$D(\lambda) = I - X_0 T_0^{-l}(W_1 \lambda^l + \cdots + W_l \lambda),$$

where $[W_1, \ldots, W_l]$ is a special left inverse of $\operatorname{col}(X_0 T_0^{-i})_{i=0}^{l-1}$, *is a greatest common divisor of $L_1(\lambda), \ldots, L_s(\lambda)$.*

The conclusions of Theorem 9.15 hold for any pair of integers (l, f) which satisfy the inequalities:

$$f \geq \min_{1 \leq i \leq s} (np_i - \operatorname{degree}(\det(L_i(\lambda)))),$$

$$l \geq n\left(\min_{1 \leq j \leq s} p_j\right) + \max_{1 \leq j \leq s} p_j, \tag{9.35}$$

where p_j is the degree of $L_j(\lambda)$, $j = 1, \ldots, s$.

Proof. Assume (9.35) holds, and let $q = l + f$. By Theorem 9.13 (taking into account the estimate (9.34)) we obtain that

$$Q(\operatorname{Ker} R_q(L_1, \ldots, L_s)) = Q(\operatorname{Im} \operatorname{col}(X_F J_F^{i-1})_{i=1}^q \dotplus \operatorname{Im} \operatorname{col}(X_\infty J_\infty^{q-i})_{i=1}^q \tag{9.36}$$

where (X_F, J_F) (resp. (X_∞, J_∞)) is the greatest common restriction of finite (resp. infinite) Jordan pairs of $L_1, \ldots, L_s$. It is easy to see that $J_\infty^f = 0$, so the right-hand side of (9.36) is equal to $\operatorname{Im} \operatorname{col}(X_F J_F^{i-1})_{i=1}^l$. Now by (9.3) (using again (9.34)) we find that the admissible pair (X_0, T_0) is similar to (X_F, J_F), i.e., (X_0, T_0) is also a greatest common restriction of the finite Jordan pairs of $L_1(\lambda), \ldots, L_s(\lambda)$. It remains to apply Theorem 9.9. □

Comments

The exposition in this chapter is based on the papers [29a, 29b, 30a].

Common multiples of monic operator polynomials in the infinite dimensional case have been studied in [30b, 70b]. The use of the Vandermonde matrix (in the infinite dimensional case) for studying operator roots of monic operator polynomials was originated in [62a, 64a, 64b]. See also [46]. Resultant matrices are well known for a pair of scalar polynomials. In the case of matrix polynomials the corresponding notion of resultant matrices was developed recently in [1, 26, 35a, 41].

Another approach to the construction of common multiples and common divisors of matrix polynomials is to be found in [6].

Behavior of monic common multiples of monic matrix polynomials, under analytic perturbation of those polynomials, is studied in [31].

Part III

Self-Adjoint Matrix Polynomials

In this part the finite dimensional space $\mathbb{C}^n$ will be regarded as equipped with the usual scalar product: $((x_1, \ldots, x_n)^T, (y_1, \ldots, y_n)^T) = \sum_{i=1}^n x_i \bar{y}_i$. Then, for a given matrix polynomial $L(\lambda) = \sum_{j=0}^l A_j \lambda^j$, define its adjoint $L^*(\lambda)$ by the formula

$$L^*(\lambda) = \sum_{j=0}^{l} A_j^* \lambda^j, \tag{10.1}$$

where A^* means the operator adjoint to A on the space $\mathbb{C}^n$; that is,

$$(A^*x, y) = (x, Ay) \qquad \text{for all} \quad x, y \in \mathbb{C}^n.$$

In the standard basis, if $A = (a_{ij})_{i,j=1}^n$, then $A^* = (\bar{a}_{ji})_{i,j=1}^n$.

We are to consider in this part an important class of monic matrix polynomials $L(\lambda)$ for which the coefficients A_j are hermitian matrices, $A_j = A_j^*$, or, what is the same, self-adjoint operators in $\mathbb{C}^n$:

$$(A_j x, y) = (x, A_j y) \qquad \text{for every} \quad x, y \in \mathbb{C}^n.$$

Such matrix polynomials will be called *self-adjoint*: $L(\lambda) = L^*(\lambda)$.

The study of self-adjoint matrix polynomials in the case $l = 2$ is motivated by the ubiquitous problem of damped oscillatory systems (mechanical and

and electrical) with a finite number of degrees of freedom, which may be described by a system of second-order differential equation with self-adjoint coefficients.

The main goal in this part is a description of the additional structure needed for the spectral theory of self-adjoint matrix polynomials. This includes the introduction of self-adjoint triples and a new invariant of a self-adjoint matrix polynomial, which we call the *sign characteristic*. Many properties of self-adjoint matrix polynomials demand some knowledge of this sign characteristic.

We remark that the requirement of monicity of the self-adjoint matrix polynomial $L(\lambda)$ is not necessary for most of the results given in this part. For instance, in many cases it can be replaced by the condition that the leading coefficient of $L(\lambda)$ is invertible, or is positive definite. However, we shall stick to the monicity requirement, and refer the reader to [34f, 34g] for the more general cases.

Chapter 10

General Theory

In this chapter we shall give an account of basic facts to be used in the analysis of self-adjoint matrix polynomials.

10.1. Simplest Properties

Consider a monic self-adjoint matrix polynomial

$$L(\lambda) = I\lambda^l + \sum_{i=0}^{l-1} A_j\lambda^j, \qquad A_i = A_i^*, \quad i = 0, \ldots, l-1. \tag{10.2}$$

Let C_1 and C_2, as defined in Section 1.1, be the first and second companion matrices of $L(\lambda)$, respectively. As in Section 2.1, define

$$B = \begin{bmatrix} A_1 & A_2 & \cdots & & I \\ A_2 & \cdots & & I & 0 \\ \vdots & & & \vdots & \vdots \\ I & 0 & \cdots & & 0 \end{bmatrix} \tag{10.3}$$

and, for a given standard triple (X, T, Y) of $L(\lambda)$, put

$$Q = Q(X, T) = \operatorname{col}(XT^i)_{i=0}^{l-1} \qquad \text{and} \qquad R = \operatorname{row}(T^iY)_{i=0}^{l-1}. \tag{10.4}$$

As we already know (cf. Section 2.1),

$$C_2 = BC_1B^{-1} = R^{-1}TR, \tag{10.5}$$

and

$$RBQ = I. \tag{10.6}$$

The crucial properties of a self-adjoint matrix polynomial $L(\lambda)$ are that $B = B^*$ and $C_1 = C_2^*$. So (10.5) becomes $BC_1 = C_1^*B$, or $[C_1x, y] = [x, C_1y]$, where the new indefinite scalar product $[x, y]$ is connected with the usual one (x, y) by the identity $[x, y] = (Bx, y)$ (see Chapter S5). This means that C_1 is self-adjoint relative to the scalar product $[x, y]$ (see Chapter S5). This property is crucial for our investigation of self-adjoint matrix polynomials.

We start with several equivalent characterizations of a self-adjoint matrix polynomial.

Theorem 10.1. *The following statements are equivalent:*

(i) $L^*(\lambda) = L(\lambda)$.

(ii) *For any standard triple* (X, T, Y) *of* $L(\lambda)$, (Y^*, T^*, X^*) *is also a standard triple for* $L(\lambda)$.

(iii) *For any standard triple* (X, T, Y) *of* $L(\lambda)$, *if* Q *is defined by* (10.4) *and* $M = Q^*BQ$, *then* $X = Y^*M$ *and* $T = M^{-1}T^*M$.

(iv) *For some standard triple* (X, T, Y) *of* $L(\lambda)$, *there is a nonsingular self-adjoint matrix* M *such that* $X = Y^*M$ *and* $T = M^{-1}T^*M$.

(v) $C_1 = B^{-1}C_1^*B$.

Proof. The line of proof follows the logical sequence (i) ⇒ (ii) ⇒ (i), (i) ⇒ (iii) ⇒ (iv) ⇒ (i), (iii) ⇒ (v) ⇒ (iv). The first implication follows immediately from Theorem 2.2.

(ii) ⇒ (i): given a standard triple for L and statement (ii) the resolvent form (2.16) can be used to write

$$L^{-1}(\lambda) = X(I\lambda - T)^{-1}Y = Y^*(I\lambda - T^*)^{-1}X^*. \tag{10.7}$$

It is immediately seen that $L^{-1}(\lambda) = (L^{-1}(\lambda))^*$ and (i) follows.

(i) ⇒ (iii): it follows from (i) that $C_1^* = C_2$. Using (10.4) and (10.5)

$$Q^{*-1}T^*Q^* = C_1^* = C_2 = R^{-1}TR$$

whence $T = (Q^*R^{-1})^{-1}T^*(Q^*R^{-1})$. But $RBQ = I$ so $R^{-1} = BQ$, and $T = M^{-1}T^*M$ where $M = Q^*BQ$.

To prove that $X = Y^*M$ we focus on the special triple.

$$X = [I \quad 0 \quad \cdots \quad 0], \qquad T = C_1, \qquad Y = \begin{bmatrix} 0 \\ \vdots \\ I \end{bmatrix}. \tag{10.8}$$

If we can establish the relation for this triple the conclusion follows for any standard triple by the similarity relationship between two standard triples (X_1, T_1, Y_1) and (X_2, T_2, Y_2) of $L(\lambda)$ established in Theorem 1.25, i.e.,

$$X_1 = X_2 S, \qquad T_1 = S^- T_2 S, \qquad Y_1 = S^{-1} Y_2, \tag{10.9}$$

where $S = [\operatorname{col}(X_2 T_2^i)_{i=0}^{l-1}]^{-1} \cdot \operatorname{col}(X_1 T_1)_{i=0}^{l-1}$.

For this choice of triple, $Q = I$ and (from (10.6)) $R = B^{-1}$. Consequently

$$XR = XB^{-1} = [0 \quad \cdots \quad 0 \quad I] = Y^*.$$

Hence $X = Y^*B = Y^*(Q^*BQ) = Y^*M$.

(iii) ⇒ (iv): requires no proof.

(iv) ⇒ (i): it is quickly verified that, from the equality $\sum_{i=0}^{l} A_i X T^i = 0$, $A_l = I$, and hypothesis (iv), $Q(Y^*, T^*) = Q(X, T)M^{-1}$ and then that Y^*, T^* form a standard pair for L. If (Y^*, T^*, Z) is the associated standard triple, then Z is defined by equations $Y^*T^{*r-1}Z = 0$ for $r = 1, \ldots, l-1$, and $Y^*T^{*l-1}Z = I$, which imply

$$Z = Q(Y^*, T^*)^{-1}\begin{bmatrix} 0 \\ \vdots \\ 0 \\ I \end{bmatrix} = MQ(X, T)^{-1}\begin{bmatrix} 0 \\ \vdots \\ 0 \\ I \end{bmatrix} = MY.$$

But $MY = X^*$ by hypothesis, so (Y^*, T^*, X^*) is a standard triple for L. Equation (10.7) can now be used again to deduce that $L^* = L$.

(iii) ⇒ (v): apply the standard triple of (10.8) in statement (iii). Then $T = C_1$, $Q = I$, $M = B$ and (v) follows.

(v) ⇒ (iv): this is obvious using the triple of (10.8) once more. □

One of our main objectives will be to find triples (X, T, Y) for the self-adjoint matrix polynomial $L(\lambda)$ such that the matrix M from Theorem 10.1 ($X = Y^*M$, $T = M^{-1}T^*M$) is as simple as possible.

EXAMPLE 10.1. Let $L(\lambda) = I\lambda - A$ be linear, $A = A^*$. In this case M can be chosen to be I. Indeed, $L(\lambda)$ has only linear elementary divisors and the spectrum $\sigma(L)$ of $L(\lambda)$ is real. Moreover, the eigenvectors of $L(\lambda)$ are orthogonal. So there exists a standard pair (X, T) of $L(\lambda)$ with unitary matrix X. With this choice of X, the matrix

$$M = Q^*BQ = X^*X$$

from Theorem 10.1(iii) is clearly I. □

EXAMPLE 10.2. Let $L(\lambda)$ be a self-adjoint matrix polynomial with *all* of its eigenvalues real and distinct, say $\lambda_1, \lambda_2, \ldots, \lambda_{ln}$. Let $x_1, \ldots, x_{ln}$ and $y_1, \ldots, y_{ln}$ be corresponding eigenvectors for $L(\lambda_i)$ and $L^*(\lambda_i)$ ($=L(\lambda_i)$ in this case), $i = 1, 2, \ldots, ln$, respectively. This means that $L(\lambda_i)x_i = L^*(\lambda_i)y_i = 0$ and $x_i, y_i \neq 0$ for $i = 1, 2, \ldots, ln$. Write $X = [x_1 \quad x_2 \quad \cdots \quad x_{ln}]$, $J = \operatorname{diag}[\lambda_1, \lambda_2, \ldots, \lambda_{ln}]$, $Y^* = [y_1 \quad \cdots \quad y_{ln}]$. The two sets of eigenvectors can be normalized in such a way that (X, J, Y) is a Jordan triple for $L(\lambda)$, so we have the resolvent representation

$$L^{-1}(\lambda) = X(I\lambda - J)^{-1}Y = \sum_{i=1}^{ln} \frac{x_i y_i^*}{\lambda - \lambda_i}, \tag{10.10}$$

and, since $J^* = J$, in this case

$$L^{-1}(\lambda) = X(I\lambda - J)^{-1}Y = Y^*(I\lambda - J)^{-1}X^*.$$

By Theorem 10.1 there exists a nonsingular self-adjoint M such that $X = Y^*M$ and $MJ = J^*M$ ($= JM$ in this special case). It follows that M must be a real diagonal matrix, say $M = \operatorname{diag}[s_1, \ldots, s_{ln}]$. Let $\alpha_i = \sqrt{|s_i|}$, $i = 1, 2, \ldots, ln$ and write

$$T = \operatorname{diag}[\alpha_1, \ldots, \alpha_{ln}], \qquad E = \operatorname{diag}[\operatorname{sgn} s_1, \ldots, \operatorname{sgn} s_{ln}].$$

Then $M = TET$.

Consider now the Jordan triple (X_1, J, Y_1) with $X_1 = XT^{-1}$ and $Y_1 = TY$. For this Jordan triple the corresponding matrix M_1 ($X_1 = Y_1^* M_1$, $M_1 J = JM_1$) is equal to E.

This is the simplest possible form of M for the self-adjoint matrix polynomial $L(\lambda)$. We shall see later that the signs $\{\operatorname{sgn} s_i\}_{i=1}^{ln}$ form the so-called sign characteristic of $L(\lambda)$ and (X_1, J, Y_1) is a self-adjoint triple. □

Let $L(\lambda)$ be a self-adjoint monic matrix polynomial. Then the spectrum of $L(\lambda)$ is symmetric with respect to the real axis, as the following lemma shows.

Lemma 10.2. *If $L = L^*$ and $\lambda_r \in \sigma(L)$ is nonreal, then $\bar{\lambda}_r$ is also an eigenvalue of L with the same partial multiplicities as λ_r. Thus, if Jordan blocks $J_{r_1}, \ldots, J_{r_t}$ are associated with λ_r, then exactly t blocks $J_{s_1}, \ldots, J_{s_t}$ are associated with $\lambda_s = \bar{\lambda}_r$ and*

$$J_{s_i} = \bar{J}_{r_i}, \qquad i = 1, 2, \ldots, t.$$

Proof. If follows from characterization (ii) of Theorem 10.1 that (Y^*, J^*, X^*) is a standard triple for L if (X, J, Y) is so. But then the matrices J and J^* are similar, and the conclusion follows. □

We conclude this section with an observation of an apparently different character. For a nonsingular self-adjoint matrix M the *signature* of M, written $\operatorname{sig} M$, is the difference between the number of positive eigenvalues and the number of negative eigenvalues (counting multiplicities, of course). Referring to the characterizations (iii) and (iv) of Theorem 10.1 of a monic self-adjoint matrix polynomial we are to show that, for the M appearing there, $\operatorname{sig} M$ depends only on the polynomial. Indeed, it will depend only on l.

Theorem 10.3. *Let $L = \sum_{j=0}^{l-1} A_j \lambda^j + I\lambda^l$ be a self-adjoint matrix polynomial on $\mathbb{C}^n$. Let (X, T, Y) be any standard triple of L. Then the signature of any self-adjoint matrix M for which $X = Y^*M$, $T = M^{-1}T^*M$ is given by*

$$\operatorname{sig} M = \begin{cases} 0 & \text{if } l \text{ is even} \\ n & \text{if } l \text{ is odd.} \end{cases} \tag{10.11}$$

Proof. The relations $X = Y^*M$, $T = M^{-1}T^*M$ imply that

$$\operatorname{col}[XT^i]_{i=0}^{l-1} = (\operatorname{col}[Y^*T^{*i}]_{i=0}^{l-1})M.$$

Thus,

$$M = (\operatorname{col}[Y^*T^{*i}]_{i=0}^{l-1})^{-1}(\operatorname{col}[XT^i]_{i=0}^{l-1}),$$

and for any other standard triple (X_1, T_1, Y_1) the corresponding matrix M_1 is given by

$$M_1 = (\operatorname{col}[Y_1^*T_1^{*i}]_{i=0}^{l-1})^{-1}(\operatorname{col}[X_1T_1^i]_{i=0}^{l-1}).$$

But there is an S for which relations (10.9) are satisfied and this leads to the conclusion $M_1 = S^*MS$. Then as is well known, sig M_1 = sig M so that the signature is independent of the choice of standard triple.

Now select the standard triple (10.8). By Theorem 10.1(v), in this case $M = B$. Thus, for any M of the theorem statement, sig M = sig B.

To compute sig B, consider the continuous family of self-ajdoint operators $B(\varepsilon) = [B_{j+k-1}(\varepsilon)]_{j,k=1}^{l}$, where $B_j(\varepsilon) = \varepsilon A_j$ for $j = 1, \dots, l-1$, $B_l(\varepsilon) = I$, $B_j(\varepsilon) = 0$ for $j > l$. Here $\varepsilon \in [0, 1]$. Then $B = B(1)$, and $B(\varepsilon)$ is nonsingular for every $\varepsilon \in [0, 1]$. Hence sig $B(\varepsilon)$ is independent of ε. Indeed, let $\lambda_1(\varepsilon) \geq \cdots \geq \lambda_{nl}(\varepsilon)$ be the eigenvalues of $B(\varepsilon)$. As $B(\varepsilon)$ is continuous in ε, so are the $\lambda_i(\varepsilon)$. Furthermore, $\lambda_i(\varepsilon) \neq 0$ for any ε in $[0, 1]$, $i = 1, \dots, nl$ in view of the nonsingularity of $B(\varepsilon)$. So the number of positive $\lambda_i(\varepsilon)$, as well as the number of negative $\lambda_i(\varepsilon)$, is independent of ε on this interval. In particular, we obtain

$$\operatorname{sig} B = \operatorname{sig} B(0) = \operatorname{sig}\begin{bmatrix} 0 & & \cdots & 0 & I \\ \vdots & & & I & 0 \\ 0 & I & \cdot^{\cdot^{\cdot}} & & \vdots \\ I & 0 & \cdots & & 0 \end{bmatrix}$$

and the verification of (10.11) is then an easy exercise. □

Theorem 10.3 implies (together with Corollary S5.2) the following unexpected result.

Theorem 10.4. *A self-adjoint monic matrix polynomial of odd degree l has at least n elementary divisors of odd degree associated with real eigenvalues.*

Proof. Indeed, Theorem 10.3 implies that

$$\operatorname{sig} B = \begin{cases} 0 & \text{if } \ l \text{ is even} \\ n & \text{if } \ l \text{ is odd.} \end{cases}$$

It remains to apply Corollary S5.2 bearing in mind that C_1 is B-self-adjoint. □

The significance of this result is brought out if we consider two familiar extreme cases: a) if $n = 1$, then the result of Theorem 10.4 says that every scalar real polynomial of odd degree has a real root; b) if $l = 1$, then Theorem 10.4 says that every self-adjoint matrix (considered in the framework of the linear self-adjoint polynomial $I\lambda - A$) has n real eigenvalues (counting multiplicities), and the elementary divisors of A are linear.

These two results are not usually seen as being closely connected, but Theorem 10.4 provides an interesting unification.

10.2. Self-Adjoint Triples: Definition

Let $L = L^*$ be a monic matrix polynomial with a Jordan triple (X, J, Y). Since (Y^*, J^*, X^*) is also a standard triple of L, we have

$$Y^* = XM, \qquad J^* = M^{-1}JM, \qquad X^* = M^{-1}Y.$$

As mentioned before, we would like to choose (X, J, Y) in such a way that the similarity matrix M is as simple as possible. In general one cannot expect $M = I$ even in the case of a scalar polynomial with all real eigenvalues, as the following example shows (see also Example 10.2).

EXAMPLE 10.3. Let $L(\lambda) = \lambda(\lambda - 1)$ be a scalar polynomial. We can take

$$J = \begin{bmatrix} 0 & 0 \\ 0 & 1 \end{bmatrix}$$

and the general forms for X, Y in the standard triple (X, J, Y) are

$$X = [x_1 \quad x_2], \qquad Y = \begin{bmatrix} -x_1^{-1} \\ x_2^{-1} \end{bmatrix},$$

where $x_1, x_2 \in \mathbb{C}\backslash\{0\}$. Thus, $Y^* = XM$, where

$$M = \begin{bmatrix} -|x_1^{-1}|^2 & 0 \\ 0 & |x_2^{-1}|^2 \end{bmatrix}.$$

So the simplest form of M is $\operatorname{diag}(-1, 1)$, which appears if x_1 and x_2 lie on the unit circle. □

We describe now the simplest structure of M. First it is necessary to specify the complete Jordan structure of X, J, and Y. Since the nonreal eigenvalues of L occur in conjugate pairs, and in view of Lemma 10.2, a Jordan form J can be associated with L (i.e., with the first companion matrix C_1) having the following form. Select a maximal set $\{\lambda_1, \ldots, \lambda_a\}$ of eigenvalues of J containing no conjugate pair, and let $\{\lambda_{a+1}, \ldots, \lambda_{a+b}\}$ be the distinct real eigenvalues of J. Put $\lambda_{a+b+j} = \bar{\lambda}_j$, for $j = 1, \ldots, a$, and let

$$J = \operatorname{diag}[J_i]_{i=1}^{2a+b} \tag{10.12}$$

where $J_i = \operatorname{diag}[J_{ij}]_{j=1}^{k_i}$ is a Jordan form with eigenvalue λ_i and Jordan blocks $J_{i,1}, \ldots, J_{i,k_i}$ of sizes $\alpha_{i,1} \geq \cdots \geq \alpha_{i,k_i}$, respectively.

Clearly, if (X, J, Y) is a Jordan triple, then there is a corresponding partitioning of X, and of Y, into blocks, each of which determines a Jordan chain. Thus, we write

$$X = [X_1 \quad X_2 \quad \cdots \quad X_{2a+b}] \tag{10.13}$$

and

$$X_i = [X_{i,1} \quad X_{i,2} \quad \cdots \quad X_{i,k_i}] \tag{10.14}$$

and

$$Y_i = \operatorname{col}[Y_{ij}]_{j=1}^{k_i}, \qquad Y = \operatorname{col}[Y_i]_{i=1}^{2a+b}. \tag{10.15}$$

A standard triple (X, J, Y) is said to be *self-adjoint* if J is Jordan and if (in the notations just introduced)

$$Y_{ij} = P_{ij} X^*_{a+b+i,j}, \qquad Y_{a+b+i,j} = P_{ij} X^*_{ij} \tag{10.16}$$

for $i = 1, 2, \ldots, a$ and $j = 1, 2, \ldots, k_i$, where P_{ij} is the $\alpha_{ij} \times \alpha_{ij}$ standard involuntary permutation (sip) matrix (see Section S5.1 for the definition of a sip matrix), and also

$$Y_{ij} = \varepsilon_{ij} P_{ij} X^*_{ij} \tag{10.17}$$

for $i = a + 1, \ldots, a + b$ and $j = 1, 2, \ldots, k_i$ where P_{ij} is the $\alpha_{ij} \times \alpha_{ij}$ sip matrix, and the numbers ε_{ij} are $+1$ or -1.

A simple analysis can convince us that in a self-adjoint triple (X, J, Y) the matrix M for which $Y^* = XM$ and $MJ = J^*M$ is of the simplest possible form. Indeed, the self-adjoint equalities (10.16) and (10.17) could be rewritten in the form $Y^* = XP_{\varepsilon,J}$, where $P_{\varepsilon,J}$ is given by the formula

$$P_{\varepsilon,J} = \begin{bmatrix} 0 & 0 & P_c \\ 0 & P_r & 0 \\ P_c & 0 & 0 \end{bmatrix}, \tag{10.18}$$

where

$$P_c = \operatorname{diag}[\operatorname{diag}[P_{ij}]_{j=1}^{k_i}]_{i=1}^{a}$$

and

$$P_r = \operatorname{diag}[\operatorname{diag}[\varepsilon_{ij} P_{ij}]_{j=1}^{k_i}]_{i=a+1}^{a+b}.$$

We remark also that $P_{\varepsilon,J} J = J^* P_{\varepsilon,J}$. Thus, for this triple we have $M = P_{\varepsilon,J}$. One cannot expect any simpler connections between X and Y in general, because the presence of blocks P_{ij} is inevitable due to the structure of J and the signs ε_{ij} appear even in the simplest examples, as shown above.

The existence of self-adjoint triples for a self-adjoint matrix polynomial will be shown in the next section and, subsequently, it will be shown that the signs ε_{ij} associated with the real eigenvalues are defined uniquely if normalized as follows: where there is more than one Jordan block of a certain fixed size associated with an eigenvalue, all the $+1$s (if any) precede all the -1s (if any). The normalized ordered set $\{\varepsilon_{ij}\}$, $i = a+1, \ldots, a+b$, $j = 1, 2, \ldots, k_i$, is called the *sign characteristic* of $L(\lambda)$.

Observe that Eq. (10.16) and (10.17) show that left Jordan chains *can* be constructed very simply from right Jordan chains. Suppose one begins with any Jordan pair (X, J) and then these equations are used to generate a matrix Y. In general (X, J, Y) will not be a self-adjoint triple because the orthogonality conditions (10.6) will not be satisfied. However, the theorem says that for a *special* choice of X, and then Y defined by (10.16) and (10.17) a self-adjoint triple can be generated. As an illustration of the notation of self-adjoint triples consider the resolvent form for self-adjoint triple (X, J, Y). First define submatrices of X, J as follows (cf. Eqs. (10.13) and (10.12)):

$$\tilde{X}_1 = [X_1 \quad \cdots \quad X_a], \qquad \tilde{X}_2 = [X_{a+1} \quad \cdots \quad X_{a+b}],$$
$$\tilde{X}_3 = [X_{a+b+1} \quad \cdots \quad X_{2a+b}]$$

and similarly for $J = \operatorname{diag}[K_1, K_2, K_3]$. Then equalities (10.16) and (10.17) imply the existence of (special) permutation matrices $\tilde{P}_1$, $\tilde{P}_2$ for which

$$Y = \begin{bmatrix} \tilde{P}_1 \tilde{X}_3^* \\ \tilde{P}_2 \tilde{X}_2^* \\ \tilde{P}_1 \tilde{X}_1^* \end{bmatrix},$$

namely, $\tilde{P}_1 = \operatorname{diag}[\operatorname{diag}(P_{ij})_{j=1}^{k_i}]_{i=1}^{a}$, $\tilde{P}_2 = \operatorname{diag}[\operatorname{diag}(\varepsilon_{ij} P_{ij})_{j=1}^{k_i}]_{i=a+1}^{a+b}$.

In this case, the resolvent form of $L(\lambda)$ becomes

$$\begin{aligned} L^{-1}(\lambda) = X(I\lambda - J)^{-1} Y &= \tilde{X}_1(I\lambda - K_1)^{-1}\tilde{P}_1\tilde{X}_2^* + \tilde{X}_2(I\lambda - K_2)^{-1}\tilde{P}_2\tilde{X}_2 \\ &\quad + \tilde{X}_3(I\lambda - \overline{K}_1)^{-1}P_1X_1^* \\ &= \tilde{X}_1(I\lambda - K_1)^{-1}\tilde{P}_1\tilde{X}_3^* + \tilde{X}_2(I\lambda - K_2)^{-1}(\tilde{P}_2\tilde{X}_2^*) \\ &\quad + (\tilde{X}_3\tilde{P}_1)(I\lambda - K_1^*)^{-1}\tilde{X}_1^*. \end{aligned}$$

This equality is characteristic for self-adjoint matrix polynomials, i.e., the following result holds.

Theorem 10.5. *A monic matrix polynomial $L(\lambda)$ is self-adjoint if and only if it admits the representation*

$$\begin{aligned} L^{-1}(\lambda) = X_1(I\lambda - J_1)^{-1} \cdot P_1 X_3^* &+ X_2(I\lambda - J_2)^{-1} \cdot P_2 X_2^* \\ &+ X_3 P_1 \cdot (I\lambda - J_1^*)^{-1} X_1^*, \end{aligned} \tag{10.19}$$

where J_1 (J_2) are Jordan matrices with $\mathscr{I}m\ \lambda_0 > 0$ $(\mathscr{I}m\ \lambda_0 = 0)$ for every eigenvalue λ_0 of J_1 (J_2), and for $i = 1, 2$

$$P_i = P_i^* = P_i^{\mathrm{T}}, \qquad P_i^2 = I, \qquad P_i J_i = J_i^{\mathrm{T}} P_i. \tag{10.20}$$

Proof. We saw above that if $L(\lambda)$ is self-adjoint, then $L^{-1}(\lambda)$ admits representation (10.19) (equalities (10.20) are checked easily for $P_i = \tilde{P}_i$ and $J_i = K_i$ in the above notation). On the other hand, if (10.19) holds, then by taking adjoints one sees easily that $L^{-1}(\lambda) = (L^{-1}(\bar{\lambda}))^*$. So $L(\lambda)$ is self-adjoint. □

We remark that the form of the resolvent representation (10.19) changes if the "monic" hypothesis is removed. More generally, it can be shown that a matrix polynomial $L(\lambda)$ with $\det L(\lambda) \not\equiv 0$ is self-adjoint if and only if it admits the representation

$$\begin{aligned} L^{-1}(\lambda) = {} & X_1(I\lambda - J_1)^{-1}P_1X_3^* + X_2(I\lambda - J_2)^{-1}P_2X_2^* \\ & + X_3P_1(I\lambda - J_1^*)^{-1}X_1^* + X_4(J_4\lambda - I)^{-1}P_4X_4^*, \end{aligned} \tag{10.21}$$

where $X_1, X_2, X_3, P_1, P_2, J_1, J_2$ are as in Theorem 10.5, J_4 is a nilpotent Jordan matrix, and

$$P_4 = P_4^* = P_4^T, \qquad P_4J_4 = J_4^{\mathrm{T}}P_4.$$

It turns out that the matrices X_1, X_2, X_3, J_1, J_2 in (10.21) are constructed using the finite Jordan pair $(X_{\mathrm{F}}, J_{\mathrm{F}})$ of $L(\lambda)$ in the same way as described above for monic polynomials. The pair (X_4, J_4) turns out to be the restriction of the infinite Jordan pair (X_∞, J_∞) of $L(\lambda)$ (as defined in Section 7.1) to some J_∞-invariant subspace. We shall not prove (10.21) in this book.

10.3. Self-Adjoint Triples: Existence

In this section the notions and results of Chapter S5 are to be applied to self-adjoint matrix polynomials and their self-adjoint triples. Some properties of these triples are deduced and an illustrative example is included. To begin with, we prove the existence of self-adjoint triples. In fact, this is included in the following more informative result. As usual, C_1 is the first companion matrix of L, and B is as defined in (10.3).

Theorem 10.6. *Let $L(\lambda) = \sum_{j=0}^{l-1} A_j\lambda^j + I\lambda^l$ be a monic self-adjoint matrix polynomial. If $P_{\varepsilon, J}$ is a C_1-canonical form of B and S is the reducing matrix $B = S^*P_{\varepsilon, J}S$, then the triple (X, J, Y) with*

$$X = [I \quad 0 \quad \cdots \quad 0]S^{-1}, \qquad J = SC_1S^{-1}, \qquad Y = S[0 \quad \cdots \quad I]^*$$

is self-adjoint.

Conversely, if for some nonsingular S the triple (X, J, Y) defined by the above relations is self-adjoint, then S is a reducing matrix of B to C_1-canonical form.

Proof. It has been shown in Theorem 10.1 that $L^* = L$ is equivalent to the statement that C_1 is B-self-adjoint. Then Theorem S5.1 implies the existence of a C_1-canonical form $P_{\varepsilon, J}$ for B. Thus, $B = S^*P_{\varepsilon, J}S$ and $SC_1S^{-1} = J$. Then it is clear that (X, J, Y) as defined in the theorem forms a Jordan triple for L (refer to Eq. (10.9)). This triple is self-adjoint if and only if $Y = P_{\varepsilon, J}X^*$. Using the definition of X and Y this relation is seen to be equivalent to

$$B\begin{bmatrix} 0 \\ \vdots \\ 0 \\ I \end{bmatrix} = \begin{bmatrix} I \\ 0 \\ \vdots \\ 0 \end{bmatrix}, \tag{10.22}$$

and this is evident in view of the definition of B.

Suppose now that

$$(X, J, Y) = ([I \quad 0 \quad \cdots \quad 0]S^{-1}, SC_1S^{-1}, S[0 \quad \cdots \quad 0 \quad I]^*)$$

is a self-adjoint triple. Then $Y = P_{\varepsilon, J}X^*$. Using the equality $P_{\varepsilon, J}J = J^*P_{\varepsilon, J}$ one checks easily that $Q^* = P_{\varepsilon, J}R$, where $Q = \operatorname{col}[XJ^i]_{i=0}^{l-1}$ and $R = \operatorname{row}[J^iY]_{i=0}^{l-1}$. By (10.6) $B = R^{-1}Q^{-1} = Q^{*-1}P_{\varepsilon, J}Q^{-1}$. It remains to note that $S = Q^{-1}$. $\square$

Now we derive some useful formulas involving self-adjoint triples.

Let (X, J) be a Jordan pair of the monic self-adjoint matrix polynomial $L(\lambda)$ constructed as in (10.12), (10.13), and (10.14). Construct the corresponding matrix $Q = \operatorname{col}[XJ^i]_{i=0}^{l-1}$ and let $U = Q^*BQ$. It is evident that U is nonsingular. It Y is the third member of a Jordan triple determined by X and J, we show first that $Y = U^{-1}X^*$. Recall that Y is determined by the orthogonality relation

$$Y = Q^{-1} \cdot \operatorname{col}(\delta_{il}I)_{i=1}^{l}. \tag{10.23}$$

We also have

$$Q(U^{-1}X^*) = B^{-1}Q^{*-1}X^* = B^{-1}\begin{bmatrix} I \\ 0 \\ \vdots \\ 0 \end{bmatrix} = \begin{bmatrix} 0 \\ \vdots \\ 0 \\ I \end{bmatrix} \tag{10.24}$$

and comparing with (10.23) it is seen that $Y = U^{-1}X^*$.

Since the columns of Q form Jordan chains for C_1, $C_1Q = QJ$ whence $Q^*(BC_1)Q = (Q^*BQ)J$. In addition $(BC_1)^* = BC_1$ and it follows that

$$J^*(Q^*BQ) = (Q^*BQ)J. \tag{10.25}$$

With a partitioning consistent with that of X in (10.13) write

$$Q = [Q_1 \quad \cdots \quad Q_{2a+b}]$$

and (10.25) can be written

$$J_r^*(Q_r^* B Q_s) = (Q_r^* B Q_s) J_s,$$

for $r, s = 1, 2, \ldots, 2a + b$. Now $\bar{\lambda}_r \neq \lambda_s$ implies that $Q_r^* B Q_s = 0$ (see Section S2.2) and it is found that

$$U = [Q_r^* B Q_s] = \begin{bmatrix} 0 & 0 & T_1^* \\ 0 & T_2 & 0 \\ T_1 & 0 & 0 \end{bmatrix} \tag{10.26}$$

where

$$T_1 = \operatorname{diag}[Q_{a+b+i}^* B Q_i]_{i=1}^a \tag{10.27}$$

and

$$T_2 = \operatorname{diag}[Q_{a+j}^* B Q_{a+j}]_{j=1}^b. \tag{10.28}$$

Suppose now that the triple (X, J, Y) is self-adjoint. Then $Q = S^{-1}$, where S is the reducing matrix to a C_1-canonical form of B (cf. Theorem 10.6). It follows that $U = P_{\varepsilon, J}$. In particular, we have for $i = a + 1, \ldots, a + b$

$$Q_i^* B Q_i = \operatorname{diag}[\varepsilon_{ij} P_{ij}]_{j=1}^{k_i}, \tag{10.29}$$

where $\varepsilon_{ij} = \pm 1$, for $i = 1, \ldots, a$,

$$Q_{a+b+i}^* B Q_i = \operatorname{diag}[P_{ij}]_{j=1}^{k_i}. \tag{10.30}$$

Note that Eqs. (10.26)–(10.28) hold for any Jordan triple of $L(\lambda)$ (not necessarily self-adjoint).

We conclude this section with an illustrative example.

EXAMPLE 10.4. Consider the fourth-degree self-adjoint matrix polynomial

$$L_4(\lambda) = \begin{bmatrix} 0 & 1 \\ 1 & -1 \end{bmatrix} \begin{bmatrix} 2(\lambda^2+1)^2 & (\lambda^2+1)^2 \\ (\lambda^2+1)^2 & \lambda^4+1 \end{bmatrix} \begin{bmatrix} 0 & 1 \\ 1 & -1 \end{bmatrix}.$$

It is found that $\det L_4(\lambda) = (\lambda^4 - 1)^2$ so the eigenvalues are the fourth roots of unity, each with multiplicity two. A Jordan pair is

$$J = \operatorname{diag}\left\{ \begin{bmatrix} i & 1 \\ 0 & i \end{bmatrix}, \begin{bmatrix} 1 & 1 \\ 0 & 1 \end{bmatrix}, \begin{bmatrix} -1 & 1 \\ 0 & -1 \end{bmatrix}, \begin{bmatrix} -i & 1 \\ 0 & -i \end{bmatrix} \right\},$$

$$X = \begin{bmatrix} 1 & 0 & -1 & 0 & -1 & 0 & 1 & 0 \\ 1 & 0 & 1 & 0 & 1 & 0 & 1 & 0 \end{bmatrix}.$$

However, it is found that this is not a self-adjoint pair. It turns out that

$$\hat{X} = \frac{\sqrt{2}}{8}\begin{bmatrix} -2 & -2i & -2 & 1 & -2 & -1 & 0 \\ -2 & -2i & 2 & -1 & 2 & 1 & 2 \end{bmatrix}$$

is such that $\hat{X}, J$ determines a self-adjoint triple. In Eq. (10.26) we then have

$$T_1 = \begin{bmatrix} 0 & 1 \\ 1 & 0 \end{bmatrix} \qquad T_2 = \operatorname{diag}\left\{\begin{bmatrix} 0 & 1 \\ 1 & 0 \end{bmatrix}, \begin{bmatrix} 0 & 1 \\ 1 & 0 \end{bmatrix}\right\}. \quad \square$$

10.4. Self-Adjoint Triples for Real Self-Adjoint Matrix Polynomials

Let $L(\lambda) = I\lambda^l + \sum_{j=0}^{l-1} A_j\lambda^j$ be a monic matrix polynomial with real and symmetric coefficients A_j: $A_j = A_j^{\mathrm{T}}$, $j = 0, \ldots, l-1$. Then clearly $L(\lambda)$ is self-adjoint and as we have already seen there exists a self-adjoint triple (X, J, Y) of $L(\lambda)$. In the case of real coefficients, however, a self-adjoint triple which enjoys special properties can be constructed, as we shall see in this section.

Let us make first a simple observation on Jordan chains for $L(\lambda)$. Let $\lambda_0 \in \sigma(L)$ be a nonreal eigenvalue of $L(\lambda)$ with corresponding Jordan chain $x_0, \ldots, x_r$. So, as defined in Section 1.4.

$$\sum_{j=0}^{k} \frac{1}{j!} L^{(j)}(\lambda_0)x_{k-j} = 0, \qquad k = 0, \ldots, r. \tag{10.31}$$

Taking complex conjugates in these equalities, (and denoting by $\bar{Z}$ the matrix whose entries are complex conjugate to the entries of the matrix Z) we obtain

$$\sum_{j=0}^{k} \frac{1}{j!} \overline{L^{(j)}(\lambda_0)}\bar{x}_{k-j} = 0, \qquad k = 0, \ldots, r.$$

But since the coefficients A_j are real, $\overline{L^{(j)}(\lambda_0)} = L^{(j)}(\bar{\lambda}_0)$; hence $\bar{x}_0, \ldots, \bar{x}_r$ is a Jordan chain of $L(\lambda)$ corresponding to the eigenvalue $\bar{\lambda}_0$. Moreover, by Proposition 1.15 it follows that a canonical set of Jordan chains for λ_0 becomes a canonical set of Jordan chains for $\bar{\lambda}_0$ under the transformation of taking complex conjugates. In other words, if (X', J') is the part of a Jordan pair (X, J) of $L(\lambda)$ corresponding to the nonreal part of $\sigma(J)$, then so is $(\bar{X}', \bar{J}')$. Let now $\lambda_0 \in \sigma(L)$ be real, and let v be the maximal partial multiplicity of $L(\lambda)$ at λ_0. Consider the system of linear homogeneous equations (10.31) with $r = v$, where the vectors $x_0, \ldots, x_v$ are regarded as unknowns. Since $L^{(j)}(\lambda_0)$ is real, there exists a basis in the linear set of solutions of (10.31), which consists of real vectors. By Proposition 1.15 this basis gives rise to a canonical set of Jordan chains for λ_0 which consists of real eigenvectors and generalized eigenvectors.

It follows from the preceding discussion that there exists a Jordan pair (X, J) of $L(\lambda)$ of the following structure:

$$X = [X^{(1)}, X^{(2)}, \overline{X^{(1)}}], \qquad J = \operatorname{diag}[J^{(1)}, J^{(2)}, \overline{J^{(1)}}], \tag{10.32}$$

where $\sigma(J^{(2)})$ is real and the spectrum of $J^{(1)}$ is nonreal and does not contain conjugate complex pairs. The matrix $X^{(2)}$ is real.

We have not yet used the condition that $A_j = A_j^{\mathrm{T}}$. From this condition it follows that there exists a self-adjoint triple of $L(\lambda)$. The next result shows that one can choose the self-adjoint triple in such a way that property (10.32) holds also.

Theorem 10.7. *There exists a self-adjoint triple (X, J, Y) of $L(\lambda)$ such that the pair (X, J) is decomposed as in (10.32).*

We need some preparations for the proof of Theorem 10.7 and first we point out the following fact.

Theorem 10.8. *For every $n \times n$ nonsingular (complex) symmetric matrix V there exists an $n \times n$ matrix U such that $V = U^{\mathrm{T}}U$.*

Proof. Consider the linear space $\mathbb{C}^n$ together with the symmetric bilinear form $[\ ,\ ]$ defined by V: $[x, y] = y^{\mathrm{T}}Vx$. Using Lagrange's alogrithm (see [22, 52c]), we reduce the form $[\ ,\ ]$ to the form $[x, y] = \sum_{i=0}^{k} x_i y_i$ for $x = (x_1, \ldots, x_n)^{\mathrm{T}}$ and $y = (y_1, \ldots, y_n)^{\mathrm{T}}$ in some basis in $\mathbb{C}^n$. Then the desired matrix U is the transformation matrix from the standard basis in $\mathbb{C}^n$ to this new basis. □

Let $K = \operatorname{diag}[K_1, \ldots, K_r]$ be a Jordan matrix with Jordan blocks $K_1, \ldots, K_r$ of sizes $m_1, \ldots, m_r$, respectively. Let P_i be the sip matrix of size m_i $(i = 1, \ldots, r)$, and put $P_K = \operatorname{diag}[P_1, \ldots, P_r]$. Denote by $\mathscr{T}$ the set of all nonsingular matrices commuting with K.

Lemma 10.9. *Let V be a nonsingular (complex) symmetric matrix such that $P_K V \in \mathscr{T}$. Then there exists $W \in \mathscr{T}$ such that $W^{\mathrm{T}}VW = P_K$. Conversely, if for some matrix V there exists $W \in \mathscr{T}$ such that $W^{\mathrm{T}}VW = P_K$, then V is nonsingular symmetric and $P_K V \in \mathscr{T}$.*

Proof. The converse statement is easily checked using the property that $P_K K = K^{\mathrm{T}}P_K$ and $P_K K^{\mathrm{T}} = KP_K$ so we focus on the direct statement of the lemma.

Consider first the case when the set $\sigma(K)$ of eigenvalues of K consists of one point and all the Jordan blocks of K have the same size:

$$K = \operatorname{diag}[K_1, \ldots, K_1],$$

where the $p \times p$ Jordan block K_1 appears r times.

It is seen from Theorem S2.2 that a matrix $U \in \mathscr{T}$ if and only if it is nonsingular and has the partitioned form $(U_{ij})_{i,j=1}^{r}$ where U_{ij} is a $p \times p$ upper triangular Toeplitz matrix, i.e., has the form

$$U_{ij} = \begin{bmatrix} t_1 & t_2 & \cdots & & t_p \\ 0 & t_1 & & \vdots & \vdots \\ \vdots & \vdots & \ddots & t_1 & t_2 \\ 0 & & & 0 & t_1 \end{bmatrix} \tag{10.33}$$

for some complex numbers $t_1, t_2, \ldots, t_p$ (dependent on i, j). It will first be proved that V admits the representation

$$U^{\mathrm{T}} V U = \operatorname{diag}[W_i]_{i=1}^{r} \tag{10.34}$$

for some $U \in \mathscr{T}$ where $W_1, \ldots, W_r$ are $p \times p$ matrices.

For brevity, introduce the class $\mathscr{A}$ of all nonsingular (complex) symmetric matrices of the form $P_K S$, $S \in \mathscr{T}$, to which V belongs. Put

$$W_0 = U^{\mathrm{T}} V U, \tag{10.35}$$

where $U \in \mathscr{T}$ is arbitrary. Then $W_0 \in \mathscr{A}$. Indeed, the symmetry of W_0 follows from the symmetry of V. Further, using equalities $UK = KU$, $P_K V K = K P_K V$, $P_K^2 = I$, and $P_K K = K^{\mathrm{T}} P_K$, we have

$$\begin{aligned} P_K W_0 K &= P_K U^{\mathrm{T}} V U K = P_K U^{\mathrm{T}} P_K P_K V K U = P_K U^{\mathrm{T}} P_K K P_K V U \\ &= P_K U^{\mathrm{T}} K^{\mathrm{T}} P_K P_K V U = P_K K^{\mathrm{T}} U^{\mathrm{T}} V U = K P_K U^{\mathrm{T}} V U = K P_K W_0, \end{aligned}$$

so that $W_0 \in \mathscr{A}$.

Now let M be the $pr \times pr$ permutation matrix which, when applied from the right moves columns $1, 2, 3, \ldots, pr$ into the positions

$$1,\quad r+1,\quad 2r+1, \ldots, (p-1)r+1,$$
$$2,\quad r+2, \ldots, (p-1)r+2, \ldots, r,\quad 2r, \ldots, pr,$$

respectively. Since $V \in \mathscr{A}$, we have $P_K V \in \mathscr{T}$, and using representation (10.33) for the blocks of $P_K V$ it is found that

$$M^{\mathrm{T}} V M = \begin{bmatrix} 0 & & & 0 & V_1 \\ \vdots & \vdots & \cdot^{\cdot^{\cdot}} & V_1 & V_2 \\ 0 & V_1 & \cdot^{\cdot^{\cdot}} & \vdots & \vdots \\ V_1 & V_2 & \cdots & & V_p \end{bmatrix} \tag{10.36}$$

where $V_i^{\mathrm{T}} = V_i$, $i = 1, 2, \ldots, p$. (Note that the effect of the permutation is to transform a partitioning of V in r^2 $p \times p$ blocks to a partitioning in p^2 blocks of size $r \times r$.) Hence for W_0 defined by (10.35) $M^T W_0 M$ has the form, analogous to (10.36), say

$$\tilde{W}_0 = M^T W_0 M = \begin{bmatrix} 0 & & & 0 & \tilde{W}_1 \\ \vdots & \vdots & \iddots & \tilde{W}_1 & \tilde{W}_2 \\ 0 & \tilde{W}_1 & & \vdots & \vdots \\ \tilde{W}_1 & \tilde{W}_2 & \cdots & & \tilde{W}_p \end{bmatrix}, \tag{10.37}$$

and furthermore, the nature of the permutation is such that W_0 is *block diagonal if and only if the blocks* $\tilde{W}_i$, $i = 1, 2, \ldots, p$, *are diagonal.*

If $U \in \mathscr{T}$, the action of the same symmetric permutation on U is

$$M^T U M = \begin{bmatrix} U_1 & U_2 & \cdots & & U_p \\ 0 & U_1 & & \vdots & \vdots \\ \vdots & \vdots & \ddots & U_1 & U_2 \\ 0 & & & 0 & U_1 \end{bmatrix}. \tag{10.38}$$

Now (using the property $MM^T = I$) we have

$$\tilde{W}_0 = M^T W_0 M = (M^T U M)^T (M^T V M)(M^T U M)$$

$$= \begin{bmatrix} U_1^T & 0 & \cdots & 0 \\ U_2^T & U_1^T & \ddots & \vdots \\ \vdots & \vdots & \ddots & 0 \\ U_p^T & & \cdots & U_1^T \end{bmatrix} \begin{bmatrix} 0 & & \cdots & 0 & V_1 \\ \vdots & \vdots & \iddots & & \vdots \\ 0 & V_1 & & & \\ V_1 & V_2 & \cdots & & V_p \end{bmatrix} \begin{bmatrix} U_1 & U_2 & \cdots & & U_p \\ 0 & U_1 & & & \vdots \\ \vdots & \vdots & \ddots & \vdots & \vdots \\ 0 & & & 0 & U_1 \end{bmatrix}$$

Multiply out these products and use (10.37) to obtain

$$\tilde{W}_i = \sum_{l=1}^{i} U_{i-l+1}^T \sum_{k=1}^{l} V_k U_{l-k+1}, \qquad i = 1, 2, \ldots, p. \tag{10.39}$$

To complete the proof of (10.34) it is to be shown that $U_1, \ldots, U_p$ can be determined in such a way that $\tilde{W}_1, \ldots, \tilde{W}_p$ are diagonal. This is done by calculating $U_1, \ldots, U_p$ successively.

First let U_1 be any nonsingular matrix for which $U_1^T V_1 U_1 = I$. Since $V_1^T = V_1$ such a U_1 exists by Lemma 10.8.

Proceeding inductively, suppose that matrices $U_1, \ldots, U_{\nu-1}$ have been found for which $\tilde{W}_1, \ldots, \tilde{W}_{\nu-1}$ are diagonal. Then it is deduced from (10.39) that

$$\tilde{W}_\nu = U_\nu^T V_1 U_1 + U_1^T V_1 U_\nu + C \tag{10.40}$$

where $C^{\mathrm{T}} = C$ and depends only on $V_1, \ldots, V_\nu$ and $U_1, \ldots, U_{\nu-1}$. Write $\hat{U}_\nu = U_1^{-1} U_\nu$, then (10.40) takes the form

$$\tilde{W}_\nu = \hat{U}_\nu^{\mathrm{T}}(U_1^{\mathrm{T}} V_1 U_1) + (U_1^{\mathrm{T}} V_1 U_1)\hat{U}_\nu + C.$$

Let $\hat{U}_\nu = [\beta_{ij}]_{i,j=1}^r$, $C = [c_{ij}]_{i,j=1}^r$. Since $U_1^{\mathrm{T}} V_1 U_1 = I$, the matrix $\tilde{W}_\nu$ is diagonal if and only if for $1 \le i \le j \le r$

$$\beta_{ji} + \beta_{ij} = -c_{ij}.$$

There certainly exist pairs β_{ji}, β_{ij} satisfying this equation. This completes the proof of (10.34).

Now each W_i from (10.34) has the form

$$W_i = \begin{bmatrix} 0 & 0 & \cdots & t_1 \\ \vdots & \vdots & \iddots & \vdots \\ 0 & t_1 & \cdots & t_{p-1} \\ t_1 & t_2 & \cdots & t_p \end{bmatrix}, \qquad t_j \in \mathbb{C}, \quad t_1 \neq 0.$$

There exists a matrix S_i of the form

$$S_i = \begin{bmatrix} s_1 & s_2 & \cdots & s_p \\ 0 & s_1 & & s_{p-1} \\ \vdots & \vdots & \ddots & \vdots \\ 0 & 0 & \cdots & s_1 \end{bmatrix}, \qquad s_i \in \mathbb{C}, \quad s_1 \neq 0$$

such that

$$S_i^{\mathrm{T}} W_i S_i = P_1 \tag{10.41}$$

the sip matrix of size $p \times p$. Indeed, (10.41) amounts to the system

$$\sum_{l=1}^{i} s_{i-l+1} \sum_{k=1}^{l} t_k s_{l-k+1} = \begin{cases} 1, & i = 1 \\ 0, & i = 2, \ldots, p. \end{cases}$$

This system can be solved for $s_1, s_2, \ldots, s_p$ successively by using the ith equation to find s_i in terms of $t_1, \ldots, t_i, s_1, \ldots, s_{i-1}$, where $s_1 = t_1^{-1/2}$. Now put $W = \operatorname{diag}[S_1, \ldots, S_r]U$ to prove Lemma 10.9 for the case when $\sigma(K)$ is a singleton and all Jordan blocks of K have the same size.

Consider now the case when $\sigma(K)$ consists of one point but the sizes of Jordan blocks are not all equal. So let

$$m_1 = \cdots = m_{k_1} > m_{k_1+1} = \cdots = m_{k_2} > \cdots > m_{k_{p-1}+1} = \cdots = m_{k_p}.$$

Let

$$V = [V_{ij}]_{i,j=1}^{k_p}, \qquad (V_{ij} \text{ an } m_i \times m_j \text{ matrix}).$$

Then the condition $P_K V \in \mathscr{T}$ means that the V_{ij} have the following structure (cf. Theorem S2.2):

$$V_{ij} = V_{ji} \quad \text{for} \quad i \neq j;$$
$$V_{ij} = [v_{r+s}^{(ij)}]_{r=1,\ldots,m_i}^{s=1,\ldots,m_j} \quad \text{with} \quad v_t^{(ij)} = 0 \quad \text{for} \quad t < \max(m_i, m_j).$$

Nonsingularity of V implies, in particular, that the $k_1 \times k_1$ matrix

$$[v_{m_1+1}^{(ij)}]_{i,j=1}^{k_1}$$

is also nonsingular. Therefore, we can reduce the problem to the case when

$$V = [V_{ij}]_{i,j=1}^{k_p} \qquad \text{and} \qquad V_{ij} = V_{ji} = 0 \qquad \text{for} \quad i = 1, \ldots, m_1, \tag{10.42}$$

by applying sequentially (a finite number of times) transformations of type $V \to E^{\mathrm{T}} V E$, where the $n \times n$ matrix E has the structure

$$E = [E_{ij}]_{i,j=1}^{k_p} \qquad (E_{ij} \text{ an } m_i \times m_j \text{ matrix}),$$

with $E_{ii} = I$ for $i = 1, \ldots, k_p$. Also.

$$E_{ij} = \begin{bmatrix} e_1 & e_2 & \cdots & e_{m_j} \\ 0 & e_1 & \cdots & e_{m_j - 1} \\ \vdots & \vdots & & \vdots \\ 0 & 0 & \cdots & e_1 \\ \vdots & \vdots & & \vdots \\ 0 & 0 & \cdots & 0 \end{bmatrix},$$

where $e_s \in \mathbb{C}$ depends on i and j, for $i = 1, \ldots, m_1$ and $j = m + 1, \ldots, m_{k_p}$, and $E_{ij} = 0$ otherwise. Then in the case (10.42) we apply induction on the number of different sizes of Jordan blocks in K to complete the proof of Lemma 10.9 in the case that $\sigma(K)$ consists of a single point.

Finally, the general case in Lemma 10.9 ($\sigma(K)$ contains more than one point) reduces to the case already considered in view of the description of the matrices from $\mathscr{T}$ (Theorem S2.2). □

The next lemma is an analog of Lemma 10.9 for real matrices. Given the Jordan matrix K and the matrices $P_1, \ldots, P_r$ as above, let

$$P_{\varepsilon, K} = \operatorname{diag}[\varepsilon_1 P_1, \ldots, \varepsilon_r P_r]$$

for a given set of signs $\varepsilon = (\varepsilon_i)_{i=1}^r$, $\varepsilon_i = \pm 1$.

Lemma 10.10. *Let V be a nonsingular real symmetric matrix such that $P_K V \in \mathscr{T}$. Then there exists a real $W \in \mathscr{T}$ such that $W^{\mathrm{T}} V W = P_{\varepsilon, K}$ for some set of signs ε. Conversely, if for some matrix V there exists real $W \in \mathscr{T}$ such that $W^{\mathrm{T}} V W = P_{\varepsilon, K}$, then V is real, invertible, and symmetric and $P_K V \in \mathscr{T}$.*

The proof of Lemma 10.10 is analogous to the proof of Lemma 10.9 (the essential difference is that, in the notation of the proof of Lemma 10.9, we now define U_1 to be a real nonsingular matrix such that

$$U_1^{\mathrm{T}} V_1 U_1 = \operatorname{diag}[\delta_1, \ldots, \delta_r],$$

for some signs $\delta_1, \ldots, \delta_r$. Also, instead of (10.41) we now prove that $S_i^{\mathrm{T}} W_i S_i = \pm P_1$, by letting $s_1 = (\pm t_1)^{-1/2}$ where the sign is chosen to ensure that s_1 is real (here $t_1 \in \mathbb{R}$)).

Proof of Theorem 10.7. We start with a Jordan triple (X, J, Y) of $L(\lambda)$ of the form

$$X = [X_1 \quad X_2 \quad \bar{X}_1], \qquad J = \begin{bmatrix} J_1 & 0 & 0 \\ 0 & J_2 & 0 \\ 0 & 0 & \bar{J}_1 \end{bmatrix}, \qquad Y = \begin{bmatrix} Y_1 \\ Y_2 \\ Y_3 \end{bmatrix}, \tag{10.43}$$

where $\operatorname{Im} \sigma(J_1) > 0$, $\sigma(J_2)$ is real, X_2 is a real matrix, and the partitions of X and Y are consistent with the partition of J. (In particular, $\operatorname{Im} \sigma(\bar{J}_1) < 0$.) The existence of such a Jordan triple was shown in the beginning of this section.

Let $Q = \operatorname{col}(XJ^i)_{i=0}^{l-1}$. Then (see (10.26))

$$U = Q^* B Q = \begin{bmatrix} 0 & 0 & T_1^* \\ 0 & T_2 & 0 \\ T_1 & 0 & 0 \end{bmatrix}, \tag{10.44}$$

where the partition is consistent with the partition in (10.43). Moreover (equality (10.25)),

$$J^* T = TJ. \tag{10.45}$$

It is easy to check (using (10.44) and (10.45)) that T_1 is nonsingular symmetric and $P_c T_1$ commutes with J_1 (where P_c is defined as in (10.18)). By Lemma 10.9 there exists a nonsingular matrix W_1 such that $W_1 J_1 = J_1 W_1$ and $W_1^{\mathrm{T}} T_1 W_1 = P_c$. Analogously, using Lemma 10.10, one shows that there exists a real nonsingular matrix W_2 such that $W_2 J_2 = J_2 W_2$ and $W_2^{\mathrm{T}} T_2 W_2 = P_{\varepsilon, J_2}$ for some set of signs ε. Denote $X_1' = X_1 W_1$ and $X_2' = X_2 W_2$. Then (cf. Theorem 1.25) the triple (X', J, Y') with

$$X' = [X_1', X_2', \bar{X}_1'] \qquad \text{and} \qquad Y' = [\operatorname{col}(X' J^i)_{i=1}^{l-1}]^{-1} \cdot \operatorname{col}(\delta_{i, l-1} I)_{i=0}^{l-1}$$

is a Jordan triple for $L(\lambda)$. Moreover, the equality

$$[\operatorname{col}(X' J^i)_{i=0}^{l-1}]^* B \operatorname{col}(X' J^i)_{i=0}^{l-1} = \begin{bmatrix} 0 & 0 & P_c \\ 0 & P_r & 0 \\ P_c & 0 & 0 \end{bmatrix}$$

holds, which means that the triple (X', J, Y') is self-adjoint (cf. (10.29) and (10.30)). □

We remark that the method employed in the proof of Theorem 10.7 can also be used to construct a self-adjoint triple for any monic self-adjoint matrix polynomial (not necessarily real), thereby proving Theorem 10.6 directly, without reference to Theorem S5.1. See [34f] for details.

The following corollary is a version of Theorem 10.5 for the case of real coefficients A_j.

Corollary 10.11. *Let $L(\lambda)$ be a monic self-adjoint matrix polynomial with real coefficients. Then $L(\lambda)$ admits the representation*

$$L^{-1}(\lambda) = 2\,\mathcal{Re}(X_1(I\lambda - J_1)^{-1}P_1X_1^{\mathrm{T}}) + X_2(I\lambda - J_2)^{-1}P_2X_2^{\mathrm{T}}, \quad (10.46)$$

where J_1 (J_2) are Jordan matrices with $\mathcal{Im}\,\lambda_0 > 0$ $(\mathcal{Im}\,\lambda_0 = 0)$ for every eigenvalue λ_0 of J_1 (J_2), the matrix X_2 is real, and for $i = 1, 2$:

$$P_i = P_i^* = P_i^{\mathrm{T}}, \qquad P_i^2 = I, \qquad P_iJ_i = J_i^{\mathrm{T}}P_i. \quad (10.47)$$

Conversely, if a monic matrix polynomial $L(\lambda)$ admits representation (10.46), *then its coefficients are real and symmetric.*

Proof. The direct part of Corollary 10.11 follows by combining Theorems 10.5 and 10.7. The converse statement is proved by direct verification that

$$L^{-1}(\lambda) = (L^{-1}(\lambda))^{\mathrm{T}} = (L^{-1}(\lambda))^* \qquad \text{for} \quad \lambda \in \mathbb{R},$$

using the equalities (10.47). □

10.5. Sign Characteristic of a Self-Adjoint Matrix Polynomial

Let $L(\lambda)$ be a monic matrix polynomial with self-adjoint coefficients. By Theorem 10.6 there exists a self-adjoint triple of $L(\lambda)$, which includes (see (10.17)) a set of signs ε_{ij}, one for every nonzero partial multiplicity α_{ij}, $j = 1, \ldots, k_i$ corresponding to every real eigenvalue λ_{a+i}, $i = 1, \ldots, b$, of $L(\lambda)$. This set of signs ε_{ij} is the *sign characteristic* of $L(\lambda)$, and it plays a central role in the investigation of self-adjoint matrix polynomials. According to this definition, the sign characteristic of $L(\lambda)$ is just the C_1-sign characteristic of B (cf. Theorem 10.6).

We begin with the following simple property of the sign characteristic.

Proposition 10.12. *Let ε_{ij}, $j = 1, \ldots, k_i$, $i = 1, \ldots, b$ be the sign characteristic of $L(\lambda) = I\lambda^l + \sum_{j=0}^{l-1} A_j\lambda^j$ with the corresponding partial multiplicities α_{ij}, $j = 1, \ldots, k_i$, $i = 1, \ldots, b$. Then*

$$\sum_{i=1}^{b}\sum_{j=1}^{k_i} \tfrac{1}{2}[1 - (-1)^{\alpha_{ij}}]\varepsilon_{ij} = \begin{cases} 0 & \text{if} \quad l \text{ is even} \\ n & \text{if} \quad l \text{ is odd.} \end{cases} \quad (10.48)$$

Proof. Let (X, J, Y) be a self-adjoint triple of $L(\lambda)$, i.e., such that $Y = P_{\varepsilon, J} X^*$, where $\varepsilon = (\varepsilon_{ij})$, $j = 1, \ldots, k_i$, $i = 1, \ldots, b$. By Theorem 10.6

$$\operatorname{sig} B = \operatorname{sig} P_{\varepsilon, J}. \tag{10.49}$$

Note that the signature of a sip matrix is 1 or 0 depending if its size is odd or even. Using this fact one deduces easily that $\operatorname{sig} P_{\varepsilon, J}$ is just the left-hand side of (10.48). Apply Theorem 10.3 to the left-hand side of (10.49) and obtain (10.48). □

In particular, Proposition 10.12 ensures that $\sum_{i=1}^{b} \sum_{j=1}^{k_i} \frac{1}{2}[1 - (-1)^{\alpha_{ij}}]\varepsilon_{ij}$ does not depend on the choice of a self-adjoint triple (X, J, Y) of $L(\lambda)$. The next result shows that in fact the signs ε_{ij} themselves do not depend on (X, J, Y), and expresses a characteristic property of the self-adjoint matrix polynomial $L(\lambda)$ (justifying the term "sign characteristic of $L(\lambda)$").

Theorem 10.13. *The sign characteristic of a monic self-adjoint matrix polynomial $L(\lambda)$ does not depend on the choice of its self-adjoint triple (X, J, Y).*

The proof is immediate: apply Theorem S5.6 and the fact that C_1 is B-self-adjoint (cf. Theorem 10.6). □

The rest of this section will be devoted to the computation of the sign characteristic and some examples.

Given a self-adjoint triple $(X, J. Y)$ of $L(\lambda)$ and given an eigenvector x from X corresponding to a real eigenvalue λ_0, denote by $v(x, \lambda_0)$ the length of the Jordan chain from X beginning with x, and by $\operatorname{sgn}(x, \lambda_0)$—the sign of the corresponding Jordan block of J in the sign characteristic of $L(\lambda)$.

Theorem 10.14. *Let (X, J, Y) be a self-adjoint triple of $L(\lambda) = I\lambda^l + \sum_{j=0}^{l-1} A_j \lambda^j$. Let x_1 and $y_1, \ldots, y_r$ be eigenvector and first vectors of a Jordan chain, respectively, from X corresponding to the same real eigenvalue λ_0. Then*

$$\left(x_1, \sum_{j=0}^{r} \frac{1}{j!} L^{(j)}(\lambda_0) y_{r+1-j}\right)$$

$$= \begin{cases} 0, & \textit{if} \quad y_1 \neq x_1 \quad \textit{or} \quad y_1 = x_1 \quad \textit{and} \quad r < v(x_1, \lambda_0) \\ \operatorname{sgn}(y_1, \lambda_0), & \textit{if} \quad y_1 = x_1 \quad \textit{and} \quad r = v(x_1, \lambda_0). \end{cases} \tag{10.50}$$

Proof. Let $Q = \operatorname{col}(XJ^i)_{i=0}^{l-1}$. Denote by $\hat{x}_1$ and $\hat{y}_1, \ldots, \hat{y}_r$ the columns in Q corresponding to the columns x_1 and $y_1, \ldots, y_r$, respectively, in X. We shall prove first that

$$\left(x_1, \sum_{j=1}^{r} \frac{1}{j!} L^{(j)}(\lambda_0) y_{r+1-j}\right) = (\hat{x}_1, B\hat{y}_r), \tag{10.51}$$

where B is given by (10.3).

The proof is a combinatorial one. Substituting $\hat{x}_1 = \operatorname{col}[\lambda_0^i x_1]_{i=0}^{l-1}$ and $\hat{y}_r = \operatorname{col}[\sum_{i=0}^{j} \binom{j}{i} \lambda_0^{j-i} y_{r-i}]_{j=0}^{l-1}$ (by definition $y_p = 0$ for $p \le 0$) in the expression $(\hat{x}_1, B\hat{y}_r)$ we deduce that

$$(\hat{x}_1, B\hat{y}_r) = \left(x_1, \sum_{i=0}^{l-1} \left(\sum_{k=0}^{l} \sum_{p=i+k}^{l} \binom{p-k}{i} \lambda_0^{p-i-1} A_p\right) y_{r-i}\right).$$

It is easy to see that

$$\sum_{k=1}^{l} \sum_{p=i+k}^{l} \binom{p-k}{i} \lambda_0^{p-i-1} A_p = \frac{1}{(i+1)!} L^{(i+1)}(\lambda_0),$$

and (10.51) follows. Now use (10.29) to obtain (10.50) from (10.51). □

Theorem 10.14 allows us to compute the sign characteristic of $L(\lambda)$ in simple cases. Let us consider some examples.

EXAMPLE 10.5. Let $L(\lambda)$ be a monic scalar polynomial with real coefficients and r different simple real roots $\lambda_1 > \cdots > \lambda_r$. It is easy to check (using Theorem 10.14) that the sign of λ_j in the sign characteristic of $L(\lambda)$ is $(-1)^{j+1}$. □

EXAMPLE 10.6. Let $L(\lambda)$ be a monic self-adjoint matrix polynomial with degree l. Let λ_1 (resp. λ_r) be the maximal (resp. minimal) real eigenvalue of $L(\lambda)$. Then $\operatorname{sgn}(x, \lambda_1) = 1$ for any eigenvector x corresponding to λ_1 with $\nu(x, \lambda_1) = 1$; and for any eigenvector y corresponding to λ_r with $\nu(y, \lambda_r) = 1$ we have $\operatorname{sgn}(y, \lambda_r) = (-1)^{l+1}$. Indeed, consider the real-valued function $f_x(\lambda) = (x, L(\lambda)x)$. Then $f_x(\lambda) > 0$ for $\lambda > \lambda_1$, because the leading coefficient of $L(\lambda)$ is positive definite. On the other hand, $f_x(\lambda_1) = 0$ (for x is an eigenvector of $L(\lambda)$ corresponding to λ_1). So $f'_x(\lambda_1) \ge 0$. But in view of Theorem 10.14, $\operatorname{sgn}(x, \lambda_1)$ is just $f'_x(\lambda_1)$, maybe multiplied by a positive constant. Hence $\operatorname{sgn}(x, \lambda_1) = 1$. The same argument is applicable to the eigenvector y of λ_r. □

These examples are extremal in the sense that the signs can be determined there using only the eigenvalues with their partial multiplicities, without knowing anything more about the matrix polynomial. The next example shows a different situation.

EXAMPLE 10.7. Let $a < b < c < d$ be real numbers, and let

$$L(\lambda) = \begin{bmatrix} (\lambda - a)(\lambda - b) & 0 \\ 0 & (\lambda - c)(\lambda - d) \end{bmatrix}.$$

Then the sign characteristic of the eigenvalues a, b, c, d is $-1, 1, -1, 1$, respectively (as follows from Example 10.5). Interchanging in $L(\lambda)$ the places of $\lambda - b$ and $\lambda - c$, we obtain a new self-adjoint matrix polynomial with the same eigenvalues and partial multiplicities, but the sign characteristic is different: $-1, -1, 1, 1$ for a, b, c, d, respectively. □

10.6. Numerical Range and Eigenvalues

For a matrix polynomial $M(\lambda)$ define the numerical range $NR(M)$ as

$$NR(M) = \{\lambda \in \mathbb{C} \mid (M(\lambda)x, x) = 0 \text{ for some } x \in \mathbb{C}^n \backslash \{0\}\} \qquad (10.52)$$

This definition is a natural generalization of the notion of numerical range for a simple matrix A, which is defined as $N_A = \{(Ax, x) \mid x \in \mathbb{C}^n, \|x\| = 1\}$. Indeed,

$$\{(Ax, x) \mid x \in \mathbb{C}^n, \|x\| = 1\} = NR(I\lambda - A).$$

It is well known that for any matrix A the set N_A is convex and compact. In general, the set $NR(M)$ is neither convex nor compact. For example, for the scalar polynomial $M(\lambda) = \lambda(\lambda - 1)$ we have $NR(M) = \{0, 1\}$, which is obviously nonconvex; for the 2×2 matrix polynomial

$$M(\lambda) = \begin{bmatrix} 0 & 1 \\ 1 & \lambda \end{bmatrix}$$

we have $NR(M) = \mathbb{R}$, which is noncompact. It is true that $NR(M)$ is closed (this fact can be checked by a standard argument observing that

$$NR(M) = \{\lambda \in \mathbb{C} \mid (M(\lambda)x, x) = 0 \text{ for some } x \in \mathbb{C}^n, \|x\| = 1\}$$

and using the compactness of the unit sphere in $\mathbb{C}^n$). Moreover, if $M(\lambda)$ is monic, then $NR(M)$ is compact. Indeed, write $M(\lambda) = I\lambda^m + \sum_{j=0}^{m-1} M_j \lambda^j$. Let $\alpha = \max\{m \cdot \max_{0 \le i \le m-1}\{|\lambda| \mid \lambda \in N_{M_i}\}, 1\}$. Clearly, α is a finite number, and for $|\lambda| > \alpha$ and $x \in \mathbb{C}^n$, $\|x\| = 1$ we have

$$\begin{aligned} |\lambda^m(x, x)| = |\lambda|^m &> m \max_{0 \le i \le m-1} \{|\lambda| \mid \lambda \in N_{M_i}\} |\lambda|^{m-1} \\ &\ge \sum_{j=0}^{m-1} |(M_j x, x)| |\lambda|^j \ge \left| \sum_{j=0}^{m-1} (\lambda^j M_j x, x) \right|, \end{aligned}$$

so $\lambda \notin NR(M)$.

We shall assume in the rest of this section that $M(\lambda) = L(\lambda)$ is a monic self-adjoint matrix polynomial, and study the set $NR(L)$ and its relationship with $\sigma(L)$. We have seen above that $NR(L)$ is a compact set; moreover, since $L(\lambda)$ is self-adjoint the coefficients of the scalar polynomial $(L(\lambda)x, x)$, $x \in \mathbb{C}^n \backslash \{0\}$ are real, and therefore $NR(L)$ is symmetric relative to the real line: if $\lambda_0 \in NR(L)$, then also $\bar{\lambda}_0 \in NR(L)$. Evidently, $NR(L) \supset \sigma(L)$.

The following result exhibits a close relationship between the eigenvalues of $L(\lambda)$ and the numerical range $NR(L)$, namely, that every real frontier (or boundary) point of $NR(L)$ is an eigenvalue of $L(\lambda)$.

Theorem 10.15. *Let $L(\lambda)$ be a monic self-adjoint matrix polynomial, and let $\lambda_0 \in NR(L) \cap \overline{(\mathbb{R} \backslash NR(L))}$. Then λ_0 is an eigenvalue of $L(\lambda)$.*

Proof. By assumption, there exists a sequence of real numbers $\lambda_1, \lambda_2, \ldots$ such that $\lim_{m\to\infty} \lambda_m = \lambda_0$ and $\lambda_i \notin NR(L)$. This means that the equality $(L(\lambda_i)x, x) = 0$ holds only for $x = 0$; in particular, either $(L(\lambda_i)x, x) \geq 0$ for all $x \in \mathbb{C}^n$, or $(L(\lambda_i)x, x) \leq 0$ for all $x \in \mathbb{C}^n$. Suppose, for instance, that the former case holds for infinitely many λ_i. Passing to the limit along the (infinite) set of such λ_i, we find that $(L(\lambda_0)x, x) \geq 0$ for all $x \in \mathbb{C}^n$. Since $\lambda_0 \in NR(L)$, we also have for some nonzero $x_0 \in \mathbb{C}^n$, $(L(\lambda_0)x_0, x_0) = 0$. It follows that the hermitian matrix $L(\lambda_0)$ is singular, i.e. $\lambda_0 \in \sigma(L)$. □

Since the number of different real eigenvalues of $L(\lambda)$ is at most nl, where l is the degree on $L(\lambda)$, the following corollary is an immediate consequence of Theorem 10.15.

Corollary 10.16. *Let $L(\lambda)$ be as in Theorem 10.15. Then the set $NR(L) \cap \mathbb{R}$ is a finite union of disjoint closed intervals and separate points:*

$$NR(L) \cap \mathbb{R} = \left(\bigcup_{i=1}^{k} [\mu_{2i-1}, \mu_{2i}]\right) \cup \left(\bigcup_{j=1}^{m} \{\nu_j\}\right), \tag{10.53}$$

where $\mu_1 < \mu_2 < \cdots < \mu_{2k}$ and $\nu_j \notin \bigcup_{i=1}^{k} [\mu_{2i-1}, \mu_{2i}]$, $\nu_{j_1} \neq \nu_{j_2}$ for $j_1 \neq j_2$. The number k of intervals and the number m of single points in (10.53) satisfy the inequality

$$2k + m \leq nl,$$

where l is the degree of $L(\lambda)$.

We point out one more corollary of Theorem 10.15.

Corollary 10.17. *A monic self-adjoint matrix polynomial $L(\lambda)$ has no real eigenvalues if and only if $NR(L) \cap \mathbb{R} = \varnothing$.*

Comments

Self-adjoint matrix polynomials of second degree appear in the theory of damped oscillatory systems with a finite number of degrees of freedom ([17, 52b]). Consideration of damped oscillatory systems with an infinite number of degrees of freedom leads naturally to self-adjoint operator polynomials (acting in infinite dimensional Hilbert space), see [51]. Self-adjoint operator polynomials (possibly with unbounded coefficients) appear also in the solution of certain partial differential equations by the Fourier method.

The results of this chapter, with the exception of Sections 10.4 and 10.6, are developed in the authors' papers [34d, 34f, 34g]. The crucial fact for the theory, that C_1 is B-self-adjoint (in the notation of Section 10.1), has been used elsewhere (see [51, 56a, 56d]). The results of Section 10.4 are new. The numerical range for operator polynomials was introduced (implicitly) in [67] and further developed in [56a]. A more general treatment appears in [44].

Chapter 11

Factorization of Self-Adjoint Matrix Polynomials

The general results on divisibility of monic matrix polynomials introduced in Part I, together with the spectral theory for self-adjoint polynomials as developed in the preceding chapter, can now be combined to prove specific theorems on the factorization of self-adjoint matrix polynomials. Chapter 11 is devoted to results of this type.

11.1. Symmetric Factorization

We begin with symmetric factorizations of the form $L = L_1^* L_3 L_1$, where $L_3^* = L_3$. Although such factorizations may appear to be of a rather special kind, it will be seen subsequently that the associated geometric properties described in the next theorem play an important role in the study of more general, nonsymmetric factorizations. As usual, C_1 stands for the first companion matrix of L, and B is defined by (10.3).

Theorem 11.1. *Let L be a monic self-adjoint polynomial of degree l with a monic right divisor L_1 of degree $k \le l/2$. Then L has a factorization $L = L_1^* L_3 L_1$ for some monic matrix polynomial L_3 if and only if*

$$(Bx, y) = 0 \qquad \text{for all} \quad x, y \in \mathcal{M}, \tag{11.1}$$

where $\mathcal{M}$ is a supporting subspace for L_1 (with respect to C_1).

We shall say that *any* subspace $\mathscr{M}$ for which (11.1) holds is *neutral with respect to B*, or is *B-neutral.* To illustrate the existence of B-neutral subspaces consider a monic right divisor L_1 for which $\sigma(L_1)$ contains no pair of complex conjugate points. If $L = L_2 L_1$, then $\sigma(L) = \sigma(L_1) \cup \sigma(L_2)$ and $\sigma(L_1) \cap \sigma(L_1^*) = \varnothing$. If L is self-adjoint, then $L = L_2 L_1 = L_1^* L_2^*$ and it follows that L_1 is a right divisor of L_2^*. Hence there is a matrix polynomial L_3 for which $L_1^* L_3 L_1 = L$. Consequently (11.1) holds.

Proof. Let $L = L_2 L_1 = L_1^* L_2^*$, and L_1 have supporting subspace $\mathscr{M}$ with respect to C_1. It follows from Theorem 3.20 that L_2^* has supporting subspace $\mathscr{M}^\perp$ with respect to C_1^*. But now, by Theorem 10.1 $C_1^* = BC_1B^{-1}$. It follows that the supporting subspace $\mathscr{N}$ for L_2^* with respect to C_1 is given by $\mathscr{N} = B^{-1}\mathscr{M}^\perp$.

Now observe that $k \leq l/2$ implies that $L = L_1^* L_3 L_1$ if and only if L_1 divides L_2^*. Thus, using Corollary 3.15, such a factorization exists if and only if $\mathscr{M} \subset \mathscr{N} = B^{-1}\mathscr{M}^\perp$, i.e., if and only if (11.1) is satisfied. □

Theorem 11.1 suggests that B-neutral subspaces play an important role in the factorization problems of self-adjoint operator polynomials. Formulas of Section 10.3 provide further examples of B-neutral subspaces. For example, the top left zero of the partitioned matrix in (10.26) demonstrates immediately that the generalized eigenspaces corresponding to the *non-real* eigenvalues $\lambda_1, \ldots, \lambda_a$ determine a B-neutral subspace. These eigenspaces are therefore natural candidates for the construction of supporting subspaces for right divisors in a symmetric factorization of L. In a natural extension of the idea of a B-neutral subspace, a subspace $\mathscr{S}$ of $\mathbb{C}^{nl}$ is said to be *B-nonnegative* (*B-nonpositive*) if $(Bx, x) \geq 0$ (≤ 0) for all $x \in \mathscr{S}$.

11.2. Main Theorem

For the construction of B-neutral subspaces (as used in Theorem 11.1) we introduce sets of nonreal eigenvalues, S, with the property that $S \cap \bar{S} = \varnothing$, i.e., such a set contains no real eigenvalues and no conjugate pairs. Call such a subset of $\sigma(L)$ a *c-set.* The main factorization theorem can then be stated as follows (the symbol $[a]$ denotes the greatest integer not exceeding a):

Theorem 11.2. *Let L be a monic matrix polynomial of degree l with $L = L^*$. Then*

(a) *There exists a B-nonnegative (B-nonpositive) invariant subspace $\mathscr{M}_1(\mathscr{M}_2)$ of C_1 such that, if $k = [\frac{1}{2}l]$,*

$$\dim \mathscr{M}_1 = \begin{cases} kn & \text{if } l \text{ is even} \\ (k+1)n & \text{if } l \text{ is odd,} \end{cases}$$

$$\dim \mathscr{M}_2 = kn$$

(b) *L has a monic right divisor $L_1(L_2)$ with supporting subspace $\mathcal{M}_1$ ($\mathcal{M}_2$) with respect to C_1 such that*

$$\deg L_1 = \begin{cases} k & \text{if} \quad l \text{ is even} \\ k+1 & \text{if} \quad l \text{ is odd,} \end{cases}$$

$$\deg L_2 = k.$$

In either case the nonreal spectrum of L_1 and of L_2 coincides with any maximal c-set of $\sigma(L)$ chosen in advance.

In order to build up the subspaces referred to in part (a) the nonreal and real parts of $\sigma(L)$ are first considered separately, and in this order.

Theorem 10.6 implies the existence of a self-adjoint triple (X, J, Y) for L, with $Y = P_{\varepsilon, J} X^*$, and the columns of X determine Jordan chains for L. Let $x_1, \ldots, x_p$ be such a Jordan chain associated with the Jordan block J_0 of J. A set of vectors $x_1, \ldots, x_r$, $r \leq p$, will be described as a set of *leading vectors* of the chain, and there is an associated $r \times r$ Jordan block which is a leading submatrix of J_0.

Now let Ξ_ε be a submatrix of X whose columns are made up of leading vectors of Jordan chains associated with a c-set. Then let J_c be the corresponding Jordan matrix (a submatrix of J) and write

$$Q_c = \operatorname{col}[\Xi_c J_c^i]_{i=0}^{l-1}. \tag{11.2}$$

The following lemma shows that $\operatorname{Im} Q_c$ is a B-neutral subspace. Maximal subspaces for L of one sign will subsequently be constructed by extending $\operatorname{Im} Q_c$.

Lemma 11.3. *With Q_c as defined in* (11.2), $Q_c^* B Q_c = 0$.

Proof. Bearing in mind the construction of the Jordan matrix J in (10.12) and in particular, the arrangement of real and nonreal eigenvalues, it can be seen that $Q_c^* B C_c$ is just a submatrix of the top-left entry in the partitioned matrix of (10.26). □

Consider now the real eigenvalues of L. Suppose that the Jordan block J_1 of J is associated with a real eigenvalue λ_0 and that J_1 is $p \times p$. Then λ_0 has an associated Jordan chain $x_1, \ldots, x_p$ with $x_1 \neq 0$, and these vectors form the columns of the $n \times p$ matrix X_1. There is a corresponding chain $\hat{x}_1, \ldots, \hat{x}_p$ associated with the eigenvalue λ_0 of C_1 and

$$Q_1 = [\hat{x}_1 \quad \cdots \quad \hat{x}_p] = \operatorname{col}[X_1 J_1^i]_{i=0}^{l-1}.$$

The definition of a self-adjoint triple shows that $x_1, \ldots, x_p$ can be chosen so that

$$Q_1^* B Q_1 = \pm P \tag{11.3}$$

where P is the $p \times p$ sip matrix. In particular, observe that $(\hat{x}_1, B\hat{x}_p) = \pm 1$. We say that *the pair* (λ_0, x_1), $\lambda_0 \in \mathbb{R}$, *is of the first (second) kind* according as $(\hat{x}_1, B\hat{x}_p) = 1$ or -1, respectively. Then let $\mathscr{S}_1$ and $\mathscr{S}_2$ be the sets of all pairs of the first and second kinds, respectively.

It follows from Theorem 10.14 that

$$(\hat{x}_1, B\hat{x}_1) = (x_1, L^{(1)}(\lambda_0)x_1) = \pm 1 \quad \text{(resp. 0)} \tag{11.4}$$

when the chain has length one (resp. more than one), i.e., when the associated elementary divisor of L is linear (resp. nonlinear). We shall describe $\hat{x}_1$ as either B-positive, B-negative, or B-neutral, with the obvious meanings.

Suppose that the pair (λ_0, x_1), $\lambda_0 \in \mathbb{R}$, determines a Jordan chain of length k, let $r = [\frac{1}{2}(k+1)]$, and consider the first r leading vectors of the chain. It is apparent from (11.3) that

$$\begin{bmatrix} \hat{x}_1^* \\ \vdots \\ \hat{x}_r^* \end{bmatrix} B[\hat{x}_1 \quad \cdots \quad \hat{x}_r] = \begin{cases} 0 & \text{for } k \text{ even} \\ \pm E & \text{for } k \text{ odd} \end{cases} \tag{11.5}$$

where $E = \operatorname{diag}[0, \ldots, 0, 1]$ and is $r \times r$.

This relation can now be used to construct subspaces of vectors which are B-nonnegative, B-nonpositive, or B-neutral, *and* are invariant under C_1. Thus, for k odd, Span$\{\hat{x}, \ldots, \hat{x}_r\}$ is B-nonnegative or B-nonpositive according as (λ_0, x_1) is a pair of the first or second kind. Furthermore, if $k > 1$, Span$\{\hat{x}_1, \ldots, \hat{x}_{r-1}\}$ is a B-neutral subspace. If k is even, then Span$\{\hat{x}_1, \ldots, \hat{x}_r\}$, $r = \frac{1}{2}k$, is B-neutral for (λ_0, x_1) of either the first or second kind.

These observations are now used to build up maximal B-nonnegative and B-nonpositive invariant subspaces for C_1 (associated with real eigenvalues only in this instance). First, basis vectors for a B-nonnegative invariant subspace $\mathscr{L}_1$ are selected as follows (here $[\hat{x}_{1,i} \quad \cdots \quad \hat{x}_{k_i,i}] = \operatorname{col}(X_i J_i^m)_{m=0}^{l-1}$, where J_i is a Jordan block of J with a real eigenvalue λ_i, and X_i is the corresponding part of X):

(a) If $(\lambda_i, x_{1,i}) \in \mathscr{S}_1$, select $\hat{x}_{1,i}, \ldots, \hat{x}_{r_i,i}$ where $r_i = [\frac{1}{2}(k_i + 1)]$.
(b) If $(\lambda_i, x_{1,i}) \in \mathscr{S}_2$ and k_i is odd, select $\hat{x}_{1,i}, \ldots, \hat{x}_{r_i-1,i}$.
(c) If $(\lambda_i, x_{1,i}) \in \mathscr{S}_2$ and k_i is even, select $\hat{x}_{1,i}, \ldots, \hat{x}_{r_i,i}$.

Second, basis vectors for a B-nonpositive invariant subspace $\mathscr{L}_2$ are selected as follows:

(a) If $(\lambda_i, x_{1,i}) \in \mathscr{S}_2$, select $\hat{x}_{1,i}, \ldots, \hat{x}_{r_i,i}$.
(b) If $(\lambda_i, x_{1,i}) \in \mathscr{S}_1$ and k_i is odd, select $\hat{x}_{1,i}, \ldots, \hat{x}_{r_i-1,i}$.
(c) If $(\lambda_i, x_{1,i}) \in \mathscr{S}_1$ and k_i is even, select $\hat{x}_{1,i}, \ldots, \hat{x}_{r_i,i}$.

Then it is easily verified that $\dim \mathscr{L}_1 + \dim \mathscr{L}_2 = \rho$, the total number of *real* eigenvalues (with multiplicities). It is also clear that $\mathscr{L}_1 \cap \mathscr{L}_2 = \{0\}$ if and only if all real eigenvalues have only linear elementary divisors.

Now let $\mathscr{L}$ represent either of the subspaces $\mathscr{L}_1$ or $\mathscr{L}_2$, and form a matrix Q_R whose columns are the basis vectors for $\mathscr{L}$ defined above. Then Q_R has full rank, $\operatorname{Im} Q_R = \mathscr{L}$, and Q_R has the form

$$Q_R = \operatorname{col}[\Xi_R J_R^i]_{i=0}^{l-1} \tag{11.6}$$

where Ξ_R is a matrix whose columns are Jordan chains (possibly truncated) for L, and J_R in the associated Jordan matrix.

It follows from these definitions and (11.5) that

$$Q_R^* B Q_R = \pm D \tag{11.7}$$

where D is a diagonal matrix of zeros and ones, each one being associated with an elementary divisor of odd degree, and the plus or minus sign is selected according as $\mathscr{L} = \mathscr{L}_1$ or $\mathscr{L} = \mathscr{L}_2$.

11.3. Proof of the Main Theorem

In general, L will have both real and nonreal eigenvalues. It has been demonstrated in the above construction that certain invariant subspaces $\mathscr{L}_1$ and $\mathscr{L}_2$ of C_1 associated with real eigenvalues can be constructed which are B-nonnegative and B-nonpositive, respectively.

However, in the discussion of nonreal eigenvalues, B-neutral invariant subspaces of C_1 were constructed (as in Lemma 11.3) using c-sets of nonreal eigenvalues. Consider now a *maximal* c-set and the corresponding matrix of Jordan chains for C_1, Q_c of Eq. (11.2). Then $\operatorname{Im} Q_c$ is B-neutral and the direct sums $\mathscr{M}_1 = \mathscr{L}_1 \dotplus \operatorname{Im} Q_c$, $\mathscr{M}_2 = \mathscr{L}_2 \dotplus \operatorname{Im} Q_c$ will generally be larger B-nonnegative, B-nonpositive invariant subspaces, respectively. Indeed, $\max(\dim \mathscr{M}_1, \dim \mathscr{M}_2) \geq kn$, where $k = [\frac{1}{2}l]$.

Construct composite matrices

$$\Xi = [\Xi_R \quad \Xi_c], \qquad K = J_R \oplus J_c, \qquad S_r = \operatorname{col}[\Xi K^j]_{j=0}^{r-1}, \tag{11.8}$$

the S_r being defined for $r = 1, 2, \ldots, l$. Then S_l is a submatrix of Q and consequently has full rank. Furthermore, the subspace

$$\mathscr{M} = \operatorname{Im} S_l = \operatorname{Im} Q_c \dotplus \operatorname{Im} Q_R = \operatorname{Im} Q_c \dotplus \mathscr{L}$$

is an invariant subspace of C_1 which is B-nonnegative or B-nonpositive according as $\operatorname{Im} Q_R = \mathscr{L}_1$ or $\operatorname{Im} Q_R = \mathscr{L}_2$. In order to prove the existence of a right divisor of L of degree $k = [\frac{1}{2}l]$ we use Theorem 3.12 and are to show that the construction of Ξ implies the invertibility of S_k. In particular, it will be shown that Ξ is an $n \times kn$ matrix, i.e., $\dim \mathscr{M} = kn$.

Our construction of Ξ ensures that S_l is a submatrix of $Q = \text{col}(XJ^i)_{i=0}^{l-1}$ and that

$$S_l^* B S_l = [\Xi^* \quad K^*\Xi^* \quad \cdots \quad K^{*l-1}\Xi^*]\begin{bmatrix} A_1 & \cdots & I \\ \vdots & & \vdots \\ I & \cdots & 0 \end{bmatrix}\begin{bmatrix} \Xi \\ \vdots \\ \Xi K^{l-1} \end{bmatrix} = \pm \hat{D} \tag{11.9}$$

where $\hat{D} = D \oplus 0$ is a diagonal matrix of zeros and ones and the plus or minus sign applies according as $\mathscr{L} = \mathscr{L}_1$ or $\mathscr{L} = \mathscr{L}_2$.

Let $x \in \text{Ker } S_k = \bigcap_{r=0}^{k-1} \text{Ker}(\Xi K^r)$ and make the unique decomposition $x = x_1 + x_2$ where $x_1 \in \text{Im } \hat{D}$, $x_2 \in \text{Ker } \hat{D}$ and, for brevity, write $y_r = \Xi K^r x$, $r = 0, 1, 2, \ldots$. Use (11.9) to write $x^* S_l^* B S_l x$ as follows:

$$\pm x_1^* \hat{D} x_1 = [0 \quad \cdots \quad 0 \quad y_k^* \quad \cdots \quad y_{l-1}^*] \times$$

$$\left[\begin{array}{ccc|ccc} A_1 & \cdots & A_k & A_{k+1} & \cdots & I \\ \vdots & & \vdots & \vdots & \iddots & \vdots \\ & & & I & & 0 \\ \hline & & I & & & \\ \vdots & \iddots & \vdots & \vdots & & \vdots \\ I & \cdots & 0 & & & 0 \end{array}\right]\begin{bmatrix} 0 \\ \vdots \\ y_k \\ \\ \vdots \\ y_{l-1} \end{bmatrix}$$

$$= \begin{cases} 0 & \text{when } l \text{ is even} \\ y_k^* y_k & \text{when } l \text{ is odd.} \end{cases} \tag{11.10}$$

Case (i), *l even.* Equation (11.10) implies $x_1 = 0$ so that $x = x_2 \in \text{Ker } \hat{D}$. (Note that if $\hat{D} = 0$ this statement is trivially true.) Consider the image of x under $S_l^* B S_l$ using (11.9), and it is found that

$$[\Xi^* \quad K^*\Xi^* \quad \cdots \quad K^{*k-1}\Xi^*]\begin{bmatrix} A_{k+1} & \cdots & I \\ \vdots & \iddots & \vdots \\ I & \cdots & 0 \end{bmatrix}\begin{bmatrix} y_k \\ \vdots \\ y_{l-1} \end{bmatrix} = 0. \tag{11.11}$$

Taking the scalar product of this vector with Kx it is deduced that $y_k^* y_k = 0$ whence $y_k = 0$ and $x \in \text{Ker}(\Xi K^k)$. Then take the scalar product of (11.11) with $K^3 x$ to find $y_{k+1}^* y_{k+1} = 0$ whence $y_{k+1} = 0$ and $x \in \text{Ker}(\Xi K^{k+1})$. Proceeding in this way it is found that $x \in \bigcap_{r=0}^{l-1} \text{Ker}(\Xi K^k) = \text{Ker } S_l$. Hence $x = 0$ so that $\text{Ker } S_k = \{0\}$. This implies that the number of columns in S_k (and hence in S_l) does not exceed kn, and hence that $\dim \mathscr{M} \le kn$. But this argument applies equally when $\mathscr{M} = \mathscr{M}_1$, or $\mathscr{M} = \mathscr{M}_2$ ($\mathscr{M}_i = \mathscr{L}_i \dotplus \text{Im } Q_c$), and we have $\dim \mathscr{M}_1 + \dim \mathscr{M}_2 = ln = 2kn$, thus it follows that $\dim \mathscr{M}_1 = \dim \mathscr{M}_2 = kn$.

Furthermore, S_k must be $kn \times kn$ and hence nonsingular (whether $\mathcal{M} = \mathcal{M}_1$ or $\mathcal{M} = \mathcal{M}_2$ is chosen). Since $[I_k \quad 0]|_{\mathcal{M}}$ has the representation S_k it follows that there is a right monic divisor L_1 of L of degree k. Then

$$L = L_2 L_1 = L_1^* L_2^*,$$

and λ is an eigenvalue of L_1 if and only if $\lambda \in \sigma(C_1|_{\mathcal{M}})$, which is the union of (any) maximal c-set of nonreal eigenvalues with certain real eigenvalues, as described in Section 11.2 (in defining the invariant subspaces $\mathcal{L}_1$, $\mathcal{L}_2$ of C_1).

Case (*ii*), *l odd.* (a) Consider $\mathcal{M}_2 = \mathcal{L}_2 \dotplus \operatorname{Im} Q_c$ which, by construction, is B-nonpositive so that the negative sign applies in (11.10). There is an immediate contradiction unless $x_1 = 0$ and $y_k = 0$, i.e., $x \in \operatorname{Ker}(\Xi K^k) \cap \operatorname{Ker} \hat{D}$. Now it is deduced from (11.9) that

$$[\Xi^* \quad \cdots \quad K^{*k-1}\Xi^*]\begin{bmatrix} A_{k+2} & & \cdots & I \\ & & \cdot^{\cdot^{\cdot}} & 0 \\ \vdots & \vdots & & \vdots \\ I & 0 & \cdots & 0 \end{bmatrix}\begin{bmatrix} y_{k+1} \\ \vdots \\ y_{l+1} \end{bmatrix} = 0. \tag{11.12}$$

Taking scalar products with $K^2x, K^4x, \ldots$, successively, it is found (as in case (i)) that $x \in \bigcap_{r=0}^{l-1} \operatorname{Ker}(\Xi K^r) = \operatorname{Ker} S_l$, and hence $x = 0$. It follows that $\dim \mathcal{M}_2 \le kn$.

(b) Since $\dim \mathcal{M}_1 + \dim \mathcal{M}_2 = ln$, we now have $\dim \mathcal{M}_1 \ge (k+1)n$. Consider the implications of (11.10) for $\mathcal{M}_1$ (when the plus sign obtains). Let $x \in \operatorname{Ker} S_{k+1} = \bigcap_{r=0}^{k} \operatorname{Ker}(\Xi K^r)$ and write $x = x_1 + x_2$, $x_1 \in \operatorname{Im} \hat{D}$, $x_2 \in \operatorname{Ker} \hat{D}$ and (11.10) reduces to

$$x_1^* \hat{D} x_1 = [0 \quad \cdots \quad 0 \quad y_{k+1}^* \quad \cdots \quad y_{l-1}^*]B\begin{bmatrix} 0 \\ \vdots \\ y_{k+1} \\ \vdots \\ y_{l-1} \end{bmatrix} = 0,$$

whence $x_1 = 0$ and $x = x_2 \in \operatorname{Ker} \hat{D}$. As in paragraph (a) above it is found that Eq. (11.12) now applies to $\mathcal{M}_1$. Consequently, as in case (a), $x \in \operatorname{Ker} S_l$, whence $x = 0$.

Thus, $\dim \mathcal{M}_1 \le (k+1)n$ and, combining with the conclusion $\dim \mathcal{M}_2 \le kn$, of part (a), we have

$$\dim \mathcal{M}_1 = (k+1)n, \qquad \dim \mathcal{M}_2 = kn$$

and S_{k+1} (defined by $\mathcal{M}_1$) and S_k (defined by $\mathcal{M}_2$) are nonsingular. Since $[I_{k+1}, 0]|_{\mathcal{M}_1} = S_{k+1}$, $[I_k, 0]|_{\mathcal{M}_2} = S_k$, it follows that $\mathcal{M}_1$, $\mathcal{M}_2$ determine monic right divisors of L of degrees $k+1, k$, respectively. This completes the proof of Theorem 11.2. $\square$

11.4. Discussion and Further Deductions

Let us summarize the elementary divisor structure of a matrix polynomial L and the factors described in Theorem 11.2. If L has the factorization $L = Q_1 L_1$ we tabulate (see Table I) the degrees of elementary divisors $(\lambda - \lambda_0)^j$ of Q_1, L_1 resulting from the presence of such a divisor for L. Consider the case in which L has supporting subspace $\mathcal{M}_1$ and an associated c-set of nonreal eigenvalues. We assume that $\mathcal{M}_1$ is constructed as in the proof of Theorem 11.2 and that S is the maximal c-set chosen in advance.

The presence of the entries in the "divisor of L_1" column is fully explained by our constructions. The first two entries in the "divisor of Q_1" column are evident. Let us check the correctness of the last three entries in the last column. To this end, recall that in view of Theorem 3.20, $Q_1(\lambda)$ is a right divisor of $L^* = L$ with supporting subspace $\mathcal{M}_1^\perp$ with respect to the standard triple

$$([0, \dots, 0, I], \quad C_1^*, \quad \operatorname{col}(\delta_{i1} I)_{i=1}^l)$$

of L; therefore, the elementary divisors of $Q_1(\lambda)$ and $I\lambda - C_1^*|_{\mathcal{M}_1^\perp}$ are the same. Now let (X, J, Y) be the self-adjoint triple of L used in the proof of Theorem 11.2. The subspace $\mathcal{M}_1^\perp$ can be easily identified in terms of X, J and the self-adjoint matrix B given by (10.3). Namely, $\mathcal{M}_1^\perp = B\mathcal{Z}$, where $\mathcal{Z}$ is the image of the $nl \times nk$ matrix (where $k = [\frac{1}{2}l]$) formed by the following selected columns of $\operatorname{col}(XJ^i)_{i=0}^{l-1}$:

(a) select the columns corresponding to Jordan blocks in J with nonreal eigenvalues outside S;

(b) in the notation introduced in Section 11.1, if $(\lambda_i, x_{1i}) \in \mathcal{S}_1$, select $\hat{x}_{1,i}, \dots, \hat{x}_{k_i - r_i, i}$;

(c) if $(\lambda_i, x_{1i}) \in \mathcal{S}_2$ and k_i is odd, select $\hat{x}_{1,i}, \dots, \hat{x}_{k_i - r_i + 1, i}$;

(d) if $(\lambda_i, x_{1i}) \in \mathcal{S}_2$ and k_i is even, select $\hat{x}_{1,i}, \dots, \hat{x}_{k_i - r_i, i}$.

It is easily seen that this rule ensures the selection of exactly nk columns of $\operatorname{col}(XJ^i)_{i=0}^{l-1}$. Furthermore, Lemma 11.3 and equality (11.7) ensure that $B\mathcal{Z}$ is orthogonal to $\mathcal{M}_1$, and since $\dim(B\mathcal{Z}) + \dim \mathcal{M}_1 = nl$, we have $B\mathcal{Z} = \mathcal{M}_1^\perp$.

TABLE I

	Divisor of L		Divisor of L_1	Divisor of Q_1
	$(\lambda - \lambda_0)^j$	$\lambda_0 \neq \bar{\lambda}_0, \lambda_0 \in S$	$(\lambda - \lambda_0)^j$	1
	$(\lambda - \lambda_0)^j$,	$\lambda_0 \neq \bar{\lambda}_0, \lambda_0 \notin S$	1	$(\lambda - \lambda_0)^j$
$r \geq 1$	$(\lambda - \lambda_0)^{2r}$,	$\lambda_0 = \bar{\lambda}_0$,	$(\lambda - \lambda_0)^r$	$(\lambda - \lambda_0)^r$
$r \geq 1$	$(\lambda - \lambda_0)^{2r-1}$,	$\lambda_0 = \bar{\lambda}_0$, 1st kind	$(\lambda - \lambda_0)^r$	$(\lambda - \lambda_0)^{r-1}$
$r \geq 1$	$(\lambda - \lambda_0)^{2r-1}$,	$\lambda_0 = \bar{\lambda}_0$, 2nd kind	$(\lambda - \lambda_0)^{r-1}$	$(\lambda - \lambda_0)^r$

Taking adjoints in

$$C_1 \operatorname{col}(XJ^i)_{i=0}^{l-1} = \operatorname{col}(XJ^i)_{i=0}^{l-1} J$$

and using the properties of the self-adjoint triple: $Y^* = XP_{\varepsilon,J}$, $P_{\varepsilon,J}J = J^*P_{\varepsilon,J}$, $P_{\varepsilon,J}J^* = JP_{\varepsilon,J}$, we obtain

$$P_{\varepsilon,J}RC_1^* = J^*P_{\varepsilon,J}R, \tag{11.13}$$

where $R = [Y, JY, \ldots, J^{l-1}Y]$. Rewrite (11.13) in the form

$$C_1^*R^{-1}P_{\varepsilon,J} = R^{-1}P_{\varepsilon,J}J^*,$$

and substitute here $R^{-1} = BQ$, where $Q = \operatorname{col}(XJ^i)_{i=0}^{l-1}$:

$$C_1^*BQP_{\varepsilon,J} = BQP_{\varepsilon,J}J^*. \tag{11.14}$$

Use equality (11.14) and the selection rule described above for $\mathscr{L}$ to verify the last three entries in the "divisor of Q_1" column.

Analogously, one computes the elementary divisors of the quotient $Q_2 = LL_2^{-1}$, where L_2 is taken from Theorem 11.2.

We summarize this discussion in the following theorem.

Theorem 11.4. *Let L be a monic self-adjoint matrix polynomial of degree l, and let $\lambda_1, \ldots, \lambda_r$ be all its different real eigenvalues with corresponding partial multiplicities $m_{i1}, \ldots, m_{i,s_i}$, $i = 1, \ldots, r$, and the sign characteristic $\varepsilon = (\varepsilon_{ij})$ where $j = 1, \ldots, s_i$, $i = 1, \ldots, r$. Let S be a maximal c-set. Then there exists a monic right divisor L_1 (L_2) of L with the following properties:*

(i) $\deg L_1 = [(l+1)/2]$, $\deg L_2 = l - [(l+1)/2]$;

(ii) *the nonreal spectrum of L_1 and of L_2 coincides with S;*

(iii) *the partial multiplicities of L_1 (resp. of the quotient $Q_1 = LL_1^{-1}$) corresponding to λ_i ($i = 1, \ldots, r$) are $\frac{1}{2}(m_{ij} + \zeta_{ij}\varepsilon_{ij})$, $j = 1, \ldots, s_i$, (resp. $\frac{1}{2}(m_{ij} - \zeta_{ij}\varepsilon_{ij})$, $j = 1, \ldots, s_i$), where $\zeta_{ij} = 0$ if m_{ij} is even, and $\zeta_{ij} = 1$ if m_{ij} is odd;*

(iv) *the partial multiplicities of L_2 (resp. of the quotient $Q_2 = LL_2^{-1}$) are $\frac{1}{2}(m_{ij} - \zeta_{ij}\varepsilon_{ij})$, $j = 1, \ldots, s_i$ (resp. $\frac{1}{2}(m_{ij} + \zeta_{ij}\varepsilon_{ij})$, $j = 1, \ldots, s_i$).*

Moreover, the supporting subspace of L_1 (resp. L_2) with respect to C_1 is B-nonnegative (resp. B-nonpositive).

Going back to the proof of Theorem 11.2, observe that the argument of case (ii) in the proof can be applied to $\mathcal{N} = \operatorname{Im} Q_c$ where Q_c corresponds to the c-set of all eigenvalues of L in the open half-plane $\mathscr{Im}\,\lambda > 0$. Since $\mathcal{N}$ is B-neutral it is also B-nonpositive and the argument referred to leads to the conclusion that $\dim \mathcal{N} \leq kn$. But this implies that, for l odd, the number of *real* eigenvalues is at least n, and thus confirms the last statement of Theorem 10.4.

Corollary 11.5. *If L has odd degree and exactly n real eigenvalues (counting multiplicities), then these eigenvalues have only linear elementary divisors and there is a factorization*

$$L(\lambda) = M^*(\lambda)(I\lambda + W)M(\lambda),$$

where M is a monic matrix polynomial, $\sigma(M)$ is in the half-plane $\mathscr{Im}\ \lambda > 0$, and $W^* = W$.

Proof. Theorem 10.4 implies that L has exactly n elementary divisors associated with real eigenvalues, and all of them are linear. By Proposition 10.2, $\sum_{ij}\varepsilon_{ij} = n$, where $\varepsilon = (\varepsilon_{ij})$ is the sign characteristic of L. Since the number of signs in the sign characteristic is exactly n, all signs must be $+1$. Now it is clear that the subspace $\mathscr{N}$ of the above discussion has dimension kn and determines a supporting subspace for L which is B-neutral. The symmetric factorization is then a deduction from Theorem 11.1. □

In the proof of Corollary 11.5 it is shown that the real eigenvalues in Corollary 11.5 are necessarily of the first kind. This can also be seen in the following way. Let λ_0 be a real eigenvalues of L with eigenvector x_0. Then $M(\lambda_0)$ is nonsingular and it is easily verified that

$$(x_0, L^{(1)}(\lambda_0)x_0) = (x_0, M^*(\lambda_0)M(\lambda_0)x_0) = (y_0, y_0)$$

where $y_0 = M(\lambda_0)x_0$. Hence $(x_0, L^{(1)}(\lambda_0)x_0) > 0$, and the conclusion follows from equation (11.4).

Corollary 11.6. *If L of Theorem* 11.2 *has even degree and the real eigenvalues of L have all their elementary divisors of even degree, then there is a matrix polynomial M such that $L(\lambda) = M^*(\lambda)M(\lambda)$.*

Proof. This is an immediate consequence of Theorems 11.1 and 11.2. The hypothesis of even degree for the real divisors means that the supporting subspace $\mathscr{M}$ constructed in Theorem 11.2 has dimension $\frac{1}{2}ln$ and is B-neutral. □

EXAMPLE 11.1. Consider the matrix polynomial L of Example 10.4. Choose the c-set, $\{i\}$ and select Ξ from the columns of matrix X of Example 10.4 as follows:

$$\Xi_R = \begin{bmatrix} -1 & -1 \\ 1 & 1 \end{bmatrix}, \qquad \Xi_c = \begin{bmatrix} 1 & 0 \\ 0 & 0 \end{bmatrix}, \qquad \Xi = [\Xi_R \quad \Xi_c]$$

and then

$$K = [1] \oplus [-1] \oplus \begin{bmatrix} i & 1 \\ 0 & i \end{bmatrix}.$$

We then have $\mathscr{M}_1 = \mathscr{L}_1 \dotplus \operatorname{Im} Q_c$, a B-neutral subspace, and

$$\mathscr{L}_1 = \operatorname{Im}\begin{bmatrix} -1 & 1 & -1 & 1 & -1 & 1 & -1 & 1 \\ -1 & 1 & 1 & -1 & -1 & 1 & 1 & -1 \end{bmatrix},$$

$$Q_c = \operatorname{Im}\begin{bmatrix} 1 & 1 & i & i & -1 & -1 & -i & -i \\ 0 & 0 & 1 & 1 & 2i & 2i & -3 & -3 \end{bmatrix}$$

and $S_2 = \operatorname{col}[\Xi K^r]_{r=0}^1$ is given by

$$S_2 = \begin{bmatrix} -1 & -1 & 1 & 0 \\ 1 & 1 & 1 & 0 \\ -1 & 1 & i & 1 \\ 1 & -1 & i & 1 \end{bmatrix} \quad \text{and} \quad S_2^{-1} = \frac{1}{4}\begin{bmatrix} -1 & 1 & -1 & 1 \\ -1 & 1 & 1 & -1 \\ 2 & 2 & 0 & 0 \\ -2i & -2i & 2 & 2 \end{bmatrix}.$$

Computing $\Xi K^2 S^{-1}$ it is then found that

$$L_1(\lambda) = \begin{bmatrix} \lambda^2 - i\lambda - 1 & -i\lambda \\ 0 & \lambda^2 - i\lambda - 1 \end{bmatrix}.$$

It is easily verified that $L(\lambda) = L_1^*(\lambda)L_1(\lambda)$. □

The special cases $l = 2, 3$ arise very frequently in practice and merit special attention. When $l = 2$ Theorem 11.2 implies that the quadratic bundle (with self-adjoint coefficients) can always be factored in the form

$$I\lambda^2 + A_1\lambda + A_0 = (I\lambda - Y)(I\lambda - Z). \tag{11.15}$$

If the additional hypothesis of Corollary 11.6 holds, then there is such a factorization with $Y = Z^*$. Further discussions of problems with $l = 2$ will be found in Chapter 13.

When $l = 3$ there are always at least n real eigenvalues and there are matrices Z, B_1, B_2 for which

$$I\lambda^3 + A_2\lambda^2 + A_1\lambda + A_0 = (I\lambda^2 + B_1\lambda + B_2)(I\lambda - Z). \tag{11.16}$$

If there are precisely n real eigenvalues, then it follows from Corollary 11.5 that there are matrices Z, M with $M^* = M$ and

$$I\lambda^3 + A_2\lambda^2 + A_1\lambda + A_0 = (I\lambda - Z^*)(I\lambda + M)(I\lambda - Z).$$

Furthermore, $\sigma(I\lambda - Z)$ is nonreal, $\sigma(I\lambda + M)$ is just the real part of $\sigma(L(\lambda))$ and the real eigenvalues necessarily have all their elementary divisors linear.

EXAMPLE 11.2. Let $l = 2$ and suppose that *each real eigenvalue has multiplicity one*. Then the following factorizations (11.15) are possible:

$$L(\lambda) = (I\lambda - Y_i)(I\lambda - Z_i),$$

$i = 1, 2$, in which $I\lambda - Z_1$ has supporting subspace which is maximal B-nonnegative and that of $I\lambda - Z_2$ is maximal B-nonpositive, while $\sigma(L)$ is the disjoint union of the spectra of $I\lambda - Z_1$ and $I\lambda - Z_2$. In this case, it follows from Theorem 2.15 that Z_1, Z_2 form a *complete pair* for L.

Comments

The presentation of this chapter is based on [34d, 34f, 34g]. Theorem 11.2 was first proved in [56d], and Corollary 11.5 is Theorem 8 in [56c]. Theorem 11.4 is new.

The notation of "kinds" of eigenvectors corresponding to real eigenvalues (introduced in Section 11.2) is basic and appears in many instances. See, for example, papers [17, 50, 51]. In [69] this idea appears in the framework of analytic operator valued functions.

Chapter 12

Further Analysis of the Sign Characteristic

In this chapter we develop further the analysis of the sign characteristic of a monic self-adjoint matrix polynomial $L(\lambda)$, begun in Chapter 10. Here we shall represent the sign characteristic as a local property. This idea eventually leads to a description of the sign characteristic in terms of eigenvalues of $L(\lambda_0)$ (as a constant matrix) for each fixed $\lambda_0 \in \mathbb{R}$ (Theorem 12.5). As an application, a description of nonnegative matrix polynomials is given in Section 12.4.

12.1. Localization of the Sign Characteristic

The idea of the sign characteristic of a self-adjoint matrix polynomial was introduced as a global notion in Chapter 10 via self-adjoint triples. In this section we give another description of the sign characteristic which ultimately shows that it can be defined locally.

Theorem 12.1. *Let $L_1(\lambda)$ and $L_2(\lambda)$ be two monic self-adjoint matrix polynomials. If $\lambda_0 \in \sigma(L_1)$ is real and*

$$L_1^{(i)}(\lambda_0) = L_2^{(i)}(\lambda_0), \qquad i = 0, 1, \ldots, \gamma,$$

where γ is the maximal length of Jordan chains of $L_1(\lambda)$ (and then also of $L_2(\lambda)$) corresponding to λ_0, then the sign characteristics of $L_1(\lambda)$ and $L_2(\lambda)$ at λ_0 are the same.

By the sign characteristic of a matrix polynomial at an eigenvalue we mean the set of signs in the sign characteristic, corresponding to the Jordan blocks with this eigenvalue. It is clear that Theorem 12.1 defines the sign characteristic at λ_0 as a local property of the self-adjoint matrix polynomials. This result will be an immediate consequence of the description of the sign characteristic to be given in Theorem 12.2 below.

Let $L(\lambda)$ be a monic self-adjoint matrix polynomial, and let λ_0 be a real eigenvalue of $L(\lambda)$. For $x \in \operatorname{Ker} L(\lambda_0)\backslash\{0\}$ denote by $v(x)$ the maximal length of a Jordan chain of $L(\lambda)$ beginning with the eigenvector x of λ_0. Let Ψ_i, $i = 1, \ldots, \gamma$ ($\gamma = \max\{v(x) \mid x \in \operatorname{Ker} L(\lambda_0)\backslash\{0\}\}$) be the subspace in $\operatorname{Ker} L(\lambda_0)$ spanned by all x with $v(x) \geq i$.

Theorem 12.2. *For $i = 1, \ldots, \gamma$, let λ_0 be a real eigenvalue of L and*

$$f_i(x, y) = \left(x, \sum_{j=1}^{i} \frac{1}{j!} L^{(j)}(\lambda_0) y^{(i+1-j)}\right), \qquad x, y \in \Psi_i,$$

where $y = y^{(1)}, y^{(2)}, \ldots, y^{(i)}$ is a Jordan chain of $L(\lambda)$ corresponding to λ_0 with eigenvector y. Then:

(i) *$f_i(x, y)$ does not depend on the choice of $y^{(2)}, \ldots, y^{(i)}$;*
(ii) *there exists a self-adjoint linear transformation $G_i\colon \Psi_i \to \Psi_i$ such that*

$$f_i(x, y) = (x, G_i y), \qquad x, y \in \Psi_i;$$

(iii) *$\Psi_{i+1} = \operatorname{Ker} G_i$ (by definition, $\Psi_{\gamma+1} = \{0\}$);*
(iv) *the number of positive (negative) eigenvalues of G_i, counting according to the multiplicities, coincides with the number of positive (negative) signs in the sign characteristic of $L(\lambda)$ corresponding to the Jordan blocks of λ_0 of size i.*

Proof. Let (X, J, Y) be a self-adjoint triple of $L(\lambda)$. Then, by Theorem 10.6 $X = [I \ 0 \ \cdots \ 0]S^{-1}$, $J = SC_1S^{-1}$, $Y = S[0 \ \cdots \ 0 \ I]^{\mathrm{T}}$, and $B = S^*P_{\varepsilon,J}S$. It is easy to see that in fact $S^{-1} = \operatorname{col}(XJ^i)_{i=0}^{l-1}$ (see, for instance, (1.67)). Recall the equivalence

$$E(\lambda)(I\lambda - C_1)F(\lambda) = L(\lambda) \oplus I_{n(l-1)} \tag{12.1}$$

(see Theorem 1.1 and its proof), where $E(\lambda)$ and $F(\lambda)$ are matrix polynomials with constant nonzero determinants, and

$$F(\lambda) = \begin{bmatrix} I & 0 & \cdots & 0 \\ I\lambda & I & \cdots & 0 \\ \vdots & \vdots & & \vdots \\ I\lambda^{l-1} & I\lambda^{l-2} & \cdots & I \end{bmatrix}$$

Equation (12.1) implies that the columns of the $n \times r$ matrix X_0 form a Jordan chain for $L(\lambda)$ corresponding to λ_0 if and only if the columns of the $ln \times r$

matrix $\operatorname{col}[X_0 J_0^j]_{j=0}^{l-1}$ form a Jordan chain for $\lambda I - C_1$, where J_0 is the $r \times r$ Jordan block with eigenvalue λ_0 (see Proposition 1.11). Now we appeal to Theorem S5.7, in view of which it remains to check that

$$\left(x, \sum_{j=1}^{r} \frac{1}{j!} L^{(j)}(\lambda_0) y^{(r+1-j)}\right) = (\hat{x}, B\hat{y}^{(r)}), \tag{12.2}$$

where $\hat{x} = \operatorname{col}(\lambda_0^j x)_{j=0}^{l-1}$, $[\hat{y}^{(1)} \quad \cdots \quad \hat{y}^{(r)}] = \operatorname{col}([y^{(1)} \quad \cdots \quad y^{(r)}] J_0^j)_{j=0}^{l-1}$,

$$B = \begin{bmatrix} A_1 & A_2 & \cdots & A_{l-1} & I \\ A_2 & & \cdot^{\cdot^{\cdot}} & \cdot^{\cdot^{\cdot}} & \\ \vdots & \cdot^{\cdot^{\cdot}} & \cdot^{\cdot^{\cdot}} & & \vdots \\ A_{l-1} & I & \cdots & & 0 \\ I & 0 & \cdots & & 0 \end{bmatrix}.$$

As usual, the self-adjoint matrices A_i stand for the coefficients of $L(\lambda)$: $L(\lambda) = I\lambda^l + \sum_{j=0}^{l-1} A_j \lambda^j$. But this is already proved in (10.51). $\square$

Let us compute now the first two linear transformations G_1 and G_2. Since $f_1(x, y) = (x, L'(\lambda_0)y)$ $(x, y \in \Psi_1 = \operatorname{Ker} L(\lambda_0))$, it is easy to see that

$$G_1 = P_1 L'(\lambda_0) P_1 |_{\Psi_1},$$

where P_1 is the orthogonal projector onto Ψ_1. Denote $L_0^{(j)} = L^{(j)}(\lambda_0)$. For $x, y \in \Psi_2 = \operatorname{Ker} G_1$, we have

$$f_2(x, y) = (x, \tfrac{1}{2} L_0'' y + L_0' y'),$$

where y, y' is a Jordan chain of $L(\lambda)$ corresponding to λ_0. Thus

$$L_0 y = L_0' y + L_0 y' = L_0' y + L_0(I - P_1) y' = 0 \tag{12.3}$$

(the last equality follows in view of $L_0 P_1 = 0$). Denote by $L_0^+ : \mathbb{C}^n \to \mathbb{C}^n$ the linear transformation which is equal to L_0^{-1} on $\Psi_1^\perp$ and zero on $\Psi_1 = \operatorname{Ker} L_0$. Then $L_0^+ L_0(I - P_1) = I - P_1$ and (12.3) gives $(I - P_1)y' = -L_0^+ L_0' y$. Now, for $x, y \in \Psi_2$,

$$\begin{aligned} (x, L_0' y') &= (x, L_0'(P_1 + (I - P_1))y') = (x, L_0' P_1 y') + (x, L_0'(I - P_1)y') \\ &= (x, G_1 P_1 y') + (x, L_0'(I - P_1)y') = (x, L_0'(I - P_1)y') \\ &= (x, -L_0' L_0^+ L_0' y), \end{aligned}$$

where the penultimate equality follows from the fact that $x \in \operatorname{Ker} G_1$ and $G_1 = G_1^*$. Thus $f_2(x, y) = (x, \tfrac{1}{2} L_0'' y - L_0' L_0^+ L_0' y)$, and

$$G_2 = P_2[\tfrac{1}{2} L_0'' - L_0' L_0^+ L_0'] P_2 |_{\Psi_2},$$

where P_2 is the orthogonal projector of $\mathbb{C}^n$ on Ψ_2. To illustrate the above construction, consider the following example.

EXAMPLE 12.1. Let

$$L(\lambda) = \begin{bmatrix} \lambda^2 & 0 & 0 \\ 0 & \lambda^2 + \lambda & \lambda \\ 0 & \lambda & \lambda^2 + \lambda \end{bmatrix}.$$

Choose $\lambda_0 = 0$ as an eigenvalue of $L(\lambda)$. Then $\operatorname{Ker} L(0) = \mathbb{C}^3$, so $\Psi_1 = \mathbb{C}^3$. Further,

$$L'(0) = \begin{bmatrix} 0 & 0 & 0 \\ 0 & 1 & 1 \\ 0 & 1 & 1 \end{bmatrix}$$

and $f_1(x, y) = (x, L'(0)y) = x_2(\bar{y}_2 + \bar{y}_3) + x_3(\bar{y}_2 + \bar{y}_3)$, where $x = (x_1, x_2, x_3)^T$, $y = (y_1, y_2, y_3)^T$. $L'(0)$ has one nonzero eigenvalue 2 and

$$\operatorname{Ker} L'(0) = \Psi_2 = \operatorname{Span}\{(1, 0, 0)^T, (0, -1, 1)^T\}.$$

Thus there exists exactly one partial multiplicity of $L(\lambda)$ corresponding to $\lambda_0 = 0$ which is equal to 1, and its sign is $+1$. It is easily see that y, 0 is a Jordan chain for any eigenvector $y \in \Psi_2$. Thus

$$f_2(x, y) = (x, \tfrac{1}{2}L''(0)y + L'(0)y') = (x, y) \qquad \text{for} \quad x, y \in \Psi_2.$$

Therefore there exist exactly two partial multiplicities of $\lambda_0 = 0$ which are equal to 2, and their signs are $+1$. □

12.2. Stability of the Sign Characteristic

In this section we describe a stability property of the sign characteristic which is closely related to its localization. It turns out that it is not only the sign characteristic of a monic self-adjoint polynomial which is determined by local properties, but also every monic self-adjoint polynomial which is sufficiently close, and has the same local Jordan structure, will have the same sign characteristic .More exactly, the following result holds.

Theorem 12.3. *Let $L(\lambda) = I\lambda^l + \sum_{j=0}^{l-1} \lambda^j A_j$ be a self-adjoint matrix polynomial, and let $\lambda_0 \in \sigma(L)$ be real. Then there exists a $\delta > 0$ with the following property: if $\tilde{L}(\lambda) = I\lambda^l + \sum_{j=0}^{l-1} \lambda^j \tilde{A}_j$ is a self-adjoint polynomial such that $\|\tilde{A}_j - A_j\| < \delta$, $j = 0, \ldots, l-1$, and if there exists a unique real eigenvalue λ_1 of $L(\lambda)$ in the disk $|\lambda_0 - \lambda| < \delta$ with the same partial multiplicities as those of $L(\lambda)$ at λ_0, then the sign characteristic of $\tilde{L}(\lambda)$ at λ_1 coincides with the sign characteristic of $L(\lambda)$ at λ_0.*

We shall not present the proof of Theorem 12.3 because it requires background material which is beyond the scope of this book. Interested readers are referred to [34f, 34g] for the proof.

12.3. A Sign Characteristic for Self-Adjoint Analytic Matrix Functions

Let L be a monic self-adjoint matrix polynomial. Then (see Theorem S6.3) for real λ the matrix $L(\lambda)$ has a diagonal form

$$L(\lambda) = U(\lambda) \cdot \operatorname{diag}[\mu_1(\lambda), \ldots, \mu_n(\lambda)] \cdot V(\lambda), \tag{12.4}$$

where $U(\lambda)$ is unitary (for real λ) and $V(\lambda) = (U(\lambda))^*$. Moreover, the functions $\mu_i(\lambda)$ and $U(\lambda)$ can be chosen to be analytic functions of the real parameter λ (but in general $\mu_i(\lambda)$ and $U(\lambda)$ are not polynomials). Our next goal will be to describe the sign characteristic of $L(\lambda)$ in terms of the functions $\mu_1(\lambda), \ldots, \mu_n(\lambda)$. Since the $\mu_i(\lambda)$ are analytic functions, we must pay some attention to the sign characteristic of analytic matrix functions in order to accomplish this goal. We shall do that in this section, which is of a preliminary nature and prepares the groundwork for the main results in the next section.

Let Ω be a connected domain in the complex plane, symmetric with respect to the real axis. An analytic $n \times n$ matrix function $A(\lambda)$ in Ω is called *self-adjoint* if $A(\lambda) = (A(\lambda))^*$ for real $\lambda \in \Omega$. We consider in what follows self-adjoint analytic functions $A(\lambda)$ with $\det A(\lambda) \not\equiv 0$ only, and shall not stipulate this condition explicitly. For such self-adjoint analytic matrix functions $A(\lambda)$ the spectrum $\sigma(A) = \{\lambda \in \Omega \,|\, \det A(\lambda) = 0\}$ is a set of isolated points, and for every $\lambda_0 \in \sigma(A)$ one defines the Jordan chains and Jordan structure of $A(\lambda)$ at λ_0 as for matrix polynomials.

We shall need the following simple fact: if $A(\lambda)$ is an analytic matrix function in Ω (not necessarily self-adjoint) and $\det A(\lambda) \not\equiv 0$, then the Jordan chains of $A(\lambda)$ corresponding to a given $\lambda_0 \in \sigma(A)$ have bounded lengths, namely, their lengths do not exceed the multiplicity of λ_0 as a zero of $\det A(\lambda)$. For completeness let us prove this fact directly. Let $\varphi_0, \ldots, \varphi_{k-1}$ be a Jordan chain of $A(\lambda)$ corresponding to λ_0; then $\varphi_0 \neq 0$,

$$A(\lambda)\varphi(\lambda) = (\lambda - \lambda_0)^k x(\lambda) \tag{12.5}$$

where $\varphi(\lambda) = \sum_{j=0}^{k-1} (\lambda - \lambda_0)^j \varphi_j$, and the vector function $x(\lambda)$ is analytic in a neighborhood of λ_0. Let $\psi_1, \ldots, \psi_{n-1}$ be vectors such that $\varphi_0, \psi_1, \ldots, \psi_{n-1}$ is a basis in $\mathbb{C}^n$, and put

$$E(\lambda) = [\varphi(\lambda), \psi_1, \ldots, \psi_{n-1}].$$

By (12.5), the analytic function $\det A(\lambda) \cdot \det E(\lambda) = \det[A(\lambda)E(\lambda)]$ has a zero of multiplicity at least k at $\lambda = \lambda_0$. Since $E(\lambda_0)$ is nonsingular by construction, the same is true for $\det A(\lambda)$, i.e., λ_0 is a zero of $\det A(\lambda)$ of multiplicity at least k.

Going back to the self-adjoint matrix function $A(\lambda)$, let $\lambda_0 \in \sigma(A)$ be real. Then there exists a monic self-adjoint matrix polynomial $L(\lambda)$ such that

$$L^{(j)}(\lambda_0) = A^{(j)}(\lambda_0), \qquad j = 1, \ldots, \gamma, \tag{12.6}$$

where γ is the maximal length of Jordan chains of $A(\lambda)$ corresponding to λ_0 (for instance, put $L(\lambda) = I(\lambda - \lambda_0)^{\gamma+1} + \sum_{j=0}^{\gamma} A^{(j)}(\lambda_0)(\lambda - \lambda_0)^j$). Bearing in mind that $x_1, \ldots, x_k$ is a Jordan chain of $L(\lambda)$ corresponding to λ_0 if and only if $x_1 \neq 0$ and

$$\begin{bmatrix} L(\lambda_0) & 0 & \cdots & 0 \\ L'(\lambda_0) & L(\lambda_0) & \cdots & 0 \\ \cdots & \cdots & \cdots & \vdots \\ \frac{1}{(k-1)!} L^{(k-1)}(\lambda_0) & \frac{1}{(k-2)!} L^{(k-2)}(\lambda_0) & \cdots & L(\lambda_0) \end{bmatrix} \begin{bmatrix} x_1 \\ x_2 \\ \vdots \\ x_k \end{bmatrix} = 0,$$

and analogously for $A(\lambda)$, we obtain from (12.6) that $L(\lambda)$ and $A(\lambda)$ have the same Jordan chains corresponding to λ_0.

In particular, the structure of Jordan blocks of $A(\lambda)$ and $L(\lambda)$ at λ_0 is the same and we define the sign characteristic of $A(\lambda)$ at λ_0 as the sign characteristic of $L(\lambda)$ at λ_0. In view of Theorem 12.1 this definition is correct (i.e., does not depend on the choice of $L(\lambda)$).

The next theorem will also be useful in the next section, but is clearly of independent interest. Given an analytic matrix function $R(\lambda)$ in Ω, let $R^*(\lambda)$ denote the analytic (in Ω) matrix function $(R(\bar{\lambda}))^*$.

Observe that the notion of a canonical set of Jordan chains extends word for word for analytic matrix functions with determinant not identically zero. As for monic matrix polynomials, the lengths of Jordan chains in a canonical set (corresponding to $\lambda_0 \in \sigma(A)$) of an analytic matrix function $A(\lambda)$, do not depend on the choice of the canonical set. The *partial multiplicities* of $A(\lambda)$ at λ_0, by definition, consist of these lengths and possibly some zeros (so that the total number of partial multiplicities is n, the size of $A(\lambda)$), cf. Proposition 1.13. Observe that for a monic matrix polynomial $L(\lambda)$ with property (12.6), the partial multiplicities of $A(\lambda)$ at λ_0 coincide with those of $L(\lambda)$ at λ_0.

Theorem 12.4. *Let A be a self-adjoint analytic matrix function in Ω, and let $\lambda_0 \in \sigma(A)$ be real. Let R be an analytic matrix function in Ω such that* $\det R(\lambda_0) \neq 0$. *Then*:

(i) *the partial multiplicities of A and R^*AR at λ_0 are the same*;
(ii) *the sign characteristics of A and R^*AR at λ_0 are the same.*

Proof. Part (i) follows from Proposition 1.11 (which extends verbatim, together with its proof, for analytic matrix functions). The rest of the proof will be devoted to part (ii).

Consider first the case that $A = L$ is a monic self-adjoint matrix polynomial. Let γ be the maximal length of Jordan chains of L corresponding to λ_0. Let $S(\lambda) = \sum_{i=0}^{\gamma+1} S_i(\lambda - \lambda_0)^i$ and $T(\lambda) = \sum_{i=0}^{\gamma+1} T_i(\lambda - \lambda_0)^j$ be monic matric polynomials of degree $\gamma + 1$ which are solutions of the following interpolation problem (cf. the construction of L in formula (12.6)):

$$S^{(j)}(\lambda_0) = R^{(j)}(\lambda_0) \qquad \text{for} \quad j = 0, 1, \ldots, \gamma,$$

$$T^{(j)}(\lambda_0) = \begin{cases} I & \text{for} \quad j = 0 \\ 0 & \text{for} \quad j = 1, \ldots, \gamma. \end{cases}$$

Then $(S^*LS)^{(j)}(\lambda_0) = (R^*LR)^{(j)}(\lambda_0)$ for $j = 0, \ldots, \gamma$, thus in view of Theorem 12.1 the sign characteristics of S^*LS and R^*LR at λ_0 are the same.

Observe that S_0 and T_0 are nonsingular. There exists a continuous matrix function $F_0(t)$, $t \in [0, 1]$, such that $F_0(0) = S_0$, $F_0(1) = T_0$ and $F_0(t)$ is nonsingular for all $t \in [0, 1]$. Indeed, let

$$U^{-1}S_0 U = \operatorname{diag}[J_1, \ldots, J_k]$$

be the Jordan form for S_0 with Jordan blocks $J_1, \ldots, J_k$ and nonsingular matrix U. Here

$$J_i = \begin{bmatrix} \mu_i & 1 & 0 & & 0 \\ & \mu_i & 1 & & \\ & & & \ddots & \\ & & & & 1 \\ & 0 & & & \mu_i \end{bmatrix}, \qquad i = 1, \ldots, k,$$

where $\mu_i \neq 0$ in view of the nonsingularity of S_0. Let $\mu_i(t)$, $t \in [0, \frac{1}{2}]$ be a continuous function such that $\mu_i(0) = \mu_i$, $\mu_i(\frac{1}{2}) = 1$ and $\mu_i(t) \neq 0$ for all $t \in [0, \frac{1}{2}]$. Finally, put

$$F_0(t) = U \operatorname{diag}(J_1(t), \ldots, J_k(t))U^{-1}, \qquad t \in [0, \tfrac{1}{2}],$$

where

$$J_i(t) = \begin{bmatrix} \mu_i(t) & 1 - 2t & & \\ & \mu_i(t) & 1 - 2t & \\ & & \ddots & \\ & 0 & & 1 - 2t \\ & & & \mu_i(t) \end{bmatrix}.$$

Clearly, $F_0(t)$ is nonsingular for all $t \in [0, \frac{1}{2}]$, $F_0(0) = S_0$, $F_0(\frac{1}{2}) = I$. Analogous construction ensures the connection between I and T_0 by means of a continuous nonsingular matrix function $F_0(t)$, $t \in [\frac{1}{2}, 1]$.

Now let $F(\lambda, t)$, $t \in [0, 1]$ be a continuous family of monic matrix polynomials of degree $\gamma + 1$ such that $F(\lambda, 0) = S(\lambda)$, $F(\lambda, 1) = T(\lambda)$ and $F(\lambda_0, t)$ is nonsingular for every $t \in [0, 1]$. (For example,

$$F(\lambda, t) = I(\lambda - \lambda_0)^{\gamma+1} + \sum_{j=1}^{\gamma} [tT_j + (1 - t)S_j](\lambda - \lambda_0)^j + F_0(t),$$

where $F_0(t)$ is constructed as above.) Consider the family

$$M(\lambda, t) = F^*(\lambda, t)L(\lambda)F(\lambda, t), \qquad t \in [0, 1],$$

of monic self-adjoint matrix polynomials. Since $\det F(\lambda_0, t) \neq 0$, it follows that $\lambda_0 \in \sigma(M(\lambda, t))$ and the partial multiplicities of $M(\lambda, t)$ at λ_0 do not depend on $t \in [0, 1]$. Furthermore,

$$M^{(j)}(\lambda_0, 0) = (R^*LR)^{(j)}(\lambda_0), \qquad M^{(j)}(\lambda_0, 1) = L^{(j)}(\lambda_0), \qquad j = 0, \ldots, \gamma. \tag{12.7}$$

Let us show that the sign characteristic of $M(\lambda, t)$ at λ_0 also does not depend on $t \in [0, 1]$. Let $\Psi_i(t)$, $i = 1, \ldots, \gamma$, $t \in [0, 1]$ be the subspace in $\operatorname{Ker} M(\lambda_0, t)$ spanned by all eigenvectors x of $M(\lambda, t)$ corresponding to λ_0 such that there exists a Jordan chain of $M(\lambda, t)$ corresponding to λ_0 of length at least i beginning with x. By Proposition 1.11, and by (12.7):

$$\Psi_i(t) = F^{-1}(\lambda_0, t)\Psi_i(1), \qquad t \in [0, 1], \quad i = 1, \ldots, \gamma. \tag{12.8}$$

For $x, y \in \Psi_i(t)$ define

$$f_i(x, y; t) = \left(x, \sum_{j=1}^{i} \frac{1}{j!} M^{(j)}(\lambda_0, t) y^{(i+1-j)}\right),$$

where $y = y^{(1)}, y^{(2)}, \ldots, y^{(i)}$ is a Jordan chain of $M(\lambda, t)$ corresponding to λ_0. According to Theorem 12.2,

$$f_i(x, y; t) = (x, G_i(t)y),$$

where $G_i(t)\colon \Psi_i(t) \to \Psi_i(t)$ is a self-adjoint linear transformation. By Theorem 12.2, in order to prove that the sign characteristic of $M(\lambda, t)$ at λ_0 does not depend on $t \in [0, 1]$, we have to show that the number of positive eigenvalues and the number of negative eigenvalues of $G_i(t)$ is constant. Clearly, it is sufficient to prove the same for the linear transformations

$$H_i(t) = [F^{-1}(\lambda_0, t)]^* G_i(t) F^{-1}(\lambda_0, t)\colon \Psi_i(1) \to \Psi_i(1) \tag{12.9}$$

(see (12.8)).

Let us prove that $H_i(t)$, $i = 1, \ldots, \gamma$, are continuous functions of $t \in [0, 1]$. Let $x_{i1}, \ldots, x_{i,k_i}$ be a basis in $\Psi_i(1)$. It is enough to prove that all scalar functions $(x_{ip}, H_i(t)x_{iq})$, $1 \leq p, q \leq k_i$ are continuous. So let us fix p and q, and let $x_{iq} = y^{(1)}, y^{(2)}, \ldots, y^{(i)}$ be a Jordan chain of $L(\lambda)$ corresponding to λ_0, which starts with x_{iq}. By Proposition 1.11, the vectors

$$z^{(k)} = \sum_{j=0}^{k-1} \frac{1}{j!} [F^{-1}(\lambda_0, t)]^{(j)} y^{(k-j)}, \qquad k = 1, \ldots, i, \tag{12.10}$$

form a Jordan chain of $M(\lambda, t)$ corresponding to λ_0 which begins with $z^{(1)} = F^{-1}(\lambda_0, t)y^{(1)}$. By Theorem 12.2, we have:

$$(x_{ip}, H_i(t)x_{iq}) = (F^{-1}(\lambda_0, t)x_{ip}, G_i(t)F^{-1}(\lambda_0, t)x_{iq})$$
$$= \left(F^{-1}(\lambda_0, t)x_{ip}, \sum_{j=1}^{i} \frac{1}{j!} M^{(j)}(\lambda_0, t)z^{(i+1-j)}\right),$$

which depends continuously on t in view of (12.10).

Now it is easily seen that the number of positive eigenvalues of $H_i(t)$ and the number of negative eigenvalues of $H_i(t)$ are constant. Indeed, write $H_i(t)$ in matrix form with respect to the decomposition $\Psi_i(1) = \Psi_{i+1}(1) \oplus [\Psi_{i+1}(1)]^{\perp}$, bearing in mind that $H_i(t)$ is self-adjoint (and therefore the corresponding matrix is hermitian) and that $\Psi_{i+1}(1) = \operatorname{Ker} H_i(t)$ (Theorem 12.2(iii)):

$$H_i(t) = \begin{bmatrix} 0 & 0 \\ 0 & \tilde{H}_i(t) \end{bmatrix}.$$

Here $\tilde{H}_i(t)$: $[\Psi_{i+1}(1)]^{\perp} \to [\Psi_{i+1}(1)]^{\perp}$ is self-adjoint, nonsingular, and continuous (nonsingularity follows from the equality $\Psi_{i+1}(1) = \operatorname{Ker} H_i(t)$, and continuity follows from the continuity of $H_i(t)$). Therefore, the number of positive eigenvalues of $\tilde{H}_i(t)$ (and, consequently, of $H_i(t)$) does not depend on t, for $i = 1, \ldots, \gamma$.

Thus, the sign characteristic of $M(\lambda, t)$ at λ_0 does not depend on $t \in [0, 1]$. In particular, the sign characteristics of S^*LS and T^*LT at λ_0 are the same. But the sign characteristic of T^*LT at λ_0 is the same as that of L and the sign characteristic of S^*LS is the same as that of R^*LR in view of Theorem 12.1.

So Theorem 12.3 is proved for the case that $A(\lambda)$ is a monic polynomial with invertible leading coefficient. The general case can be easily reduced to this. Namely, let $L(\lambda)$ be a monic self-adjoint matrix polynomial satisfying (12.6). Let $M(\lambda)$ be a matrix polynomial such that

$$M^{(j)}(\lambda_0) = R^{(j)}(\lambda_0), \qquad j = 0, \ldots, \gamma.$$

By the definition, the sign characteristic of R^*AR at λ_0 is defined by M^*LM, and that of A is defined by L. But in view of the already proved case, the sign characteristics of L and M^*LM at λ_0 are the same. □

12.4. Third Description of the Sign Characteristic

Let $L(\lambda)$ be a monic self-adjoint matrix polynomial. The sign characteristic of $L(\lambda)$ was defined via self-adjoint triples for $L(\lambda)$. Here we shall describe the same sign characteristic in a completely different way, namely, from the point of view of perturbation theory.

We shall consider $L(\lambda)$ as a hermitian matrix which depends on the real parameter λ. Let $\mu_1(\lambda), \ldots, \mu_n(\lambda)$ be the eigenvalues of $L(\lambda)$, when $L(\lambda)$ is considered as a constant matrix for every fixed real λ. In other words, $\mu_i(\lambda)$ $(i = 1, \ldots, n)$ are the roots of the equation

$$\det(I\mu - L(\lambda)) = 0. \tag{12.11}$$

From Theorem S6.3 it follows that $\mu_i(\lambda)$ are real analytic functions of the real parameter λ (when enumerated properly), and $L(\lambda)$ admits the representation (12.4).

It is easy to see that $\lambda_0 \in \sigma(L)$ if and only if at least one of the $\mu_j(\lambda_0)$ is zero. Moreover, dim Ker $L(\lambda_0)$ is exactly the number of indices $(1 \leq j \leq n)$ such that $\mu_j(\lambda_0) = 0$. As the next theorem shows, the partial multiplicities of $L(\lambda)$ at λ_0 coincide with the multiplicities of λ_0 as a zero of the analytic functions $\mu_1(\lambda), \ldots, \mu_n(\lambda)$. Moreover, we can describe the sign characteristics of $L(\lambda)$ at λ_0 in terms of the functions $\mu_j(\lambda)$.

The following description of the sign characteristic is one of the main results of this chapter.

Theorem 12.5. *Let $L = L^*$ be monic, and let $\mu_1(\lambda), \ldots, \mu_n(\lambda)$ be real analytic functions of real λ such that $\det(I\mu_j - L(\lambda)) = 0$, $j = 1, \ldots, n$. Let $\lambda_1 < \cdots < \lambda_r$ be the different real eigenvalues of $L(\lambda)$. For every $i = 1, \ldots, r$, write $\mu_j(\lambda) = (\lambda - \lambda_i)^{m_{ij}} v_{ij}(\lambda)$, where $v_{ij}(\lambda_i) \neq 0$ is real. Then the nonzero numbers among $m_{i1}, \ldots, m_{in}$ are the partial multiplicites of L associated with λ_i, and* sgn *$v_{ij}(\lambda_i)$ (for $m_{ij} \neq 0$) is the sign attached to the partial multiplicity m_{ij} of $L(\lambda)$ at λ_i in its (possibly nonnormalized) sign characteristic.*

Proof. Equation (12.4), which a priori holds only for real λ in a neighborhood of λ_i, can be extended to complex λ which are close enough to λ_i, so that $U(\lambda)$, $\mu_j(\lambda)$, and $V(\lambda)$ can be regarded as analytic functions in some complex neighborhood of λ_i in $\mathbb{C}$. This is possible since $U(\lambda)$, $\mu_j(\lambda)$, and $V(\lambda)$ can be expressed as convergent series in a real neighborhood of λ_i; therefore these series converge also in some complex neighborhood of λ_i. (But then of course it is no longer true that $U(\lambda)$ is unitary and $V(\lambda) = (U(\lambda))^*$.) Now the first assertion of Theorem 12.5 follows from Theorem 12.4(i).

Further, in view of Theorem 12.4(ii) the sign characteristics of L and $\operatorname{diag}[\mu_j(\lambda)]_{j=1}^n$ at λ_i are the same. Let us compute the latter. Choose scalar polynomials $\tilde{\mu}_1(\lambda), \ldots, \tilde{\mu}_n(\lambda)$ of the same degree with real coefficients and with the properties

$$\tilde{\mu}_j^{(k)}(\lambda_i) = \mu_j^{(k)}(\lambda_i)$$

for $k = 0, \ldots, m_{ij}$, $i = 1, \ldots, r$, $j = 1, \ldots, n$. (Here and further we write $\mu^{(k)}(\lambda)$ for the kth derivative of μ with respect to λ.) By definition, the sign characteristics of $\operatorname{diag}[\tilde{\mu}_j(\lambda)]_{j=1}^n$ and $\operatorname{diag}[\mu_j(\lambda)]_{j=1}^n$ are the same. Using the description of the sign characteristic of $\operatorname{diag}[\tilde{\mu}_j(\lambda)]_{j=1}^n$ given in Theorem 12.2 it is easy to see that the first nonzero derivative $\tilde{\mu}_j^{(k)}(\lambda_i)$ (for fixed i and j) is positive or negative depending on whether the sign of the Jordan block corresponding to the Jordan chain $(0, \ldots, 0, 1, 0, \ldots, 0)^T, 0, \ldots, 0$ ("1" in the jth place) of λ_i is $+1$ or -1. Thus, the second assertion of Theorem 12.5 follows. □

As a corollary to Theorem 12.5 we obtain the next interesting result.

Note that in this theorem and elsewhere, as appropriate, the notation $\operatorname{sig} L(\lambda)$ (where λ is real and $\lambda \notin \sigma(L)$), is used to denote the signature of the $n \times n$ self-adjoint matrix $L(\lambda)$, in the classical sense, i.e., the difference between the number of positive and the number of negative roots of the equation $\det(I\mu - L(\lambda)) = 0$.

Theorem 12.6. *Let $L = L^*$ be monic, and let $\lambda_0 \in \sigma(L)$ be real. Let ξ be the number of odd partial multiplicities of L at λ_0, having the signs $\varepsilon_1, \ldots, \varepsilon_\xi$ in the sign characteristic. Then*

$$\operatorname{sig} L(\lambda_0 + 0) - \operatorname{sig} L(\lambda_0 - 0) = 2 \sum_{i=1}^{\xi} \varepsilon_i. \tag{12.12}$$

Proof. Consider the eigenvalues $\mu_1(\lambda), \ldots, \mu_n(\lambda)$ of $L(\lambda)$. Let $\alpha_1, \ldots, \alpha_n$ (≥ 0) be the partial multiplicities of $L(\lambda)$ at λ_0 arranged in such an order that $\mu_j^{(i)}(\lambda_0) = 0$ for $i = 0, 1, \ldots, \alpha_j - 1$, and $\mu_j^{(\alpha_j)}(\lambda_0) \neq 0$ ($j = 1, \ldots, n$). This is possible by Theorem 12.5. Then for α_j even, $\mu_j(\lambda_0 + 0)$ has the same sign as $\mu_j(\lambda_0 - 0)$; for α_j odd, the signs of $\mu_j(\lambda_0 + 0)$ and $\mu_j(\lambda_0 - 0)$ are different, and $\mu_j(\lambda_0 + 0)$ is >0 or <0 according as $\mu_j^{(\alpha_j)}(\lambda_0)$ is >0 or <0. Bearing in mind that $\operatorname{sig} L(\lambda)$ is the difference between the number of positive and the number of negative $\mu_j(\lambda)$, we obtain now the assertion of Theorem 12.6 by using Theorem 12.5. □

Corollary 12.7. *Let L, λ_0 and ξ be as in Theorem* 12.6. *Then*

$$\operatorname{sig} L(\lambda_0 + 0) = \operatorname{sig} L(\lambda_0 - 0) \tag{12.13}$$

if and only if ξ is even and exactly $\xi/2$ of the corresponding signs are $+1$.

In particular, (12.13) holds whenever all the partial multiplicities of L at λ_0 are even.

We conclude this section with an illustrative example.

EXAMPLE 12.2. Consider the polynomial

$$L(\lambda) = \begin{bmatrix} \lambda^2 - 2 & 1 \\ 1 & \lambda^2 - 2 \end{bmatrix}.$$

The polynomial $L(\lambda)$ has 4 spectral points: ± 1, $\pm\sqrt{3}$, each one with multiplicity 1. It is easy to check that

$$\operatorname{sig} L(\lambda) = \begin{cases} 2 & \text{for } |\lambda| > \sqrt{3} \\ 0 & \text{for } 1 < |\lambda| < \sqrt{3} \\ -2 & 0 \leq |\lambda| < 1. \end{cases}$$

Now it is possible to compute the sign characteristic of $L(\lambda)$ by using Theorem 12.6. It turns out that

$$\varepsilon_{-\sqrt{3}} = \varepsilon_{-1} = -1, \qquad \varepsilon_{\sqrt{3}} = \varepsilon_1 = 1,$$

where by ε_{λ_0} we denote the sign corresponding to the eigenvalue λ_0. □

12.5. Nonnegative Matrix Polynomials

A matrix polynomial L of size $n \times n$ is called *nonnegative* if

$$(L(\lambda)f, f) \geq 0$$

for all real λ and all $f \in \mathbb{C}^n$. Clearly, every nonnegative polynomial is self-adjoint. In this section we give a description of monic nonnegative polynomials.

Theorem 12.8. *For a monic self-adjoint matrix polynomial $L(\lambda)$ the following statements are equivalent:*

(i) *$L(\lambda)$ is nonnegative;*
(ii) *$L(\lambda)$ admits a representation*

$$L(\lambda) = M^*(\lambda)M(\lambda), \tag{12.14}$$

where $M(\lambda)$ is a monic matrix polynomial;
(iii) *$L(\lambda)$ admits the representation* (12.14) *with $\sigma(M)$ in the closed upper half-plane;*
(iv) *the partial multiplicities of $L(\lambda)$ for real points of spectrum are all even;*
(v) *the degree of $L(\lambda)$ is even, and the sign characteristic of $L(\lambda)$ consists only of $+1$s.*

Proof. Implications (iii) ⇒ (ii) ⇒ (i) are evident.

(i) ⇒ (iv). Observe that the analytic eigenvalues $\mu_1(\lambda), \ldots, \mu_n(\lambda)$ of $L(\lambda)$ (defined in the preceding section) are nonnegative (for real λ), therefore the multiplicities of each real eigenvalue as a zero of $\mu_1(\lambda), \ldots, \mu_n(\lambda)$ are even. Now apply Theorem 12.5.

(i) ⇒ (v). Since $L(\lambda)$ is nonnegative, we have the $\mu_j(\lambda) > 0$ for real $\lambda \notin \sigma(L)$, $j = 1, \ldots, n$. Let $\lambda_0 \in \sigma(L)$. Then clearly the first nonzero derivative

$\mu_j^{(\alpha)}(\lambda_0), j = 1, \ldots, n$, is positive. So the signs are $+1$ in view of Theorem 12.5.

(v) $\Rightarrow$ (i). It is sufficient to show that $\mu_j(\lambda) \geq 0$ for real λ, $j = 1, \ldots, n$. But this can easily be deduced by using Theorem 12.5 again.

To complete the proof of Theorem 12.8 it remains to prove the implication (iv) $\Rightarrow$ (iii). First let us check that the degree l of $L(\lambda)$ is even. Indeed, otherwise in view of Theorem 10.4 $L(\lambda)$ will have at least one odd partial multiplicity corresponding to a real eigenvalue, which is impossible in view of (iv). Now let $L_1(\lambda)$ be the right divisor of $L(\lambda)$ constructed in Theorem 11.2, such that its c-set lies in the open upper half-plane. From the proof of Theorem 11.2 it is seen that the supporting subspace of L_1 is B-neutral (because the partial multiplicities are even). By Theorem 11.1.

$$L = L_1^* L_1,$$

so (iii) holds (with $M = L_1$). □

We can say more about the partial multiplicities of $L(\lambda) = M^*(\lambda)M(\lambda)$.

Theorem 12.9. *Let $M(\lambda)$ be a monic matrix polynomial. Let $\lambda_0 \in \sigma(M)$ be real, and let $\alpha_1 \geq \cdots \geq \alpha_r$ be the partial multiplicities of $M(\lambda)$ at λ_0. Then the partial multiplicities of $L(\lambda) = M^*(\lambda)M(\lambda)$ at λ_0 are $2\alpha_1, \ldots, 2\alpha_r$.*

Proof. Let (X, J, Y) be a Jordan triple for M. By Theorem 2.2, (Y^*, J^*, X^*) is a standard triple for $M^*(\lambda)$, and in view of the multiplication theorem (Theorem 3.1) the triple

$$[X \quad 0], \begin{bmatrix} J & YY^* \\ 0 & J^* \end{bmatrix}, \begin{bmatrix} 0 \\ X^* \end{bmatrix}$$

is a standard triple of $L(\lambda)$. Let J_0 be the part of J corresponding to λ_0, and and let Y_0 be the corresponding part of Y. We have to show that the elementary divisors of

$$\lambda I - \begin{bmatrix} J_0 & Y_0 Y_0^* \\ 0 & J_0^{\mathrm{T}} \end{bmatrix}$$

are $2\alpha_1, \ldots, 2\alpha_r$. Note that the rows of Y_0 corresponding to some Jordan block in J_0 and taken in the reverse order form a left Jordan chain of $M(\lambda)$ (or, what is the same, their transposes form a usual Jordan chain of $M^{\mathrm{T}}(\lambda)$ corresponding to λ_0). Let $y_1, \ldots, y_k$ be the left eigenvectors of $M(\lambda)$ (corresponding to λ^0). Then the desired result will follow from Lemma 3.4 if we show first that the matrix $A = \mathrm{col}[y_i]_{i=1}^k \cdot \mathrm{row}[y_i^*]_{i=1}^k$ is nonsingular and can be decomposed into a product of lower and upper triangular matrices.

Since $(Ax, x) \geq 0$ for all $x \in \mathbb{C}^n$ and $y_1^*, \ldots, y_k^*$ are linearly independent, the matrix A is positive definite. It is well known (and easily proved by induction on the size of A) that such a matrix can be represented as a product $A_1 A_2$ of the lower triangular matrix A_1 and the upper triangular matrix A_2

(the Cholesky factorization). So indeed we can apply Lemma 3.4 to complete the proof of Theorem 12.9. □

We remark that Theorem 12.9 holds for regular matrix polynomials $M(\lambda)$ as well. The proof in this case can be reduced to Theorem 12.9 by considering the monic matrix polynomial $\tilde{M}(\lambda) = \lambda^l M(\lambda^{-1} + a)M(a)^{-1}$, where l is the degree of $M(\lambda)$ and $a \notin \sigma(M)$.

Comments

The results presented in Sections 12.1–12.4 are from [34f, 34g], and we refer to these papers for additional information.

Factorizations of matrix polynomials of type (12.14) are well known in linear systems theory; see, for instance, [15a, 34h, 45], where the more general case of factorization of a self-adjoint polynomial $L(\lambda)$ with constant signature for all real λ is considered.

Chapter 13

Quadratic Self-Adjoint Polynomials

This is a short chapter in which we illustrate the concepts of spectral theory for self-adjoint matrix polynomials in the case of matrix polynomials of the type

$$L(\lambda) = I\lambda^2 + B\lambda + C, \tag{13.1}$$

where B and C are positive definite matrices. Such polynomials occur in the theory of damped oscillatory systems, which are governed by the system of equations

$$L\left(\frac{d}{dt}\right)x = \frac{d^2x}{dt^2} + B\frac{dx}{dt} + Cx = f, \tag{13.2}$$

Note that the numerical range and, in particular, all the eigenvalues of $L(\lambda)$, lie in the open left half-plane. Indeed, let

$$(L(\lambda)f, f) = \lambda^2(f, f) + \lambda(Bf, f) + (Cf, f) = 0$$

for some $\lambda \in \mathbb{C}$ and $f \neq 0$. Then

$$\lambda = \frac{-(Bf, f) \pm \sqrt{(Bf, f)^2 - 4(f, f)(Cf, f)}}{2(f, f)}, \tag{13.3}$$

and since $(Bf, f) > 0$, $(Cf, f) > 0$, the real part of λ is negative. This fact reflects dissipation of energy in damped oscillatory systems.

13.1. Overdamped Case

We consider first the case when the system governed by Eqs. (13.2) is *overdamped*, i.e., the inequality

$$(Bf,f)^2 - 4(f,f)(Cf,f) > 0 \tag{13.4}$$

holds for every $f \in \mathbb{C}^n \setminus \{0\}$. Equation (13.3) shows that, in the over-damped case, the numerical range of $L(\lambda)$ is real and, consequently, so are all the eigenvalues of $L(\lambda)$ (refer to Section 10.6).

Note that from the point of view of applications the overdamped systems are not *very* interesting. However, this case is relatively simple, and a fairly complete description of spectral properties of the quadratic self-adjoint polynomial (13.1) is available, as the following theorem shows.

Theorem 13.1. *Let system* (13.2) *be overdamped. Then*

(i) *all the eigenvalues of* $L(\lambda)$ *are real and nonpositive*;

(ii) *all the elementary divisors of* $L(\lambda)$ *are linear*;

(iii) *there exists a negative number* q *such that* n *eigenvalues* $\lambda_1^{(1)}, \ldots, \lambda_n^{(1)}$ *are less than* q *and* n *eigenvalues* $\lambda_1^{(2)}, \ldots, \lambda_n^{(2)}$ *are greater than* q;

(iv) *for* $i = 1$ *and* 2, *the sign of* $\lambda_j^{(i)}$ $(j = 1, \ldots, n)$ *in the sign characteristic of* $L(\lambda)$ *is* $(-1)^i$;

(v) *the eigenvectors* $u_1^{(i)}, \ldots, u_n^{(i)}$ *of* $L(\lambda)$, *corresponding to the eigenvalues* $\lambda_1^{(i)}, \ldots, \lambda_n^{(i)}$, *respectively, are linearly independent for* $i = 1, 2$.

Proof. Property (i) has been verified already. Let us prove (ii). If (ii) is false, then according to Theorem 12.2 there exists an eigenvalue λ_0 of $L(\lambda)$ and a corresponding eigenvector f_0 such that

$$(L'(\lambda_0)f_0, f_0) = 2\lambda_0(f_0, f_0) + (Bf_0, f_0) = 0.$$

But we also have

$$\lambda_0^2(f_0, f_0) + \lambda_0(Bf_0, f_0) + (Cf_0, f_0) = 0,$$

which implies

$$2\lambda_0(f_0, f_0) + (Bf_0, f_0) = \pm\sqrt{(Bf_0, f_0)^2 - 4(f_0, f_0)(Cf_0, f_0)} = 0,$$

a contradiction with (13.4); so (ii) is proved.

This argument shows also that for every eigenvalue λ_0 of $L(\lambda)$, the bilinear form $(L'(\lambda_0)f, f)$, $f \in \operatorname{Ker} L(\lambda_0)$ is positive definite or negative definite. By Theorem 12.2 this means that all the signs associated with a fixed eigenvalue λ_0 are the same (all $+1$ or all -1). By Proposition 10.12 (taking into account that by (ii) all partial multiplicities corresponding to the (real) eigenvalues of $L(\lambda)$ are equal to one) the number of $+1$s in the sign characteristic of $L(\lambda)$

is equal to the number of -1s, and since all the elementary divisors of $L(\lambda)$ are linear, the number of $+1$s (or -1s) is equal to n. Let $\lambda_i^{(1)}$, $i = 1, \ldots, n$, be the eigenvalues of $L(\lambda)$ (not necessarily different) having the sign -1, and let $\lambda_i^{(2)}$, $i = 1, \ldots, n$, be the eigenvalues of $L(\lambda)$ with the sign $+1$. According to the construction employed in the proof of Theorem 11.2 (Sections 11.2, 11.3) the eigenvectors $u_1^{(i)}, \ldots, u_n^{(i)}$ corresponding to $\lambda_1^{(i)}, \ldots, \lambda_n^{(i)}$, respectively, are linearly independent, for $i = 1, 2$. This proves (v).

Observe that for $i = 1, 2$ and $j = 1, \ldots, n$, we have

$$(L'(\lambda_j^{(i)})u_j^{(i)}, u_j^{(i)}) = 2\lambda_j^{(i)}(u_j^{(i)}, u_j^{(i)}) + (Bu_j^{(i)}, u_j^{(i)})$$

and

$$\lambda_j^{(i)} = \frac{-(Bu_j^{(i)}, u_j^{(i)}) \pm \sqrt{(Bu_j^{(i)}, u_j^{(i)})^2 - 4(u_j^{(i)}, u_j^{(i)})(Cu_j^{(i)}, u_j^{(i)})}}{2(u_j^{(i)}, u_j^{(i)})}. \tag{13.5}$$

So by Theorem 12.2, the sign in (13.5) is $+$ if $i = 2$, and $-$ if $i = 1$. We shall use this observation to prove (iii) and (iv).

For the proof of (iii) and (iv) it remains to show that

$$\lambda_i^{(1)} < \lambda_j^{(2)}, \qquad \text{for every} \quad i \text{ and } j \quad (1 \le i, j \le n). \tag{13.6}$$

Define two functions $\rho_1(x)$ and $\rho_2(x)$, for every $x \in \mathbb{C}^n \backslash \{0\}$, as follows:

$$\rho_1(x) = \frac{-(Bx, x) - d(x)}{2(x, x)}, \qquad \rho_2(x) = \frac{-(Bx, x) + d(x)}{2(x, x)},$$

where $d(x) = \sqrt{(Bx, x)^2 - 4(x, x)(Cx, x)}$. Let us show that

$$\{\rho_1(x) \mid x \in \mathbb{C}^n \backslash \{0\}\} < \min\{\lambda_1^{(2)}, \ldots, \lambda_n^{(2)}\}. \tag{13.7}$$

Suppose the contrary; then for some j we have:

$$\rho_1(y) \ge \lambda_j^{(2)} = \rho_2(x), \tag{13.8}$$

where x is some eigenvector corresponding to $\lambda_j^{(2)}$, and $y \in \mathbb{C}^n \backslash \{0\}$. Clearly, $\rho_2(y) > \rho_1(y)$ so that $\rho_2(y) \ne \rho_2(x)$, and since $\rho_2(\alpha x) = \rho_2(x)$ for every $\alpha \in \mathbb{C} \backslash \{0\}$, we deduce that x and y are linearly independent. Define

$$v = \mu y + (1 - \mu)x,$$

where μ is real. Then (as $L(\lambda_j^{(2)})x = 0$)

$$\begin{aligned}(\lambda_j^{(2)})^2(v, v) &+ \lambda_j^{(2)}(Bv, v) + (Cv, v) \\ &= \mu^2\{(\lambda_j^{(2)})^2(y, y) + \lambda_j^{(2)}(By, y) + (Cy, y)\}.\end{aligned} \tag{13.9}$$

However, $\rho_1(y) \ge \lambda_j^{(2)}$ implies that the right-hand side of (13.9) is non-negative and consequently,

$$(\lambda_j^{(2)})^2(v, v) + \lambda_j^{(2)}(Bv, v) + (Cv, v) \ge 0 \tag{13.10}$$

for all μ. Define $g(\mu) = 2\lambda_j^{(2)}(v, v) + (Bv, v)$; then $g(1) < 0$ by (13.8) and $g(0) > 0$ by (13.5) (with $i = 2$ and $u_j^{(i)} = x$; so the sign $+$ appears in (13.5)). Furthermore, $g(\mu)$ is a continuous function of μ, and hence there exists a $\mu_0 \in (0, 1)$ such that $g(\mu_0) = 2\lambda_j^{(2)}(v_0, v_0) + (Bv_0, v_0) = 0$, $v_0 = v(\mu_0)$. Together with (13.10) this implies $(Bv_0, v_0)^2 \leq 4(v_0, v_0)(Cv_0, v_0)$; since the system is assumed to be overdamped, this is possible only if $v_0 = 0$. This contradicts the fact that x and y are linearly independent. So (13.7) is proved.

Since

$$\lambda_i^{(1)} \in \{\rho_1(x) \mid x \in \mathbb{C}^n \backslash \{0\}\},$$

inequalities (13.6) follow immediately from (13.7). □

Observe that using Theorem 12.5 one can easily deduce that

$$\max\{\lambda_1^{(1)}, \ldots, \lambda_n^{(1)}\} < \lambda_0 < \min\{\lambda_1^{(2)}, \ldots, \lambda_n^{(2)}\}$$

if and only if $L(\lambda_0)$ is negative definite (assuming of course that the system is overdamped). In particular, the set $\{\lambda \in \mathbb{R} \mid L(\lambda) \text{ is negative definite}\}$ is not void, and the number q satisfies the requirements of Theorem 13.1 if and only if $L(q)$ is negative definite.

We shall deduce now general formulas for the solutions of the homogeneous equation corresponding to (13.2), as well as for the two-point boundary problem.

Theorem 13.2. *Let system* (13.2) *be overdamped, and let* $\lambda_j^{(i)}$ *be as in Theorem* 13.1. *Let* Γ_+ *(resp.* Γ_-*) be a contour in the complex plane containing no points of* $\sigma(L)$, *and such that* $\lambda_1^{(2)}, \ldots, \lambda_n^{(2)}$ *(resp.* $\lambda_1^{(1)}, \ldots, \lambda_n^{(1)}$*) are the only eigenvalues of* $L(\lambda)$ *lying inside* Γ_+ *(resp.* Γ_-*). Then*

(i) *the matrices* $\int_{\Gamma_+} L^{-1}(\lambda)\, d\lambda$ *and* $\int_{\Gamma_-} L^{-1}(\lambda)\, d\lambda$ *are nonsingular;*
(ii) *a general solution of* $L(d/dt)x = 0$ *has the form*

$$x(t) = e^{X_+ t}c_1 + e^{X_- t}c_2$$

for some $c_1, c_2 \in \mathbb{C}^n$, *where*

$$X_{\pm} = \left(\int_{\Gamma_{\pm}} L^{-1}(\lambda)\, d\lambda\right)^{-1} \int_{\Gamma_{\pm}} \lambda L^{-1}(\lambda)\, d\lambda; \tag{13.11}$$

(iii) *the matrices* X_+ *and* X_- *form a complete pair of solutions of the matrix equation* $X^2 + BX + C = 0$, *i.e.,* $X_+ - X_-$ *is nonsingular;*

(iv) *if* $b - a > 0$ *is large enough, then the matrix* $e^{X_+(b-a)} - e^{X_-(b-a)}$ *is nonsingular;*

(v) *for* $b - a > 0$ *large enough, the two-point boundary value problem*

$$L\left(\frac{d}{dt}\right)x(t) = f(t), \qquad x(a) = x(b) = 0$$

has the unique solution

$$x(t) = \int_a^b G_0(t, \tau) f(\tau)\, d\tau - [e^{X_+(b-a)}, e^{X_-(b-a)}] W \int_a^b \begin{bmatrix} e^{X_+(a-\tau)} \\ e^{X_-(a-\tau)} \end{bmatrix} Z f(\tau)\, d\tau,$$

where $Z = (X_+ - X_-)^{-1}$,

$$W = \begin{bmatrix} I & I \\ e^{X_+(b-a)} & e^{X_-(b-a)} \end{bmatrix}^{-1},$$

and

$$G_0(t, \tau) = \begin{cases} e^{X_+(t-\tau)} Z, & a \le t \le \tau \\ e^{X_-(t-\tau)} Z, & \tau \le t \le b. \end{cases}$$

Proof. By Theorem 11.2 and its proof, the matrix polynomial $L(\lambda)$ admits decompositions

$$L(\lambda) = (I\lambda - Y_+)(I\lambda - X_+) = (I\lambda - Y_-)(I\lambda - X_-),$$

where $\sigma(X_-) = \{\lambda_1^{(1)}, \ldots, \lambda_n^{(1)}\}$, $\sigma(X_+) = \{\lambda_1^{(2)}, \ldots, \lambda_n^{(2)}\}$. In view of Theorem 13.1(iii), $I\lambda - X_+$ and $I\lambda - Y_-$ are right and left Γ_+-spectral divisors of $L(\lambda)$, respectively; also $I\lambda - X_-$ and $I\lambda - Y_+$ are right and left Γ_--spectral divisors of $L(\lambda)$, respectively. So (i) follows from Theorem 4.2. By the same theorem, $X_\pm$ are given by formulas (13.11). Now by Theorem 2.16, (ii) and (v) follow from (iii) and (iv). Property (iii) follows immediately from part (a) of Theorem 2.16; so it remains to prove (iv).

Write

$$e^{X_+(b-a)} - e^{X_-(b-a)} = e^{X_+(b-a)} \cdot (I - e^{-X_+(b-a)} e^{X_-(b-a)}). \tag{13.12}$$

Now

$$\sigma(X_+) = \{\lambda_1^{(2)}, \ldots, \lambda_n^{(2)}\} \quad \text{and} \quad \sigma(X_-) = \{\lambda_1^{(1)}, \ldots, \lambda_n^{(1)}\}.$$

Therefore, we can estimate

$$\|e^{-X_+(b-a)}\| \le K_1 e^{-\lambda_0^{(2)}(b-a)}, \tag{13.13}$$

$$\|e^{X_-(b-a)}\| \le K_2 e^{\lambda_0^{(1)}(b-a)}, \tag{13.14}$$

where $\lambda_0^{(2)} = \min_{1 \le i \le n} \lambda_i^{(2)}$, $\lambda_0^{(1)} = \max_{1 \le i \le n} \lambda_i^{(1)}$, and K_1 and K_2 are constants. Indeed, by Theorem 13.1 (ii), $X_+ = S^{-1} \cdot \operatorname{diag}[\lambda_1^{(2)}, \ldots, \lambda_n^{(2)}] \cdot S$ for some nonsingular matrix S; so $e^{-X_+(b-a)} = S^{-1} \cdot \operatorname{diag}[e^{-\lambda_n^{(2)}(b-a)}, \ldots, e^{-\lambda_n^{(2)}(b-a)}] \cdot S$, and (13.13) follows. The proof of (13.14) is analogous. So

$$\|e^{-X_+(b-a)} e^{X_-(b-a)}\| \le K_1 K_2 \exp[(b-a)(\lambda_0^{(1)} - \lambda_0^{(2)})], \tag{13.15}$$

and since $\lambda_0^{(1)} < \lambda_0^{(2)}$, expression (13.15) can be made less than 1 for $b - a$ large enough. Hence the nonsingularity of (13.12) follows for such a choice of $b - a$, and (iv) is proved. □

13.2. Weakly Damped Case

Consider now the case when the system (13.2) satisfies the condition

$$(Bf,f)^2 - 4(f,f)(Cf,f) < 0, \qquad f \in \mathbb{C}^n \backslash \{0\}. \tag{13.16}$$

The system (13.2) for which (13.16) holds will be called *weakly damped*. Physically, condition (13.16) means that the free system (i.e., without external forces) can have only oscillatory solutions, whatever the initial values may be.

For a weakly damped system we have $(L(\lambda)f,f) > 0$ for every real λ and every $f \neq 0$; in particular, the polynomial $L(\lambda)$ is nonnegative and has no real eigenvalues. According to Theorem 12.8, $L(\lambda)$ therefore admits the factorization

$$L(\lambda) = (I\lambda - Z^*)(I\lambda - Z),$$

where $\sigma(Z)$ coincides with a maximal c-set S of eigenvalues of $L(\lambda)$ chosen in advance. (Recall that the set of eigenvalues S is called a c-set if $S \cap \{\bar{\lambda} \mid \lambda \in S\} = \varnothing$.) In particular, as a maximal c-set one may wish to choose the eigenvalues lying in the open upper half-plane. In such a case, by Theorem 2.15, the matrices Z and Z^* form a complete pair of solutions of the matrix equation $X^2 + BX + C = 0$.

If, in addition, $e^{Z(b-a)} - e^{Z^*(b-a)}$ is nonsingular, then we are led to explicit formulas for the general solution of the homogeneous equation, and the solution of the two-point boundary value problem for (13.2), analogous to those of Theorem 13.2 for overdamped systems. The reader will be able to derive these formulas, as required.

Note also that the weak damping hypothesis does not yield a result analogous to part (ii) of Theorem 13.1 for overdamped systems, i.e., nonlinear elementary divisors may arise under the weak damping hypothesis. This occurs in the following example.

EXAMPLE 13.1. Let

$$L(\lambda) = \begin{bmatrix} 1 & 0 \\ 0 & 1 \end{bmatrix}\lambda^2 + \begin{bmatrix} 1 & \sqrt{3}/2 \\ \sqrt{3}/2 & 2 \end{bmatrix}\lambda + \begin{bmatrix} 1 & 0 \\ 0 & 4 \end{bmatrix}.$$

Note that the matrices

$$B = \begin{bmatrix} 1 & \sqrt{3}/2 \\ \sqrt{3}/2 & 2 \end{bmatrix}, \qquad C = \begin{bmatrix} 1 & 0 \\ 0 & 4 \end{bmatrix}$$

are positive definite. The eigenvalues of $L(\lambda)$ are $\frac{1}{4}(-3 \pm i\sqrt{23})$ each taken twice. It is easily seen that $L(\lambda_0) \neq 0$ for $\lambda_0 = \frac{1}{4}(-3 \pm i\sqrt{23})$; so for each eigenvalue there is only one eigenvector (up to multiplication by a nonzero number). So the elementary divisors of $L(\lambda)$ must be quadratic:

$$(\lambda - \tfrac{1}{4}(-3 + i\sqrt{23}))^2 \qquad (\lambda - \tfrac{1}{4}(-3 - i\sqrt{23}))^2. \quad \square$$

Comments

The results of this chapter are in general not new. The finite-dimensional overdamped case was first studied in [17] (see also [52b]). As has been remarked earlier, this case has generated many generalizations almost all of which are beyond the subject matter of this book. Results in the infinite-dimensional setting are to be found in [51]. Example 13.1 is in [52d].

Part IV

Supplementary Chapters in Linear Algebra

In order to make this book more self-contained, several topics of linear algebra are presented in Part IV which are called upon in the first three parts. None of this subject matter is new. However, some of the topics are not readily accessible in standard texts on linear algebra and matrix theory. In particular, the self-contained treatment of Chapter S5 cannot be found elsewhere in such a form. Other topics (as in Chapter S1) are so important for the theory of matrix polynomials that advantage can be gained from an exposition consistent with the style and objectives of the whole book. Standard works on linear algebra and matrix theory which will be useful for filling any gaps in our exposition include [22], [23], and [52c].

Chapter S1

The Smith Form and Related Problems

This chapter is devoted to the Smith form for matrix polynomials (which is sometimes called a canonical form, or diagonal form of λ-matrices) and its applications. As a corollary of this analysis we derive the Jordan normal form for matrices. A brief introduction to functions of matrices is also presented.

S1.1. The Smith Form

Let $L(\lambda)$ be a matrix polynomial of type $\sum_{j=0}^{s} A_j \lambda^j$, where A_j are $m \times n$ matrices whose entries are complex numbers (so that we admit the case of rectangular matrices A_j). The main result, in which the Smith form of such a polynomial is described, is as follows.

Theorem S1.1. *Every $m \times n$ matrix polynomial $L(\lambda)$ admits the representation*

$$L(\lambda) = E(\lambda)D(\lambda)F(\lambda), \tag{S1.1}$$

where

$$D(\lambda) = \begin{bmatrix} d_1(\lambda) & & & & 0 \\ & \ddots & & & \\ & & d_r(\lambda) & & \vdots \\ & & & 0 & \\ 0 & & \cdots & & \ddots \; 0 \end{bmatrix} \tag{S1.2}$$

is a diagonal polynomial matrix with monic scalar polynomials $d_i(\lambda)$ such that $d_i(\lambda)$ is divisible by $d_{i-1}(\lambda)$; $E(\lambda)$ and $F(\lambda)$ are matrix polynomials of sizes $m \times m$ and $n \times n$, respectively, with constant nonzero determinants.

The proof of Theorems 1.1 will be given later.

Representation (S1.1) as well as the diagonal matrix $D(\lambda)$ from (S1.2) is called the Smith form of the matrix polynomial $L(\lambda)$ and plays an important role in the analysis of matrix polynomials. Since $\det E(\lambda) \equiv \text{const} \neq 0$, $\det F(\lambda) \equiv \text{const} \neq 0$, where $E(\lambda)$ and $F(\lambda)$ are taken from the Smith form (S1.1), the inverses $(E(\lambda))^{-1}$ and $(F(\lambda))^{-1}$ are also matrix polynomials with constant nonzero determinant. The matrix polynomials $E(\lambda)$ and $F(\lambda)$ are not defined uniquely, but the diagonal matrix $D(\lambda)$ is unique. This follows from the fact that the entries of $D(\lambda)$ can be expressed in terms of the original matrix itself, as the next theorem shows.

We shall need the definition of minors of the matrix polynomial $L(\lambda) = (a_{ij}(\lambda))_{i=1,\dots,m}^{j=1,\dots,n}$. Choose k rows, $1 \leq i_1 < \cdots < i_k \leq m$, and k columns, $1 \leq j_1 < \cdots < j_k \leq n$, in $L(\lambda)$, and consider the determinant $\det(a_{i_l j_m})_{l,m=1}^{k}$ of the $k \times k$ submatrix of $L(\lambda)$ formed by the rows and columns. This determinant is called a minor of $L(\lambda)$. Loosely speaking, we shall say that this minor is of order k and is composed of the rows $i_1, \dots, i_k$ and columns $j_1, \dots, j_k$ of L. Taking another set of columns and/or rows, we obtain another minor of order k of $L(\lambda)$. The total number of minors of order k is $\binom{n}{k}\binom{m}{k}$.

Theorem S1.2. *Let $L(\lambda)$ be an $m \times n$ matrix polynomial. Let $p_k(\lambda)$ be the greatest common divisor (with leading coefficient 1) of the minors of $L(\lambda)$ of order k, if not all of them are zeros, and let $p_k(\lambda) \equiv 0$ if all the minors of order k of $L(\lambda)$ are zeros. Let $p_0(\lambda) = 1$ and $D(\lambda) = \operatorname{diag}[d_1(\lambda), \dots, d_r(\lambda), 0, \dots, 0]$ be the Smith form of $L(\lambda)$. Then r is the maximal integer such that $p_r(\lambda) \not\equiv 0$, and*

$$d_i(\lambda) = p_i(\lambda)/p_{i-1}(\lambda), \qquad i = 1, \dots, r. \tag{S1.3}$$

Proof. Let us show that if $L_1(\lambda)$ and $L_2(\lambda)$ are matrix polynomials of the same size such that

$$L_1(\lambda) = E(\lambda)L_2(\lambda)F(\lambda),$$

where $E^{\pm 1}(\lambda)$ and $F^{\pm 1}(\lambda)$ are square matrix polynomials, then the greatest common divisors $p_{k,1}(\lambda)$ and $p_{k,2}(\lambda)$ of the minors of order k of $L_1(\lambda)$ and $L_2(\lambda)$, respectively, are equal. Indeed, apply the Binet–Cauchy formula (see, for instance, [22] or [52c]) twice to express a minor of $L_1(\lambda)$ of order k as a linear combination of minors of $L_2(\lambda)$ of the same order. It therefore follows that $p_{k,2}(\lambda)$ is a divisor of $p_{k,1}(\lambda)$. But the equation

$$L_2(\lambda) = E^{-1}(\lambda)L_1(\lambda)F^{-1}(\lambda)$$

implies that $p_{k,1}(\lambda)$ is a divisor of $p_{k,2}(\lambda)$. So $p_{k,1}(\lambda) = p_{k,2}(\lambda)$. In the same way one shows that the maximal integer r_1 such that $p_{r_1,1}(\lambda) \not\equiv 0$, coincides with the maximal integer r_2 such that $p_{r_2,2}(\lambda) \not\equiv 0$.

Now apply this observation for the matrix polynomials $L(\lambda)$ and $D(\lambda)$. It follows that we have to prove Theorems 1.2 only in the case that $L(\lambda)$ itself is in the diagonal form $L(\lambda) = D(\lambda)$. From the structure of $D(\lambda)$ it is clear that

$$d_1(\lambda)d_2(\lambda)\cdots d_s(\lambda), \qquad s = 1, \ldots, r,$$

is the greatest common divisor of the minors of $D(\lambda)$ of order s. So $p_s(\lambda) = d_1(\lambda)\cdots d_s(\lambda)$, $s = 1, \ldots, r$, and (1.3) follows. □

To prove Theorem S1.1, we shall use the following elementary transformations of a matrix polynomial $L(\lambda)$ of size $m \times n$: (1) interchange two rows, (2) add to some row another row multiplied by a scalar polynomial, and (3) multiply a row by a nonzero complex number, together with the three corresponding operations on columns.

Note that each of these transformations is equivalent to the multiplication of $L(\lambda)$ by an invertible matrix as follows:

Interchange of rows (columns) i and j in $L(\lambda)$ is equivalent to multiplication on the left (right) by

$$\begin{array}{r} \\ \\ i\rightarrow \\ \\ j\rightarrow \\ \\ \\ \end{array}\begin{bmatrix} 1 & & & & & \cdots & 0 \\ \vdots & \ddots & & & & & \vdots \\ & & 0 & \cdots & 1 & \cdots & \\ \vdots & & \vdots & & \vdots & & \vdots \\ & \cdots & 1 & \cdots & 0 & & \\ \vdots & & & & & \ddots & \vdots \\ 0 & \cdots & & & & & 1 \end{bmatrix}. \qquad \text{(S1.4)}$$

Adding to the ith row of $L(\lambda)$ the jth row multiplied by the polynomial $f(\lambda)$ is equivalent to multiplication on the left by

$$
\begin{array}{c}
\\ \\ \\ \\ i\rightarrow \\ \\ \\ \\ \\ \\ \\
\end{array}
\begin{bmatrix}
1 & & & & & & & & \overset{j}{\downarrow} & & \\
& \ddots & & & & & & & \vdots & & \\
& & 1 & & & & & & & & \\
& & & \ddots & & & & & & & \\
& & \cdots & & 1 & & \cdots & & f(\lambda) & & \\
& & & & & \ddots & & & \vdots & & \\
& & & & & & 1 & & & & \\
& & & & & & & \ddots & & & \\
& & & & & & & & 1 & & \\
& & & & & & & & & \ddots & \\
& & & & & & & & & & 1
\end{bmatrix}; \qquad \text{(S1.5)}
$$

the same operation for columns is equivalent to multiplication on the right by the matrix

$$
\begin{array}{c}
\\ \\ \\ \\ i\rightarrow \\ \\ \\ \\ \\ \\ j\rightarrow \\ \\ \\
\end{array}
\begin{bmatrix}
1 & & & & & & & & & & & & \\
& \ddots & & & & & & & & & & & \\
& & 1 & & \vdots & & & & & & & & \\
& & & \ddots & & & & & & & & & \\
& & \cdots & & 1 & & & & & & & & \\
& & & & & \ddots & & & & & & & \\
& & & & & & 1 & & & & & & \\
& & & & \vdots & & & \ddots & & & & & \\
& & & & & & & & 1 & & & & \\
& & & & & & & & & \ddots & & & \\
& & \cdots & & f(\lambda) & \cdots & & \cdots & & & 1 & & \\
& & & & & & & & & & & \ddots & \\
& & & & & & & & & & & & 1
\end{bmatrix}. \qquad \text{(S1.6)}
$$

Finally multiplication of the ith row (column) in $L(\lambda)$ by a number $a \neq 0$ is equivalent to the multiplication on the left (right) by

$$
i\rightarrow \begin{bmatrix} 1 & & & & & & \\ & \ddots & & & & & \\ & & 1 & & & & \\ & & & \ddots & & & \\ \cdots & \cdots & & & a & & & \\ & & & & & \ddots & & \\ & & & & & & 1 & \\ & & & & & & & \ddots & \\ & & & & & & & & 1 \end{bmatrix} \tag{S1.7}
$$

(Empty places in (S1.4)–(S1.7) are assumed to be zeros.) Matrices of the form (S1.4)–(S1.7) will be called *elementary*. It is apparent that the determinant of any elementary matrix is a nonzero constant.

Proof of Theorem S1.1. Since the determinant of any elementary matrix is a nonzero constant, it is sufficient to prove that by applying a sequence of elementary transformations every matrix polynomial $L(\lambda)$ can be reduced to a diagonal form: $\operatorname{diag}[d_1(\lambda), \ldots, d_r(\lambda), 0, \ldots, 0]$, (the zeros on the diagonal can be absent, in which case $r = \min(m, n)$), where $d_1(\lambda), \ldots, d_r(\lambda)$ are scalar polynomials such that $d_i(\lambda)/d_{i-1}(\lambda)$, $i = 1, 2, \ldots, r - 1$, are also scalar polynomials. We shall prove this statement by induction on m and n. For $m = n = 1$ it is evident.

Consider now the case $m = 1$, $n > 1$, i.e.,

$$L(\lambda) = [a_1(\lambda) \quad a_2(\lambda) \quad \cdots \quad a_n(\lambda)].$$

If all $a_j(\lambda)$ are zeros, there is nothing to prove. Suppose that not all of $a_j(\lambda)$ are zeros, and let $a_{j_0}(\lambda)$ be a polynomial of minimal degree among the nonzero entries of $L(\lambda)$. We can suppose that $j_0 = 1$ (otherwise interchange columns in $L(\lambda)$). By elementary transformations it is possible to replace all the other entries in $L(\lambda)$ by zeros. Indeed, let $a_j(\lambda) \neq 0$. Divide $a_j(\lambda)$ by $a_1(\lambda)$: $a_j(\lambda) = b_j(\lambda)a_1(\lambda) + r_j(\lambda)$, where $r_j(\lambda)$ is the remainder and its degree is less than the degree of $a_1(\lambda)$, or $r_j(\lambda) \equiv 0$. Add to the jth column the first column multiplied by $-b_j(\lambda)$. Then in the place j in the new matrix will be $r_j(\lambda)$. If $r_j(\lambda) \not\equiv 0$, then put $r_j(\lambda)$ in the place 1, and if there still exists a nonzero entry (different from $r_j(\lambda)$), apply the same argument again. Namely, divide this (say, the kth) entry by $r_j(\lambda)$ and add to the kth column the first multiplied by minus the quotient of the division, and so on. Since the degrees of the remainders decrease, after a finite number (not more than the degree of $a_1(\lambda)$) of steps we find that all the entries in our matrix, except the first, are zeros.

This proves Theorem S1.1 in the case $m = 1, n > 1$. The case $m > 1, n = 1$ is treated in a similar way.

Assume now $m, n > 1$, and assume that the theorem is proved for $m - 1$ and $n - 1$. We can suppose that the (1, 1) entry of $L(\lambda)$ is nonzero and has the minimal degree among the nonzero entries of $L(\lambda)$ (indeed, we can reach this condition by interchanging rows and/or columns in $L(\lambda)$ if $L(\lambda)$ contains a nonzero entry; if $L(\lambda) \equiv 0$, Theorem S1.1 is trivial). With the help of the procedure described in the previous paragraph (applied for the first row and the first column of $L(\lambda)$), by a finite number of elementary transformations we reduce $L(\lambda)$ to the form

$$L_1(\lambda) = \begin{bmatrix} a_{11}^{(1)}(\lambda) & 0 & \cdots & 0 \\ 0 & a_{22}^{(1)}(\lambda) & \cdots & a_{2n}^{(1)}(\lambda) \\ \vdots & \vdots & & \vdots \\ 0 & a_{m2}^{(1)}(\lambda) & \cdots & a_{mn}^{(1)}(\lambda) \end{bmatrix}.$$

Suppose that some $a_{ij}^{(1)}(\lambda) \not\equiv 0$ $(i, j > 1)$ is not divisible by $a_{11}^{(1)}(\lambda)$ (without remainder). Then add to the first row the ith row and apply the above arguments again. We obtain a matrix polynomial of the structure

$$L_2(\lambda) = \begin{bmatrix} a_{11}^{(2)}(\lambda) & 0 & \cdots & 0 \\ 0 & a_{22}^{(2)}(\lambda) & \cdots & a_{2n}^{(2)}(\lambda) \\ \vdots & \vdots & & \vdots \\ 0 & a_{m2}^{(2)}(\lambda) & \cdots & a_{mn}^{(2)}(\lambda) \end{bmatrix},$$

where the degree of $a_{11}^{(2)}(\lambda)$ is less than the degree of $a_{11}^{(1)}(\lambda)$. If there still exists some entry $a_{ij}^{(2)}(\lambda)$ which is not divisible by $a_{11}^{(2)}(\lambda)$, repeat the same procedure once more, and so on. After a finite number of steps we obtain the matrix

$$L_3(\lambda) = \begin{bmatrix} a_{11}^{(3)}(\lambda) & 0 & \cdots & 0 \\ 0 & a_{22}^{(3)}(\lambda) & \cdots & a_{2n}^{(3)}(\lambda) \\ \vdots & \vdots & & \vdots \\ 0 & a_{m2}^{(3)}(\lambda) & \cdots & a_{mn}^{(3)}(\lambda) \end{bmatrix},$$

where every $a_{ij}^{(3)}(\lambda)$ is divisible by $a_{11}^{(3)}(\lambda)$. Multiply the first row (or column) by a nonzero constant to make the leading coefficient of $a_{11}^{(3)}(\lambda)$ equal to 1. Now write

$$L_4(\lambda) = \frac{1}{a_{11}^{(3)}(\lambda)} \begin{bmatrix} a_{22}^{(3)}(\lambda) & \cdots & a_{2n}^{(3)}(\lambda) \\ \vdots & & \vdots \\ a_{m2}^{(3)}(\lambda) & \cdots & a_{mn}^{(3)}(\lambda) \end{bmatrix},$$

(here $L_4(\lambda)$ is an $(m - 1) \times (n - 1)$ matrix), and apply the induction hypothesis for $L_4(\lambda)$ to complete the proof of Theorem S1.1. □

The existence of the Smith form allows us to prove easily the following fact.

Corollary S1.3. *An $n \times n$ matrix polynomial $L(\lambda)$ has constant nonzero determinant if and only if $L(\lambda)$ can be represented as a product $L(\lambda) = F_1(\lambda)F_i(\lambda) \cdots F_p(\lambda)$ of a finite number of elementary matrices $F_i(\lambda)$.*

Proof. Since $\det F_i(\lambda) \equiv \text{const} \neq 0$, obviously any product of elementary matrices has a constant nonzero determinant. Conversely, suppose $\det L(\lambda) \equiv \text{const} \neq 0$. Let

$$L(\lambda) = E(\lambda)D(\lambda)F(\lambda) \tag{S1.8}$$

be the Smith form of $L(\lambda)$. Note that by definition of a Smith form $E(\lambda)$ and $F(\lambda)$ are products of elementary matrices. Taking determinants in (S1.8), we obtain

$$\det L(\lambda) = \det E(\lambda) \cdot \det D(\lambda) \cdot \det F(\lambda),$$

and, consequently, $\det D(\lambda) \equiv \text{const} \neq 0$. This happens if and only if $D(\lambda) = I$. But then $L(\lambda) = E(\lambda)F(\lambda)$ is a product of elementary matrices. □

S1.2. Invariant Polynomials and Elementary Divisors

The diagonal elements $d_1(\lambda), \ldots, d_r(\lambda)$ in the Smith form are called the *invariant polynomials* of $L(\lambda)$. The number r of invariant polynomials can be defined as

$$r = \max_{\lambda \in C}\{\text{rank } L(\lambda)\}. \tag{S1.9}$$

Indeed, since $E(\lambda)$ and $F(\lambda)$ from (S1.1) are invertible matrices for every λ, we have $\text{rank } L(\lambda) = \text{rank } D(\lambda)$ for every $\lambda \in \mathbb{C}$. On the other hand, it is clear that $\text{rank } D(\lambda) = r$ if λ is not a zero of one of the invariant polynomials, and $\text{rank } D(\lambda) < r$ otherwise. So (S1.9) follows.

Represent each invariant polynomial as a product of linear factors

$$d_i(\lambda) = (\lambda - \lambda_{i1})^{\alpha_{i1}} \cdots (\lambda - \lambda_{i,k_i})^{\alpha_{ik_i}}, \qquad i = 1, \ldots, r,$$

where $\lambda_{i1}, \ldots, \lambda_{i,k_i}$ are different complex numbers and $\alpha_{i1}, \ldots, \alpha_{ik_i}$ are positive integers. The factors $(\lambda - \lambda_{ij})^{\alpha_{ij}}, j = 1, \ldots, k_i, i = 1, \ldots, r$, are called the *elementary divisors* of $L(\lambda)$. An elementary divisor is said to be *linear* or *nonlinear* according as $\alpha_{ij} = 1$ or $\alpha_{ij} > 1$.

Some different elementary divisors may contain the same polynomial $(\lambda - \lambda_0)^\alpha$ (this happens, for example, in case $d_i(\lambda) = d_{i+1}(\lambda)$ for some i); the total number of elementary divisors of $L(\lambda)$ is therefore $\sum_{i=1}^r k_i$.

Consider a simple example.

EXAMPLE S1.1. Let

$$L(\lambda) = \begin{bmatrix} \lambda(\lambda-1) & 1 \\ 0 & \lambda(\lambda-1) \end{bmatrix}$$

First let us find the Smith form for $L(\lambda)$ (we shall not mention explicitly the elementary transformations):

$$\begin{bmatrix} \lambda(\lambda-1) & 1 \\ 0 & \lambda(\lambda-1) \end{bmatrix} \to \begin{bmatrix} 0 & 1 \\ -\lambda^2(\lambda-1)^2 & \lambda(\lambda-1) \end{bmatrix} \to$$

$$\begin{bmatrix} 0 & 1 \\ -\lambda^2(\lambda-1)^2 & 0 \end{bmatrix} \to \begin{bmatrix} 1 & 0 \\ 0 & \lambda^2(\lambda-1)^2 \end{bmatrix}.$$

Thus the elementary divisors are λ^2 and $(\lambda-1)^2$. □

The degrees α_{ij} of the elementary divisors form an important characteristic of the matrix polynomial $L(\lambda)$. As we shall see later, they determine, in particular, the Jordan structure of $L(\lambda)$. Here we mention only the following simple property of the elementary divisors, whose verification is left to the reader.

Proposition S1.4. *Let $L(\lambda)$ be an $n \times n$ matrix polynomial such that* $\det L(\lambda) \not\equiv 0$. *Then the sum $\sum_{i=1}^{r} \sum_{j=1}^{k_i} \alpha_{ij}$ of degrees of its elementary divisors $(\lambda - \lambda_{ij})^{\alpha_{ij}}$ coincides with the degree of* $\det L(\lambda)$.

Note that the knowledge of the elementary divisors of $L(\lambda)$ and of the number r of its invariant polynomials $d_1(\lambda), \ldots, d_r(\lambda)$ is sufficient to construct $d_1(\lambda), \ldots, d_r(\lambda)$. In this construction we use the fact that $d_i(\lambda)$ is divisible by $d_{i-1}(\lambda)$. Let $\lambda_1, \ldots, \lambda_p$ be all the different complex numbers which appear in the elementary divisors, and let $(\lambda - \lambda_i)^{\alpha_{i1}}, \ldots, (\lambda - \lambda_i)^{\alpha_{i,k_i}}$ $(i = 1, \ldots, p)$ be the elementary divisors containing the number λ_i, and ordered in the descending order of the degrees $\alpha_{i1} \geq \cdots \geq \alpha_{i,k_i} > 0$. Clearly, the number r of invariant polynomials must be greater than or equal to $\max\{k_1, \ldots, k_p\}$. Under this condition, the invariant polynomials $d_1(\lambda), \ldots, d_r(\lambda)$ are given by the formulas

$$d_j(\lambda) = \prod_{i=1}^{p} (\lambda - \lambda_i)^{\alpha_{i,r+1-j}}, \qquad j = 1, \ldots, r,$$

where we put $(\lambda - \lambda_i)^{\alpha_{ij}} = 1$ for $j > k_i$.

The following property of the elementary divisors will be used subsequently:

Proposition S1.5. *Let $A(\lambda)$ and $B(\lambda)$ be matrix polynomials, and let $C(\lambda) =$ $\operatorname{diag}[A(\lambda), B(\lambda)]$, a block diagonal matrix polynomial. Then the set of elementary divisors of $C(\lambda)$ is the union of the elementary divisors of $A(\lambda)$ and $B(\lambda)$.*

Proof. Let $D_1(\lambda)$ and $D_2(\lambda)$ be the Smith forms of $A(\lambda)$ and $B(\lambda)$, respectively. Then clearly

$$C(\lambda) = E(\lambda)\begin{bmatrix} D_1(\lambda) & 0 \\ 0 & D_2(\lambda) \end{bmatrix} F(\lambda)$$

for some matrix polynomials $E(\lambda)$ and $F(\lambda)$ with constant nonzero determinant. Let $(\lambda - \lambda_0)^{\alpha_1}, \ldots, (\lambda - \lambda_0)^{\alpha_p}$ and $(\lambda - \lambda_0)^{\beta_1}, \ldots, (\lambda - \lambda_0)^{\beta_q}$ be the elementary divisors of $D_1(\lambda)$ and $D_2(\lambda)$, respectively, corresponding to the same complex number λ_0. Arrange the set of exponents $\alpha_1, \ldots, \alpha_p, \beta_1, \ldots, \beta_q$, in a nonincreasing order: $\{\alpha_1, \ldots, \alpha_p, \beta_1, \ldots, \beta_q\} = \{\gamma_1, \ldots, \gamma_{p+q}\}$, where $0 < \gamma_1 \leq \cdots \leq \gamma_{p+q}$. Using Theorem S1.2 it is clear that in the Smith form $D = \operatorname{diag}[d_1(\lambda), \ldots, d_r(\lambda), 0, \ldots, 0]$ of $\operatorname{diag}[D_1(\lambda), D_2(\lambda)]$, the invariant polynomial $d_r(\lambda)$ is divisible by $(\lambda - \lambda_0)^{\gamma_{p+q}}$ but not by $(\lambda - \lambda_0)^{\gamma_{p+q}+1}$, $d_{r-1}(\lambda)$ is divisible by $(\lambda - \lambda_0)^{\gamma_{p+q-1}}$ but not by $(\lambda - \lambda_0)^{\gamma_{p+q-1}+1}$, and so on. It follows that the elementary divisors of

$$\begin{bmatrix} D_1(\lambda) & 0 \\ 0 & D_2(\lambda) \end{bmatrix}$$

(and therefore also those of $C(\lambda)$) corresponding to λ_0, are just $(\lambda - \lambda_0)^{\gamma_1}, \ldots, (\lambda - \lambda_0)^{\gamma_{p+q}}$, and Proposition S1.5 is proved. □

S1.3. Application to Differential Equations with Constant Coefficients

Consider the system of homogeneous differential equations

$$A_l \frac{d^l x}{dt^l} + \cdots + A_1 \frac{dx}{dt} + A_0 x = 0, \tag{S1.10}$$

where $x = x(t)$ is a vector function (differentiable l times) of the real argument t with values in $\mathbb{C}^n$, and $A_l, \ldots, A_0$ are constant $m \times n$ matrices.

Introduce the following matrix polynomial connected with system (S1.9)

$$A(\lambda) = \sum_{j=0}^{l} A_j \lambda^j. \tag{S1.11}$$

The properties of the solutions of (S1.10) are closely related to the spectral properties of the polynomial $A(\lambda)$. This connection appears many times in the book. Here we shall only mention the possibility of reducing system (S1.10) to n (or less) independent scalar equations, using the Smith form of the matrix polynomial (S1.11).

Given matrix polynomial $B(\lambda) = \sum_{j=0}^{p} B_j \lambda^j$ and a p-times differentiable vector function $y = y(t)$, denote for brevity

$$B\left(\frac{d}{dt}\right)y = \sum_{j=1}^{p} B_j \frac{d^j y}{dt^j}.$$

Thus, $B(d/dt)y$ is obtained if we write formally $(\sum_{j=1}^{p} B_j \lambda^j)y$ and then replace λ by d/dt; for example, if

$$B(\lambda) = \begin{bmatrix} \lambda^2 & \lambda + 2 \\ 0 & \lambda^3 + 1 \end{bmatrix} \quad \text{and} \quad y = \begin{bmatrix} y_1(t) \\ y_2(t) \end{bmatrix},$$

then

$$B\left(\frac{d}{dt}\right)y = \begin{bmatrix} y_1'' + y_2' + 2y_2 \\ y_2''' + y_2 \end{bmatrix}.$$

Note the following simple property. If $B(\lambda) = B_1(\lambda)B_2(\lambda)$ is a product of two matrix polynomials $B_1(\lambda)$ and $B_2(\lambda)$, then

$$B\left(\frac{d}{dt}\right)y = B_1\left(\frac{d}{dt}\right)\left(B_2\left(\frac{d}{dt}\right)y\right). \tag{S1.12}$$

This property follows from the facts that for $\alpha_1, \alpha_2 \in \mathbb{C}$

$$\frac{d}{dt}(\alpha_1 y_1 + \alpha_2 y_2) = \alpha_1 \frac{d}{dt} y_1 + \alpha_2 \frac{d}{dt} y_2,$$

and

$$\frac{d^j y}{dt^j} = \frac{d^i}{dt^i}\left(\frac{d^{j-i} y}{dt^{j-i}}\right), \qquad 0 \leq i \leq j.$$

Let us go back to the system (S1.10) and the corresponding matrix polynomial (S1.11). Let $D(\lambda) = \operatorname{diag}[d_1(\lambda), \ldots, d_r(\lambda), 0 \cdots 0]$ be the Smith form of $A(\lambda)$:

$$A(\lambda) = E(\lambda)D(\lambda)F(\lambda), \tag{S1.13}$$

where $E(\lambda)$ and $F(\lambda)$ are matrix polynomials with constant nonzero determinant. According to (S1.12), the system (S1.10) can be written in the form

$$E\left(\frac{d}{dt}\right)D\left(\frac{d}{dt}\right)F\left(\frac{d}{dt}\right)x(t) = 0. \tag{S1.14}$$

Denote $y = F(d/dt)x(t)$. Multiplying (S1.14) on the left by $E^{-1}(d/dt)$, we obtain the system

$$\begin{bmatrix} d_1\left(\frac{d}{dt}\right) & \cdots & & & \cdots & 0 \\ & & \vdots & & & 0 \\ & \ddots & & & & \\ \vdots & \vdots & d_r\left(\frac{d}{dt}\right) & \vdots & & \vdots \\ & & & 0 & & \\ & & \vdots & & \ddots & \\ 0 & 0 & & & & 0 \end{bmatrix} \begin{bmatrix} y_1 \\ \vdots \\ y_r \\ \vdots \\ y_n \end{bmatrix} = 0, \tag{S1.15}$$

($y = (y_1, \ldots, y_n)^{\mathrm{T}}$), which is equivalent to (S1.14). System (S1.15) splits into r independent scalar equations

$$d_i\left(\frac{d}{dt}\right)y_i(t) = 0, \qquad i = 1, \ldots, r, \tag{S1.16}$$

(the last $n - r$ equations are identities: $0 \cdot y_i = 0$ for $i = r + 1, \ldots, n$). As is well known, solutions of the ith equation from (S1.16) form a m_i-dimensional linear space Π_i, where m_i is the degree of $d_i(t)$. It is clear then that every solution of (S1.15) has the form $(y_1(t), \ldots, y_n(t))$ with $y_i(t) \in \Pi_i(t)$, $i = 1, \ldots, r$, and $y_{r+1}(t), \ldots, y_n(t)$ are arbitrary l-times differentiable functions. The solutions of (S1.10) are now given by the formulas

$$x = \begin{bmatrix} x_1 \\ \vdots \\ x_n \end{bmatrix} = F^{-1}\left(\frac{d}{dt}\right)\begin{bmatrix} y_1 \\ \vdots \\ y_n \end{bmatrix}$$

where $y_i \in \Pi_i$, $i = 1, \ldots, r$, and $y_{r+1}, \ldots, y_n$ are arbitrary l-times differentiable functions. As a particular case of this fact we obtain the following result.

Theorem S1.6. *Let $A(\lambda) = \sum_{j=0}^{l} A_j\lambda^j$ be an $n \times n$ matrix polynomial such that* $\det A(\lambda) \not\equiv 0$. *Then the dimension of the solution space of* (S1.10) *is equal to the degree of* $\det L(\lambda)$.

Indeed, under the condition $\det A(\lambda) \not\equiv 0$, the Smith form of $A(\lambda)$ does not contain zeros on the main diagonal. Thus, the dimension of the solution space is just $\dim \Pi_1 + \cdots + \dim \Pi_n$, which is exactly degree($\det A(\lambda)$) (Proposition S1.4). Theorem S1.6 is often referred to as Chrystal's theorem.

Let us illustrate this reduction to scalar equations in an example.

EXAMPLE S1.2. Consider the system

$$\frac{d}{dt}x_1 + x_2 = 0, \qquad x_1 + \frac{d^2}{dt^2}x_2 = 0. \tag{S1.17}$$

The corresponding matrix polynomial is

$$A(\lambda) = \begin{bmatrix} \lambda & 1 \\ 1 & \lambda^2 \end{bmatrix}.$$

Compute the Smith form for $A(\lambda)$:

$$A(\lambda) = \begin{bmatrix} 1 & 0 \\ \lambda^2 & 1 \end{bmatrix}\begin{bmatrix} 1 & 0 \\ 0 & \lambda^3 - 1 \end{bmatrix}\begin{bmatrix} \lambda & 1 \\ -1 & 0 \end{bmatrix}.$$

Equations (S1.15) take the form

$$y_1 = 0, \qquad \frac{d^3 y_2}{dt^3} - y_2 = 0.$$

Linearly independent solutions of the last equation are

$$y_{2,1} = e^t, \qquad y_{2,2} = \exp\left[\frac{-1 + i\sqrt{3}}{2}t\right], \qquad y_{2,3} = \exp\left[\frac{-1 - i\sqrt{3}}{2}t\right]. \tag{S1.18}$$

Thus, linearly independent solutions of (S1.17) are

$$\begin{bmatrix} x_1 \\ x_2 \end{bmatrix} = F^{-1}\left(\frac{d}{dt}\right)\begin{bmatrix} y_1 \\ y_2 \end{bmatrix} = \begin{bmatrix} 0 & -1 \\ 1 & \dfrac{d}{dt} \end{bmatrix}\begin{bmatrix} 0 \\ y_2 \end{bmatrix} = \begin{bmatrix} -y_2 \\ y_2' \end{bmatrix},$$

where $y_2 = y_2(t)$ can be each one of the three functions determined in (S1.18). □

In an analogous fashion a reduction to the scalar case can be made for the nonhomogeneous equation

$$A_l \frac{d^l x}{dt^l} + \cdots + A_1 \frac{dx}{dt} + A_0 x = f. \tag{S1.19}$$

Indeed, this equation is equivalent to

$$E\left(\frac{d}{dt}\right)D\left(\frac{d}{dt}\right)F\left(\frac{d}{dt}\right)x(t) = f(t),$$

where $f(t)$ is a given function (we shall suppose for simplicity that $f(t)$ has as many derivatives as necessary).

Here $D(\lambda)$ is the Smith form for $A(\lambda) = \sum_{j=0}^{l} A_j\lambda^j$, and $E(\lambda)$ and $F(\lambda)$ are defined by (S1.13). Let $y = F(d/dt)x$, $g = E^{-1}(d/dt)f(t)$; then the preceding equation takes the form

$$D\left(\frac{d}{dt}\right)y(t) = g(t); \qquad \text{(S1.20)}$$

thus, it is reduced to a system of n independent scalar equations.

Using this reduction it is not hard to prove the following result.

Lemma S1.7. *The system* (S1.19) (*where* $n = m$) *has a solution for every right-hand side* $f(t)$ (*which is sufficiently many-times differentiable*) *if and only if* $\det A(\lambda) \not\equiv 0$.

Proof. If $\det A(\lambda) \not\equiv 0$, then none of the entries on the main diagonal in $D(\lambda)$ is zero, and therefore every scalar equation in (S1.20) has a solution. If $\det A(\lambda) \equiv 0$, then the last entry on the main diagonal in $D(\lambda)$ is zero. Hence, if $f(t)$ is chosen in such a way that the last entry in $E^{-1}(d/dt)f(t)$ is not zero, the last equation in (S1.20) does not hold, and (S1.19) has no solution for this choice of $f(t)$. □

Of course, it can happen that for some special choice of $f(t)$ Eq. (S1.19) still has a solution even when $\det A(\lambda) \equiv 0$.

S1.4. Application to Difference Equations

Let $A_0, \ldots, A_l$ be $m \times n$ matrices with complex entries. Consider the system of difference equations

$$A_0x_k + A_1x_{k+1} + \cdots + A_lx_{k+l} = y_k, \qquad k = 0, 1, 2, \ldots, \qquad \text{(S1.21)}$$

where $(y_0, y_1, \ldots,)$ is a given sequence of vectors in $\mathbb{C}^m$ and $(x_0, x_1, \ldots,)$ is a sequence in $\mathbb{C}^n$ to be found.

For example, such systems appear if we wish to solve approximately a system of differential equations

$$B_2\frac{d^2}{dt^2}x(t) + B_1\frac{dx(t)}{dt} + B_0x(t) = y(t), \qquad 0 < t < \infty, \qquad \text{(S1.22)}$$

(here $y(t)$ is a given function and $x(t)$ is unknown), by replacing each derivative by a finite difference approximations as follows: let h be a positive number; denote $t_j = jh$, $j = 0, 1, \ldots$. Given the existence of a solution $x(t)$, consider (S1.22) only at the points t_j:

$$B_2\frac{d^2x(t_j)}{dt^2} + B_1\frac{dx(t_j)}{dt} + B_0x(t_j) = y(t_j), \qquad j = 0, 1, \ldots, \qquad \text{(S1.23)}$$

and replace each derivative $(d^l/dt^l)x(t_j)$ by a finite difference approximation. For example, if x_j is an approximation for $x(t_j)$, $j = 0, 1, 2, \ldots$, then one may use the central difference approximations

$$\frac{dx(t_j)}{dt} \doteq \frac{x_{j+1} - x_{j-1}}{2h}, \tag{S1.24}$$

$$\frac{d^2x(d_j)}{dt^2} \doteq \frac{x_{j+1} - 2x_j + x_{j-1}}{h^2}. \tag{S1.25}$$

Inserting these expressions in (S1.23) gives

$$B_2 \frac{x_{j+1} - 2x_j + x_{j-1}}{h^2} + B_1 \frac{x_{j+1} - x_{j-1}}{2h} + B_0 x_j = y(t_j), \qquad j = 1, 2, \ldots,$$

or

$$\begin{aligned}(B_2 + \tfrac{1}{2}B_1 h)x_{j+1} - (2B_2 - B_0 h^2)x_j + (B_2 - \tfrac{1}{2}B_1 h)x_{j-1} \\ = h^2 y(t_j), \qquad j = 1, 2, \ldots,\end{aligned} \tag{S1.26}$$

which is a difference equation of type (S1.21) with $y_k = h^2 y(t_{k+1})$, $k = 0, 1, 2, \ldots$. Some other device may be needed to provide the approximate values of $x(0)$ and $x(h)$ (since, generally speaking, the unique solution of (S1.26) is determined by the values of $x(0)$ and $x(h)$). These can be approximated if, for instance, initial values $x(0)$ and $dx(0)/dt$ are given; then the approximation x_1 for $x(h)$ could be

$$x_1 = x(0) + h\frac{dx(0)}{dt}.$$

This technique can, of course, be extended to approximate by finite differences the equation of lth order

$$B_l \frac{d^l x(t)}{dt^l} + \cdots + B_0 x(t) = y(t).$$

The approximation by finite differences is also widely used in the numerical solution of partial differential equations. In all these cases we arrive at a system of difference equations of type (S1.21).

The extent to which solutions of the difference equation give information about solutions of the original differential equation is, of course, quite another question. This belongs to the study of numerical analysis and will not be pursued here.

To study the system (S1.21), introduce operator $\mathscr{E}$ (the shift operator), which acts on the set of all sequences $(x_1, x_2, \ldots,)$ of n-dimensional vectors x_i as follows: $\mathscr{E}(x_1, x_2, \ldots,) = (x_2, x_3, \ldots,)$. For difference equations, the operator $\mathscr{E}$ plays a role analogous to the operator d/dt in Section S1.3. The system (S1.21) can be rewritten in the form

$$(A_0 + A_1\mathscr{E} + \cdots + A_l\mathscr{E}^l)x = f, \tag{S1.27}$$

where $f = (f_0, f_1, \ldots,)$ is a given sequence of n-dimensional vectors, and $x = (x_0, x_1, \ldots,)$ is a sequence to be found. Now it is quite clear that we have to consider the matrix polynomial $A(\lambda) = \sum_{j=0}^{l} A_j\lambda^j$ connected with (S1.21).

Let $D(\lambda)$ be the Smith form of $A(\lambda)$ and

$$A(\lambda) = E(\lambda)D(\lambda)F(\lambda), \tag{S1.28}$$

where $\det E(\lambda) \equiv \text{const} \neq 0$, $\det F(\lambda) \equiv \text{const} \neq 0$. Replace λ in (S1.28) by $\mathscr{E}$ and substitute in (S1.17) to obtain

$$D(\mathscr{E})y = g, \tag{S1.29}$$

where $y = (y_0, y_1, \ldots,) = F(\mathscr{E})(x_0, x_1, \ldots,)$ and

$$g = (g_0, g_1, \ldots,) = (E(\mathscr{E}))^{-1}(f_0, f_1, \ldots,).$$

Equation (S1.29) splits into m independent scalar equations, and using the results concerning scalar difference equations (for the convenience of the reader they are given in the appendix below), we obtain the following result analogous to Theorem S1.6.

Theorem S1.8. *Let $A(\lambda) = \sum_{j=0}^{l} A_j\lambda^j$ be an $n \times n$ matrix polynomial with* $\det A(\lambda) \not\equiv 0$. *Then the dimension of the solution space of the homogeneous equation*

$$A_0x_k + A_1x_{k+1} + \cdots + A_lx_{k+l} = 0, \qquad k = 0, 1, 2, \ldots,$$

is the degree of $\det A(\lambda)$.

Appendix. Scalar Difference Equations

In this appendix we shall give a short account of the theory of homogeneous scalar difference equations with constant coefficients.

Consider the scalar homogeneous difference equation

$$a_0x_k + a_1x_{k+1} + \cdots + a_lx_{k+l} = 0, \qquad k = 0, 1, \ldots \tag{S1.30}$$

where a_i are complex numbers and $a_l \neq 0$; $(x_0, x_1, \ldots,)$ is a sequence of complex numbers to be found. It is clear that the solutions of (S1.30) form a linear space, i.e., the sum of two solutions is again a solution, as well as a scalar multiple of a solution.

Proposition S1.9. *The linear space of all solutions of Eq.* (S1.30) *has dimension l.*

Proof. Suppose that $(x_0^{(j)}, x_1^{(j)}, \ldots,)$, $j = 1, \ldots, l+1$ are nonzero solutions of (S1.30). In particular, we have

$$a_0 x_0^{(j)} + a_1 x_1^{(j)} + \cdots + a_l x_l^{(j)} = 0, \qquad j = 1, \ldots, l+1. \tag{S1.31}$$

Consider each equality as a linear homogeneous equation with unknowns $x_0^{(j)}, \ldots, x_l^{(j)}$. From the general theory of linear homogeneous equations it follows that (S1.31) has exactly l linearly independent solutions. This means that there exists a relationship

$$\sum_{j=1}^{l+1} \alpha_j x_k^{(j)} = 0, \qquad k = 0, 1, \ldots, l, \tag{S1.32}$$

where $\alpha_0, \ldots, \alpha_l$ are complex numbers and not all of them are zeros. It follows that the same relationship holds for the complete sequences

$$x^{(j)} = (x_0^{(j)}, x_1^{(j)}, \ldots,), \qquad j = 1, \ldots, l+1:$$

$$\sum_{j=1}^{l+1} \alpha_j x_k^{(j)} = 0, \qquad k = 0, 1, \ldots, \tag{S1.33}$$

or, in other words,

$$\sum_{j=1}^{l+1} \alpha_j x^{(j)} = 0. \tag{S1.34}$$

Indeed, we shall prove (S1.33) by induction on k. For $k = 0, \ldots, l$, (S1.33) is exactly (S1.32). Suppose (S1.33) holds for all $k \leq k_0$, where k_0 is some integer greater than or equal to l. Using the equalities (S1.33), we obtain

$$x_{k_0+1}^{(j)} = -a_l^{-1}(a_{l-1} x_{k_0}^{(j)} + \cdots + a_0 x_{k_0-l}^{(j)}), \qquad j = 1, \ldots, l+1.$$

So

$$\sum_{j=1}^{l+1} \alpha_j x_{k_0+1}^{(j)} = -a_l^{-1}\left(a_{l-1} \sum_{j=1}^{l+1} \alpha_j x_{k_0}^{(j)} + \cdots + \alpha_0 \sum_{j=1}^{l+1} \alpha_j x_{k_0-l}^{(j)}\right),$$

and by the induction hypothesis,

$$\sum_{j=1}^{l+1} \alpha_j x_{k_0+1}^{(j)} = -a_l^{-1}(a_{l-1} \cdot 0 + \cdots + a_0 \cdot 0) = 0.$$

So (S1.33) follows. Now it is clear from (S1.34) that the solutions $x^{(1)}, \ldots, x^{(l+1)}$ are linearly dependent. We have proved therefore that any $l+1$ solutions of (S1.30) are linearly dependent; so the dimension of the solution space is less than or equal to l.

To prove that the dimension of the solution space is exactly l, we construct now a linearly independent set of l solutions of (S1.30). Let us seek for a solution of (S1.30) in the form of a geometric sequence $x_k = q^k$, $k = 0, 1, \ldots$.

Inserting in (S1.30) we see that $(1, q, q^2, \ldots,)$ is a solution of (S1.30) if and only if q is a root of the *characteristic polynomial*

$$f(\lambda) = a_0 + a_1\lambda + \cdots + a_l\lambda^l$$

of Eq. (S1.30). Let $q_1, \ldots, q_s$ be the different roots of $f(\lambda)$, and let $\alpha_1, \ldots, \alpha_s$ be the multiplicity of the roots $q_1, \ldots, q_s$, respectively. So

$$f(\lambda) = \prod_{j=1}^{s} (\lambda - q_j)^{\alpha_j} \qquad \text{and} \quad \alpha_1 + \cdots + \alpha_s = l.$$

It is not hard to see that the solutions

$$x_k^{(j)} = q_j^k, \qquad k = 0, 1, \ldots, \quad j = 1, \ldots, s,$$

are linearly independent. Indeed, suppose

$$\sum_{j=1}^{s} \alpha_j x_k^{(j)} = \sum_{j=1}^{s} \alpha_j q_j^k = 0 \qquad \text{for} \quad k = 0, 1, \ldots. \tag{S1.35}$$

Consider the equations for $k = 0, \ldots, s - 1$ as a linear system with unknowns $\alpha_1, \ldots, \alpha_s$. The matrix of this system is

$$V = \begin{bmatrix} 1 & 1 & \cdots & 1 \\ q_1 & q_2 & \cdots & q_s \\ \vdots & \vdots & & \vdots \\ q_1^{s-1} & q_2^{s-1} & \cdots & q_s^{s-1} \end{bmatrix},$$

and it is invertible, since $q_1, \ldots, q_s$ are all different and $\det V = \Pi_{i>j}(q_i - q_j)$ (Vandermonde determinant, see, for instance, [65]). So (S1.35) has only the trivial solution, i.e., $\alpha_1 = \cdots = \alpha_s = 0$. Thus, the solutions $x_k^{(j)} = q_j^{(k)}$, $k = 0, 1, \ldots, j = 1, \ldots, s$ are linearly independent.

In the case that $s = l$, i.e., all the roots of $f(\lambda)$ are simple, we have finished: a set of l independent solutions of (S1.30) is found. If not all the roots of $f(\lambda)$ are simple, then $s < l$, and the s independent solutions constructed above do not span the linear space of all solutions. In this case we have to find additional $l - s$ solutions of (S1.30), which form, together with the above s solutions, a linearly independent set. This can be done in the following way: let q_j be a root of $f(\lambda)$ of multiplicity $\alpha_j \geq 1$. Put $x_k = q_j^k y_k$, $k = 0, 1, \ldots$, and substitute in (S1.30)

$$a_0 q_j^k y_k + a_1 q_j^{k+1} y_{k+1} + \cdots + a_l q_j^{k+l} y_{k+l} = 0, \qquad k = 0, 1, \ldots,. \tag{S1.36}$$

Introduce now finite differences of the sequence $y_0, y_1, \ldots,$:

$$\delta y_k = y_{k+1} - y_k,$$
$$\delta^2 y_k = y_{k+2} - 2y_{k+1} + y_k,$$

and in general

$$\delta^m y_k = \sum_{i=0}^{m} \binom{m}{i} (-1)^{m-i} y_{k+i}, \qquad m = 1, \ldots, l.$$

It is easily seen that the sequence $y_0, y_1, \ldots,$ can be expressed through its finite differences as follows:

$$y_{k+m} = \sum_{i=0}^{m} \binom{m}{i} \delta^i y_k, \qquad k = 0, 1, \ldots, \quad m = 0, 1, \ldots, l,$$

where $\delta^0 y_k = y_k$ by definition. Substituting these expressions in (S1.36) and rearranging terms gives

$$q_j^k \left(f(q_j) y_k + q_j f^{(1)}(q_j) \delta y_k + \cdots + q_j^l \frac{f^{(l)}(q_j)}{l!} \delta^l y_k \right) = 0, \qquad k = 0, 1, \ldots. \tag{S1.37}$$

Note that $f^{(i)}(q_j) = 0$ for $i = 0, \ldots, \alpha_j - 1$. Therefore Eq. (S1.37) will be satisfied for every sequence $(y_0, y_1, \ldots,)$ such that

$$\delta^i y_k = 0 \qquad \text{for} \quad i \geq \alpha_j. \tag{S1.38}$$

It is not hard to see that (S1.38) is satisfied for every polynomial sequence $y_k = \sum_{m=0}^{\alpha_j - 1} \beta_m k^m$, $k = 0, 1, \ldots,$ where $\beta_m \in \mathbb{C}$. Indeed, $\delta y_k = \sum_{m=0}^{\alpha_j - 2} \beta'_m k^m$ for some coefficients $\beta'_m \in \mathbb{C}$. Now

$$\delta^{\alpha_j} y_k = \underbrace{\delta(\delta \cdots (\delta y_k)}_{\alpha_j \text{ times}} = 0,$$

since operator δ when applied to a polynomial sequence reduces its degree at least by 1.

So we have obtained α_j solutions of (S1.30) connected with the root q_j of $f(\lambda)$ of multiplicity α_j:

$$x_k^{(j,m)} = k^m q_j^k, \qquad k = 0, 1, \ldots,, \quad m = 0, \ldots, \alpha_j - 1. \tag{S1.39}$$

It turns out that the $l = \sum_{j=1}^{s} \alpha_j$ solutions of (S1.30) given by (S1.39) for $j = 1, \ldots, s$, are linearly independent.

Thus, we have proved Proposition S1.9 and have constructed also a basis in the solution space of (S1.30). □

S1.5. Local Smith Form and Partial Multiplicities

It is well known that for any scalar polynomial $a(\lambda)$ and any $\lambda_0 \in \mathbb{C}$ the following representation holds:

$$a(\lambda) = (\lambda - \lambda_0)^\alpha b(\lambda), \tag{S1.40}$$

where α is a nonnegative integer and $b(\lambda)$ is a scalar polynomial such that $b(\lambda_0) \neq 0$. (If $a(\lambda_0) \neq 0$, then put $\alpha = 0$ and $b(\lambda) = a(\lambda)$.) In this section we shall obtain an analogous representation for matrix polynomials. We restrict ourselves to the case when the matrix polynomial $A(\lambda)$ is square (of size $n \times n$) and $\det A(\lambda) \not\equiv 0$. It turns out that a representation of type (S1.40) for matrices is defined by n nonnegative integers (instead of one nonnegative integer α in (S1.40)) which are called the partial multiplicities.

Theorem S1.10. *Let $A(\lambda)$ be an $n \times n$ matrix polynomial with* $\det A(\lambda) \not\equiv 0$. *Then for every $\lambda_0 \in \mathbb{C}$, $A(\lambda)$ admits the representation*

$$A(\lambda) = E_{\lambda_0}(\lambda)\begin{bmatrix} (\lambda - \lambda_0)^{\kappa_1} & & 0 \\ & \ddots & \\ 0 & & (\lambda - \lambda_0)^{\kappa_n} \end{bmatrix} F_{\lambda_0}(\lambda), \qquad \text{(S1.41)}$$

where $E_{\lambda_0}(\lambda)$ and $F_{\lambda_0}(\lambda)$ are matrix polynomials invertible at λ_0, and $\kappa_1 \leq \cdots \leq \kappa_n$ are nonnegative integers, which coincide (after striking off zeros) with the degrees of the elementary divisors of $A(\lambda)$ corresponding to λ_0 (i.e., of the form $(\lambda - \lambda_0)^n$).

In particular, the integers $\kappa_1 \leq \cdots \leq \kappa_n$ from Theorem S1.10 are uniquely determined by $A(\lambda)$ and λ_0; they are called the *partial multiplicities* of $A(\lambda)$ at λ_0. The representation (S1.41) will be referred to as a *local Smith form* of $A(\lambda)$ at λ_0.

Proof. The existence of representation (S1.41) follows easily from the Smith form. Namely, let $D(\lambda) = \operatorname{diag}(d_1(\lambda), \ldots, d_n(\lambda))$ be the Smith form of $A(\lambda)$ and let

$$A(\lambda) = E(\lambda)D(\lambda)F(\lambda), \qquad \text{(S1.42)}$$

where $\det E(\lambda) \equiv \text{const} \neq 0$, $\det F(\lambda) \equiv \text{const} \neq 0$. Represent each $d_i(\lambda)$ as in (S1.40):

$$d_i(\lambda) = (\lambda - \lambda_0)^{\kappa_i}\tilde{d}_i(\lambda), \qquad i = 1, \ldots, n,$$

where $\tilde{d}_i(\lambda_0) \neq 0$ and $\kappa_i \geq 0$. Since $d_i(\lambda)$ is divisible by $d_{i-1}(\lambda)$, we have $\kappa_i \geq \kappa_{i-1}$. Now (S1.41) follows from (S1.42), where

$$E_{\lambda_0}(\lambda) = E(\lambda)\operatorname{diag}(\tilde{d}_1(\lambda), \ldots, \tilde{d}_n(\lambda)), \qquad F_{\lambda_0}(\lambda) = F(\lambda).$$

It remains to show that κ_i coincide (after striking off zeros) with the degrees of elementary divisors of $A(\lambda)$ corresponding to λ_0. To this end we shall show that any factorization of $A(\lambda)$ of type (S1.41) with $\kappa_1 \leq \cdots \leq \kappa_n$ implies that κ_j is the multiplicity of λ_0 as a zero of $d_j(\lambda)$, $j = 1, \ldots, n$, where $D(\lambda) = \operatorname{diag}(d_1(\lambda), \ldots, d_n(\lambda))$ is the Smith form of $A(\lambda)$. Indeed, let

$$A(\lambda) = E(\lambda)D(\lambda)F(\lambda),$$

where $E(\lambda)$ and $F(\lambda)$ are matrix polynomials with constant nonzero determinants. Comparing with (S1.41), write

$$\begin{bmatrix} d_1(\lambda) & & \cdots & 0 \\ & d_2(\lambda) & & \vdots \\ & & \ddots & \\ 0 & & \cdots & d_n(\lambda) \end{bmatrix}$$
$$= \tilde{E}_{\lambda_0}(\lambda) \begin{bmatrix} (\lambda - \lambda_0)^{\kappa_1} & 0 & \cdots & 0 \\ 0 & (\lambda - \lambda_0)^{\kappa_2} & & \\ \vdots & \vdots & \ddots & \vdots \\ 0 & 0 & \cdots & (\lambda - \lambda_0)^{\kappa_n} \end{bmatrix} \tilde{F}_{\lambda_0}(\lambda), \qquad \text{(S1.43)}$$

where $\tilde{E}_{\lambda_0}(\lambda) = (E(\lambda))^{-1} E_{\lambda_0}(\lambda)$, $\tilde{F}_{\lambda_0}(\lambda)(F(\lambda))^{-1}$ are matrix polynomials invertible at λ_0. Applying the Binet–Cauchy formula for minors of products of matrices, we obtain

$$d_1(\lambda) d_2(\lambda) \cdots d_{i_0}(\lambda) = \sum_{i,j,k} m_{i,\tilde{E}}(\lambda) \cdot m_{j,D_{\lambda_0}}(\lambda) \cdot m_{k,\tilde{F}}(\lambda), \qquad i_0 = 1, 2, \ldots, n \qquad \text{(S1.44)}$$

where $m_{i,\tilde{E}}(\lambda)$ (resp. $m_{j,D_{\lambda_0}}(\lambda)$, $m_{k,\tilde{F}}(\lambda)$) is a minor of order i_0 of $\tilde{E}(\lambda)$ (resp. $\operatorname{diag}((\lambda - \lambda_0)^{\kappa_1}, \ldots, (\lambda - \lambda_0)^{\kappa_n})$, $\tilde{F}(\lambda)$), and thc sum in (S1.44) is taken over certain set of triples (i, j, k). It follows from (S1.44) and the condition $\kappa_1 \leq \cdots \leq \kappa_n$, that λ_0 is a zero of the product $d_1(\lambda) d_2(\lambda) \cdots d_{i_0}(\lambda)$ of multiplicity at least $\kappa_1 + \kappa_2 + \cdots + \kappa_{i_0}$. Rewrite (S1.43) in the form

$$(\tilde{E}_{\lambda_0}(\lambda))^{-1} \begin{bmatrix} d_1(\lambda) & 0 & \cdots & 0 \\ 0 & d_2(\lambda) & \cdots & 0 \\ \vdots & \vdots & & \vdots \\ 0 & 0 & \cdots & d_n(\lambda) \end{bmatrix} (\tilde{F}_{\lambda_0}(\lambda))^{-1}$$
$$= \begin{bmatrix} (\lambda - \lambda_0)^{\kappa_1} & 0 & \cdots & 0 \\ 0 & (\lambda - \lambda^0)^{\kappa_2} & \cdots & 0 \\ \vdots & \vdots & \ddots & \vdots \\ 0 & 0 & \cdots & (\lambda - \lambda_0)^{\kappa_n} \end{bmatrix}$$

and apply the Binet–Cauchy formula again. Using the fact that $(\tilde{E}_{\lambda_0}(\lambda))^{-1}$ and $(\tilde{F}_{\lambda_0}(\lambda))^{-1}$ are rational matrix functions which are defined and invertible at $\lambda = \lambda_0$, and using the fact that $d_i(\lambda)$ is a divisor of $d_{i+1}(\lambda)$, we deduce that

$$(\lambda - \lambda_0)^{\kappa_1 + \cdots + \kappa_{i_0}} = d_1(\lambda) d_2(\lambda) \cdots d_{i_0}(\lambda) \Phi_{i_0}(\lambda),$$

where $\Phi_{i_0}(\lambda)$ is a rational function defined at $\lambda = \lambda_0$ (i.e., λ_0 is not a pole of $\Phi_{i_0}(\lambda)$). It follows that λ_0 is a zero of $d_1(\lambda)d_2(\lambda)\cdots d_{i_0}(\lambda)$ of multiplicity exactly $\kappa_1 + \kappa_2 + \cdots + \kappa_{i_0}$, $i = 1, \ldots, n$. Hence κ_i is exactly the multiplicity of λ_0 as a zero of $d_i(\lambda)$ for $i = 1, \ldots, n$. □

Note that according to the definition, the partial multiplicities are all zero for every λ_0 which is not a zero of an invariant polynomial of $A(\lambda)$. For instance, in Example S1.1, the partial multiplicities for $\lambda_0 = 0$ and $\lambda_0 = 1$ are 0 and 2; for $\lambda_0 \notin \{0, 1\}$, the partial multiplicities are zeros.

S1.6. Equivalence of Matrix Polynomials

Two matrix polynomials $A(\lambda)$ and $B(\lambda)$ of the same size are called *equivalent* if

$$A(\lambda) = E(\lambda)B(\lambda)F(\lambda)$$

for some matrix polynomials $E(\lambda)$ and $F(\lambda)$ with constant nonzero determinants. For this equivalence relation we shall use the symbol $\sim$: $A(\lambda) \sim B(\lambda)$ means $A(\lambda)$ and $B(\lambda)$ are equivalent.

It is easy to see that $\sim$ is indeed an equivalence relation, i.e.,

(1) $A(\lambda) \sim A(\lambda)$ for every matrix polynomial $A(\lambda)$;
(2) $A(\lambda) \sim B(\lambda)$ implies $B(\lambda) \sim A(\lambda)$;
(3) $A(\lambda) \sim B(\lambda)$ and $B(\lambda) \sim C(\lambda)$ implies $A(\lambda) \sim C(\lambda)$.

Let us check the last assertion, for example. We have

$$A(\lambda) = E_1(\lambda)B(\lambda)F_1(\lambda), \quad B(\lambda) = E_2(\lambda)C(\lambda)F_2(\lambda),$$

where $E_1(\lambda)$, $E_2(\lambda)$, $F_1(\lambda)$, $F_2(\lambda)$ have constant nonzero determinants. Then $A(\lambda) = E(\lambda)C(\lambda)F(\lambda)$ with $E(\lambda) = E_1(\lambda)E_2(\lambda)$ and $F(\lambda) = F_1(\lambda)F_2(\lambda)$. So $A(\lambda) \sim C(\lambda)$.

The uniqueness of the Smith form allows us to give the following criterion for the equivalence relation between matrix polynomials.

Theorem S1.11. *$A(\lambda) \sim B(\lambda)$ if and only if the invariant polynomials of $A(\lambda)$ and $B(\lambda)$ are the same.*

Proof. Suppose the invariant polynomials of $A(\lambda)$ and $B(\lambda)$ are the same. Then their Smith forms are equal:

$$A(\lambda) = E_1(\lambda)D(\lambda)F_1(\lambda), \qquad B(\lambda) = E_2(\lambda)D(\lambda)F_2(\lambda),$$

where $\det E_i(\lambda) \equiv \text{const} \neq 0$, $\det F_i(\lambda) \equiv \text{const} \neq 0$, $i = 1, 2$. Consequently,

$$(E_1(\lambda))^{-1}A(\lambda)(F_1(\lambda))^{-1} = (E_2(\lambda))^{-1}B(\lambda)(F_2(\lambda))^{-1}(=D(\lambda))$$

and

$$A(\lambda) = E(\lambda)B(\lambda)F(\lambda),$$

where $E(\lambda) = E_1(\lambda)(E_2(\lambda))^{-1}$, $F(\lambda) = F_1(\lambda)(F_2(\lambda))^{-1}$. Since $E_2(\lambda)$ and $F_2(\lambda)$ are matrix polynomials with constant nonzero determinants, the same is true for $E^{-1}(\lambda)$ and $F^{-1}(\lambda)$, and, consequently, for $E(\lambda)$ and $F(\lambda)$. So $A(\lambda) \sim B(\lambda)$.

Conversely, suppose $A(\lambda) = E(\lambda)B(\lambda)F(\lambda)$, where $\det E(\lambda) \equiv \text{const} \neq 0$, $\det F(\lambda) \equiv \text{const} \neq 0$. Let $D(\lambda)$ be the Smith form for $B(\lambda)$:

$$B(\lambda) = E_1(\lambda)D(\lambda)F_1(\lambda).$$

Then $D(\lambda)$ is also the Smith form for $A(\lambda)$:

$$A(\lambda) = E(\lambda)E_1(\lambda)D(\lambda)F_1(\lambda)F(\lambda).$$

By the uniqueness of the Smith form for $A(\lambda)$ (more exactly, by the uniqueness of the invariant polynomials of $A(\lambda)$) it follows that the invariant polynomials of $A(\lambda)$ are the same as those of $B(\lambda)$. □

Of special interest is the equivalence of $n \times n$ linear matrix polynomials of the form $I\lambda - A$. It turns out that for linear polynomials the concept of equivalence is closely related to the concept of similarity between matrices. Recall, that matrices A and B are *similar* if $A = SBS^{-1}$ for some nonsingular matrix S. The following theorem clarifies this relationship.

Theorem S1.12. *$I\lambda - A \sim I\lambda - B$ if and only if A and B are similar.*

To prove this theorem, we have to introduce division of matrix polynomials.

We restrict ourselves to the case when the dividend is a general matrix polynomial $A(\lambda) = \sum_{j=0}^{l} A_j\lambda^j$, and the divisor is a matrix polynomial of type $I\lambda + X$, where X is a constant $n \times n$ matrix. In this case the following representation holds:

$$A(\lambda) = Q_r(\lambda)(I\lambda + X) + R_r, \tag{S1.45}$$

where $Q_r(\lambda)$ is a matrix polynomial, which is called the right quotient, and R_r is a constant matrix, which is called the right remainder, on division of $A(\lambda)$ by $I\lambda + X$;

$$A(\lambda) = (I\lambda + X)Q_l(\lambda) + R_l, \tag{S1.46}$$

where $Q_l(\lambda)$ is the left quotient, and R_l is the left remainder (R_l is a constant matrix).

Let us check the existence of representation (S1.45) ((S1.46) can be checked analogously). If $l = 0$ (i.e., $A(\lambda)$ is constant), put $Q_r(\lambda) \equiv 0$ and

$R_r = A(\lambda)$. So we can suppose $l \geq 1$; write $Q_r(\lambda) = \sum_{j=0}^{l-1} Q_j^{(r)}\lambda^j$. Comparing the coefficients of the same degree of λ in right and left hand sides of (S1.45), this relation can be rewritten as follows:

$$A_l = Q_{l-1}^{(r)}, \qquad A_{l-1} = Q_{l-2}^{(r)} + Q_{l-1}^{(r)}X, \qquad \cdots \quad A_1 = Q_0^{(r)} + Q_1^{(r)}X,$$
$$A_0 = Q_0^{(r)}X + R_r.$$

Clearly, these equalities define $Q_{l-1}^{(r)}, \ldots, Q_1^{(r)}, Q_0^{(r)}$, and R_r, sequentially.

It follows from this argument that the left and right quotient and remainder are uniquely defined.

Proof of Theorem S1.12. In one direction this result is immediate: if $A = SBS^{-1}$ for some nonsingular S, then the equality $I\lambda - A = S(I\lambda - B)S^{-1}$ proves the equivalence of $I\lambda - A$ and $I\lambda - B$. Conversely, suppose $I\lambda - A \sim I\lambda - B$. Then for some matrix polynomials $E(\lambda)$ and $F(\lambda)$ with constant nonzero determinant we have

$$E(\lambda)(I\lambda - A)F(\lambda) = I\lambda - B.$$

Suppose that division of $(E(\lambda))^{-1}$ on the left by $I\lambda - A$ and of $F(\lambda)$ on the right by $I\lambda - B$ yield

$$(E(\lambda))^{-1} = (I\lambda - A)S(\lambda) + E_0, \tag{S1.47}$$
$$F(\lambda) = T(\lambda)(I\lambda - B) + F_0.$$

Substituting in the equation

$$(E(\lambda))^{-1}(I\lambda - B) = (I\lambda - A)F(\lambda),$$

we obtain

$$\{(I\lambda - A)S(\lambda) + E_0\}(I\lambda - B) = (I\lambda - A)\{T(\lambda)(I\lambda - B) + F_0\},$$

whence

$$(I\lambda - A)(S(\lambda) - T(\lambda))(I\lambda - B) = (I\lambda - A)F_0 - E_0(I\lambda - B).$$

Since the degree of the matrix polynomial in the right-hand side here is 1 it follows that $S(\lambda) = T(\lambda)$; otherwise the degree of the matrix polynomial on the left is at least 2. Hence,

$$(I\lambda - A)F_0 = E_0(\lambda I - B),$$

so that

$$F_0 = E_0, \qquad AF_0 = E_0B, \qquad \text{and} \qquad AE_0 = E_0B_0.$$

It remains only to prove that E_0 is nonsingular. To this end divide $E(\lambda)$ on the left by $I\lambda - B$:

$$E(\lambda) = (I\lambda - B)U(\lambda) + R_0. \tag{S1.48}$$

Then using Eqs. (S1.47) and (S1.48) we have

$$\begin{aligned}I &= (E(\lambda))^{-1}E(\lambda) = \{(I\lambda - A)S(\lambda) + E_0\}\{(I\lambda - B)U(\lambda) + R_0\} \\ &= (I\lambda - A)\{S(\lambda)(I\lambda - B)U(\lambda)\} + (I\lambda - A)F_0 U(\lambda) \\ &\quad + (I\lambda - A)S(\lambda)R_0 + E_0 R_0 \\ &= (I\lambda - A)[S(\lambda)(I\lambda - B)U(\lambda) + F_0 U(\lambda) + S(\lambda)R_0] + E_0 R_0.\end{aligned}$$

Hence the matrix polynomial in the square brackets is zero, and $E_0 R_0 = I$, i.e., E_0 is nonsingular. □

S1.7. Jordan Normal Form

We prove here (as an application of the Smith form) the theorem on the existence of a Jordan form for every square matrix.

Let us start with some definitions. A square matrix of type

$$\begin{bmatrix} \lambda_0 & 1 & \cdots & 0 \\ 0 & \lambda_0 & \ddots & 0 \\ & & \ddots & \vdots \\ \vdots & \vdots & & 1 \\ 0 & 0 & \cdots & \lambda_0 \end{bmatrix}, \qquad \lambda_0 \in \mathbb{C},$$

is called a Jordan block, and λ_0 is its eigenvalue. A matrix J is called *Jordan*, if it is a block diagonal matrix formed by Jordan blocks on the main diagonal:

$$J = \begin{bmatrix} J_1 & 0 & \cdots & 0 \\ 0 & J_2 & \cdots & 0 \\ \vdots & \vdots & & \vdots \\ 0 & 0 & \cdots & J_k \end{bmatrix},$$

where J_i are Jordan blocks (their eigenvalues may be arbitrary).

The theorem on existence and uniqueness of a Jordan form can now be stated as follows:

Theorem S1.13. *Every $n \times n$ matrix B is similar to a Jordan matrix. This Jordan matrix is unique up to permutation of some Jordan blocks.*

Before we start to prove this theorem, let us compute the elementary divisors of $I\lambda - J$ where J is a Jordan matrix.

Consider first the matrix polynomial $I\lambda - J_0$, where J_0 is a Jordan block of size k and eigenvalue λ_0. Clearly, $\det(I\lambda - J_0) = (\lambda - \lambda_0)^k$. On the other hand, there exists a minor (namely, composed by rows $1, \ldots, k - 1$ and columns

$2, \dots, k$) of order $k - 1$ which is 1. So the greatest common divisor of the minors of order $k - 1$ is also 1, and by Theorem 1.2 the Smith form of $I\lambda - J_0$ is

$$\begin{bmatrix} 1 & 0 & \cdots & 0 \\ 0 & 1 & \cdots & 0 \\ \vdots & \vdots & \ddots & \vdots \\ 0 & 0 & \cdots & (\lambda - \lambda_0)^k \end{bmatrix},$$

i.e., with the single elementary divisor $(\lambda - \lambda_0)^k$.

Now let $\lambda_1, \dots, \lambda_p$ be different complex numbers and let J be the Jordan matrix consisting of Jordan blocks of sizes $\alpha_{i1}, \dots, \alpha_{i,k_i}$ and eigenvalue λ_i, $i = 1, \dots, p$. Using Proposition S1.5 and the assertion proved in the preceding paragraph, we see that the elementary divisors of $I\lambda - J$ are just $(\lambda - \lambda_i)^{\alpha_{ij}}$, $j = 1, \dots, k_i$, $i = 1, \dots, p$.

Proof of Theorem S1.13. The proof is based on reduction to the equivalence of matrix polynomials. Namely, B is similar to a Jordan matrix J if and only if $I\lambda - B \sim I\lambda - J$, and the latter condition means that the invariant polynomials of $I\lambda - B$ and $I\lambda - J$ are the same. From the investigation of the Smith form of $I\lambda - J$, where J is a Jordan matrix, it is clear how to construct a Jordan matrix J such that $I\lambda - J$ and $I\lambda - B$ have the same elementary divisors. Namely, J contains exactly one Jordan block of size r with eigenvalue λ_0 for every elementary divisor $(\lambda - \lambda_0)^r$ of $I\lambda - B$. The size of J is then equal to the sum of degrees of all elementary divisors of $I\lambda - B$, which in turn is equal to the degree of $\det(I\lambda - B)$, i.e., to the size n of B. Thus, the sizes of J and B are the same, and Theorem S1.13 follows. □

S1.8. Functions of Matrices

Let T be an $n \times n$ matrix. We would like to give a meaning to $f(T)$, as a matrix-valued function, where $f(\lambda)$ is some scalar function of a complex variable. We shall restrict our attention here to functions f which are analytic in a neighborhood of the spectrum $\sigma(T)$ of T (this case is sufficient for our purposes).

So let $f(\lambda)$ be a scalar-valued function which is analytic in some open set $U \subset \mathbb{C}$ such that $\sigma(T) \subset U$. Let $\Gamma \subset U$ be a rectifiable contour (depending on $f(\lambda)$) such that $\sigma(T)$ is inside Γ; the contour Γ may consist of several simple contours (i.e., without self-intersections). For example, let $\sigma(T) = \{\lambda_1, \dots, \lambda_r\}$ and let $\Gamma = \bigcup_{i=1}^r \{\lambda \in \mathbb{C} \mid |\lambda - \lambda_i| = \delta_i\}$, where $\delta_i > 0$ are chosen small enough to ensure that $\Gamma \subset U$. Put, by definition

$$f(T) = \frac{1}{2\pi i} \int_\Gamma f(\lambda)(I\lambda - T)^{-1}\, d\lambda. \tag{S1.49}$$

In view of the Cauchy integral formula, this definition does not depend on the choice of Γ (provided the above requirements on Γ are met). Note that according to (S1.49), $f(S^{-1}TS) = S^{-1}f(T)S$ for any nonsingular matrix S.

The following proposition shows that the definition (S1.49) is consistent with the expected value of $f(T) = T^i$ for the analytic functions $f(\lambda) = \lambda^i$, $i = 0, 1, \ldots$.

Proposition S1.14.

$$\frac{1}{2\pi i}\int_\Gamma \lambda^j(\lambda I - T)^{-1}\,d\lambda = T^j, \qquad j = 0, 1, \ldots. \tag{S1.50}$$

Proof. Suppose first that T is a Jordan block with eigenvalue $\lambda = 0$:

$$T = \begin{bmatrix} 0 & 1 & & \cdots & 0 \\ 0 & 0 & 1 & \cdots & 0 \\ & & & \ddots & \vdots \\ \vdots & \vdots & & & 1 \\ 0 & 0 & & \cdots & 0 \end{bmatrix}. \tag{S1.51}$$

Then

$$(\lambda I - T)^{-1} = \begin{bmatrix} \lambda^{-1} & \lambda^{-2} & \cdots & \lambda^{-n} \\ 0 & \lambda^{-1} & \cdots & \lambda^{-n+1} \\ \vdots & \vdots & \ddots & \\ 0 & 0 & \cdots & \lambda^{-1} \end{bmatrix} \tag{S1.52}$$

(recall that n is the size of T). So

$$\frac{1}{2\pi i}\int_\Gamma \lambda^j(\lambda I - T)^{-1}\,d\lambda = \frac{1}{2\pi i}\int_\Gamma \begin{bmatrix} \lambda^{j-1} & \lambda^{j-2} & \cdots & \lambda^{j-n} \\ 0 & \lambda^{j-1} & \cdots & \lambda^{j-n+1} \\ \vdots & \vdots & \ddots & \vdots \\ 0 & 0 & \cdots & \lambda^{j-1} \end{bmatrix} d\lambda$$

$$= \begin{bmatrix} 0 & \cdots & \overset{\text{place } j+1}{1} & \cdots & 0 \\ & & & \ddots & \vdots \\ \vdots & & \vdots & & 1 \\ & & & & \vdots \\ 0 & \cdots & & \cdots & 0 \end{bmatrix} = T^j.$$

It is then easy to verify (S1.50) for a Jordan block T with eigenvalue λ_0 (not necessarily 0). Indeed, $T - \lambda_0 I$ has an eigenvalue 0, so by the case already already considered,

$$\frac{1}{2\pi i}\int_{\Gamma} \lambda^j (I\lambda - (T - \lambda_0 I))^{-1}\, d\lambda = (T - \lambda_0 I)^j, \qquad j = 0, 1, \ldots,$$

where $\Gamma_0 = \{\lambda - \lambda_0 \mid \lambda \in \Gamma\}$. The change of variables $\mu = \lambda + \lambda_0$ in the left-hand side leads to

$$\frac{1}{2\pi i}\int_{\Gamma} (\mu - \lambda_0)^j (I\mu - T)^{-1}\, d\mu = (T - \lambda_0 I)^j, \qquad j = 0, 1, \ldots. \tag{S1.53}$$

Now

$$\begin{aligned}\frac{1}{2\pi i}\int_{\Gamma} \mu^j (I\mu - T)^{-1}\, d\mu &= \sum_{p=0}^{j} \binom{j}{p} \lambda_0^{j-p} \frac{1}{2\pi i}\int_{\Gamma} (\mu - \lambda_0)^p (I\mu - T)^{-1}\, d\mu \\ &= \sum_{p=0}^{j} \binom{j}{p} \lambda_0^{j-p} (T - \lambda_0 I)^p = T^j,\end{aligned}$$

so (S1.50) holds.

Applying (S1.50) separately for each Jordan block, we can carry (S1.50) further for arbitrary Jordan matrices T. Finally, for a given matrix T there exists a Jordan matrix J and an invertible matrix S such that $T = S^{-1}JS$. Since (S1.50) is already proved for J, we have

$$\frac{1}{2\pi i}\int_{\Gamma} \lambda^j (I\lambda - T)^{-1}\, d\lambda = S^{-1} \cdot \frac{1}{2\pi i}\int_{\Gamma} \lambda^j (I\lambda - J)^{-1}\, d\lambda \cdot S = S^{-1} J^j S = T^j.$$

□

The definition of this function of T establishes the map $\varphi: A \to \mathbb{C}^{n\times n}$ from the set A of all functions analytic in a neighborhood of $\sigma(T)$ (each function in its own neighborhood) to the set $\mathbb{C}^{n\times n}$ of all $n \times n$ matrices:

$$\varphi(f) = f(T) = \frac{1}{2\pi i}\int_{\Gamma} f(\lambda)(I\lambda - T)^{-1}\, d\lambda, \qquad f(\lambda) \in A. \tag{S1.54}$$

Observe that the set A is a commutative algebra with the natural definitions: $(f \cdot g)(\lambda) = f(\lambda) \cdot g(\lambda) \in A$ for every pair of functions $f(\lambda), g(\lambda) \in A$, $(\alpha f)(\lambda) = \alpha \cdot f(\lambda)$ for $f(\lambda) \in A$ and $\alpha \in \mathbb{C}$.

Proposition S1.14 is a special case of the following theorem (where we use the notions introduced above):

Theorem S1.15. *The map $\varphi: A \to \mathbb{C}^{n \times n}$ is a homomorphism of algebra A into the algebra of all $n \times n$ matrices. In other words,*

$$\varphi(f \cdot g) = \varphi(f) \cdot \varphi(g), \tag{S1.55}$$

$$\varphi(\alpha f + \beta g) = \alpha\varphi(f) + \beta\varphi(g), \tag{S1.56}$$

where $f(\lambda), g(\lambda) \in A$ and $\alpha, \beta \in \mathbb{C}$.

Proof. As in the proof of Proposition S1.14 the general case is reduced to the case when T is a single Jordan block with eigenvalue zero as in (S1.51). Let $f(\lambda) = \sum_{j=0}^{\infty} \lambda^j f_j$, $g(\lambda) = \sum_{j=0}^{\infty} \lambda^j g_j$ be the developments of $f(\lambda)$ and $g(\lambda)$ into power series in a neighborhood of zero. Then, using (S1.52) we find that

$$\begin{aligned}\varphi(f) &= \frac{1}{2\pi i}\int_\Gamma \sum_{j=0}^{\infty} \lambda^j f_j \begin{bmatrix} \lambda^{-1} & \lambda^{-2} & \cdots & \lambda^{-n} \\ 0 & \lambda^{-1} & \cdots & \lambda^{-n+1} \\ \vdots & \vdots & & \vdots \\ 0 & 0 & \cdots & \lambda^{-1} \end{bmatrix} d\lambda \\ &= \begin{bmatrix} f_0 & f_1 & \cdots & f_{n-1} \\ 0 & f_0 & \ddots & \vdots \\ \vdots & \vdots & \ddots & f_1 \\ 0 & 0 & \cdots & f_0 \end{bmatrix} \end{aligned} \tag{S1.57}$$

Analogously,

$$\varphi(g) = \begin{bmatrix} g_0 & g_1 & \cdots & g_{n-1} \\ 0 & g_0 & \ddots & \vdots \\ \vdots & \vdots & \ddots & g_1 \\ 0 & 0 & \cdots & g_1 \end{bmatrix}, \qquad \varphi(f \cdot g) = \begin{bmatrix} h_0 & h_1 & \cdots & h_{n-1} \\ 0 & h_0 & \ddots & \vdots \\ \vdots & \vdots & \ddots & h_1 \\ & 0 & \cdots & h_0 \end{bmatrix}, \tag{S1.58}$$

where the coefficients h_j are taken from the development $f(\lambda)g(\lambda) = \sum_{j=0}^{\infty} \lambda^j h_j$. So $h_k = \sum_{j=0}^{k} f_j g_{k-j}$, $k = 0, 1, \ldots$, and direct computation of the product $\varphi(f) \cdot \varphi(g)$, using (S1.57) and (S1.58), shows that $\varphi(fg) = \varphi(f) \cdot \varphi(g)$, i.e., (S1.55) holds. Since the equality (S1.56) is evident, Theorem S1.15 follows. □

We note the following formula, which was established in the proof of Theorem S1.15: if J is the Jordan block of size $k \times k$ with eigenvalue λ_0, then

$$f(J) = \begin{bmatrix} f(\lambda_0) & \frac{1}{1!} f'(\lambda_0) & \cdots & \frac{1}{(k-1)!} f^{(k-1)}(\lambda_0) \\ 0 & f(\lambda_0) & & \vdots \\ \vdots & \vdots & \ddots & \\ 0 & 0 & \cdots & f(\lambda_0) \end{bmatrix} \tag{S1.59}$$

for any analytic function $f(\lambda)$ defined in a neighborhood of λ_0 ($f^{(i)}(\lambda_0)$ means the ith derivative of $f(\lambda)$ evaluated at λ_0).

As an immediate consequence of Theorem S1.15 we obtain the following corollary.

Corollary S1.16. *Let $f(\lambda)$ and $g(\lambda)$ be analytic functions in a neighborhood of $\sigma(T)$. Then*

$$f(T) \cdot g(T) = g(T) \cdot f(T),$$

i.e., the matrices $f(T)$ and $g(T)$ commute.

Comments

A good source for the theory of partial multiplicities and the local Smith form of operator valued analytic functions in infinite dimensional Banach space is [38]; see also [2, 37d]. For further development of the notion of equivalence see [28, 70c].

A generalization of the global Smith form for operator-valued analytic functions is obtained in [57a, 57b]; see also [37d].

Chapter S2

The Matrix Equation $AX - XB = C$

In this chapter we consider the equation

$$AX - XB = C, \tag{S2.1}$$

where A, B, C are given matrices and X is a matrix to be found. We shall assume that A and B are square, but not necessarily of the same size: the size of A is $r \times r$, and the size of B is $s \times s$; then C, as well as X, is of size $r \times s$. In the particular case of Eq. (S2.1) with $A = B$ and $C = 0$, the solutions of (S2.1) are exactly the matrices which commute with A. This case is considered in Section S2.2.

S2.1. Existence of Solutions of $AX - XB = C$

The main result on existence of solutions of (S2.1) reads as follows.

Theorem S2.1. *Equation* (S2.1) *has a solution X if and only if the matrices*

$$\begin{bmatrix} A & 0 \\ 0 & B \end{bmatrix} \quad \textit{and} \quad \begin{bmatrix} A & C \\ 0 & B \end{bmatrix}$$

are similar.

Proof. If (S2.1) has a solution X, then direct multiplication shows that

$$\begin{bmatrix} I & -X \\ 0 & I \end{bmatrix}\begin{bmatrix} A & 0 \\ 0 & B \end{bmatrix}\begin{bmatrix} I & X \\ 0 & I \end{bmatrix} = \begin{bmatrix} A & C \\ 0 & B \end{bmatrix};$$

but

$$\begin{bmatrix} I & -X \\ 0 & I \end{bmatrix} = \begin{bmatrix} I & X \\ 0 & I \end{bmatrix}^{-1},$$

so

$$\begin{bmatrix} A & 0 \\ 0 & B \end{bmatrix} \quad \text{and} \quad \begin{bmatrix} A & C \\ 0 & B \end{bmatrix}$$

are similar with the similarity matrix

$$\begin{bmatrix} I & X \\ 0 & I \end{bmatrix}.$$

Suppose now that the matrices

$$\begin{bmatrix} A & 0 \\ 0 & B \end{bmatrix} \quad \text{and} \quad \begin{bmatrix} A & C \\ 0 & B \end{bmatrix}$$

are similar:

$$\begin{bmatrix} A & 0 \\ 0 & B \end{bmatrix} = S^{-1}\begin{bmatrix} A & C \\ 0 & B \end{bmatrix}S \tag{S2.2}$$

for some nonsingular matrix S. Define the linear transformations Ω_1 and Ω_2 on the set $\mathcal{C}_{r+s}^{r+s}$ of all $(r + s) \times (r + s)$ complex matrices (considered as an $(r + s)^2$-dimensional vector space) as follows:

$$\Omega_1(Z) = \begin{bmatrix} A & 0 \\ 0 & B \end{bmatrix}Z - Z\begin{bmatrix} A & 0 \\ 0 & B \end{bmatrix},$$

$$\Omega_2(Z) = \begin{bmatrix} A & C \\ 0 & B \end{bmatrix}Z - Z\begin{bmatrix} A & 0 \\ 0 & B \end{bmatrix},$$

where $Z \in \mathcal{C}_{r+s}^{r+s}$. Equality (S2.2) implies that

$$\Omega_1(Z) = S^{-1} \cdot \Omega_2(SZ), \qquad Z \in \mathcal{C}_{r+s}^{r+s}.$$

In particular,

$$\operatorname{Ker} \Omega_2 = \{SZ \mid Z \in \mathcal{C}_{r+s}^{r+s} \quad \text{and} \quad Z \in \operatorname{Ker} \Omega_1\}.$$

Consequently,

$$\dim \operatorname{Ker} \Omega_1 = \dim \operatorname{Ker} \Omega_2. \tag{S2.3}$$

One checks easily that

$$\operatorname{Ker}\Omega_1 = \left\{\begin{bmatrix} T & U \\ V & W \end{bmatrix} \middle| AT = TA, AU = UB, BV = VA, BW = WB\right\}$$

and

$$\operatorname{Ker}\Omega_2 = \left\{\begin{bmatrix} T & U \\ V & W \end{bmatrix} \middle| AT + CV = TA, AU + CW = UB, BV = VA, BW = WB\right\}.$$

It is sufficient to find a matrix in $\operatorname{Ker}\Omega_2$ of the form

$$\begin{bmatrix} T & U \\ 0 & -I \end{bmatrix}, \tag{S2.4}$$

(indeed, in view of the equality $AU + C(-I) = UB$, the matrix U is a solution of (S2.1)). To this end introduce the set R consisting of all pairs of complex matrices (V, W), where V is of size $s \times r$ and W is of size $s \times s$ such that $BV = VA$ and $BW = WB$. Then R is a linear space with multiplication by a complex number and addition defined in a natural way:

$$\alpha(V, W) = (\alpha V, \alpha W), \qquad \alpha \in \mathbb{C},$$

$$(V_1, W_1) + (V_2, W_2) = (V_1 + V_2, W_1 + W_2).$$

Define linear transformations $\Phi_i\colon \operatorname{Ker}\Omega_i \to R$, $i = 1, 2$ by the formula

$$\Phi_i\begin{bmatrix} T & U \\ V & W \end{bmatrix} = (V, W).$$

Then

$$\operatorname{Ker}\Phi_1 = \operatorname{Ker}\Phi_2 = \left\{\begin{bmatrix} T & U \\ 0 & 0 \end{bmatrix} \middle| AT = TA, AU = UB\right\}. \tag{S2.5}$$

In addition, we have

$$\operatorname{Im}\Phi_1 = \operatorname{Im}\Phi_2 = R. \tag{S2.6}$$

To see this, observe that $\operatorname{Im}\Phi_1 = R$, because if $BV = VA$ and $BW = WB$, then

$$\begin{bmatrix} 0 & 0 \\ V & W \end{bmatrix} \in \operatorname{Ker}\Omega_1 \qquad \text{and} \qquad \Phi_1\begin{bmatrix} 0 & 0 \\ V & W \end{bmatrix} = (V, W).$$

Therefore

$$\operatorname{Im}\Phi_2 \subset \operatorname{Im}\Phi_1. \tag{S2.7}$$

On the other hand, by the well-known property of linear transformations,

$$\dim \operatorname{Ker} \Phi_i + \dim \operatorname{Im} \Phi_i = \dim \operatorname{Ker} \Omega_i, \qquad i = 1, 2.$$

In view of (S2.3) and (S2.5), $\dim \operatorname{Im} \Phi_1 = \dim \operatorname{Im} \Phi_2$; now (S2.6) follows from (S2.7).

Consider the pair

$$(0, -I) = \Phi_1 \begin{bmatrix} I & 0 \\ 0 & -I \end{bmatrix}, \qquad \text{where} \quad \begin{bmatrix} I & 0 \\ 0 & -I \end{bmatrix} \in \operatorname{Ker} \Omega_1.$$

By (S2.6), there exists a matrix

$$\begin{bmatrix} T_0 & U_0 \\ V_0 & W_0 \end{bmatrix} \in \operatorname{Ker} \Omega_2$$

such that

$$(0, -I) = \Phi_2 \begin{bmatrix} T_0 & U_0 \\ V_0 & W_0 \end{bmatrix}.$$

This means that $V_0 = 0$, $W_0 = -I$; so we have found a matrix in $\operatorname{Ker} \Omega_2$ of the form (S2.4). □

S2.2. Commuting Matrices

Matrices A and B (both of the same size $n \times n$) are said to *commute* if $AB = BA$. We shall describe the set of all matrices which commute with a given matrix A. In other words, we wish to find all the solutions of the equation

$$AX = XA, \tag{S2.8}$$

where X is an $n \times n$ matrix to be found. Recall that (S2.8) is a particular case of Eq. (S2.1) (with $B = A$ and $C = 0$).

We can restrict ourselves to the case that A is in the Jordan form. Indeed, let $J = S^{-1}AS$ be a Jordan matrix for some nonsingular matrix S. Then X is a solution of (S2.8) if and only if $Z = S^{-1}XS$ is a solution of

$$JZ = ZJ. \tag{S2.9}$$

So we shall assume that $A = J$ is in the Jordan form. Write

$$J = \operatorname{diag}(J_1, \ldots, J_u),$$

where J_α ($\alpha = 1, \ldots, u$) is a Jordan block of size $m_\alpha \times m_\alpha$, $J_\alpha = \lambda_\alpha I_\alpha + H_\alpha$, where I_α is the unit matrix of size $m_\alpha \times m_\alpha$, and

$$H_\alpha = \begin{bmatrix} 0 & 1 & & 0 \\ & 0 & \ddots & \\ \vdots & & \ddots & \\ & & & 1 \\ 0 & & \cdots & 0 \end{bmatrix}.$$

Let Z be a matrix which satisfies (S2.9). Write

$$Z = (Z_{\alpha\beta})_{\alpha,\beta=1}^{u},$$

where $Z_{\alpha\beta}$ is a $m_\alpha \times m_\beta$ matrix. Rewrite Eq. (S2.9) in the form

$$(\lambda_\alpha - \lambda_\beta)Z_{\alpha\beta} = Z_{\alpha\beta}H_\beta - H_\alpha Z_{\alpha\beta}, \qquad 1 \le \alpha, \beta \le u. \tag{S2.10}$$

Two cases can occur:

(1) $\lambda_\alpha \neq \lambda_\beta$. We show that in this case $Z_{\alpha\beta} = 0$. Indeed, multiply the left-hand side of (S2.10) by $\lambda_\alpha - \lambda_\beta$ and in each term in the right-hand side replace $(\lambda_\alpha - \lambda_\beta)Z_{\alpha\beta}$ by $Z_{\alpha\beta}H_\beta - H_\alpha Z_{\alpha\beta}$. We obtain

$$(\lambda_\alpha - \lambda_\beta)^2 Z_{\alpha\beta} = Z_{\alpha\beta}H_\beta^2 - 2H_\alpha Z_{\alpha\beta}H_\beta + H_\alpha^2 Z_{\alpha\beta}.$$

Repeating this process, we obtain for every $p = 1, 2, \ldots,$

$$(\lambda_\alpha - \lambda_\beta)^p Z_{\alpha\beta} = \sum_{q=0}^{p} (-1)^q \binom{p}{q} H_\alpha^q Z_{\alpha\beta} H_\beta^{p-q}. \tag{S2.11}$$

Choose p large enough so that either $H_\alpha^q = 0$ or $H_\beta^{p-q} = 0$ for every $q = 0, \ldots, p$. Then the right-hand side of (S2.11) is zero, and since $\lambda_\alpha \neq \lambda_\beta$, we obtain that $Z_{\alpha\beta} = 0$.

(2) $\lambda_\alpha = \lambda_\beta$. Then

$$Z_{\alpha\beta}H_\beta = H_\alpha Z_{\alpha\beta}. \tag{S2.12}$$

From the structure of H_α and H_β it follows that the product $H_\alpha Z_{\alpha\beta}$ is obtained from $Z_{\alpha\beta}$ by shifting all the rows one place upwards and filling the last row with zeros; similarly, $Z_{\alpha\beta}H_\beta$ is obtained from $Z_{\alpha\beta}$ by shifting all the columns one place to the right and filling the first column with zeros. So Eq. (S2.12) gives (where ζ_{ik} is the (i, k)th entry in $Z_{\alpha\beta}$, which depends, of course, on α and β):

$$\zeta_{i+1,k} = \zeta_{i,k-1}, \qquad i = 1, \ldots, m_\alpha, \quad k = 1, \ldots, m_\beta,$$

where by definition $\zeta_{i0} = \zeta_{m_\alpha+1,k} = 0$. These equalities mean that the matrix $Z_{\alpha\beta}$ has the structure

(1) for $m_\alpha = m_\beta$:

$$Z_{\alpha\beta} = \begin{bmatrix} c_{\alpha\beta}^{(1)} & c_{\alpha\beta}^{(2)} & \cdots & c_{\alpha\beta}^{(m_\alpha)} \\ 0 & c_{\alpha\beta}^{(1)} & \cdots & c_{\alpha\beta}^{(m_\alpha - 1)} \\ \vdots & \vdots & & \vdots \\ 0 & 0 & \cdots & c_{\alpha\beta}^{(1)} \end{bmatrix} \stackrel{\text{def}}{=} T_{m_\alpha} \qquad (c_{\alpha\beta}^{(i)} \in \mathbb{C}); \tag{S2.13}$$

(2) for $m_\alpha < m_\beta$: $Z_{\alpha\beta} = [\overbrace{\ 0\ }^{m_\beta - m_\alpha} \quad T_{m_\alpha}]$; (S2.14)

(3) for $m_\alpha > m_\beta$: $Z_{\alpha\beta} = \begin{bmatrix} T_{m_\beta} \\ 0 \end{bmatrix} \}_{m_\alpha - m_\beta}$ (S2.15)

Matrices of types (S2.13)–(S2.15) will be referred to as *upper triangular Toeplitz matrices*. So we have proved the following result.

Theorem S2.2. *Let $J = \operatorname{diag}[J_1, \ldots, J_k]$ be an $n \times n$ Jordan matrix with Jordan blocks $J_1, \ldots, J_k$ and eigenvalues $\lambda_1, \ldots, \lambda_k$, respectively. Then an $n \times n$ matrix Z commutes with J if and only if $Z_{\alpha\beta} = 0$ for $\lambda_\alpha \neq \lambda_\beta$ and $Z_{\alpha\beta}$ is an upper triangular Toeplitz matrix for $\lambda_\alpha = \lambda_\beta$, where $Z = (Z_{\alpha\beta})_{\alpha,\beta=1,\ldots,k}$ is the partition of Z consistent with the partition of J into Jordan blocks.*

We repeat that Theorem S2.2 gives, after applying a suitable similarity transformation, a description of all matrices commuting with a fixed matrix A.

Corollary S2.3. *If $\sigma(A) \cap \sigma(B) = \varnothing$, then for all C the equation $AX - XB = C$ has a unique solution.*

Proof. Since Eq. (S2.1) is linear, it is sufficient to show that $AX - XB = 0$ has only the zero solution. Indeed, let X be a solution of this equation; then $\left[\begin{smallmatrix} I & X \\ 0 & I \end{smallmatrix}\right]$ commutes with $\left[\begin{smallmatrix} A & 0 \\ 0 & B \end{smallmatrix}\right]$. However, since $\sigma(A) \cap \sigma(B) = \varnothing$, Theorem S2.2 implies that $X = 0$. $\square$

Comments

Theorem S2.1 was proved in [72]. The proof presented here is from [18]. For a comprehensive treatment of commuting matrices see [22, Chapter VIII]. Formulas for the solution of Eq. (S2.1) when $\sigma(A) \cap \sigma(B) \neq \varnothing$, as well as a treatment of more general equations in infinite dimensional spaces can be found in [16a].

Chapter S3

One-Sided and Generalized Inverses

In the main text we use some well-known facts about one-sided and generalized inverses. These facts, in an appropriate form, are presented here with their proofs.

Let A be an $m \times n$ matrix with complex entries. A is called *left invertible* (resp. *right invertible*) if there exists an $n \times m$ complex matrix A^{I} such that $A^{\mathrm{I}}A = I$ (resp. $AA^{\mathrm{I}} = I$). In this case A^{I} is called a left (resp. right) inverse of A.

The notion of one-sided invertibility is a generalization of the notion of invertibility (nonsingularity) of square matrices. If A is a square matrix and $\det A \neq 0$, then A^{-1} is the unique left and right inverse of A.

One-sided invertibility is easily characterized in other ways. Thus, in the case of left invertibility, the following statements are equivalent;

(i) *the* $m \times n$ *matrix* A *is left invertible*;
(ii) *the columns of* A *are linearly independent in* $\mathbb{C}^m$ (*in particular*, $m \geq n$);
(iii) $\operatorname{Ker} A = \{0\}$, *where* A *is considered as a linear transformation from* $\mathbb{C}^n$ *to* $\mathbb{C}^m$.

Let us check this assertion. Assume A is left invertible, and assume that $Ax = 0$ for some $x \in \mathbb{C}^n$. Let A^{I} be a left inverse of A. Then $x = A^{\mathrm{I}}Ax = 0$ i.e., $\operatorname{Ker} A = \{0\}$, so (i) $\Rightarrow$ (iii) follows.

On the other hand, if $\operatorname{Ker} A = \{0\}$, then the columns $f_1, \ldots, f_n$ of A are linearly independent. For equality $\sum_{i=0}^{n} \alpha_i f_i = 0$, for some complex numbers $\alpha_1, \ldots, \alpha_n$, yields

$$A\begin{bmatrix}\alpha_1\\ \alpha_2\\ \vdots\\ \alpha_n\end{bmatrix} = 0, \qquad \text{so} \qquad \begin{bmatrix}\alpha_1\\ \alpha_2\\ \vdots\\ \alpha_n\end{bmatrix} \in \operatorname{Ker} A = \{0\}.$$

Finally, we shall construct a left inverse of A provided the columns of A are linearly independent. In this case rank $A = n$, so we can choose a set of n linearly independent rows of A, and let A_0 be the square $n \times n$ submatrix of A formed by these rows. Since the rows of A_0 are linearly independent, A_0 is nonsingular, so there exists an inverse A_0^{-1}. Now construct a left inverse A^{I} of A as follows: A^{I} is of size $n \times m$; the columns of A^{I} corresponding to the chosen linearly independent set of rows in A, form the matrix A_0^{-1}; all other columns of A^{I} are zeros. By a straightforward multiplication on checks that $A^{\mathrm{I}}A = I$, and (ii) $\Rightarrow$ (i) is proved.

The above construction of a left inverse for the left invertible matrix A can be employed further to obtain a general formula for the left inverse A^{I}. Namely, assuming for simplicity that the nonsingular $n \times n$ submatrix A_0 of A occupies the top n rows, write

$$A = \begin{bmatrix}A_0\\ A_1\end{bmatrix}.$$

Then a straightforward calculation shows that

$$A^{\mathrm{I}} = [A_0^{-1} - BA_1A_0^{-1}, B] \tag{S3.1}$$

is a left inverse for A for every $n \times (m - n)$ matrix B. Conversely, if A^{I} is a left inverse of A, it can be represented in the form (S3.1). In particular, a left inverse is unique if and only if the left invertible matrix A is square (and then necessarily nonsingular).

In the case of right invertibility, we have equivalence of the following:

(iv) *the $m \times n$ matrix A is right invertible*;
(v) *the rows of A are linearly independent in $\mathbb{C}^n$ (in particular, $m \le n$)*;
(vi) $\operatorname{Im} A = \mathbb{C}^m$.

This assertion can be obtained by arguments like those used above (or by using the equivalence of (i), (ii), and (iii) for the transposed matrix A^{T}).

In an analogous way, the $m \times n$ matrix A is right invertible if and only if there exists a nonsingular $m \times m$ submatrix A_0 of A (this can be seen from (v)). Assuming for simplicity of notation that A_0 occupies the leftmost columns of A, and writing $A = [A_0, A_1]$, we obtain the general formula for a right inverse A^{I} (analogous to (S3.1)):

$$A^{\mathrm{I}} = \begin{bmatrix} A_0^{-1} - A_0^{-1} A_1 C \\ C \end{bmatrix} \tag{S3.2}$$

where C is an arbitrary $(n - m) \times m$ matrix.

Right invertibility of an $m \times n$ matrix A is closely related to the equation

$$Ax = y, \tag{S3.3}$$

where $y \in \mathbb{C}^m$ is given and $x \in \mathbb{C}^n$ is to be found. Assume A is right invertible; then obviously

$$x = A^{\mathrm{I}}y, \tag{S3.4}$$

where A^{I} is any right inverse of A, is a solution of (S3.3). It turns out that formula (S3.4) gives the general solution of (S3.3) provided $y \neq 0$. Indeed, write $A = [A_0, A_1]$ with nonsingular matrix A_0, and let $x = \left[\begin{smallmatrix} x_1 \\ x_2 \end{smallmatrix}\right]$ be a solution of (S3.3), where the partition is consistent with that of A. Since $y \neq 0$, there exists an $(n - m) \times m$ matrix C such that $Cy = x_2$. Then using (S3.3),

$$(A_0^{-1} - A_0^{-1}A_1C)y = A_0^{-1}y - A_0^{-1}A_1x_2 = A_0^{-1}y - A_0^{-1}(y - A_0x_1) = x_1.$$

In other words, $A^{\mathrm{I}}y = x$, where A^{I} is given by (S3.2). Thus, for $y \neq 0$, every solution of (S3.3) is in the form (S3.4).

We point out that one-sided invertibility is stable under small perturbations. Namely, let A be a one-sided (left or right) invertible matrix. Then there exists $\varepsilon > 0$ such that every matrix B with $\|A - B\| < \varepsilon$ is also one-sided invertible, and from the same side as A. (Here and in what follows the norm $\|M\|$ of the matrix M is understood as follows: $\|M\| = \sup_{\|x\|=1} \|Mx\|$, where the norms of vectors x and Mx are euclidean.) This property follows from (ii) (in case of left invertibility) and from (v) (in case of right invertibility). Indeed, assume for example that A is left invertible (of size $m \times n$). Let A_0 be a nonsingular $n \times n$ submatrix of A. The corresponding $n \times n$ submatrix B_0 of B is as close as we wish to A_0, provided $\|A - B\|$ is sufficiently small. Since nonsingularity of a matrix is preserved under small perturbations, B_0 will be nonsingular as well if $\|A - B\|$ is small enough. But this implies linear independence of the columns of B, and therefore B is left invertible.

One can take $\varepsilon = \|A^{\mathrm{I}}\|^{-1}$ in the preceding paragraph, where A^{I} is some right or left inverse of A. Indeed, suppose for definiteness that A is right invertible, $AA^{\mathrm{I}} = I$. Let B be such that $\|A - B\| < \|A^{\mathrm{I}}\|^{-1}$. Then $BA^{\mathrm{I}} = AA^{\mathrm{I}} + (B - A)A^{\mathrm{I}} = I + S$, where $S = (B - A)A^{\mathrm{I}}$. By the conditions $\|S\| < 1$, $I + S$ is invertible and

$$(I + S)^{-1} = I - S + S^2 - S^3 + \cdots,$$

where the series converges absolutely (because $\|S\| < 1$). Thus, $BA^{\mathrm{I}}(I + S)^{-1} = I$, i.e., B is right invertible.

We conclude this chapter with the notion of generalized inverse. An $n \times m$ matrix A^{I} is called a *generalized inverse* of the $m \times n$ matrix A if the following equalities hold:

$$AA^{\mathrm{I}}A = A, \qquad A^{\mathrm{I}}AA^{\mathrm{I}} = A^{\mathrm{I}}.$$

This notion incorporates as special cases the notions of one-sided (left or right) inverses. A generalized inverse of A exists always (in contrast with one-sided inverses which exist if and only if A has full rank). One of the easiest ways to verify the existence of a generalized inverse is by using the representation

$$A = B_1 \begin{bmatrix} I & 0 \\ 0 & 0 \end{bmatrix} B_2, \tag{S3.5}$$

where B_1 and B_2 are nonsingular matrices of sizes $m \times m$ and $n \times n$, respectively. Representation (S3.5) may be achieved by performing elementary transformations of columns and rows of A. For A in the form (S3.5), a generalized inverse can be found easily:

$$A^{\mathrm{I}} = B_2^{-2} \begin{bmatrix} I & 0 \\ 0 & 0 \end{bmatrix} B_1^{-1}.$$

Using (S3.5), one can easily verify that a generalized inverse is unique if and only if A is square and nonsingular.

We shall also need the following fact: if $\mathscr{M} \subset \mathbb{C}^m$ is a direct complement to $\operatorname{Im} A$, and $\mathscr{N} \subset \mathbb{C}^n$ is a direct complement to $\operatorname{Ker} A$, then there exists a generalized inverse A^{I} of A such that $\operatorname{Ker} A^{\mathrm{I}} = \mathscr{M}$ and $\operatorname{Im} A^{\mathrm{I}} = \mathscr{N}$. (For the definition of direct complements see Section S4.1.) Indeed, let us check this assertion first for

$$A = \begin{bmatrix} I_r & 0 \\ 0 & 0 \end{bmatrix}. \tag{S3.6}$$

The conditions on $\mathscr{M}$ and $\mathscr{N}$ imply that

$$\mathscr{M} = \operatorname{Im} \begin{bmatrix} X \\ I_{m-r} \end{bmatrix} \qquad \mathscr{N} = \operatorname{Im} \begin{bmatrix} I_r \\ Y \end{bmatrix}$$

for some $r \times (m - r)$ matrix X and $(n - r) \times r$ matrix Y. Then

$$A^{\mathrm{I}} = \begin{bmatrix} I_r & -X \\ Y & -YX \end{bmatrix}$$

is a generalized inverse of A with the properties that $\operatorname{Ker} A^{\mathrm{I}} = \mathscr{M}$ and $\operatorname{Im} A^{\mathrm{I}} = \mathscr{N}$. The general case is easily reduced to the case (S3.6) by means of representation (S3.5), using the relations

$$\operatorname{Ker} A = B_2^{-1}(\operatorname{Ker} D), \qquad \operatorname{Im} A = B_1(\operatorname{Im} D),$$

where

$$D = \begin{bmatrix} I & 0 \\ 0 & 0 \end{bmatrix} \quad \text{and} \quad A = B_1 D B_2 .$$

Chapter S4

Stable Invariant Subspaces

In Chapters 3 and 7 it is proved that the description of divisors of a matrix polynomial depends on the structure of invariant subspaces for its linearization. So the properties of the invariant subspaces for a given matrix (or linear transformation) play an important role in the spectral analysis of matrix polynomials. Consequently, we consider in this chapter invariant subspaces of a linear transformation acting in a finite dimensional linear vector space $\mathscr{X}$. In the main, attention is focused on a certain class of invariant subspaces which will be called stable. Sections S4.3, S4.4, and S4.5 are auxiliary, but the results presented there are also used in the main text. We shall often assume that $\mathscr{X} = \mathbb{C}^n$, where $\mathbb{C}^n$ is considered as the linear space of n-dimensional column vectors together with the customary scalar product (x, y) defined in $\mathbb{C}^n$ by

$$(x, y) = \sum_{i=1}^{n} x_i \overline{y_i},$$

where $x = (x_1, \ldots, x_n)^{\mathrm{T}}$, $y = (y_1, \ldots, y_n)^{\mathrm{T}} \in \mathbb{C}^n$. We may also assume that linear transformations are represented by matrices in the standard orthonormal basis in $\mathbb{C}^n$.

S4.1. Projectors and Subspaces

A linear transformation $P: \mathscr{X} \to \mathscr{X}$ is called a *projector* if $P^2 = P$. It follows immediately from the definition that if P is a projector, so is $S^{-1}PS$

for any invertible transformation S. The important feature of projectors is that there exists a one-to-one correspondence between the set of all projectors and the set of all pairs of complementary subspaces in $\mathscr{X}$. This correspondence is described in Theorem S4.1.

Recall first that, if $\mathscr{M}$, $\mathscr{L}$ are subspaces of $\mathscr{X}$, then $\mathscr{M} + \mathscr{L} = \{z \in \mathscr{X} \mid z = x + y, x \in \mathscr{M}, y \in \mathscr{L}\}$. This sum is said to be *direct* if $\mathscr{M} \cap \mathscr{L} = \{0\}$ in which case we write $\mathscr{M} \dotplus \mathscr{L}$ for the sum. The subspaces $\mathscr{M}$, $\mathscr{L}$ are *complementary* (*are direct complements of each other*) *if* $\mathscr{M} \cap \mathscr{L} = \{0\}$ and $\mathscr{M} \dotplus \mathscr{L} = \mathscr{X}$.

Subspaces $\mathscr{M}$, $\mathscr{L}$ are *orthogonal* if for each $x \in \mathscr{M}$ and $y \in \mathscr{L}$ we have $(x, y) = 0$ and they are *orthogonal complements* if, in addition, they are complementary. In this case, we write $\mathscr{M} = \mathscr{L}^{\perp}$, $\mathscr{L} = \mathscr{M}^{\perp}$.

Theorem S4.1. *Let P be a projector. Then* (Im P, Ker P) *is a pair of complementary subspaces in $\mathscr{X}$. Conversely, for every pair* $(\mathscr{L}_1, \mathscr{L}_2)$ *of complementary subspaces in $\mathscr{X}$, there exists a unique projector P such that* Im $P = \mathscr{L}_1$, Ker $P = \mathscr{L}_2$.

Proof. Let $x \in \mathscr{X}$. Then

$$x = (x - Px) + Px.$$

Clearly, $Px \in \operatorname{Im} P$ and $x - Px \in \operatorname{Ker} P$ (because $P^2 = P$). So $\operatorname{Im} P + \operatorname{Ker} P = \mathscr{X}$. Further, if $x \in \operatorname{Im} P \cap \operatorname{Ker} P$, then $x - Py$ for some $y \in \mathscr{X}$ and $Px = 0$. So

$$x = Py = P^2 y = P(Py) = Px = 0,$$

and $\operatorname{Im} P \cap \operatorname{Ker} P = \{0\}$. Hence Im P and Ker P are indeed complementary subspaces.

Conversely, let $\mathscr{L}_1$ and $\mathscr{L}_2$ be a pair of complementary subspaces. Let P be the unique linear transformation in $\mathscr{X}$ such that $Px = x$ for $x \in \mathscr{L}_1$ and $Px = 0$ for $x \in \mathscr{L}_2$. Then clearly $P^2 = P$, $\mathscr{L}_1 \subset \operatorname{Im} P$, and $\mathscr{L}_2 \subset \operatorname{Ker} P$. But we already know from the first part of the proof that $\operatorname{Im} P \dotplus \operatorname{Ker} P = \mathscr{X}$. By dimensional considerations, we have, consequently, $\mathscr{L}_1 = \operatorname{Im} P$ and $\mathscr{L}_2 = \operatorname{Ker} P$. So P is a projector with the desired properties. The uniqueness of P follows from the property that $Px = x$ for every $x \in \operatorname{Im} P$ (which in turn is a consequence of the equality $P^2 = P$). □

We say that P is the projector *on* $\mathscr{L}_1$ *along* $\mathscr{L}_2$ if $\operatorname{Im} P = \mathscr{L}_1$, $\operatorname{Ker} P = \mathscr{L}_2$. A projector P is called *orthogonal* if $\operatorname{Ker} P = (\operatorname{Im} P)^{\perp}$. Orthogonal projectors are particularly important and can be characterized as follows:

Proposition S4.2. *A projector P is orthogonal if and only if P is self-adjoint, i.e., $P^* = P$.*

Recall that for a linear transformation $T: \mathbb{C}^n \to \mathbb{C}^n$ the *adjoint* $T^*: \mathbb{C}^n \to \mathbb{C}^n$ is defined by the relation $(Tx, y) = (x, T^*y)$ for all $x, y \in \mathbb{C}^n$. A transformation T is *self-adjoint* if $T = T^*$; in such a case it is represented by a hermitian matrix in the standard orthonormal basis.

Proof. Suppose $P^* = P$, and let $x \in \operatorname{Im} P$, $y \in \operatorname{Ker} P$. Then $(x, y) = (Px, y) = (x, Py) = (x, 0) = 0$, i.e., $\operatorname{Ker} P$ is orthogonal to $\operatorname{Im} P$. Since by Theorem S4.1 $\operatorname{Ker} P$ and $\operatorname{Im} P$ are complementary, it follows that in fact $\operatorname{Ker} P = (\operatorname{Im} P)^\perp$.

Conversely, let $\operatorname{Ker} P = (\operatorname{Im} P)^\perp$. In order to prove that $P^* = P$, we have to check the equality

$$(Px, y) = (x, Py) \qquad \text{for all} \quad x, y \in \mathbb{C}^n. \tag{S4.1}$$

Because of the sesquilinearity of the function (Px, y) in the arguments $x, y \in \mathscr{X}$, and in view of Theorem S4.1, it is sufficient to prove (S4.1) for the following 4 cases: (1) $x, y \in \operatorname{Im} P$; (2) $x \in \operatorname{Ker} P$, $y \in \operatorname{Im} P$; (3) $x \in \operatorname{Im} P$, $y \in \operatorname{Ker} P$; (4) $x, y \in \operatorname{Ker} P$. In case (4), the equality (4.1) is trivial because both sides are 0. In case (1) we have

$$(Px, y) = (Px, Py) = (x, Py),$$

and (S4.1) follows. In case (2), the left-hand side of (S4.1) is zero (since $x \in \operatorname{Ker} P$) and the right-hand side is also zero in view of the orthogonality $\operatorname{Ker} P = (\operatorname{Im} P)^\perp$. In the same way, one checks (S4.1) in case (3).

So (S4.1) holds, and $P^* = P$. $\square$

Note that if P is a projector, so is $I - P$. Indeed, $(I - P)^2 = I - 2P + P^2 = I - 2P + P = I - P$. Moreover, $\operatorname{Ker} P = \operatorname{Im}(I - P)$ and $\operatorname{Im} P = \operatorname{Ker}(I - P)$. It is natural to call the projectors P and $I - P$ *complementary projectors*.

We shall now give useful representations of a projector with respect to a decomposition of $\mathscr{X}$ into a sum of two complementary subspaces. Let $T: \mathscr{X} \to \mathscr{X}$ be a linear transformation and let $\mathscr{L}_1, \mathscr{L}_2$ be a pair of complementary subspaces in $\mathscr{X}$. Denote $m_i = \dim \mathscr{L}_i$ $(i = 1, 2)$; then $m_1 + m_2 = n$ $(= \dim \mathscr{X})$. The transformation T may be written as a 2×2 block matrix with respect to the decomposition $\mathscr{L}_1 \dotplus \mathscr{L}_2 = \mathscr{X}$:

$$T = \begin{bmatrix} T_{11} & T_{12} \\ T_{21} & T_{22} \end{bmatrix}. \tag{S4.2}$$

Here T_{ij} $(i, j = 1, 2)$ is an $m_i \times m_j$ matrix which represents in some basis the linear transformation $P_i T|_{\mathscr{L}_j}: \mathscr{L}_j \to \mathscr{L}_i$, where P_i is the projector on $\mathscr{L}_i$ along $\mathscr{L}_{3-i}$ (so $P_1 + P_2 = I$).

Suppose now that $T = P$ is a projector on $\mathscr{L}_1 = \operatorname{Im} P$. Then representation (S4.2) takes the form

$$P = \begin{bmatrix} I & X \\ 0 & 0 \end{bmatrix} \tag{S4.3}$$

for some matrix X. In general, $X \neq 0$. One can check easily that $X = 0$ if and only if $\mathscr{L}_2 = \operatorname{Ker} P$. Analogously, if $\mathscr{L}_1 = \operatorname{Ker} P$, then (S4.2) takes the form

$$P = \begin{bmatrix} 0 & Y \\ 0 & I \end{bmatrix}, \tag{S4.4}$$

and $Y = 0$ if and only if $\mathscr{L}_2 = \operatorname{Im} P$. By the way, the direct multiplication $P \cdot P$, where P is given by (S4.3) or (S4.4), shows that P is indeed a projector: $P^2 = P$.

S4.2. Spectral Invariant Subspaces and Riesz Projectors

Let $A: \mathscr{X} \to \mathscr{X}$ be a linear transformation on a linear vector space $\mathscr{X}$ with $\dim \mathscr{X} = n$.

Recall that a subspace $\mathscr{L}$ is *invariant* for A (or *A-invariant*) if $A\mathscr{L} \subset \mathscr{L}$. The subspace consisting of the zero vector and $\mathscr{X}$ itself is invariant for every linear transformation. A less trivial example is a one-dimensional subspace spanned by an eigenvector. An A-invariant subspace $\mathscr{L}$ is called *A-reducing* if there exists a direct complement $\mathscr{L}'$ to $\mathscr{L}$ in $\mathscr{X}$ which is also A-invariant. In this case we shall say also that the pair of subspaces $(\mathscr{L}, \mathscr{L}')$ is A-reducing. Not every A-invariant subspace is A-reducing: for example, if A is a Jordan block of size k, considered as a linear transformation of $\mathbb{C}^k$ to $\mathbb{C}^k$, then there exists a sequence of A-invariant subspaces $\mathbb{C}^k \supset \mathscr{L}_{k-1} \supset \mathscr{L}_{k-2} \supset \cdots \supset \mathscr{L}_1 \supset \{0\}$, where $\mathscr{L}_i$ is the subspace spanned by the first i unit coordinate vectors. It is easily seen that none of the invariant subspaces $\mathscr{L}_i$ $(i = 1, \ldots, k-1)$ is A-reducing (because $\mathscr{L}_i$ is the single A-invariant subspace of dimension i).

From the definition it follows immediately that $\mathscr{L}$ is an A-invariant (resp. A-reducing) subspace if and only if $S\mathscr{L}$ is an invariant (resp. reducing) subspace for the linear transformation SAS^{-1}, where $S: \mathscr{X} \to \mathscr{X}$ is invertible. This simple observation allows us to use the Jordan normal form of A in many questions concerning invariant and reducing subspaces, and we frequently take advantage of this in what follows.

Invariant and reducing subspaces can be characterized in terms of projectors, as follows: let $A: \mathscr{X} \to \mathscr{X}$ be a linear transformation and let $P: \mathscr{X} \to \mathscr{X}$ be a projector. Then

(i) the subspace Im P is A-invariant if and only if

$$PAP = AP;$$

(ii) the pair of subspaces Im P, Ker P is A-reducing if and only if

$$AP = PA.$$

Indeed, write A as a 2×2 block matrix with respect to the decomposition Im $P \dotplus$ Ker $P = \mathscr{X}$:

$$A = \begin{bmatrix} A_{11} & A_{12} \\ A_{21} & A_{22} \end{bmatrix}$$

(so, for instance, $A_{11} = PA|_{\operatorname{Im} P}$: Im $P \to$ Im P). The projector P has the corresponding representation (see (S4.3))

$$P = \begin{bmatrix} I & 0 \\ 0 & 0 \end{bmatrix}.$$

Now $PAP = AP$ means that $A_{21} = 0$. But $A_{21} = 0$ in turn means that Im P is A-invariant, and (i) is proved. Further, $AP = PA$ means that $A_{21} = 0$ and $A_{12} = 0$. But clearly, the condition that Im P, Ker P is an A-reducing pair is the same. So (ii) holds as well.

We consider now an important class of reducing subspaces, namely, spectral invariant subspaces. Let $\sigma(A)$ be the spectrum of A, i.e., the set of all eigenvalues. Among the A-invariant subspaces of special interest are spectral invariant subspaces. An A-invariant subspace $\mathscr{L}$ is called *spectral with respect to a subset* $\sigma \subset \sigma(A)$, if $\mathscr{L}$ is a maximal A-invariant subspace with the property that $\sigma(A|_{\mathscr{L}}) \subset \sigma$. If $\sigma = \{\lambda_0\}$ consists of only one point λ_0, then a spectral subspace with respect to σ is just the root subspace of A corresponding to λ_0.

To describe spectral invariant subspaces, it is convenient to use Riesz projectors. Consider the linear transformation

$$P_\sigma = \frac{1}{2\pi i} \int_{\Gamma_0} (I\lambda - A)^{-1}\, d\lambda, \tag{S4.5}$$

where Γ_σ is a simple rectifiable contour such that σ is inside Γ_σ and $\sigma(A)\backslash\sigma$ is outside Γ_σ. The matrix P_σ given by (S4.5) is called the *Riesz projector* of A corresponding to the subset $\sigma \subset \sigma(A)$.

We shall list now some simple properties of Riesz projectors. First, let us check that (S4.5) is indeed a projector. To this end let $f(\lambda)$ be the function defined as follows: $f(\lambda) = 1$ inside and on Γ_σ, $f(\lambda) = 0$ outside Γ_σ. Then we can rewrite (S4.5) in the form

$$P_\sigma = \frac{1}{2\pi i} \int_{\Gamma} f(\lambda)(I\lambda - A)^{-1}\, d\lambda,$$

where Γ is some contour such that $\sigma(A)$ is inside Γ. Since $f(\lambda)$ is analytic in a neighborhood of $\sigma(A)$, by Theorem S1.15 we have

$$\begin{aligned} P_\sigma^2 &= \frac{1}{2\pi i}\int_\Gamma f(\lambda)(I\lambda - A)^{-1}\,d\lambda \cdot \frac{1}{2\pi i}\int_\Gamma f(\lambda)(I\lambda - A)^{-1}\,d\lambda \\ &= \frac{1}{2\pi i}\int_\Gamma (f(\lambda))^2(I\lambda - A)^{-1}\,d\lambda \\ &= \frac{1}{2\pi i}\int_\Gamma f(\lambda)(I\lambda - A)^{-1}\,d\lambda = P_\sigma \end{aligned}$$

and we have used the fact that $(f(\lambda))^2 = f(\lambda)$. So P_σ is a projector. By Corollary S1.16, $P_\sigma A = AP_\sigma$, so Im P_σ is a reducing A-invariant subspace.

Proposition S4.3 below shows that Im P_σ is in fact the spectral A-invariant subspace with respect to σ.

Another way to show that P_σ is a projector, is to consider the linear transformation A in a *Jordan basis.* By a Jordan basis we mean the basis in which A is represented by a matrix in a Jordan normal form. The existence of a Jordan basis follows from Theorem S1.13, if we take into account the fact that similarity of two matrices means that they represent the same linear transformation, but in different bases.

So, identifying A with its matrix representation in a Jordan basis, we can write

$$A = \begin{bmatrix} J_\sigma & 0 \\ 0 & J_{\bar\sigma} \end{bmatrix} \tag{S4.6}$$

where J_σ (resp. $J_{\bar\sigma}$) is a Jordan matrix with spectrum σ (resp. $\bar\sigma = \sigma(A)\setminus\sigma$). From formula (S4.6) we deduce then that

$$P_\sigma = \begin{bmatrix} I & 0 \\ 0 & 0 \end{bmatrix}$$

(in the matrix form with respect to the same basis as in (S4.6)).

Using (S4.6) one can prove the following characterization of spectral invariant subspaces.

Proposition S4.3. *The subspace* Im P_σ *is the maximal A-invariant subspace such that*

$$\sigma(A|_{\operatorname{Im} P_\sigma}) \subset \sigma.$$

Moreover, Im P_σ *is characterized by this property, i.e., if $\mathscr{L}$ is a maximal A-invariant subspace such that $\sigma(A|_{\mathscr{L}}) \subset \sigma$, then $\mathscr{L}$ =* Im P_σ.

In particular, a spectral A-invariant subspace $\mathscr{L}_\sigma$ (with respect to a given set $\sigma \subset \sigma(A)$) is unique and reduces A; namely, the spectral A-invariant subspace $\mathscr{L}_{\bar{\sigma}}$ (where $\bar{\sigma} = \sigma(A) \backslash \sigma$) is a direct complement to $\mathscr{L}_\sigma$ in $\mathscr{X}$.

Let $\lambda_1, \ldots, \lambda_k$ be all the different eigenvalues of A, and let

$$\mathscr{W}_i = \operatorname{Ker}(A - \lambda_i I)^n, \qquad i = 1, \ldots, k,$$

where $n = \dim \mathscr{X}$. The subspace $\mathscr{W}_i$ is the *root subspace* associated with eigenvalue λ_i. Writing A as a Jordan matrix (in a Jordan basis) it is easily seen that $\mathscr{W}_i$ coincides with the spectral subspace corresponding to $\sigma = \{\lambda_i\}$. Clearly,

$$\mathscr{X} = \mathscr{W}_1 \dotplus \cdots \dotplus \mathscr{W}_k \tag{S4.7}$$

decomposes $\mathscr{X}$ into the direct sum of root subspaces. Moreover, the following proposition shows that an analogous decomposition holds for every A-invariant subspace.

Proposition S4.4. *Let $\mathscr{L} \subset \mathscr{X}$ be an A-invariant subspace. Then*

$$\mathscr{L} = (\mathscr{L} \cap \mathscr{W}_1) \dotplus \cdots \dotplus (\mathscr{L} \cap \mathscr{W}_k), \tag{S4.8}$$

where $\mathscr{W}_1, \ldots, \mathscr{W}_k$ are the root subspaces of A.

Proof. Let P_i be the Riesz projector corresponding to λ_i; then $\mathscr{W}_i = \operatorname{Im} P_i$, $i = 1, \ldots, k$. On the other hand, $P_1 + P_2 + \cdots + P_k = I$, so any $x \in \mathscr{L}$ can be written in the form

$$x = P_1 x + P_2 x + \cdots + P_k x. \tag{S4.9}$$

From formula (S4.5) for Riesz projectors and the fact that $\mathscr{L}$ is A-invariant, it follows that $P_i x \in \mathscr{L}$ for $x = 1, \ldots, k$. Indeed, let Γ_i be a contour such that λ_i is inside Γ_i and λ_j, for $j \neq i$, is outside Γ_i. For every $\lambda \in \Gamma_i$, the linear transformation $I\lambda - A$ is invertible and $(I\lambda - A)x \in \mathscr{L}$ for every $x \in \mathscr{L}$. Therefore,

$$(I\lambda - A)^{-1} x \in \mathscr{L} \qquad \text{for every} \quad x \in \mathscr{L}. \tag{S4.10}$$

Now for fixed $x \in \mathscr{L}$, the integral

$$\left(\frac{1}{2\pi i} \int_{\Gamma_i} (I\lambda - A)^{-1}\, d\lambda\right) x$$

is a limit of Riemannian sums, all of them belonging to $\mathscr{L}$ in view of (S4.10). Hence the integral, as a limit, also belongs to $\mathscr{L}$, i.e., $P_i x \in \mathscr{L}$.

We have proved that $P_i x \in \mathscr{L} \cap \mathscr{W}_i$ for any $x \in \mathscr{L}$. Now formula (S4.9) gives the inclusion $\mathscr{L} \subset (\mathscr{L} \cap \mathscr{W}_1) + \cdots + (\mathscr{L} \cap \mathscr{W}_k)$. Since the opposite inclusion is obvious, and the sum in the right hand side of (S4.8) is direct (because (S4.7) is a direct sum), Proposition S4.4 follows. □

Proposition S4.4 often allows us to reduce consideration of invariant subspaces of A to the case when $\sigma(A)$ consists only of a single point (namely, by considering the restrictions $A|_{\mathscr{W}_i}$ to the root subspaces).

S4.3. The Gap between Subspaces

This section, and the next, are of an auxiliary character and are devoted to the basic properties of the set of all subspaces in $\mathscr{X} = \mathbb{C}^n$. These properties will be needed in Section S4.5 to prove results on stable invariant subspaces.

As usual, we assume that $\mathbb{C}^n$ is endowed with the scalar product $(x, y) = \sum_{i=1}^n x_i \bar{y}_i$; $x = (x_1, \ldots, x_n)^T$, $y = (y_1, \ldots, y_n)^T \in \mathbb{C}^n$ and the corresponding norm

$$\|x\| = \left(\sum_{i=1}^n |x_i|^2\right)^{1/2}, \qquad x = (x_1, \ldots, x_n)^T \in \mathbb{C}^n.$$

The norm for an $n \times n$ matrix A (or a linear transformation $A: \mathbb{C}^n \to \mathbb{C}^n$) is defined accordingly:

$$\|A\| = \max_{x \subset \mathbb{C}^n \setminus \{0\}} \|Ax\|/\|x\|.$$

The *gap* between subspaces $\mathscr{L}$ and $\mathscr{M}$ (in $\mathbb{C}^n$) is defined as

$$\theta(\mathscr{L}, \mathscr{M}) = \|P_{\mathscr{M}} - P_{\mathscr{L}}\|, \tag{S4.11}$$

where $P_{\mathscr{L}}(P_{\mathscr{M}})$ is the orthogonal projector on $\mathscr{L}(\mathscr{M})$ (refer to Section S4.1). It is clear from the definition that $\theta(\mathscr{L}, \mathscr{M})$ is a *metric* in the set of all subspaces in $\mathbb{C}^n$, i.e., $\theta(\mathscr{L}, \mathscr{M})$ enjoys the following properties:

(i) $\theta(\mathscr{L}, \mathscr{M}) > 0$ if $\mathscr{L} \neq \mathscr{M}$, $\theta(\mathscr{L}, \mathscr{L}) = 0$;
(ii) $\theta(\mathscr{L}, \mathscr{M}) = \theta(\mathscr{M}, \mathscr{L})$;
(iii) $\theta(\mathscr{L}, \mathscr{M}) \leq \theta(\mathscr{L}, \mathscr{N}) + \theta(\mathscr{N}, \mathscr{M})$.

Note also that $\theta(\mathscr{L}, \mathscr{M}) \leq 1$ (this property follows immediately from the characterization (S4.12) below).

In what follows we shall denote by $S_{\mathscr{L}}$ the unit sphere in a subspace $\mathscr{L} \subset \mathbb{C}^n$, i.e., $S_{\mathscr{L}} = \{x \mid x \in \mathscr{L}, \|x\| = 1\}$. We shall also need the concept of the distance of $d(x, Z)$ from $x \in \mathbb{C}^n$ to a set $Z \subset \mathbb{C}^n$. This is defined by $d(x, Z) = \inf_{t \in Z} \|x - t\|$.

Theorem S4.5. *Let $\mathscr{M}$, $\mathscr{L}$ be subspaces in $\mathbb{C}^n$. Then*

$$\theta(\mathscr{L}, \mathscr{M}) = \max\left\{\sup_{x \in S_{\mathscr{M}}} d(x, \mathscr{L}), \sup_{x \in S_{\mathscr{L}}} d(x, \mathscr{M})\right\}. \tag{S4.12}$$

If $P_1(P_2)$ are projectors with Im $P_1 = \mathscr{L}$ (Im $P_2 = \mathscr{M}$), *not necessarily orthogonal, then*

$$\theta(\mathscr{L}, \mathscr{M}) \leq \|P_1 - P_2\|. \tag{S4.13}$$

Proof. For every $x \in S_{\mathscr{L}}$ we have

$$\|x - P_2 x\| = \|(P_1 - P_2)x\| \leq \|P_1 - P_2\|,$$

therefore

$$\sup_{x \in S_{\mathscr{L}}} d(x, \mathscr{M}) \leq \|P_1 - P_2\|.$$

Analogously, $\sup_{x \in S_{\mathscr{M}}} d(x, \mathscr{L}) \leq \|P_1 - P_2\|$; so

$$\max\{\rho_{\mathscr{L}}, \rho_{\mathscr{M}}\} \leq \|P_1 - P_2\|, \tag{S4.14}$$

where $\rho_{\mathscr{L}} = \sup_{x \in S_{\mathscr{L}}} d(x, \mathscr{M})$, $\rho_{\mathscr{M}} = \sup_{x \in S_{\mathscr{M}}} d(x, \mathscr{L})$.

Observe that $\rho_{\mathscr{L}} = \sup_{x \in S_{\mathscr{L}}} \|(I - P_{\mathscr{M}})x\|$, $\rho_{\mathscr{M}} = \sup_{x \in S_{\mathscr{M}}} \|(I - P_{\mathscr{L}})x\|$. Consequently, for every $x \in \mathbb{C}^n$ we have

$$\|(I - P_{\mathscr{L}})P_{\mathscr{M}} x\| \leq \rho_{\mathscr{M}} \|P_{\mathscr{M}} x\|, \qquad \|(I - P_{\mathscr{M}})P_{\mathscr{L}} x\| \leq \rho_{\mathscr{L}} \|P_{\mathscr{L}} x\|. \tag{S4.15}$$

Now

$$\begin{aligned} \|P_{\mathscr{M}}(I - P_{\mathscr{L}})x\|^2 &= ((I - P_{\mathscr{L}})P_{\mathscr{M}}(I - P_{\mathscr{L}})x, \quad (I - P_{\mathscr{L}})x) \\ &\leq \|(I - P_{\mathscr{L}})P_{\mathscr{M}}(I - P_{\mathscr{L}})x\| \cdot \|(I - P_{\mathscr{L}})x\|; \end{aligned}$$

hence by (S4.15)

$$\|P_{\mathscr{M}}(I - P_{\mathscr{L}})x\|^2 \leq \rho_{\mathscr{M}} \|P_{\mathscr{M}}(I - P_{\mathscr{L}})x\| \cdot \|(I - P_{\mathscr{L}})x\|,$$

$$\|P_{\mathscr{M}}(I - P_{\mathscr{L}})x\| \leq \rho_{\mathscr{M}} \|(I - P_{\mathscr{L}})x\|. \tag{S4.16}$$

On the other hand, using the relation

$$P_{\mathscr{M}} - P_{\mathscr{L}} = P_{\mathscr{M}}(I - P_{\mathscr{L}}) - (I - P_{\mathscr{M}})P_{\mathscr{L}}$$

and the orthogonality of $P_{\mathscr{M}}$, we obtain

$$\|(P_{\mathscr{M}} - P_{\mathscr{L}})x\|^2 = \|P_{\mathscr{M}}(I - P_{\mathscr{L}})x\|^2 + \|(I - P_{\mathscr{M}})P_{\mathscr{L}} x\|^2.$$

Combining with (S4.15) and (S4.16) gives

$$\|(P_{\mathscr{M}} - P_{\mathscr{L}})x\|^2 \leq \rho_{\mathscr{M}}^2 \|(I - P_{\mathscr{L}})x\|^2 + \rho_{\mathscr{L}}^2 \|P_{\mathscr{L}} x\|^2 \leq \max\{\rho_{\mathscr{M}}^2, \rho_{\mathscr{L}}^2\} \|x\|^2.$$

So

$$\|P_{\mathscr{M}} - P_{\mathscr{L}}\| \leq \max\{\rho_{\mathscr{L}}, \rho_{\mathscr{M}}\}.$$

Using (S4.14) (with $P_1 = P_{\mathscr{L}}$, $P_2 = P_{\mathscr{M}}$) we obtain (S4.12). The inequality (S4.13) follows now from (S4.14). □

An important property of the metric $\theta(\mathscr{L}, \mathscr{M})$ is that in a neighborhood of every subspace $\mathscr{L} \subset \mathbb{C}^n$ all the subspaces have the same dimension (equal to $\dim \mathscr{L}$). This is a consequence of the following theorem.

Theorem S4.6. *If* $\theta(\mathscr{L}, \mathscr{M}) < 1$, *then* $\dim \mathscr{L} = \dim \mathscr{M}$.

Proof. Condition $\theta(\mathscr{L}, \mathscr{M}) < 1$ implies that $\mathscr{L} \cap \mathscr{M}^\perp = \{0\}$ and $\mathscr{L}^\perp \cap \mathscr{M} = \{0\}$. Indeed, suppose the contrary, and assume, for instance, that $\mathscr{L} \cap \mathscr{M}^\perp \neq \{0\}$. Let $x \in S_{\mathscr{L}} \cap \mathscr{M}^\perp$. Then $d(x, \mathscr{M}) = 1$, and by (S4.12), $\theta(\mathscr{L}, \mathscr{M}) \geq 1$, a contradiction. Now $\mathscr{L} \cap \mathscr{M}^\perp = \{0\}$ implies that $\dim \mathscr{L} \leq \dim \mathscr{M}$ and $\mathscr{L}^\perp \cap \mathscr{M} = \{0\}$ implies that $\dim \mathscr{L} \geq \dim \mathscr{M}$. $\square$

It also follows directly from this proof that the hypothesis $\theta(\mathscr{L}, \mathscr{M}) < 1$ implies $\mathbb{C}^n = \mathscr{L} \dotplus \mathscr{M}^\perp = \mathscr{L}^\perp \dotplus \mathscr{M}$. In addition, we have

$$P_{\mathscr{M}}(\mathscr{L}) = \mathscr{M}, \qquad P_{\mathscr{L}}(\mathscr{M}) = \mathscr{L}.$$

To see the first of these, for example, observe that for any $x \in \mathscr{M}$ there is the unique decomposition $x = y + z$, $y \in \mathscr{L}$, $z \in \mathscr{M}^\perp$. Hence, $x = P_{\mathscr{M}}x = P_{\mathscr{M}}y$ so that $\mathscr{M} \subset P_{\mathscr{M}}(\mathscr{L})$. But the reverse inclusion is obvious and so we must have equality.

We conclude this section with the following result, which makes precise the idea that direct sum decompositions of $\mathbb{C}^n$ are stable under small perturbations of the subspaces in the gap metric.

Theorem S4.7. *Let* $\mathscr{M}, \mathscr{M}_1 \subset \mathbb{C}^n$ *be subspaces such that*

$$\mathscr{M} \dotplus \mathscr{M}_1 = \mathbb{C}^n.$$

If $\mathscr{N}$ *is a subspace in* $\mathbb{C}^n$ *such that* $\theta(\mathscr{M}, \mathscr{N})$ *is sufficiently small, then*

$$\mathscr{N} \dotplus \mathscr{M}_1 = \mathbb{C}_n \tag{S4.17}$$

and

$$\theta(\mathscr{M}, \mathscr{N}) \leq \|\tilde{P}_{\mathscr{M}} - \tilde{P}_{\mathscr{N}}\| \leq C\theta(\mathscr{M}, \mathscr{N}), \tag{S4.18}$$

where $\tilde{P}_{\mathscr{M}}(\tilde{P}_{\mathscr{N}})$ *projects* $\mathbb{C}^n$ *onto* $\mathscr{M}$ *(onto* $\mathscr{N}$*) along* $\mathscr{M}_1$ *and* C *is a constant depending on* $\mathscr{M}$ *and* $\mathscr{M}_1$ *but not on* $\mathscr{N}$.

Proof. Let us prove first that the sum $\mathscr{N} \dotplus \mathscr{M}_1$ is indeed direct. The condition that $\mathscr{M} \dotplus \mathscr{M}_1 = \mathbb{C}^n$ is a direct sum implies that $\|x - y\| \geq \delta > 0$ for every $x \in S_{\mathscr{M}_1}$ and every $y \in \mathscr{M}$. Here δ is a fixed positive constant. Take $\mathscr{N}$ so close to $\mathscr{M}$ that $\theta(\mathscr{M}, \mathscr{N}) \leq \delta/2$. Then $\|z - y\| \leq \delta/2$ for every $z \in S_{\mathscr{N}}$, where $y = y(z)$ is the orthogonal projection of z on $\mathscr{M}$. Thus for $x \in S_{\mathscr{M}_1}$ and $z \in S_{\mathscr{N}}$ we have

$$\|x - z\| \geq \|x - y\| - \|z - y\| \geq \delta/2,$$

so $\mathcal{N} \cap \mathcal{M}_1 = \{0\}$. By Theorem S4.6 $\dim \mathcal{N} = \dim \mathcal{M}$ if $\theta(\mathcal{M}, \mathcal{N}) < 1$, so the dimensional consideration tells us that $\mathcal{N} + \mathcal{M}_1 = \mathbb{C}^n$ for $\theta(\mathcal{M}, \mathcal{N}) < 1$, and (S4.17) follows.

To establish the right-hand inequality in (S4.18) two preliminary remarks are needed. First note that for any $x \in \mathcal{M}$, $y \in \mathcal{M}_1$, $x = \tilde{P}_{\mathcal{M}}(x + y)$ so that

$$\|x + y\| \geq \|\tilde{P}_{\mathcal{M}}\|^{-1}\|x\|. \tag{S4.19}$$

It is claimed that, for $\theta(\mathcal{M}, \mathcal{N})$ small enough,

$$\|z + y\| \geq \tfrac{1}{2}\|\tilde{P}_{\mathcal{M}}\|^{-1}\|z\| \tag{S4.20}$$

for all $z \in \mathcal{N}$ and $y \in \mathcal{M}_1$.

Without loss of generality assume $\|z\| = 1$. Suppose $\theta(\mathcal{M}, \mathcal{N}) < \delta$ and let $x \in \mathcal{M}$. Then, using (S4.19) we obtain

$$\|z + y\| \geq \|x + y\| - \|z - x\| \geq \|\tilde{P}_{\mathcal{M}}\|^{-1}\|x\| - \delta.$$

But then $x = (x - z) + z$ implies $\|x\| \geq 1 - \delta$ and so

$$\|z + y\| \geq \|\tilde{P}_{\mathcal{M}}\|^{-1}(1 - \delta) - \delta$$

and, for δ small enough, (S4.20) is established.

The second remark is that, for any $x \in \mathbb{C}^n$

$$\|x - \tilde{P}_{\mathcal{M}}x\| \leq C_0 d(x, \mathcal{M}) \tag{S4.21}$$

for some constant C_0. To establish (S4.21), it is sufficient to consider the case that $x \in \operatorname{Ker} \tilde{P}_{\mathcal{M}}$ and $\|x\| = 1$. But then, obviously, we can take $C_0 = \max_{x \in \operatorname{Ker} \tilde{P}_{\mathcal{M}},\, \|x\| = 1} \{d(x, \mathcal{M})^{-1}\}$.

Now for any $x \in S_{\mathcal{N}}$, by use of (S4.21) and (S4.12),

$$\|(\tilde{P}_{\mathcal{M}} - \tilde{P}_{\mathcal{N}})x\| = \|x - \tilde{P}_{\mathcal{M}}x\| \leq C_0 d(x, \mathcal{M}) \leq C_0 \theta(\mathcal{M}, \mathcal{N}).$$

Then, if $w \in \mathbb{C}^N$, $\|w\| = 1$, and $w = y + z$, $y \in \mathcal{N}$, $z \in \mathcal{M}_1$,

$$\|(\tilde{P}_{\mathcal{M}} - \tilde{P}_{\mathcal{N}})w\| = \|(\tilde{P}_{\mathcal{M}} - \tilde{P}_{\mathcal{N}})y\| \leq \|y\|C_0\theta(\mathcal{M}, \mathcal{N}) \leq 2C_0\|\tilde{P}_{\mathcal{M}}\|\theta(\mathcal{M}, \mathcal{N})$$

and the last inequality follows from (S4.20). This completes the proof of the theorem. □

S4.4. The Metric Space of Subspaces

Consider the metric space $\mathcal{G}_n$ of all subspaces in $\mathbb{C}^n$ (considered in the gap metric $\theta(\mathcal{L}, \mathcal{M})$). In this section we shall study some basic properties of $\mathcal{G}_n$.

We remark that the results of the preceding section can be extended to subspaces in infinite-dimensional Hilbert space (instead of $\mathbb{C}^n$) and, with some modifications, to the case of infinite-dimensional Banach spaces. The following fundamental theorem is, in contrast, essentially "finite dimensional" and cannot be extended to the framework of subspaces in infinite dimensional space.

Theorem S4.8. *The metric space $\mathcal{G}_n$ is compact and, therefore, complete (as a metric space).*

Recall that compactness of $\mathcal{G}_n$ means that for every sequence $\mathcal{L}_1\ \mathcal{L}_2, \ldots$ of subspaces in $\mathcal{G}_n$ there exists a converging subsequence $\mathcal{L}_{i_1}, \mathcal{L}_{i_2}, \ldots$, i.e., such that

$$\lim_{k\to\infty} \theta(\mathcal{L}_{i_k}, \mathcal{L}_0) = 0$$

for some $\mathcal{L}_0 \in \mathcal{G}_n$. Completeness of $\mathcal{G}_n$ means that every sequence of subspaces $\mathcal{L}_i$, $i = 1, 2, \ldots$, for which $\lim_{i,j\to\infty} \theta(\mathcal{L}_i, \mathcal{L}_j) = 0$ is convergent.

Proof. In view of Theorem S4.6 the metric space $\mathcal{G}_n$ is decomposed into components $\mathcal{G}_{n,m}$, $m = 0, \ldots, n$, where $\mathcal{G}_{n,m}$ is a closed and open set in $\mathcal{G}_n$ consisting of all m-dimensional subspaces in $\mathcal{C}^n$.

Obviously, it is sufficient to prove the compactness of each $\mathcal{G}_{n,m}$. To this end consider the set $\$_{n,m}$ of all orthonormal systems $u - \{u_k\}_{k=1}^m$ consisting of m vectors $u_1, \ldots, u_m$ in $\mathcal{C}^n$.

For $u = \{u_k\}_{k=1}^m$, $v = \{v_k\}_{k=1}^m \in \$_{n,m}$ define

$$\delta(u, v) = \left(\sum_{k=1}^m \|u_k - v_k\|^2\right)^{1/2}.$$

It is easily seen that $\delta(u, v)$ is a metric in $\$_{n,m}$, so turning $\$_{n,m}$ into a metric space. For each $u = \{u_k\}_{k=1}^m \in \$_{n,m}$ define $A_m u = \operatorname{Span}\{u_1, \ldots, u_m\} \in \mathcal{G}_{n,m}$. In this way we obtain a map $A_m: \$_{n,m} \to \mathcal{G}_{n,m}$ of metric spaces $\$_{n,m}$ and $\mathcal{G}_{n,m}$.

We prove now that the map A_m is continuous. Indeed, let $\mathcal{L} \in \mathcal{G}_{n,m}$ and let $v_1, \ldots, v_m$ be an orthonormal basis in $\mathcal{L}$. Pick some $u = \{u_k\}_{k=1}^m \in \$_{n,m}$ (which is supposed to be in a neighborhood of $v = \{v_k\}_{k=1}^m \in \$_{n,m}$). For v_i, $i = 1, \ldots, m$, we have (where $\mathcal{M} = A_m u$ and $P_{\mathcal{N}}$ stands for the orthogonal projector on the subspace $\mathcal{N}$):

$$\begin{aligned} \|(P_{\mathcal{M}} - P_{\mathcal{L}})v_i\| &= \|P_{\mathcal{M}} v_i - v_i\| \le \|P_{\mathcal{M}}(v_i - u_i)\| + \|u_i - v_i\| \\ &\le \|P_{\mathcal{M}}\|\,\|v_i - u_i\| + \|u_i - v_i\| \le 2\delta(u, v), \end{aligned}$$

and therefore for $x = \sum_{i=1}^m \alpha_i v_i \in S_{\mathcal{L}}$

$$\|(P_{\mathcal{M}} - P_{\mathcal{L}})x\| \le 2 \sum_{i=1}^m |\alpha_i| \delta(u, v).$$

Now, since $\|x\| = \sum_{i=1}^{m}|\alpha_i|^2 = 1$, we obtain that $|\alpha_i| \le 1$ and $\sum_{i=1}^{m}|\alpha_i| \le m$, and so

$$\|(P_{\mathscr{M}} - P_{\mathscr{L}})|_{\mathscr{L}}\| \le 2m\delta(u, v). \tag{S4.22}$$

Fix now some $y \in S_{\mathscr{L}^\perp}$. We wish to evaluate $P_{\mathscr{M}}y$. For every $x \in \mathscr{L}$, write

$$(x, P_{\mathscr{M}}y) = (P_{\mathscr{M}}x, y) = ((P_{\mathscr{M}} - P_{\mathscr{L}})x, y) + (x, y) = ((P_{\mathscr{M}} - P_{\mathscr{L}})x, y),$$

and

$$|(x, P_{\mathscr{M}}y)| \le 2m\|x\|\,\delta(u, v) \tag{S4.23}$$

by (S4.22). On the other hand, write

$$P_{\mathscr{M}}y = \sum_{i=1}^{m} \alpha_i u_i;$$

then for every $z \in \mathscr{L}^\perp$

$$(z, P_{\mathscr{M}}y) = \left(z, \sum_{i=1}^{m}\alpha_i(u_i - v_i)\right) + \left(z, \sum_{i=1}^{m}\alpha_i v_i\right) = \left(z, \sum_{i=1}^{m}\alpha_i(u_i - v_i)\right),$$

and

$$|(z, P_{\mathscr{M}}y)| \le \|z\|\|\sum_{i=1}^{m}\alpha_i(u_i - v_i)\| \le \|z\|m \max_{1\le i\le m}|\alpha_i|\,\|u_i - v_i\|.$$

But $\|y\| = 1$ implies that $\sum_{i=1}^{m}|\alpha_i|^2 \le 1$, so $\max_{1\le i\le m}|\alpha_i| \le 1$. Hence,

$$|(z, P_{\mathscr{M}}y)| \le \|z\|m\delta(u, v). \tag{S4.24}$$

Combining (S4.23) and (S4.24) we obtain that

$$|(t, P_{\mathscr{M}}y)| \le 3m\,\delta(u, v)$$

for every $t \in \mathbb{C}^n$ with $\|t\| = 1$. Thus,

$$\|P_{\mathscr{M}}y\| \le 3m\delta(u, v). \tag{S4.25}$$

Now we can easily prove the continuity of A_m. Pick $x \in \mathbb{C}^n$, $\|x\| = 1$. Then, using (S4.22), (S4.25), we have

$$\|(P_{\mathscr{M}} - P_{\mathscr{L}})x\| \le \|(P_{\mathscr{M}} - P_{\mathscr{L}})P_{\mathscr{L}}x\| + \|P_{\mathscr{M}}(x - P_{\mathscr{L}}x)\| \le 5m \cdot \delta(u, v),$$

so

$$\theta(\mathscr{M}, \mathscr{L}) = \|P_{\mathscr{M}} - P_{\mathscr{L}}\| \le 5m\delta(u, v),$$

which obviously implies the continuity of A_m.

It is easily seen that $\mathscr{S}_{n,m}$ is compact. Since $A_m: \mathscr{S}_{n,m} \to \mathscr{G}_{n,m}$ is a continuous map onto $\mathscr{G}_{n,m}$, the latter is also compact.

Finally, let us prove the completeness of $\mathcal{G}_{n,m}$. Let $\mathcal{L}_1, \mathcal{L}_2, \dots$ be a Cauchy sequence in $\mathcal{G}_{n,m}$, i.e., $\theta(\mathcal{L}_i, \mathcal{L}_j) \to 0$ as $i, j \to 0$. By compactness, there exists a partial sequence $\mathcal{L}_{i_k}$ such that $\lim_{k\to\infty} \theta(\mathcal{L}_{i_k}, \mathcal{L}) = 0$ for some $\mathcal{L} \in \mathcal{G}_{n,m}$. But then it is easily seen that in fact $\mathcal{L} = \lim_{i\to\infty} \mathcal{L}_i$. □

S4.5. Stable Invariant Subspaces: Definition and Main Result

Let $A: \mathbb{C}^n \to \mathbb{C}^n$ be a linear transformation. If λ_0 is an eigenvalue of A, the corresponding root space $\operatorname{Ker}(\lambda_0 I - A)^n$ will be denoted by $\mathcal{N}(\lambda_0)$.

An A-invariant subspace $\mathcal{N}$ is called *stable* if given $\varepsilon > 0$ there exists $\delta > 0$ such that $\|B - A\| < \delta$ for a linear transformation $B: \mathbb{C}^n \to \mathbb{C}^n$ implies that B has an invariant subspace $\mathcal{M}$ with $\|P_{\mathcal{M}} - P_{\mathcal{N}}\| < \varepsilon$. Here $P_{\mathcal{M}}$ denotes the orthogonal projector with $\operatorname{Im} P = \mathcal{M}$. The same definition applies for matrices.

This concept is particularly important from the point of view of numerical computation. It is generally true that the process of finding a matrix representation for a linear transformation and then finding invariant subspaces can only be performed approximately. Consequently, the stable invariant subspaces will generally be the only ones amenable to numerical computation.

Suppose $\mathcal{N}$ is a direct sum of root subspaces of A. Then $\mathcal{N}$ is a stable invariant subspace for A. This follows from the fact that $\mathcal{N}$ appears as the image of a Riesz projector

$$R_A = \frac{1}{2\pi i}\int_\Gamma (I\lambda - A)^{-1}\, d\lambda,$$

where Γ is a suitable contour in $\mathbb{C}$ (see Proposition S4.3). Indeed, this formula ensures that $\|R_B - R_A\|$ is arbitrarily small if $\|B - A\|$ is small enough. Let $\mathcal{M} = \operatorname{Im} R_B$; since $P_{\mathcal{N}}$ and $P_{\mathcal{M}}$ are *orthogonal* projectors with the same images as R_A and R_B respectively, the following inequality holds (see Theorem S4.5)

$$\|P_{\mathcal{N}} - P_{\mathcal{M}}\| \le \|R_B - R_A\|, \tag{S4.26}$$

so $\|P_{\mathcal{N}} - P_{\mathcal{M}}\|$ is small together with $\|R_B - R_A\|$.

It will turn out, however, that in general not every stable invariant subspace is spectral. On the other hand, if $\dim \operatorname{Ker}(\lambda_j I - A) > 1$ and $\mathcal{N}$ is a one-dimensional subspace of $\operatorname{Ker}(\lambda_j I - A)$, it is intuitively clear that a small perturbation of A can result in a large change in the gap between invariant subspaces. The following simple example provides such a situation. Let A be the 2×2 zero matrix, and let $\mathcal{N} = \operatorname{Span}\{\left[\begin{smallmatrix}1\\1\end{smallmatrix}\right]\} \subset \mathbb{C}^2$. Clearly, $\mathcal{N}$ is A-invariant; but $\mathcal{N}$ is unstable. Indeed, let $B = \operatorname{diag}[0, \varepsilon]$, where $\varepsilon \neq 0$ is close enough to zero. The only one-dimensional B-invariant subspaces are

$\mathcal{M}_1 = \operatorname{Span}\{[\begin{smallmatrix}1\\0\end{smallmatrix}]\}$ and $\mathcal{M}_2 = \operatorname{Span}\{[\begin{smallmatrix}0\\1\end{smallmatrix}]\}$, and both are far from $\mathcal{N}$: computation shows that

$$\|P_{\mathcal{N}} - P_{\mathcal{M}_i}\| = 1/\sqrt{2}, \qquad i = 1, 2.$$

The following theorem gives the description of all stable invariant subspaces.

Theorem S4.9. *Let $\lambda_1, \ldots, \lambda_r$ be the different eigenvalues of the linear transformation A. A subspace $\mathcal{N}$ of $\mathbb{C}^n$ is A-invariant and stable if and only if $\mathcal{N} = \mathcal{N}_1 \dotplus \cdots \dotplus \mathcal{N}_r$, where for each j the space $\mathcal{N}_j$ is an arbitrary A-invariant subspace of $\mathcal{N}(\lambda_j)$ whenever* $\dim \operatorname{Ker}(\lambda_j I - A) = 1$, *while otherwise $\mathcal{N}_j = \{0\}$ or $\mathcal{N}_j = \mathcal{N}(\lambda_j)$.*

The proof of Theorem S4.9 will be based on a series of lemmas and an auxiliary theorem which is of some interest in itself.

In the proof of Theorem S4.9 we will use the following observation which follows immediately from the definition of a stable subspace: the A-invariant subspace $\mathcal{N}$ is stable if and only if the SAS^{-1}-invariant subspace $S\mathcal{N}$ is stable. Here $S: \mathbb{C}^n \to \mathbb{C}^n$ is an arbitrary invertible linear transformation.

S4.6. Case of a Single Eigenvalue

The results presented in this subsection will lead to the proof of Theorem S4.9 for the case when A has one eigenvalue only. To state the next theorem we need the following notion: a chain $\mathcal{M}_1 \subset \mathcal{M}_2 \subset \cdots \subset \mathcal{M}_{n-1}$ of A-invariant subspaces is said to be *complete* if $\dim \mathcal{M}_j = j$ for $j = 1, \ldots, n-1$.

Theorem S4.10. *Given $\varepsilon > 0$, there exists $\delta > 0$ such that the following holds true: If B is a linear transformation with $\|B - A\| < \delta$ and $\{\mathcal{M}_j\}$ is a complete chain of B-invariant subspaces, then there exists a complete chain $\{\mathcal{N}_i\}$ of A-invariant subspaces such that $\|P_{\mathcal{N}_j} - P_{\mathcal{M}_j}\| < \varepsilon$ for $j = 1, \ldots, n-1$.*

In general the chain $\{\mathcal{N}_j\}$ for A will depend on the choice of B. To see this, consider

$$A = \begin{bmatrix} 0 & 0 \\ 0 & 0 \end{bmatrix}, \qquad B_v = \begin{bmatrix} 0 & 0 \\ v & 0 \end{bmatrix}, \qquad B'_v = \begin{bmatrix} 0 & v \\ 0 & 0 \end{bmatrix},$$

where $v \in \mathbb{C}$. Observe that for $v \neq 0$ the only one dimensional invariant subspace of B_v is $\operatorname{Span}\{(0, 1)^T\}$ and for B'_v, $v \neq 0$, the only one-dimensional invariant subspace is $\operatorname{Span}\{(1, 0)^T\}$.

Proof. Assume that the conclusion of the theorem is not correct. Then there exists $\varepsilon > 0$ with the property that for every positive integer m there exists a linear transformation B_m satisfying $\|B_m - A\| < 1/m$ and a complete chain $\{\mathcal{M}_{mj}\}$ of B_m-invariant subspaces such that for every complete chain $\{\mathcal{N}_j\}$ of A-invariant subspaces:

$$\max_{1 \le j \le k-1} \|P_{\mathcal{N}_j} - P_{\mathcal{M}_{mj}}\| \ge \varepsilon \qquad m = 1, 2, \ldots. \tag{S4.27}$$

Denote for brevity $P_{mj} = P_{\mathcal{M}_{mj}}$.

Since $\|P_{mj}\| = 1$, there exists a subsequence $\{m_i\}$ of the sequence of positive integers and linear transformations $P_1, \ldots, P_{n-1}$ on $\mathbb{C}^n$, such that

$$\lim_{i \to \infty} P_{m_i, j} = P_j, \qquad j = 1, \ldots, n-1. \tag{S4.28}$$

Observe that $P_1, \ldots, P_{n-1}$ are orthogonal projectors. Indeed, passing to the limit in the equalities $P_{m_i, j} = (P_{m_i, j})^2$, we obtain that $P_j = P_j^2$. Further (S4.28) combined with $P_{m_i, j}^* = P_{m_i, j}$ implies that $P_j^* = P_j$; so P_j is an orthogonal projector (Proposition S4.2).

Further, the subspace $\mathcal{N}_j = \operatorname{Im} P_j$ has dimension j, $j = 1, \ldots, n-1$. This is a consequence of Theorem S4.6.

By passing to the limits it follows from $B_m P_{mj} = P_{mj} B_m P_{mj}$ that $AP_j = P_j A P_j$. Hence $\mathcal{N}_j$ is A-invariant. Since $P_{mj} = P_{m, j+1} P_{mj}$ we have $P_j = P_{j+1} P_j$, and thus $\mathcal{N}_j \subset \mathcal{N}_{j+1}$. It follows that $\mathcal{N}_j$ is a complete chain of A-invariant subspaces. Finally, $\theta(\mathcal{N}_j, \mathcal{M}_{m_i, j}) = \|P_j - P_{m_i j}\| \to 0$. But this contradicts (S4.27), and the proof is complete. □

Corollary S4.11. *If A has only one eigenvalue, λ_0 say, and if* $\dim \operatorname{Ker}(\lambda_0 I - A) = 1$, *then each invariant subspace of A is stable.*

Proof. The conditions on A are equivalent to the requirement that for each $1 \le j \le k-1$ the operator A has only one j-dimensional invariant subspace and the nontrivial invariant subspaces form a complete chain. So we may apply the previous theorem to get the desired result. □

Lemma S4.12. *If A has only one eigenvalue, λ_0 say, and if* $\dim \operatorname{Ker}(\lambda_0 I - A) \ge 2$, *then the only stable A-invariant subspaces are* $\{0\}$ *and* $\mathbb{C}^n$.

Proof. Let $J = \operatorname{diag}(J_1, \ldots, J_s)$ be a Jordan matrix for A. Here J_i is a simple Jordan block with λ_0 on the main diagonal and of size κ_i, say. As $\dim \operatorname{Ker}(\lambda_0 I - A) \ge 2$ we have $s \ge 2$. By similarity, it suffices to prove that J has no nontrivial stable invariant subspace.

Let $e_1, \ldots, e_k$ be the standard basis for $\mathbb{C}^n$. Define the linear transformation T_ε on $\mathbb{C}^n$ by setting

$$T_\varepsilon e_i = \begin{cases} \varepsilon e_{i-1} & \text{if } i = \kappa_1 + \cdots + \kappa_j + 1, \quad j = 1, \ldots, s-1 \\ 0 & \text{otherwise,} \end{cases}$$

and put $B_\varepsilon = J + T_\varepsilon$. Then $\|B_\varepsilon - J\|$ tends to 0 as $\varepsilon \to 0$. For $\varepsilon \neq 0$ the linear transformation B_ε has exactly one j-dimensional invariant subspace namely, $\mathcal{N}_j = \operatorname{Span}\{e_1, \ldots, e_j\}$. Here $1 \leq j \leq k - 1$. It follows that $\mathcal{N}_j$ is the only candidate for a stable J-invariant subspace of dimension j.

Now consider $\tilde{J} = \operatorname{diag}(J_s, \ldots, J_1)$. Repeating the argument of the previous paragraph for $\tilde{J}$ instead of J, we see that $\mathcal{N}_j$ is the only candidate for a stable $\tilde{J}$-invariant subspace of dimension j. But $J = S\tilde{J}S^{-1}$, where S is the similarity transformation that reverses the order of the blocks in J. It follows that $S\mathcal{N}_j$ is the only candidate for a stable J-invariant subspace of dimension j. However, as $s \geq 2$, we have $S\mathcal{N}_j \neq \mathcal{N}_j$ for $1 \leq j \leq k - 1$, and the proof is complete. □

Corollary S4.1 and Lemma S4.12 together prove Theorem S4.9 for the case when A has one eigenvalue only.

S4.7. General Case of Stable Invariant Subspaces

The proof of Theorem S4.9 in the general case will be reduced to the case of one eigenvalue which was considered in Section S4.6. To carry out this reduction we will introduce some additional notions.

We begin with the following notion of minimal opening. For two subspaces $\mathcal{M}$ and $\mathcal{N}$ of $\mathbb{C}^n$ the number

$$\eta(\mathcal{M}, \mathcal{N}) = \inf\{\|x + y\| \mid x \in \mathcal{M}, y \in \mathcal{N}, \max(\|x\|, \|y\|) = 1\}$$

is called the *minimal opening* between $\mathcal{M}$ and $\mathcal{N}$. Note that always $0 \leq \eta(\mathcal{M}, \mathcal{N}) \leq 1$, except when both $\mathcal{M}$ and $\mathcal{N}$ are the zero subspace, in which case $\eta(\mathcal{M}, \mathcal{N}) = \infty$. It is easily seen that $\eta(\mathcal{M}, \mathcal{N}) > 0$ if and only if $\mathcal{M} \cap \mathcal{N} = \{0\}$.

The following result shows that there exists a close connection between the minimal opening and gap $\theta(\mathcal{M}, \mathcal{N})$ between subspaces.

Proposition S4.13. *Let $\mathcal{M}_m$, $m = 1, 2, \ldots$, be a sequence of subspaces in $\mathbb{C}^n$. If $\lim_{n\to\infty} \theta(\mathcal{M}_m, \mathcal{L}) = 0$ for some subspace $\mathcal{L}$, then*

$$\eta(\mathcal{M}_m, \mathcal{N}) \to \eta(\mathcal{L}, \mathcal{N}) \tag{S4.29}$$

for every subspace $\mathcal{N}$.

Proof. Note that the set of pairs $\{(x, y) \mid x \in \mathcal{M}, y \in \mathcal{N}, \max(\|x\|, \|y\|) = 1\} \subset \mathbb{C}^n \times \mathbb{C}^n$ is compact for any pair of subspaces $\mathcal{M}, \mathcal{N}$. Since $\|x + y\|$ is a continuous function of x and y, in fact

$$\eta(\mathcal{M}, \mathcal{N}) = \|x + y\|$$

for some vectors $x \in \mathscr{M}$, $y \in \mathscr{N}$ such that $\max(\|x\|, \|y\|) = 1$. So we can choose $x_m \in \mathscr{M}_m$, $y_m \in \mathscr{N}$, $m = 1, 2, \ldots$, such that $\max(\|x_m\|, \|y_m\|) = 1$ and $\eta(\mathscr{M}_m, \mathscr{N}) = \|x_m + y_m\|$. Pick convergent subsequences $x_{m_k} \to x_0 \in \mathbb{C}^n$, $y_{m_k} \to y_0 \in \mathscr{N}$. Clearly, $\max(\|x_0\|, \|y_0\|) = 1$ and $\eta(\mathscr{M}_{m_k}, \mathscr{N}) \to \|x_0 + y_0\|$. It is easy to verify (for instance by reductio ad absurdum) that $x_0 \in \mathscr{L}$. Thus

$$\eta(\mathscr{M}_{m_k}, \mathscr{N}) \to \|x_0 + y_0\| \geq \eta(\mathscr{L}, \mathscr{N}). \tag{S4.30}$$

On the other hand

$$\lim_{m\to\infty} \sup \eta(\mathscr{M}_m, \mathscr{N}) \leq \eta(\mathscr{L}, \mathscr{N}). \tag{S4.31}$$

Indeed, let $x \in \mathscr{L}$, $y \in \mathscr{N}$ be such that $\max(\|x\|, \|y\|) = 1$ and

$$\eta(\mathscr{L}, \mathscr{N}) = \|x + y\|.$$

Assume first that $x \neq 0$. For any given $\varepsilon > 0$ there exists m_0 such that $d(x, \mathscr{M}_m) < \varepsilon$ for $m \geq m_0$. Let $x_m \in \mathscr{M}_m$ be such that $d(x, \mathscr{M}_m) = \|x - x_m\|$. Then $z_m \overset{\text{def}}{=} (\|x\| x_m)/\|x_m\| \in \mathscr{M}_m$, $\max(\|z_m\|, \|y\|) = 1$ and for $m \geq m_0$ we obtain

$$\begin{aligned}\|z_m + y\| &\leq \|x + y\| + \left\| z_m - \frac{\|x\|}{\|x_m\|} \cdot x \right\| + \|x\| \left| 1 - \frac{\|x\|}{\|x_m\|} \right| \\ &\leq \|x + y\| + \frac{\|x\|}{\|x_m\|} \varepsilon + \frac{|\|x_m\| - \|x\||}{\|x_m\|}.\end{aligned}$$

Taking $\varepsilon < \|x\|/2$, we obtain that $\|z_m + y\| \leq \|x + y\| + 2\varepsilon + 5\varepsilon$, so

$$\eta(\mathscr{M}_m, \mathscr{N}) \leq \|z_m + y\| \to \|x + y\| = \eta(\mathscr{L}, \mathscr{N}),$$

and (S4.31) follows.

Combining (S4.30) and (S4.31) we see that for some subsequence m_k we have $\lim_{k\to\infty} \eta(\mathscr{M}_{m_k}, \mathscr{N}) = \eta(\mathscr{L}, \mathscr{N})$. Now by a standard argument one proves (S4.29). Indeed, if (S4.29) were not true, then we could pick a subsequence m_k' such that

$$\lim_{k\to\infty} \eta(\mathscr{M}_{m_k'}, \mathscr{N}) \neq \eta(\mathscr{L}, \mathscr{N}). \tag{S4.32}$$

But then we repeat the above argument replacing $\mathscr{M}_m$ by $\mathscr{M}_{m_k'}$. So it is possible to pick a subsequence m_k'' from m_k' such that $\lim_{k\to\infty} \eta(\mathscr{M}_{m_k''}, \mathscr{N}) = \eta(\mathscr{L}, \mathscr{N})$, a contradiction with (S.4.32). □

Let us introduce some terminology and notation which will be used in the next two lemmas and their proofs. We use the shorthand $A_m \to A$ for $\lim_{m\to\infty} \|A_m - A\| = 0$, where A_m, $m = 1, 2, \ldots$, and A are linear transformations on $\mathbb{C}^n$. Note that $A_m \to A$ if and only if the entries of the matrix representations of A_m (in some fixed basis) converge to the corresponding entries of A (represented as a matrix in the same basis). We say that a simple rectifiable contour Γ *splits the spectrum* of a linear transformation T if $\sigma(T) \cap \Gamma = \varnothing$. In that case we can associate with T and Γ the Riesz projector

$$P(T; \Gamma) = \frac{1}{2\pi i} \int_{\Gamma} (I\lambda - T)^{-1} \, d\lambda.$$

The following observation will be used subsequently. If T is a linear transformation for which Γ splits the spectrum, then Γ splits the spectrum for every linear transformation S which is sufficiently close to T (i.e., $\|S - T\|$ is close enough to zero). Indeed, if it were not so, we have $\det(\lambda_m I - S_m) = 0$ for some sequence $\lambda_m \in \Gamma$ and $S_m \to T$. Pick a subsequence $\lambda_{m_k} \to \lambda_0$ for some $\lambda_0 \in \Gamma$; passing to the limit in the equality $\det(\lambda_{m_k} I - S_{m_k}) = 0$ (here we use the matrix representations of S_{m_k} and T in some fixed basis, as well as the continuous dependence of $\det A$ on the entries of A) we obtain $\det(\lambda_0 I - T) = 0$, a contradiction with the splitting of $\sigma(T)$.

We shall also use the notion of the angular operator. If P is a projector of $\mathbb{C}^n$ and $\mathcal{M}$ is a subspace of $\mathbb{C}^n$ with $\operatorname{Ker} P \dotplus \mathcal{M} = \mathbb{C}^n$, then there exists a unique linear transformation R from $\operatorname{Im} P$ into $\operatorname{Ker} P$ such that $\mathcal{M} = \{Rx + x \mid x \in \operatorname{Im} P\}$. This transformation is called the *angular operator* of $\mathcal{M}$ with respect to the projector P. We leave to the reader the (easy) verification of existence and uniqueness of the angular operator (or else see [3c, Chapter 5]).

Lemma S4.14. *Let Γ be a simple rectifiable contour that splits the spectrum of T, let T_0 be the restriction of T to* $\operatorname{Im} P(T; \Gamma)$ *and let $\mathcal{N}$ be a subspace of* $\operatorname{Im} P(T; \Gamma)$. *Then $\mathcal{N}$ is a stable invariant subspace for T if and only if $\mathcal{N}$ is a stable invariant subspace for T_0.*

Proof. Suppose $\mathcal{N}$ is a stable invariant subspace for T_0, but not for T. Then one can find $\varepsilon > 0$ such that for every positive integer m there exists a linear transformation S_m such that

$$\|S_m - T\| < 1/m \tag{S4.33}$$

and

$$\theta(\mathcal{N}, \mathcal{M}) \geq \varepsilon, \qquad \mathcal{M} \in \Omega_m. \tag{S4.34}$$

Here Ω_m denotes the collection of all invariant subspaces for S_m. From (S4.33) it is clear that $S_m \to T$. By assumption Γ splits the spectrum of T. Thus, for m sufficiently large, the contour T will split the spectrum of S_m. Moreover, $P(S_m; \Gamma) \to P(T; \Gamma)$ and hence $\operatorname{Im} P(S_m; \Gamma)$ tends to $\operatorname{Im} P(T; \Gamma)$ in the gap topology. But then, for m sufficiently large,

$$\operatorname{Im} P(T; \Gamma) \dotplus \operatorname{Im} P(S_m; \Gamma) = \mathbb{C}^n$$

(cf. Theorem S4.7).

Let R_m be the angular operator of Im $P(S_m; T)$ with respect to $P(T; \Gamma)$. Here, as in what follows, m is supposed to be sufficiently large. As $P(S_m; \Gamma) \to P(T; \Gamma)$, we have $R_m \to 0$. Put

$$E_m = \begin{bmatrix} I & R_m \\ 0 & I \end{bmatrix}$$

where the matrix representation corresponds to the decomposition

$$\mathbb{C}^n = \operatorname{Ker} P(T; \Gamma) \dotplus \operatorname{Im} P(T; \Gamma). \tag{S4.35}$$

Then E_m is invertible with inverse

$$E_m^{-1} = \begin{bmatrix} I & -R_m \\ 0 & I \end{bmatrix},$$

$E_m \operatorname{Im} P(T; \Gamma) = \operatorname{Im} P(S_m; \Gamma)$, and $E_m \to I$.

Put $T_m = E_m^{-1} S_m E_m$. Then $T_m \operatorname{Im} P(T; \Gamma) \subset \operatorname{Im} P(T; \Gamma)$ and $T_m \to T$. Let T_{m0} be the restriction of T_m to Im $P(T; \Gamma)$. Then $T_{m0} \to T_0$. As $\mathcal{N}$ is a stable invariant subspace for T_0 there exists a sequence $\{\mathcal{N}_m\}$ of subspaces of Im $P(T; \Gamma)$ such that $\mathcal{N}_m$ is T_{m0}-invariant and $\theta(\mathcal{N}_m, \mathcal{N}) \to 0$. Note that $\mathcal{N}_m$ is also T_m-invariant.

Now put $\mathcal{M}_m = E_m \mathcal{N}_m$. Then $\mathcal{M}_m$ is an invariant subspace for S_m. Thus $\mathcal{M}_m \in \Omega_m$. From $E_m \to I$ one can easily deduce that $\theta(\mathcal{M}_m, \mathcal{N}_m) \to 0$. Together with $\theta(\mathcal{N}_m, \mathcal{N}) \to 0$ this gives $\theta(\mathcal{M}_m, \mathcal{N}) \to 0$, which contradicts (S4.34).

Next assume that $\mathcal{N} \subset \operatorname{Im} P(T; \Gamma)$ is a stable invariant subspace for T, but not for T_0. Then one can find $\varepsilon > 0$ such that for every positive integer m there exists a linear transformation S_{m0} on Im $P(T; \Gamma)$ satisfying

$$\|S_{m0} - T_0\| < 1/m \tag{S4.36}$$

and

$$\theta(\mathcal{N}, \mathcal{M}) \geq \varepsilon, \qquad \mathcal{N} \in \Omega_{m0}. \tag{S4.37}$$

Here Ω_{m0} denotes the collection of all invariant subspaces of S_{m0}. Let T_0 be the restriction of T to Ker $P(T; \Gamma)$ and write

$$S_m = \begin{bmatrix} T_1 & 0 \\ 0 & S_{m0} \end{bmatrix}$$

where the matrix representation corresponds to the decomposition (S4.35). From (S4.36) it is clear that $S_m \to T$. Hence, as $\mathcal{N}$ is a stable invariant subspace for T, there exists a sequence $\{\mathcal{N}_m\}$ of subspaces of $\mathbb{C}^n$ such that $\mathcal{N}_m$ is S_m-invariant and $\theta(\mathcal{N}_m, \mathcal{N}) \to 0$. Put $\mathcal{M}_m = P(T; \Gamma)\mathcal{N}_m$. Since $P(T; \Gamma)$ commutes with S_m, then $\mathcal{M}_m$ is an invariant subspace for S_{m0}. We shall now prove that $\theta(\mathcal{M}_m, \mathcal{N}) \to 0$, thus obtaining a contradiction to (S4.37).

Take $y \in \mathscr{M}_m$ with $\|y\| \leq 1$. Then $y = P(T;\Gamma)x$ for some $x \in \mathscr{M}_m$. As

$$\|y\| = \|P(T;\Gamma)x\| \geq \inf\{\|x - u\| \mid u \in \operatorname{Ker} P(T;\Gamma)\}$$
$$\geq \eta(\mathscr{N}_m, \operatorname{Ker} P(T;\Gamma)) \cdot \|x\|, \tag{S4.38}$$

where $\eta(\cdot,\cdot)$ is the minimal opening. By Proposition S4.13, $\theta(\mathscr{N}_m, \mathscr{N}) \to 0$ implies that $\eta(\mathscr{N}_m, \operatorname{Ker} P(T;\Gamma)) \to \eta_0$, where $\eta_0 = \eta(\mathscr{N}, \operatorname{Ker} P(T;\Gamma))$. So, for m sufficiently large, $\eta(\mathscr{N}_m, \operatorname{Ker} P(T;\Gamma)) \geq \frac{1}{2}\eta_0$. Together with (S4.38), this gives

$$\|y\| \geq \tfrac{1}{2}\eta_0\|x\|,$$

for m sufficiently large. Using this it is not difficult to deduce that

$$\theta(\mathscr{M}_m, \mathscr{N}) \leq (1 + 2/\eta_0)\|P(T;\Gamma)\|\theta(\mathscr{N}_m, \mathscr{N})$$

for m sufficiently large. We conclude that $\theta(\mathscr{M}_m, \mathscr{N}) \to 0$, and the proof is complete. □

Lemma S4.15. *Let $\mathscr{N}$ be an invariant subspace for T, and assume that the contour Γ splits the spectrum of T. If $\mathscr{N}$ is stable for T, then $P(T;\Gamma)\mathscr{N}$ is a stable invariant subspace for the restriction T_0 of T to* $\operatorname{Im} P(T;\Gamma)$.

Proof. It is clear that $\mathscr{M} = P(T;\Gamma)\mathscr{N}$ is T_0-invariant.

Assume that $\mathscr{M}$ is not stable for T_0. Then $\mathscr{M}$ is not stable for T either, by Lemma S4.14. Hence there exist $\varepsilon > 0$ and a sequence $\{S_m\}$ such that $S_m \to T$ and

$$\theta(\mathscr{L}, \mathscr{M}) \geq \varepsilon, \qquad \mathscr{L} \in \Omega_m, \quad m = 1, 2, \ldots, \tag{S4.39}$$

where Ω_m denotes the set of all invariant subspaces of S_m.

As $\mathscr{N}$ is stable for T, one can find a sequence of subspaces $\{\mathscr{N}_m\}$ such that $S_m\mathscr{N}_m \subset \mathscr{N}_m$ and $\theta(\mathscr{N}_m, \mathscr{N}) \to 0$. Further, since Γ splits the spectrum of T and $S_m \to T$, the contour Γ will split the spectrum of S_m for m sufficiently large. But then, without loss of generality, we may assume that Γ splits the spectrum of each S_m. Again using $S_m \to T$, it follows that $P(S_m;\Gamma) \to P(T;\Gamma)$.

Let $\mathscr{Z}$ be a direct complement of $\mathscr{N}$ in $\mathbb{C}^n$. As $\theta(\mathscr{N}_m, \mathscr{N}) \to 0$, we have $\mathbb{C}^n = \mathscr{Z} \dotplus \mathscr{N}_m$ for m sufficiently large (Theorem S4.7). So, without loss of generality, we may assume that $\mathbb{C}^n = \mathscr{Z} \dotplus \mathscr{N}_m$ for each m. Let R_m be the angular operator of $\mathscr{N}_m$ with respect to the projector of $\mathbb{C}^n$ along $\mathscr{Z}$ onto $\mathscr{N}$, and put

$$E_m = \begin{bmatrix} I & R_m \\ 0 & I \end{bmatrix}$$

where the matrix corresponds to the decomposition $\mathbb{C}^n = \mathscr{Z} \dotplus \mathscr{N}$. Note that $T_m = E_m^{-1}S_mE_m$ leaves $\mathscr{N}$ invariant. Because $R_m \to 0$ we have $E_m \to I$, and so $T_m \to T$.

Clearly, Γ splits the spectrum of $T|_{\mathcal{N}}$. As $T_m \to T$ and $\mathcal{N}$ is invariant for T_m, the contour Γ will split the spectrum of $T_m|_{\mathcal{N}}$ too, provided m is sufficiently large. But then we may assume that this happens for all m. Also, we have

$$\lim_{m\to\infty} P(T_m|_{\mathcal{N}};\Gamma) \to P(T|_{\mathcal{N}};\Gamma).$$

Hence $\mathcal{M}_m = \operatorname{Im} P(T_m|_{\mathcal{N}};\Gamma) \to \operatorname{Im} P(T|_{\mathcal{N}};\Gamma) = \mathcal{M}$ in the gap topology.

Now consider $\mathcal{L}_m = E_m\mathcal{M}_m$. Then $\mathcal{L}_m$ is an S_m-invariant subspace. From $E_m \to I$ it follows that $\theta(\mathcal{L}_m, \mathcal{M}_m) \to 0$. This together with $\theta(\mathcal{M}_m, \mathcal{M}) \to 0$, gives $\theta(\mathcal{L}_m, \mathcal{M}) \to 0$. So we arrive at a contradiction to (S4.39), and the proof is complete. □

After this long preparation we are now able to give a short proof for Theorem S4.9.

Proof of Theorem S4.9. Suppose $\mathcal{N}$ is a stable invariant subspace for A. Put $\mathcal{N}_j = P_j\mathcal{N}$. Then $\mathcal{N} = \mathcal{N}_1 \dotplus \cdots \dotplus \mathcal{N}_r$. By Lemma S4.15 the space $\mathcal{N}_j$ is a stable invariant subspace for the restriction A_j of A to $\mathcal{N}(\lambda_j)$. But A_j has one eigenvalue only, namely, λ_j. So we may apply Lemma S4.12 to prove that $\mathcal{N}_j$ has the desired form.

Conversely, assume that each $\mathcal{N}_j$ has the desired form, and let us prove that $\mathcal{N} = \mathcal{N}_1 \dotplus \cdots \dotplus \mathcal{N}_r$ is a stable invariant subspace for A. By Corollary S4.11 the space $\mathcal{N}_j$ is a stable invariant subspace for the restriction A_j of A to $\operatorname{Im} P_j$. Hence we may apply Lemma S4.14 to show that each $\mathcal{N}_j$ is a stable invariant subspace for A. But then the same is true for the direct sum $\mathcal{N} = \mathcal{N}_1 \dotplus \cdots \dotplus \mathcal{N}_r$. □

Comments

Sections S4.3 and S4.4 provide basic notions and results concerning the set of subspaces (in finite dimensional space); these topics are covered in [32c, 48] in the infinite-dimensional case. The proof of Theorem S4.5 is from [33, Chapter IV] and Theorem S4.7 appears in [34e].

The results of Sections 4.5–4.7 are proved in [3b, 3c]; see also [12]. The book [3c] also contains additional information on stable invariant subspaces, as well as applications to factorization of matrix polynomials and rational functions, and to the stability of solutions of the algebraic Riccati equation. See also [3c, 4] for additional information concerning the notions of minimal opening and angular operator.

We mention that there is a close relationship between solutions of matrix Riccati equations and invariant subspaces with special properties of a certain linear transformation. See [3c, 12, 15b, 16, 53] for detailed information.

Chapter S5

Indefinite Scalar Product Spaces

This chapter is devoted to some basic properties of finite-dimensional spaces with an indefinite scalar product. The results presented here are used in Part III. Attention is focused on the problem of description of all indefinite scalar products in which a given linear transformation is self-adjoint. A special canonical form is used for this description. First, we introduce the basic definitions and conventions.

Let $\mathscr{X}$ be a finite-dimensional vector space with scalar product (x, y), $x, y \in \mathscr{X}$, and let H be a self-adjoint linear transformation in $\mathscr{X}$; i.e., $(Hx, y) = (x, Hy)$ for all $x, y \in \mathscr{X}$. This property allows us to define a new scalar product $[x, y]$, $x, y \in \mathscr{X}$ by the formula

$$[x, y] = (Hx, y)$$

The scalar product $[\ ,\ ]$ has all the properties of the usual scalar product, except that $[x, x]$ may be positive, negative, or zero. More exactly, $[x, y]$ has the following properties:

(1) $[\alpha_1 x_1 + \alpha_2 x_2, y] = \alpha_1[x_1, y] + \alpha_2[x_2, y]$;
(2) $[x, \alpha_1 y_1 + \alpha_2 y_2] = \bar{\alpha}_1[x, y_1] + \bar{\alpha}_2[x, y_2]$;
(3) $[x, y] = \overline{[y, x]}$;
(4) $[x, x]$ is real for every $x \in \mathscr{X}$.

Because of the lack of positiveness (in general) of the form $[x, x]$, the scalar product $[\ ,\ ]$ is called indefinite, and $\mathscr{X}$ endowed with the indefinite scalar product $[\ ,\ ]$ will be called an indefinite scalar product space.

We are particularly interested in the case when the scalar product $[\ ,\ ]$ is *nondegenerate*, i.e., $[x, y] = 0$ for all $y \in \mathscr{X}$ implies $x = 0$. This happens if and only if the underlying self-adjoint linear transformation H is invertible. This condition will be assumed throughout this chapter.

Often we shall identify $\mathscr{X}$ with $\mathbb{C}^n$. In this case the standard scalar product is given by $(x, y) = \sum_{i=1}^{r} x_i \bar{y}_i$, $x = (x_1, \ldots, x_n)^{\mathrm{T}} \in \mathbb{C}^n$, $y = (y_1, \ldots, y_n)^{\mathrm{T}} \in \mathbb{C}^n$, and an indefinite scalar product is determined by an $n \times n$ nonsingular hermitian (or self-adjoint) matrix H: $[x, y] = (Hx, y)$ for all $x, y \in \mathbb{C}^n$.

Finally, a linear transformation $A: \mathscr{X} \to \mathscr{X}$ is called *self-adjoint* with respect to H (or H-self-adjoint) if $[Ax, y] = [x, Ay]$ for all $x, y \in \mathscr{X}$, where $[x, y] = (Hx, y)$. This means $HA = A^*H$.

S5.1. Canonical Form of a Self-Adjoint Matrix and the Indefinite Scalar Product

Consider the following problem: given a fixed $n \times n$ matrix A, describe all the self-adjoint and nonsingular matrices H such that A is self-adjoint in the indefinite scalar product $[x, y] = (Hx, y)$, i.e., $[Ax, y] = [x, Ay]$, or, what is the same, $HA = A^*H$. Clearly, in order that such an H exist, it is necessary that A is similar to A^*. We shall see later (Theorem S5.1 below) that this condition is also sufficient. For the spectral properties of A this condition means that the elementary divisors of $I\lambda - A$ are symmetric relative to the real line; i.e., if $\lambda_0 (\neq \bar{\lambda}_0)$ is an eigenvalue of $I\lambda - A$ with the corresponding elementary divisors $(\lambda - \lambda_0)^{\alpha_i}$, $i = 1, \ldots, k$, then $\bar{\lambda}_0$ is also an eigenvalue of $I\lambda - A$ with the elementary divisor $(\lambda - \bar{\lambda}_0)^{\alpha_i}$, $i = 1, \ldots, k$. Now the problem posed above can be reformulated as follows: given a matrix A similar to A^*, describe and classify all the self-adjoint nonsingular matrices which carry out this similarity.

Consider first the case that $A = I$. Then any self-adjoint nonsingular matrix H is such that I is H-self-adjoint. Thus, the problem of description and classification of such matrices H is equivalent to the classical problem of reduction of H to the form S^*PS for some nonsingular matrix S and diagonal matrix P such that $P^2 = I$. Then the equation $H = S^*PS$ is nothing more than the reduction of the quadratic form (Hx, x) to a sum of squares.

To formulate results for any A we need a special construction connected with a Jordan matrix J which is similar to J^*. Similarity between J and J^* means that the number and the sizes of Jordan blocks in J corresponding to some eigenvalue $\lambda_0 (\neq \bar{\lambda}_0)$ and those corresponding to $\bar{\lambda}_0$ are the same. We now fix the structure of J in the following way: select a maximal set $\{\lambda_1, \ldots, \lambda_a\}$ of eigenvalues of J containing no conjugate pair, and let $\{\lambda_{a+1}, \ldots, \lambda_{a+b}\}$ be the distinct real eigenvalues of J. Put $\lambda_{a+b+j} = \bar{\lambda}_j$, for $j = 1, \ldots, a$, and let

$$J = \operatorname{diag}[J_i]_{i=1}^{2a+b} \tag{S5.1}$$

where $J_i = \operatorname{diag}[J_{ij}]_{j=1}^{k_i}$ is a Jordan form with eigenvalue λ_i and Jordan blocks, $J_{i,1}, \ldots, J_{i,k_i}$ of sizes $\alpha_{i,1} \geq \cdots \geq \alpha_{i,k_i}$, respectively.

An $\alpha \times \alpha$ matrix whose (p, q) entry is 1 or 0 according as $p + q = \alpha + 1$ or $p + q \neq \alpha + 1$ will be called a *sip matrix* (standard involuntary permutation).

An important role will be played in the sequel by the matrix $P_{\varepsilon, J}$ connected with J as follows:

$$P_{\varepsilon, J} = \begin{bmatrix} 0 & 0 & P_c \\ 0 & P_r & 0 \\ P_c & 0 & 0 \end{bmatrix} \tag{S5.2}$$

where

$$P_c = \operatorname{diag}[\operatorname{diag}[P_{ij}]_{j=1}^{k_i}]_{i=1}^{a}$$

and

$$P_r = \operatorname{diag}[\operatorname{diag}[\varepsilon_{ij} P_{ij}]_{j=1}^{k_i}]_{i=a+1}^{a+b}$$

with sip matrices P_{st} of sizes $\alpha_{st} \times \alpha_{st}$ $(s = 1, \ldots, a + b,\ t = 1, \ldots, k_s)$ and the ordered set of signs $\varepsilon = (\varepsilon_{ij})$, $i = a + 1, \ldots, a + b$, $j = 1, \ldots, k_i$, $\varepsilon_{ij} = \pm 1$.

Using these notations the main result, which solves the problem stated at the beginning of this section, can bow be formulated.

Theorem S5.1. *The matrix A is self-adjoint relative to the scalar product (Hx, y) (where $H = H^*$ is nonsingular) if and only if*

$$T^*HT = P_{\varepsilon, J}, \qquad T^{-1}AT = J \tag{S5.3}$$

for some invertible T, a matrix J in Jordan form, and a set of signs ε.

The matrix $P_{\varepsilon, J}$ of (S5.3) will be called an *A-canonical form* of H with *reducing transformation* $S = T^{-1}$. It will be shown later (Theorem S5.6) that the set of signs ε is uniquely defined by A and H, up to certain permutations. The proof of Theorem S5.1 is quite long and needs some auxiliary results. This will be the subject matter of the next section.

Recall that the *signature* (denoted sig H) of a self-adjoint matrix H is the difference between the number of positive eigenvalues and the number of negative eigenvalues of H (in both cases counting with multiplicities). It coincides with the difference between the number of positive squares and the number of negative squares in the canonical quadratic form determined by H.

The following corollary connects the signature of H with the structure of real eigenvalues of H-self-adjoint matrices.

Corollary S5.2. *Let A be an H-self-adjoint matrix, and let s be the signature of H. Then the real eigenvalues of A have at least $|s|$ associated elementary divisors of odd degree.*

Proof. It follows from the relation (S5.3) that s is also the signature of $P_{\varepsilon, J}$. It is readily verified that each conjugate pair of nonreal eigenvalues makes no contribution to the signature; neither do sip matrices associated with even degree elementary divisors and real eigenvalues. A sip matrix associated with an odd degree elementary divisor of a real eigenvalue contributes "+1" or "−1" to s depending on the corresponding sign in the sequence ε. The conclusion then follows. □

Another corollary of Theorem S5.1 is connected with the simultaneous reduction of a pair of hermitian matrices A and B to a canonical form, where at least one of them is nonsingular.

Corollary S5.3. *Let A, B be $n \times n$ hermitian matrices, and suppose that B is invertible. Then there exists a nonsingular matrix X such that*

$$X^*AX = P_{\varepsilon, J}J, \qquad X^*BX = P_{\varepsilon, J},$$

where J is the Jordan normal form of $B^{-1}A$, and $P_{\varepsilon, J}$ is defined as above.

Proof. Observe that the matrix $B^{-1}A$ is B-self-adjoint. By Theorem S5.1, we have

$$X^*BX = P_{\varepsilon, J}, \qquad X^{-1}B^{-1}AX = J$$

for some nonsingular matrix X. So $A = BXJX^{-1}$ and

$$X^*AX = X^*BXJ = P_{\varepsilon, J}J. \quad \square$$

In the case when B is positive definite (or negative definite) and A is hermitian one can prove that the Jordan form of $B^{-1}A$ is actually a diagonal matrix with real eigenvalues. In this case Corollary S5.3 reduces to the well-known result on simultaneous reduction of a pair of quadratic forms, when one of them is supposed to be definite (positive or negative), to the sum of squares. See, for instance, [22, Chapter X] for details.

S5.2. Proof of Theorem S5.1

The line of argument used to prove Theorem S5.1 is the successive reduction of the problem to the study of restrictions of A and H to certain invariant subspaces of A. This approach is justified by the results of the next two lemmas. A subspace $\mathcal{M}$ of $\mathbb{C}^n$ is said to be *nondegenerate* (with respect to H) if, for every $x \in \mathcal{M}\backslash\{0\}$ there is a $y \in \mathcal{M}$ such that $(Hx, y) \neq 0$. For any subspace $\mathcal{M}$ the *orthogonal companion* of $\mathcal{M}$ with respect to H is

$$\mathcal{M}^{\Delta} = \{x \in \mathbb{C}^n \,|\, (Hx, y) = 0, \ y \in \mathcal{M}\}.$$

The first lemma says that, for a nondegenerate subspace, the orthogonal companion is also complementary, together with a converse statement.

Lemma S5.4. *If $\mathcal{M}$ is a nondegenerate subspace of $\mathbb{C}^n$, then $\mathbb{C}^n = \mathcal{M} \dotplus \mathcal{M}^{\Delta}$ and conversely, if $\mathcal{M}$ is a subspace and $\mathbb{C}^n = \mathcal{M} \dotplus \mathcal{M}^{\Delta}$, then $\mathcal{M}$ is nondegenerate.*

Proof. Note first that $\mathcal{M}^{\Delta}$ is the image under H^{-1} of the usual orthogonal complement (in the sense of the usual scalar product (,)) to $\mathcal{M}$. So

$$\dim \mathcal{M}^{\Delta} + \dim \mathcal{M} = n.$$

Hence it is sufficient to prove that $\mathcal{M} \cap \mathcal{M}^{\Delta} = \{0\}$ if and only if $\mathcal{M}$ is nondegenerate. But this follows directly from the definition of a nondegenerate subspace. □

In the next lemma a matrix A is considered as a linear transformation with respect to the standard basis in $\mathbb{C}^n$.

Lemma S5.5. *Let A be self-adjoint with respect to H and let $\mathcal{M}$ be a non-degenerate invariant subspace of A. Then*

(a) *$\mathcal{M}^{\Delta}$ is an invariant subspace of A.*

(b) *If Q is the orthogonal projector onto $\mathcal{M}$* (i.e., $Q^2 = Q$, $Q^* = Q$, $\operatorname{Im} Q = \mathcal{M}$), *then $A|_{\mathcal{M}}$ defines a linear transformation on $\mathcal{M}$ which is self-adjoint with respect to the invertible self-adjoint linear transformation $QH|_{\mathcal{M}}: \mathcal{M} \to \mathcal{M}$.*

Proof. Part (a) is straightforward and is left to the reader. For part (b) note first that $QH|_{\mathcal{M}}$ is self-adjoint since $H = H^*$, $Q = Q^*$, and $QH|_{\mathcal{M}} = QHQ|_{\mathcal{M}}$. Further, $QH|_{\mathcal{M}}$ is invertible in view of the invertibility of H and nondegeneracy of $\mathcal{M}$. Then $HA = A^*H$ implies $Q(HA)Q = Q(A^*H)Q$ and, since $AQ = QAQ$,

$$(QHQ)(QAQ) = (QA^*Q)(QHQ),$$

from which the result follows. □

Now we can start with the proof of Theorem S5.1.

If (S5.3) holds, then one can easily check that A is self-adjoint relative to H. Indeed, the equalities $P_{\varepsilon,J}J = J^*P_{\varepsilon,J}$ are verified directly. Then

$$\begin{aligned} HA &= T^{*-1}P_{\varepsilon,J}T^{-1} \cdot TJT^{-1} = T^{*-1}P_{\varepsilon,J}JT^{-1} = T^{*-1}J^*P^*_{\varepsilon,J}T^{-1} \\ &= T^{*-1}J^*T^*T^{*-1}P_{\varepsilon,J}T^{-1} = A^*H, \end{aligned}$$

i.e., A is self-adjoint relative to H.

Now let A be self-adjoint relative to the scalar product $[x, y] = (Hx, y)$. Decompose $\mathcal{C}^n$ into a direct sum

$$\mathcal{C}^n = \mathscr{X}_1 \dotplus \cdots \dotplus \mathscr{X}_a \dotplus \mathscr{X}_{a+1} \dotplus \cdots \dotplus \mathscr{X}_{a+b}, \tag{S5.4}$$

where $\mathscr{X}_i$ is the sum of the root subspaces of A corresponding to the eigenvalues λ_i and $\bar{\lambda}_i$, $i = 1, \ldots, a$. For $i = a + 1, \ldots, a + b$, $\mathscr{X}_i$ is the root subspace of A corresponding to the eigenvalue λ_i.

We show first that $\mathscr{X}_i$ and $\mathscr{X}_j$ ($i \neq j$, $i, j = 1, \ldots, a + b$) are orthogonal with respect to H, i.e., $[x, y] = 0$ for every $x \in \mathscr{X}_i$ and $y \in \mathscr{X}_j$. Indeed, let $x \in \mathscr{X}_i$, $y \in \mathscr{X}_j$ and assume first that $(A - \mu_i I)^s x = (A - \mu_j I)^t y = 0$, where μ_i (resp. μ_j) is equal to either λ_i or $\bar{\lambda}_i$ (resp. λ_j or $\bar{\lambda}_j$), for some nonnegative integers s and t. We have to show that $[x, y] = 0$. It is easily verified that this is so when $s = t = 1$. Now assume inductively that $[x', y'] = 0$ for all pairs $x' \in \mathscr{X}_i$ and $y' \in \mathscr{X}_j$ for which

$$(A - \mu_i I)^{s'} x' = (A - \mu_j I)^{t'} y' = 0, \qquad \text{with} \quad s' + t' < s + t.$$

Consider $x' = (A - \mu_i I)x$, $y' = (A - \mu_j I)y$. Then, by the induction hypothesis,

$$[x', y] = [x, y'] = 0$$

or

$$\mu_i[x, y] = [Ax, y], \qquad \bar{\mu}_j[x, y] = [x, Ay].$$

But $[Ax, y] = [x, Ay]$ (since A is self-adjoint with respect to H); so $(\mu_i - \bar{\mu}_j)[x, y] = 0$. By the choice of $\mathscr{X}_i$ and $\mathscr{X}_j$ we have $\mu_i \neq \bar{\mu}_j$; so $[x, y] = 0$. As

$$\mathscr{X}_i = \{x_1 + x_2 \mid (A - \mu_i I)^{s_1} x_1 = (A - \bar{\mu}_i I)^{s_2} x_2 = 0 \text{ for some } s_1 \text{ and } s_2\},$$

with an analogous representation for $\mathscr{X}_j$, it follows from the above that $[x, y] = 0$ for all $x \in \mathscr{X}_i$, $y \in \mathscr{X}_j$, where $i \neq j$. It then follows from (S5.4) and Lemma S5.4 that each $\mathscr{X}_i$ is nondegenerate.

Consider fixed $\mathscr{X}_i$, where $i = 1, \ldots, a$ (i.e., $\lambda_i \neq \bar{\lambda}_i$). Then $\mathscr{X}_i = \mathscr{X}_i' \dotplus \mathscr{X}_i''$, where $\mathscr{X}_i'$ (resp. $\mathscr{X}_i''$) is the root subspace of A corresponding to λ_i (resp. $\bar{\lambda}_i$). As above, it follows that the subspaces $\mathscr{X}_i'$ and $\mathscr{X}_i''$ are *isotropic* with respect to H, i.e., $[x, y] = 0$ for either $x, y \in \mathscr{X}_i'$ or for $x, y \in \mathscr{X}_i''$.

There exists an integer m with the properties that $(A - \lambda_i I)^m|_{\mathscr{X}_i'} = 0$, but $(A - \lambda_i I)^{m-1} a_1 \neq 0$ for some $a_1 \in \mathscr{X}_i'$. Since $\mathscr{X}_i$ is nondegenerate and $\mathscr{X}_i'$ is isotropic, there exists a $b_1 \in \mathscr{X}_i''$ such that $[(A - \lambda_i I)^{m-1} a_1, b_1] = 1$. Define sequences $a_1, \ldots, a_m \in \mathscr{X}_i'$ and $b_1, \ldots, b_m \in \mathscr{X}_i''$ by

$$a_j = (A - \lambda_i I)^{j-1} a_1, \qquad b_j = (A - \bar{\lambda}_i I)^{j-1} b_1, \qquad j = 1, \ldots, m.$$

Observe that $[a_1, b_m] = [a_1, (A - \bar{\lambda}_i I)^{m-1} b_1] = [(A - \lambda_i I)^{m-1} a_1, b_1] = 1$, in particular, $b_m \neq 0$. Further, for every $x \in \mathscr{X}_i'$ we have

$$[x, (A - \bar{\lambda}_i I) b_m] = [x, (A - \bar{\lambda}_i I)^m b_1] = [(A - \lambda_i I)^m x, b_1] = 0;$$

so the vector $(A - \bar{\lambda}_i I)b_m$ is H-orthogonal to $\mathscr{X}'_i$. In view of (S5.4) we deduce that $(A - \bar{\lambda}_i I)b_m$ is H-orthogonal to $\mathbb{C}^n$, and hence

$$(A - \bar{\lambda}_i I)b_m = 0.$$

Then clearly $a_m, \ldots, a_1$ (resp. $b_m, \ldots, b_1$) is a Jordan chain of A corresponding to λ_i (resp. $\bar{\lambda}_i$), i.e., for $j = 1, 2, \ldots, m - 1$,

$$Aa_j - \lambda_i a_j = a_{j+1} \qquad \text{and} \qquad Aa_m = \lambda_i a_m,$$

with analogous relations for the b_j (replacing λ_i by $\bar{\lambda}_i$). For $j + k = m + 1$ we have

$$[a_j, b_k] = [(A - \lambda_i I)^{j-1}a_1, (A - \bar{\lambda}_i I)^{k-1}b_1] = [(A - \lambda_i I)^{j+k-2}a_1, b_1] = 1; \tag{S5.5}$$

and analogously

$$[a_j, b_k] = 0 \qquad \text{for} \quad j + k > m + 1. \tag{S5.6}$$

Now put

$$c_1 = a_1 + \sum_{j=2}^{m} \alpha_j a_j, \qquad c_{j+1} = (A - \lambda_i I)c_j, \quad j = 1, \ldots, m - 1,$$

where $\alpha_2, \ldots, \alpha_m$ are chosen so that

$$[c_1, b_{m-1}] = [c_1, b_{m-2}] = \cdots = [c_1, b_1] = 0.$$

Such a choice is possible, as can be checked easily using (S5.5) and (S5.6). Now for $j + k \le m$

$$[c_j, b_k] = [(A - \lambda_i I)^{j-1}c_1, b_k] = [c_1, (A - \bar{\lambda}_i I)^{j-1}b_k] = [c_1, b_{k+j-1}] = 0,$$

and for $j + k \ge m + 1$ we obtain, using $(A - \lambda_i I)^m a_1 = 0$ together with (S5.5), (S5.6):

$$\begin{aligned}[c_j, b_k] &= [(A - \lambda_i I)^{j-1}c_1, (A - \bar{\lambda}_i I)^{k-1}b_1] \\ &= [(A - \lambda_i I)^{j+k-2}c_1, b_1] = [(A - \lambda_i I)^{j+k-2}a_1, b_1] \\ &= \begin{cases} 1, & \text{for} \quad j + k = m + 1 \\ 0, & \text{for} \quad j + k > m + 1. \end{cases}\end{aligned}$$

Let $\mathcal{N}_1 = \operatorname{Span}\{c_1, \ldots, c_m, b_1, \ldots, b_m\}$. The relations above show that $A|_{\mathcal{N}_1} = J_1 \oplus \bar{J}_1$ in the basis $c_1, \ldots, c_m, b_1, \ldots, b_m$, where J_1 is the Jordan block of size m with eigenvalue λ_i;

$$[x, y] = y^* \begin{bmatrix} 0 & P_1 \\ P_1 & 0 \end{bmatrix} x, \qquad x, y \in \mathcal{N}_1$$

in the same basis, and P_1 is the sip matrix of size m. We see from this representation that $\mathcal{N}_1$ is nondegenerate. By Lemma S5.4, $\mathcal{C}^n = \mathcal{N}_1 \dotplus \mathcal{N}_1^{\Delta}$, and by Lemma S5.5, N_1^{Δ} is an invariant subspace for A. If $A|_{\mathcal{N}_1^{\Delta}}$ has nonreal eigenvalues, apply the same procedure to construct a subspace $\mathcal{N}_2 \subset \mathcal{N}_1^{\Delta}$ with basis $c_1', \ldots, c_{m'}', b_1', \ldots, b_{m'}'$ such that in this basis $A|_{\mathcal{N}_2} = J_2 \oplus \bar{J}_2$, where J_2 is the Jordan block of size m' with nonreal eigenvalue, and

$$[x, y] = y^* \begin{bmatrix} 0 & P_2 \\ P_2 & 0 \end{bmatrix} x, \qquad x, y \in \mathcal{N}_2$$

with the sip matrix P_2 of size m'. Continue this procedure until the nonreal eigenvalues of A are exhausted.

Consider now a fixed $\mathcal{X}_i$, where $i = a + 1, \ldots, a + b$, so that λ_i is real. Again, let m be such that $(A - \lambda_i I)^m|_{\mathcal{X}_i} = 0$ but $(A - \lambda_i I)^{m-1}|_{\mathcal{X}_i} \neq 0$. Let Q_i be the orthogonal projector on $\mathcal{X}_i$ and define $F: \mathcal{X}_i \to \mathcal{X}_i$ by

$$F = Q_i H(A - \lambda_i I)^{m-1}.$$

Since λ_i is real, it is easily seen that F is self-adjoint. Moreover, $F \neq 0$; so there is a nonzero eigenvalue of F (necessarily real) with an eigenvector a_1. Normalize a_1 so that

$$(Fa_1, a_1) = \varepsilon, \qquad \varepsilon = \pm 1.$$

In other words,

$$[(A - \lambda_i I)^{m-1} a_1, a_1] = \varepsilon. \tag{S5.7}$$

Let $a_j = (A - \lambda_i I)^{j-1} a_1, j = 1, \ldots, m$. It follows from (S5.7) that for $j + k = m + 1$

$$[a_j, a_k] = [(A - \lambda_i I)^{j-1} a_1, (A - \lambda_i I)^{k-1} a_1] = [(A - \lambda_i I)^{m-1} a_1, a_1] = \varepsilon. \tag{S5.8}$$

Moreover, for $j + k > m + 1$ we have:

$$[a_j, a_k] = [(A - \lambda_i I)^{j+k-2} a_1, a_1] = 0 \tag{S5.9}$$

in view of the choice of m. Now put

$$b_1 = a_1 + \alpha_2 a_2 + \cdots + \alpha_m a_m, \qquad b_j = (A - \lambda_i I)^{j-1} b_1, \quad j = 1, \ldots, m,$$

and choose α_i so that

$$[b_1, b_1] = [b_1, b_2] = \cdots = [b_1, b_{m-1}] = 0.$$

Such a choice of α_i is possible. Indeed, equality $[b_1, b_j] = 0 (j = 1, \ldots, m - 1)$ gives, in view of (S5.8) and (S5.9),

$$\begin{aligned} 0 &= [a_1 + \alpha_2 a_2 + \cdots + \alpha_m a_m, a_j + \alpha_2 a_{j+1} + \cdots + \alpha_{m-j+1} \alpha_m] \\ &= [a_1, a_j] + 2\varepsilon \alpha_{m-j+1} + (\text{terms in } \alpha_2, \ldots, \alpha_{m-j}). \end{aligned}$$

Evidently, these equalities determine unique numbers $\alpha_2, \alpha_3, \ldots, \alpha_m$ in succession.

Let $\mathcal{N} = \operatorname{Span}\{b_1, \ldots, b_m\}$. In the basis $b_1, \ldots, b_m$ the linear transformation $A|_{\mathcal{N}}$ is represented by the single Jordan block with eigenvalue λ_i, and

$$[x, y] = y^* \varepsilon P_0 x, \qquad x, y \in \mathcal{N},$$

where P_0 is the sip matrix of size m.

Continue the procedure on the orthogonal companion to $\mathcal{N}$, and so on.

Applying this construction, we find a basis $f_1, \ldots, f_n$ in $\mathbb{C}^n$ such that A is represented by the Jordan matrix J of (5.1) in this basis and, with $P_{\varepsilon, J}$ as defined in (S5.2),

$$[x, y] = y^* P_{\varepsilon, J} x, \qquad x, y \in \mathbb{C}^n,$$

where x and y are represented by their coordinates in the basis $f_1, \ldots, f_n$. Let T be the $n \times n$ invertible matrix whose ith column is formed by the coordinates of f_i (in the standard orthonormal basis), $i = 1, \ldots, n$. For such a T, the relation $T^{-1}AT = J$ holds because $f_1, \ldots, f_n$ is a Jordan basis for A, and equality $T^*HT = P_{\varepsilon, J}$ follows from (S5.5), (S5.6), and (S5.8). So (S5.3) holds. Theorem S5.1 is proved completely. □

S5.3. Uniqueness of the Sign Characteristic

Let H be an $n \times n$ hermitian nonsingular matrix, and let A be some H-self-adjoint matrix. If J is a normal form of A, then in view of Theorem S5.1 H admits an A-canonical form $P_{\varepsilon, J}$. We suppose that the order of Jordan blocks in J is fixed as explained in the first section. The set of signs ε in $P_{\varepsilon, J}$ will be called the *A-sign characteristic* of H. The problem considered in this section is that of uniqueness of ε.

Recall that $\varepsilon = \{\varepsilon_{a+j, i}\}$, $i = 1, 2, \ldots, k_j$, and $j = 1, 2, \ldots, b$. in the notation of Section S5.1. Two sets of signs $\varepsilon^{(r)} = \{\varepsilon^{(r)}_{a+j, i}\}$, $r = 1, 2$, will be said to be *equivalent* if one can be obtained from the other by permutation of signs within subsets corresponding to Jordan blocks of J having the same size and the same real eigenvalue.

Theorem S5.6. *Let A be an H-self-adjoint matrix. Then the A-sign characteristic of H is defined uniquely up to equivalence.*

Note that in the special case $A = I$, this theorem states that in any reduction of the quadratic form (Hx, x) to a sum of squares, the number of positive coefficients, and of negative coefficients is invariant, i.e., the classical inertia law.

Proof. Without loss of generality a Jordan form for A can be fixed, with the conventions of Eq. (S5.1). It is then to be proved that if $P_{\varepsilon,J}$, $P_{\delta,J}$ are A-canonical forms for H, then the sets of signs ε and δ are equivalent. It is evident from the definition of an A-canonical form (Eqs. (S5.1) and (S5.2)) that the conclusion of the theorem follows if it can be established for an A having just one real eigenvalue, say α.

Thus, the Jordan form J for A is assumed to have k_i blocks of size m_i, $i = 1, 2, \ldots, t$, where $m_1 > m_2 > \cdots > m_t$. Thus, we may write

$$J = \operatorname{diag}[\operatorname{diag}[J_j]_{i=1}^{k_i}]_{j=1}^{t}$$

and

$$P_{\varepsilon,J} = \operatorname{diag}[\operatorname{diag}[\varepsilon_{j,i}P_j]_{i=1}^{k_i}]_{j=1}^{t}$$

with a similar expression for $P_{\delta,J}$, replacing $\varepsilon_{j,i}$ by signs $\delta_{j,i}$.

It is proved in Theorem S5.1 that, for some nonsingular S, $H = S^*P_{\varepsilon,J}S$ and $A = S^{-1}JS$. It follows that for any nonnegative integer k,

$$H(I\alpha - A)^k = S^*P_{\varepsilon,J}(I\alpha - J)^kS, \tag{S5.10}$$

and is a relation between hermitian matrices.

Observe that $(I\alpha - J_i)^{m_1-1} = 0$ for $i = 2, 3, \ldots, t$, and $(I\alpha - J_1)^{m_1-1}$ has all entries zeros except for the entry in the right upper corner, which is 1. Consequently,

$$P_{\varepsilon,J}(I\alpha - J)^{m_1-1} = \operatorname{diag}[\varepsilon_{1,1}P_1(I\alpha - J_1)^{m_1-1}, \ldots, \varepsilon_{1,k_1}P_1(I\alpha - J_1)^{m_1-1}, 0, \ldots, 0].$$

Using this representation in (S5.10) with $k = m_1 - 1$, we conclude that $\sum_{i=1}^{k_1} \varepsilon_{1,i}$ coincides with the signature of $H(I\alpha - A)^{m_1-1}$.

But precisely the same argument applies using the A-canonical form $P_{\delta,J}$ for H and it is concluded that

$$\sum_{i=1}^{k_1} \varepsilon_{1,i} = \sum_{i=1}^{k_1} \delta_{1,i}. \tag{S5.11}$$

Consequently, the subsets $\{\varepsilon_{1,1}, \ldots, \varepsilon_{1,k_1}\}$ and $\{\delta_{1,1}, \ldots, \delta_{1,k_1}\}$ of ε, δ are equivalent.

Now examine the hermitian matrix $P_{\varepsilon,J}(I\alpha - J)^{m_2-1}$. This is found to be block diagonal with nonzero blocks of the form

$$\varepsilon_{1,i}P_1(I\alpha - J_1)^{m_2-1}, \qquad i = 1, 2, \ldots, k_1$$

and

$$\varepsilon_{2,j}P_2(I\alpha - J_2)^{m_2-1}, \qquad j = 1, 2, \ldots, k_2.$$

Consequently, using (S5.10) with $k = m_2 - 1$, the signature of $H(I\alpha - A)^{m_2-1}$ is given by

$$\left(\sum_{i=1}^{k_1} \varepsilon_{1,i}\right)(\text{sig}[P_1(I\alpha - J_1)^{m_2-1}]) + \sum_{j=1}^{k_2} \varepsilon_{2,j}.$$

But again this must be equal to the corresponding expression formulated using δ instead of ε. Hence, using (S5.11) it is found that

$$\sum_{j=1}^{k_2} \varepsilon_{2,j} = \sum_{j=1}^{k_2} \delta_{2,j}$$

and the subsets $\{\varepsilon_{2,1}, \ldots, \varepsilon_{2,k_2}\}$ and $\{\delta_{2,1}, \ldots, \delta_{2,k_2}\}$ of ε and δ are equivalent.

Now it is clear that the argument can be continued for t steps after which the equivalence of ε and δ is established. □

It follows from Theorem S5.6 that the A-sign characteristic of H is uniquely defined if we apply the following normalization rule: in every subset of signs corresponding to the Jordan blocks of the same size and the same eigenvalue, $+1$s (if any) precede -1s (if any).

We give a simple example for illustration.

EXAMPLE S5.1 Let

$$J = \text{diag}\left\{\begin{bmatrix} 0 & 1 \\ 0 & 0 \end{bmatrix}, \begin{bmatrix} 1 & 1 \\ 0 & 1 \end{bmatrix}\right\}.$$

Then

$$P_{\varepsilon,J} = \text{diag}\left\{\begin{bmatrix} 0 & \varepsilon_1 \\ \varepsilon_1 & 0 \end{bmatrix}, \begin{bmatrix} 0 & \varepsilon_2 \\ \varepsilon_2 & 0 \end{bmatrix}\right\} \quad \text{for} \quad \varepsilon = (\varepsilon_1, \varepsilon_2), \quad \varepsilon_i = \pm 1.$$

According to Theorem S5.1 and Theorem S5.6, the set $\Omega = \Omega_J$ of all invertible and self-adjoint matrices H such that J is H-self-adjoint, splits into 4 disjoint sets $\Omega_1, \Omega_2, \Omega_3, \Omega_4$ corresponding to the sets of signs $(+1, +1)$, $(+1, -1)$, $(-1, +1)$, and $(-1, -1)$, respectively:

$$\Omega = \Omega_1 \cup \Omega_2 \cup \Omega_3 \cup \Omega_4.$$

An easy computation shows that each set Ω_i consists of all matrices H of the form

$$H = \text{diag}\left\{\begin{bmatrix} 0 & a_1\varepsilon_1 \\ a_1\varepsilon_1 & b_1 \end{bmatrix}, \begin{bmatrix} 0 & a_2\varepsilon_2 \\ a_2\varepsilon_2 & b_2 \end{bmatrix}\right\},$$

where a_1, a_2 are positive and b_1, b_2 are real parameters; $\varepsilon_1, \varepsilon_2$ are ± 1 depending on the set Ω_i. Note also that each set Ω_i $(i = 1, 2, 3, 4)$ is connected. □

S5.4. Second Description of the Sign Characteristic

Let H be an $n \times n$ hermitian nonsingular matrix, and let A be H-self-adjoint. In the main text we make use of another description of the A-sign characteristic of H, which is given below. In view of Theorem S5.1, H admits an A-canonical form $P_{\varepsilon, J}$ in some orthonormal basis, where J is the Jordan form of A:

$$H = S^* P_{\varepsilon, J} S. \tag{S5.12}$$

Here S is a nonsingular matrix such that $SA = JS$ and ε is the A-sign characteristic of H. We can suppose that the basis is the standard orthonormal basis in $\mathbb{C}^n$.

Let λ_0 be a fixed real eigenvalue of A, and let $\Psi_1 \subset \mathbb{C}^n$ be the subspace spanned by the eigenvectors of $I\lambda - A$ corresponding to λ_0. For $x \in \Psi_1 \backslash \{0\}$ denote by $v(x)$ the maximal length of a Jordan chain beginning with the eigenvector x. Let Ψ_i, $i = 1, 2, \ldots, \gamma$, $(\gamma = \max\{v(x) | x \in \Psi_1 \backslash \{0\}\})$ be the subspace of Ψ_1 spanned by all $x \in \Psi_1$ with $v(x) \geq i$. Then

$$\operatorname{Ker}(I\lambda_0 - A) = \Psi_1 \supset \Psi_2 \supset \cdots \supset \Psi_\gamma.$$

The following result describes the A-sign characteristic of H in terms of certain bilinear forms defined on Ψ_i.

Theorem S5.7. *For $i = 1, \ldots, \gamma$, let*

$$f_i(x, y) = (x, Hy^{(i)}), \qquad x, y \in \Psi_i,$$

where $y = y^{(1)}, y^{(2)}, \ldots, y^{(i)}$, is a Jordan chain of $I\lambda - A$ corresponding to real λ_0 with the eigenvector y. Then

(i) *$f_i(x, y)$ does not depend on the choice of $y^{(2)}, \ldots, y^{(k)}$;*
(ii) *for some self-adjoint linear transformation $G_i \colon \Psi_i \to \Psi_i$,*

$$f_i(x, y) = (x, G_i y), \qquad x, y \in \Psi_i;$$

(iii) *for G_i of* (ii), *$\Psi_{i+1} = \operatorname{Ker} G_i$ (by definition $\Psi_{\gamma+1} = \{0\}$);*
(vi) *the number of positive (negative) eigenvalues of G_i, counting multiplicities, coincides with the number of positive (negative) signs in the A-sign characteristic of H corresponding to the Jordan blocks of J with eigenvalue λ_0 and size i.*

Proof. By (S5.12) we have $f_i(x, y) = (Sx, P_{\varepsilon, J} Sy^{(i)})$, $x, y \in \Psi_i$. Clearly, Sx and $Sy^{(1)}, \ldots, Sy^{(i)}$ are eigenvector and Jordan chain, respectively, of $I\lambda - J$ corresponding to λ_0. Thus, the proof is reduced to the case $A = J$ and $H = P_{\varepsilon, J}$. But in this case the assertions (i)–(iv) can be checked without difficulties. $\square$

Comments

The presentation in this chapter follows a part of the authors' paper [34f]. Indefinite scalar products in infinite-dimensional spaces have been studied extensively; see [8, 43, 51] for the theory of indefinite scalar product spaces and some of its applications. Results close to Theorem S5.1 appear in [59, Chapter 7], and [77].

Applications of the theory of indefinite scalar product spaces to solutions of algebraic Riccati equations and related problems in linear control theory are found in [53, 70d, 70f].

Chapter S6

Analytic Matrix Functions

The main result in this chapter is the perturbation theorem for self-adjoint matrices (Theorem S6.3). The proof is based on auxiliary results on analytic matrix functions from Section S6.1, which is also used in the main text.

S6.1. General Results

Here we consider problems concerning existence of analytic bases in the image and in the kernel of an analytic matrix-valued function. The following theorem provides basic results in this direction.

Theorem S6.1. *Let $A(\varepsilon)$, $\varepsilon \in \Omega$, be an $n \times n$ complex matrix-valued function which is analytic in a domain $\Omega \subset \mathcal{C}$. Let $r = \max_{\varepsilon \in \Omega} \operatorname{rank} A(\varepsilon)$. Then there exist n-dimensional analytic (in Ω) vector-valued functions $y_1(\varepsilon), \ldots, y_n(\varepsilon)$ with the following properties:*

(i) $y_1(\varepsilon), \ldots, y_r(\varepsilon)$ *are linearly independent for every* $\varepsilon \in \Omega$;
(ii) $y_{r+1}(\varepsilon), \ldots, y_n(\varepsilon)$ *are linearly independent for every* $\varepsilon \in \Omega$;

(iii)
$$\operatorname{Span}\{y_1(\varepsilon), \ldots, y_r(\varepsilon)\} = \operatorname{Im} A(\varepsilon) \tag{S6.1}$$

and

$$\operatorname{Span}\{y_{r+1}(\varepsilon), \ldots, y_n(\varepsilon)\} = \operatorname{Ker} A(\varepsilon) \tag{S6.2}$$

for every $\varepsilon \in \Omega$, *except for a set of isolated points which consists exactly of those* $\varepsilon_0 \in \Omega$ *for which* rank $A(\varepsilon_0) < r$. *For such exceptional* $\varepsilon_0 \in \Omega$, *the inclusions*

$$\operatorname{Span}\{y_1(\varepsilon_0), \ldots, y_r(\varepsilon_0)\} \supset \operatorname{Im} A(\varepsilon_0) \tag{S6.3}$$

and

$$\operatorname{Span}\{y_{r+1}(\varepsilon_0), \ldots, y_r(\varepsilon_0)\} \subset \operatorname{Ker} A(\varepsilon_0) \tag{S6.4}$$

hold.

We remark that Theorem S6.1 includes the case in which $A(\varepsilon)$ is an analytic function of the *real* variable ε, i.e., in a neighborhood of every real point ε_0 the entries of $A(\varepsilon) = (a_{ik}(\varepsilon))_{i,k=1}^{n}$ can be expressed as power series in $\varepsilon - \varepsilon_0$:

$$a_{ik}(\varepsilon) = \sum_{j=0}^{\infty} \alpha_j(\varepsilon - \varepsilon_0)^j \tag{S6.5}$$

where α_j are complex numbers depending on i, k, and ε, and the power series converge in some real neighborhood of ε_0. (Indeed, the power series (S6.5) converges also in some *complex* neighborhood of ε_0, so in fact $A(\varepsilon)$ is analytic in some complex neighborhood Ω of the real line.) Theorem S6.1 will be used in this form in the next section.

The proof of Theorem S6.1 is based on the following lemma.

Lemma S6.2. *Let* $x_1(\varepsilon), \ldots, x_r(\varepsilon)$, $\varepsilon \in \Omega$ *be n-dimensional vector-valued functions which are analytic in a domain* Ω *in the complex plane. Assume that for some* $\varepsilon_0 \in \Omega$, *the vectors* $x_1(\varepsilon_0), \ldots, x_r(\varepsilon_0)$ *are linearly independent. Then there exist n-dimensional vector functions* $y_1(\varepsilon), \ldots, y_r(\varepsilon)$, $\varepsilon \in \Omega$, *with the following properties*:

(i) $y_1(\varepsilon), \ldots, y_r(\varepsilon)$ *are analytic on* Ω;
(ii) $y_1(\varepsilon), \ldots, y_r(\varepsilon)$ *are linearly independent for every* $\varepsilon \in \Omega$;
(iii) $\operatorname{Span}\{y_1(\varepsilon), \ldots, y_r(\varepsilon)\} = \operatorname{Span}\{x_1(\varepsilon), \ldots, x_r(\varepsilon)\}$ $(\subset \mathbb{C}^n)$

for every $\varepsilon \in \Omega \backslash \Omega_0$, *where* $\Omega_0 = \{\varepsilon \in \Omega \mid x_1(\varepsilon), \ldots, x_r(\varepsilon)$ *are linearly dependent*$\}$.

Proof. We shall proceed by induction on r. Consider first the case $r = 1$. Let $g(\varepsilon)$ be an analytic scalar function in Ω with the property that every zero of $g(\varepsilon)$ is also a zero of $x_1(\varepsilon)$ having the same multiplicity, and vice versa. The existence of such a $g(\varepsilon)$ is ensured by the Weierstrass theorem (concerning construction of an analytic function given its zeros with corresponding multiplicities); see, for instance, [63, Vol. III, Chapter 3].

Put $y_1(\varepsilon) = (g(\varepsilon))^{-1}x_1(\varepsilon)$ to prove Lemma S6.2 in the case $r = 1$. We pass now to the general case. Using the induction assumption, we can suppose that $x_1(\varepsilon), \ldots, x_{r-1}(\varepsilon)$ are linearly independent for every $\varepsilon \in \Omega$. Let $X_0(\varepsilon)$ be an $r \times r$ submatrix of the $n \times r$ matrix $[x_1(\varepsilon), \ldots, x_r(\varepsilon)]$ such that $\det X_0(\varepsilon_0) \neq 0$. It is well known in the theory of analytic functions that the set of zeros of the not identically zero analytic function $\det X_0(\varepsilon)$ is discrete, i.e., it is either empty or consists of isolated points. Since $\det X_0(\varepsilon) \neq 0$ implies that the vectors $x_1(\varepsilon), \ldots, x_r(\varepsilon)$ are linearly independent, it follows that the set

$$\Omega_0 = \{\varepsilon \in \Omega \mid x_1(\varepsilon), \ldots, x_r(\varepsilon) \text{ are linearly dependent}\}$$

is also discrete. Disregarding the trivial case when Ω_0 is empty, we can write $\Omega_0 = \{\zeta_1, \zeta_2, \ldots\}$, where $\zeta_i \in \Omega$, $i = 1, 2, \ldots,$ is a finite or countable sequence with no limit points inside Ω.

Let us show that for every $j = 1, 2, \ldots,$ there exist a positive integer s_j and scalar functions $a_{1j}(\varepsilon), \ldots, a_{r-1,j}(\varepsilon)$ which are analytic in a neighborhood of ζ_j such that the system of n-dimensional analytic vector functions in Ω

$$x_1(\varepsilon), \ldots, x_{r-1}(\varepsilon), (\varepsilon - \zeta_j)^{-sj}\left[x_r(\varepsilon) + \sum_{i=1}^{r-1} a_{ij}(\varepsilon)x_i(\varepsilon)\right] \tag{S6.6}$$

has the following properties: for $\varepsilon \neq \zeta_j$ it is linearly equivalent to the system $x_1(\varepsilon), \ldots, x_r(\varepsilon)$ (i.e., both systems span the same subspace in $\mathbb{C}^n$); for $\varepsilon = \zeta_j$ it is linearly independent. Indeed, consider the $n \times r$ matrix $B(\varepsilon)$ whose columns are formed by $x_1(\varepsilon), \ldots, x_r(\varepsilon)$. By the induction hypothesis, there exists an $(r-1) \times (r-1)$ submatrix $B_0(\varepsilon)$ in the first $r-1$ columns of $B(\varepsilon)$ such that $\det B_0(\zeta_j) \neq 0$. For simplicity of notation suppose that $B_0(\varepsilon)$ is formed by the first $r-1$ columns and rows in $B(\lambda)$; so

$$B(\varepsilon) = \begin{bmatrix} B_0(\varepsilon) & B_1(\varepsilon) \\ B_2(\varepsilon) & B_3(\varepsilon) \end{bmatrix}$$

where $B_1(\varepsilon)$, $B_2(\varepsilon)$, and $B_3(\varepsilon)$ are of sizes $(r-1) \times 1$, $(n-r+1) \times (r-1)$, and $(n-r+1) \times 1$, respectively. Since $B_0(\varepsilon)$ is invertible in a neighborhood of ζ_j we can write

$$B(\varepsilon) = \begin{bmatrix} I & 0 \\ B_2(\varepsilon)B_0^{-1}(\varepsilon) & I \end{bmatrix}\begin{bmatrix} B_0(\varepsilon) & 0 \\ 0 & Z(\varepsilon) \end{bmatrix}\begin{bmatrix} I & B_0^{-1}(\varepsilon)B_1(\varepsilon) \\ 0 & I \end{bmatrix}, \tag{S6.7}$$

where $Z(\varepsilon) = B_3(\varepsilon) - B_2(\varepsilon)B_0^{-1}(\varepsilon)B_1(\varepsilon)$ is an $(n-r+1) \times 1$ matrix. Let s_j be the multiplicity of ζ_j as a zero of the vector function $Z(\varepsilon)$. Consider the matrix function

$$\tilde{B}(\varepsilon) = \begin{bmatrix} I & 0 \\ B_2(\varepsilon)B_0^{-1}(\varepsilon) & I \end{bmatrix}\begin{bmatrix} B_0(\varepsilon) & 0 \\ 0 & (\varepsilon - \zeta_j)^{-s_j}Z(\varepsilon) \end{bmatrix}.$$

Clearly, the columns $b_1(\varepsilon), \ldots, b_r(\varepsilon)$ of $\tilde{B}(\varepsilon)$ are analytic and linearly independent vector functions in a neighborhood $V(\zeta_j)$ of ζ_j. From formula (S6.7) it is clear that $\operatorname{Span}\{x_1(\varepsilon), \ldots, x_r(\varepsilon)\} = \operatorname{Span}\{b_1(\varepsilon), \ldots, b_r(\varepsilon)\}$ for $\varepsilon \in U(\zeta_j)\backslash\zeta_j$. Further, from (S6.7) we obtain

$$\tilde{B}(\varepsilon) = \begin{bmatrix} B_0(\varepsilon) & 0 \\ B_2(\varepsilon) & (\varepsilon - \zeta_j)^{-s_j} Z(\varepsilon) \end{bmatrix}$$

and

$$\begin{bmatrix} 0 \\ (\varepsilon - \zeta_j^{-s_j} Z(\varepsilon) \end{bmatrix} = (\varepsilon - \zeta_j)^{-s_j} B(\varepsilon) \begin{bmatrix} -B_0^{-1}(\varepsilon) B_1(\varepsilon) \\ I \end{bmatrix}.$$

So the columns $b_1(\varepsilon), \ldots, b_r(\varepsilon)$ of $\tilde{B}(\varepsilon)$ have the form (S6.6), where $a_{ij}(\varepsilon)$ are analytic scalar functions in a neighborhood of ζ_j.

Now choose $y_1(\varepsilon), \ldots, y_r(\varepsilon)$ in the form

$$y_1(\varepsilon) = x_1(\varepsilon), \ldots, y_{r-1}(\varepsilon) = x_{r-1}(\varepsilon), \qquad y_r(\varepsilon) = \sum_{i=1}^{r} g_i(\varepsilon) x_i(\varepsilon),$$

where the scalar functions $g_i(\varepsilon)$ are constructed as follows:

(a) $g_r(\varepsilon)$ is analytic and different from zero in Ω except for the set of poles $\zeta_1, \zeta_2, \ldots,$ with corresponding multiplicities $s_1, s_2, \ldots$;

(b) the functions $g_i(\varepsilon)$ (for $i = 1, \ldots, r - 1$) are analytic in Ω except for the poles $\zeta_1, \zeta_2, \ldots,$ and the singular part of $g_i(\varepsilon)$ at ζ_j (for $j = 1, 2, \ldots$) is equal to the singular part of $a_{ij}(\varepsilon) g_r(\varepsilon)$ at ζ_j.

Let us check the existence of such functions $g_i(\varepsilon)$. Let $g_r(\varepsilon)$ be the inverse of an analytic function with zeros at $\zeta_1, \zeta_2, \ldots,$ with corresponding multiplicities $s_1, s_2, \ldots,$ (such an analytic function exists by the Weierstrass theorem mentioned above). The functions $g_1(\varepsilon), \ldots, g_{r-1}(\varepsilon)$ are constructed using the Mittag-Leffler theorem (see [63, Vol. III, Chapter 3].

Property (a) ensures that $y_1(\varepsilon), \ldots, y_r(\varepsilon)$ are linearly independent for every $\varepsilon \in \Omega\backslash\{\zeta_1, \zeta_2, \ldots\}$. In a neighborhood of each ζ_j we have

$$\begin{aligned} y_r(\varepsilon) &= \sum_{i=1}^{r-1} (g_i(\varepsilon) - a_{ij}(\varepsilon) g_r(\varepsilon)) x_i(\varepsilon) + g_r(\varepsilon)\Big(x_r(\varepsilon) + \sum_{i=1}^{r-1} a_{ij}(\varepsilon) x_i(\varepsilon)\Big) \\ &= (\varepsilon - \zeta_j)^{-s_j} \left[x_r(\varepsilon) + \sum_{i=1}^{r-1} a_{ij}(\varepsilon) x_i(\varepsilon) \right] \\ &\qquad + \{\text{linear combination of } x_1(\zeta_j), \ldots, x_{r-1}(\zeta_j)\} + \cdots, \end{aligned} \tag{S6.8}$$

where the final ellipsis denotes an analytic (in a neighborhood of ζ_j) vector function which assumes the value zero at ζ_j. Formula (S6.8) and the linear independence of vectors (S6.6) for $\varepsilon = \zeta_j$ ensures that $y_1(\zeta_j), \ldots, y_r(\zeta_j)$ are linearly independent. □

The proof of Lemma S6.2 shows that if for some s $(\leq r)$ the vector functions $x_1(\varepsilon), \ldots, x_s(\varepsilon)$ are linearly independent for all $\varepsilon \in \Omega$, then $y_i(\varepsilon)$, $i = 1, \ldots, r$, can be chosen in such a way that (i)–(iii) hold, and moreover, $y_1(\varepsilon) = x_1(\varepsilon), \ldots, y_s(\varepsilon) = x_s(\varepsilon)$ for all $\varepsilon \in \Omega$. We shall use this observation in the proof of Theorem S6.1.

Proof of Theorem S6.1. Let $A_0(\varepsilon)$ be an $r \times r$ submatrix of $A(\varepsilon)$, which is nonsingular for some $\tilde{\varepsilon} \in \Omega$, i.e., $\det A_0(\tilde{\varepsilon}) \neq 0$. So the set Ω_0 of zeros of the analytic function $\det A_0(\varepsilon)$ is either empty or consists of isolated points. In what follows we assume for simplicity that $A_0(\varepsilon)$ is located in the top left corner of size $r \times r$ or $A(\varepsilon)$.

Let $x_1(\varepsilon), \ldots, x_r(\varepsilon)$ be the first r columns of $A(\varepsilon)$, and let $y_1(\varepsilon), \ldots, y_r(\varepsilon)$ be the vector functions constructed in Lemma S6.2. Then for each $\varepsilon \in \Omega \backslash \Omega_0$ we have

$$\operatorname{Span}\{y_1(\varepsilon), \ldots, y_r(\varepsilon)\} = \operatorname{Span}\{x_1(\varepsilon), \ldots, x_r(\varepsilon)\} = \operatorname{Im} A(\varepsilon) \qquad \text{(S6.9)}$$

(the last equality follows from the linear independence of $x_1(\varepsilon), \ldots, x_r(\varepsilon)$ for $\varepsilon \in \Omega \backslash \Omega_0$). We shall prove now that

$$\operatorname{Span}\{y_1(\varepsilon), \ldots, y_r(\varepsilon)\} \supset \operatorname{Im} A(\varepsilon), \qquad \varepsilon \in \Omega. \qquad \text{(S6.10)}$$

Equality (S6.9) means that for every $\varepsilon \in \Omega \backslash \Omega_0$ there exists an $r \times r$ matrix $B(\varepsilon)$ such that

$$Y(\varepsilon)B(\varepsilon) = A(\varepsilon), \qquad \varepsilon \in \Omega \backslash \Omega_0, \qquad \text{(S6.11)}$$

where $Y(\varepsilon) = [y_1(\varepsilon), \ldots, y_r(\varepsilon)]$. Note that $B(\varepsilon)$ is necessarily unique (indeed, if $B'(\varepsilon)$ also satisfies (S6.11), we have $Y(\varepsilon)(B(\varepsilon) - B'(\varepsilon)) = 0$, and, in view of the linear independence of the columns of $Y(\varepsilon)$, $B(\varepsilon) = B'(\varepsilon)$). Further, $B(\varepsilon)$ is analytic in $\Omega \backslash \Omega_0$. To check this, pick an arbitrary $\varepsilon' \in \Omega \backslash \Omega_0$,and let $Y_0(\varepsilon)$ be an $r \times r$ submatrix of $Y(\varepsilon)$ such that $\det(Y_0(\varepsilon')) \neq 0$ (for simplicity of notation assume that $Y_0(\varepsilon)$ occupies the top r rows of $Y(\varepsilon)$). Then $\det(Y_0(\varepsilon)) \neq 0$ in some neighborhood V of ε', and $(Y_0(\varepsilon))^{-1}$ is analytic on $\varepsilon \in V$. Now $Y(\varepsilon)^{\mathrm{I}} \overset{\text{def}}{=} [(Y_0(\varepsilon))^{-1}, 0]$ is a left inverse of $Y(\varepsilon)$; premultiplying (S6.11) by $Y(\varepsilon)^{\mathrm{I}}$ we obtain

$$B(\varepsilon) = Y(\varepsilon)^{\mathrm{I}} A(\varepsilon), \qquad \varepsilon \in V. \qquad \text{(S6.12)}$$

So $B(\varepsilon)$ is analytic on $\varepsilon \in V$; since $\varepsilon' \in \Omega \backslash \Omega_0$ was arbitrary, $B(\varepsilon)$ is analytic in $\Omega \backslash \Omega_0$.

Moreover, $B(\varepsilon)$ admits analytic continuation to the whole of Ω, as follows. Let $\varepsilon_0 \in \Omega_0$, and let $Y(\varepsilon)^{\mathrm{I}}$ be a left inverse of $Y(\varepsilon)$, which is analytic in a neighborhood V_0 of ε_0 (the existence of such $Y(\varepsilon)$ is proved as above).

Define $B(\varepsilon)$ as $Y(\varepsilon)^{\mathrm{I}}A(\varepsilon)$ for $\varepsilon \in V_0$. Clearly, $B(\varepsilon)$ is analytic in V_0, and for $\varepsilon \in V_0\backslash\{\varepsilon_0\}$, this definition coincides with (S6.12) in view of the uniqueness of $B(\varepsilon)$. So $B(\varepsilon)$ is analytic in Ω.

Now it is clear that (S6.11) holds also for $\varepsilon \in \Omega_0$, which proves (S6.10). Consideration of dimensions shows that in fact we have an equality in (S6.10), unless rank $A(\varepsilon) < r$. Thus (S6.1) and (S6.3) are proved.

We pass now to the proof of existence of $y_{r+1}(\varepsilon), \dots, y_n(\varepsilon)$ such that (ii), (S6.2), and (S6.4) hold. Let $a_1(\varepsilon), \dots, a_r(\varepsilon)$ be the first r rows of $A(\varepsilon)$. By assumption $a_1(\tilde{\varepsilon}), \dots, a_r(\tilde{\varepsilon})$ are linearly independent for some $\varepsilon \in \Omega$. Apply Lemma S6.2 to construct n-dimensional analytic row functions $b_1(\varepsilon), \dots, b_r(\varepsilon)$ such that for all $\varepsilon \in \Omega$ the rows $b_1(\varepsilon), \dots, b_r(\varepsilon)$ are independent, and

$$\operatorname{Span}\{b_1(\varepsilon)^{\mathrm{T}}, \dots, b_r(\varepsilon)^{\mathrm{T}}\} = \operatorname{Span}\{a_1(\varepsilon)^{\mathrm{T}}, \dots, a_r(\varepsilon)^{\mathrm{T}}\}, \qquad \varepsilon \in \Omega\backslash\Omega_0. \tag{S6.13}$$

Fix $\varepsilon_0 \in \Omega$, and let $b_{r+1}, \dots, b_r$ be n-dimensional rows such that the vectors $b_1(\varepsilon_0)^{\mathrm{T}}, \dots, b_r(\varepsilon_0)^{\mathrm{T}}, b_{r+1}^{\mathrm{T}}, \dots, b_n^{\mathrm{T}}$ form a basis in $\mathbb{C}^n$. Applying Lemma S6.2 again (for $x_1(\varepsilon) = b_1(\varepsilon)^{\mathrm{T}}, \dots, x_r(\varepsilon) = b_r(\varepsilon)^{\mathrm{T}}, x_{r+1}(\varepsilon) = b_{r+1}^{\mathrm{T}}, \dots, x_n(\varepsilon) = b_n^{\mathrm{T}}$) and the remark after the proof of Lemma S6.2, we construct n-dimensional analytic row functions $b_{r+1}(\varepsilon), \dots, b_n(\varepsilon)$ such that the $n \times n$ matrix

$$B(\varepsilon) = \begin{bmatrix} b_1(\varepsilon) \\ b_2(\varepsilon) \\ \vdots \\ b_n(\varepsilon) \end{bmatrix}$$

is nonsingular for all $\varepsilon \in \Omega$. Then the inverse $B(\varepsilon)^{-1}$ is analytic on $\varepsilon \in \Omega$. Let $y_{r+1}(\varepsilon), \dots, y_n(\varepsilon)$ be the last $(n - r)$ columns of $B(\varepsilon)^{-1}$. We claim that (ii), (S6.2), and (S6.4) are satisfied with this choice.

Indeed, (ii) is evident. Take $\varepsilon \in \Omega\backslash\Omega_0$; from (S6.13) and the construction of $y_i(\varepsilon)$, $i = r + 1, \dots, n$, it follows that

$$\operatorname{Ker} \begin{bmatrix} a_1(\varepsilon) \\ a_2(\varepsilon) \\ \vdots \\ a_r(\varepsilon) \end{bmatrix} \supset \operatorname{Span}\{y_{r+1}(\varepsilon), \dots, y_n(\varepsilon)\}.$$

But since $\varepsilon \notin \Omega_0$, every row of $A(\varepsilon)$ is a linear combination of the first r rows; so in fact

$$\operatorname{Ker} A(\varepsilon) \supset \operatorname{Span}\{y_{r+1}(\varepsilon), \dots, y_n(\varepsilon)\}. \tag{S6.14}$$

Now (S6.14) implies

$$A(\varepsilon)[y_{r+1}(\varepsilon), \dots, y_n(\varepsilon)] = 0, \qquad \varepsilon \in \Omega\backslash\Omega_0. \tag{S6.15}$$

Passing to the limit when ε approaches a point from Ω_0, we obtain that (S6.15), as well as the inclusion (S6.14), holds for every $\varepsilon \in \Omega$. Consideration of dimensions shows that the equality holds in (S6.14) if and only if rank $A(\varepsilon) = r$, $\varepsilon \in \Omega$. $\square$

S6.2. Analytic Perturbations of Self-Adjoint Matrices

In this section we shall consider eigenvalues and eigenvectors of a self-adjoint matrix which is an analytic function of a parameter. It turns out that the eigenvalues and eigenvectors are also analytic. This result is used in Chapter 11.

Consider an $n \times n$ complex matrix function $A(\varepsilon)$ which depends on the real parameter ε. We impose the following conditions:

(i) $A(\varepsilon)$ is self-adjoint for every $\varepsilon \in \mathbb{R}$; i.e., $A(\varepsilon) = (A(\varepsilon))^*$, where star denotes the conjugate transpose matrix;

(ii) $A(\varepsilon)$ is an analytic function of the *real* variable ε.

Such a matrix function $A(\varepsilon)$ can be diagonalized simultaneously for all real ε. More exactly, the following result holds.

Theorem S6.3. *Let $A(\varepsilon)$ be a matrix function satisfying conditions* (i) *and* (ii). *Then there exist scalar functions $\mu_1(\varepsilon), \ldots, \mu_n(\varepsilon)$ and a matrix-valued function $U(\varepsilon)$, which are analytic for real ε and possess the following properties for every $\varepsilon \in \mathbb{R}$:*

$$A(\varepsilon) = (U(\varepsilon))^{-1} \operatorname{diag}[\mu_1(\varepsilon), \ldots, \mu_n(\varepsilon)]U(\varepsilon), \qquad U(\varepsilon)(U(\varepsilon))^* = I.$$

Proof. Consider the equation

$$\det(I\lambda - A(\varepsilon)) = \lambda^n + a_1(\varepsilon)^{n-1} + \cdots + a_{n-1}(\varepsilon)\lambda + a_n(\varepsilon) = 0, \quad \text{(S6.16)}$$

which $a_i(\varepsilon)$ are scalar analytic functions of the real variable ε. In general, the solutions $\mu(\varepsilon)$ of (S6.16) can be chosen (when properly ordered) as power series in $(\varepsilon - \varepsilon_0)^{1/p}$ in a real neighborhood of every $\varepsilon_0 \in \mathbb{R}$ (Puiseux series):

$$\mu(\varepsilon) = \mu_0 + b_1(\varepsilon - \varepsilon_0)^{1/p} + b_2(\varepsilon - \varepsilon_0)^{2/p} + \cdots, \qquad \text{(S6.17)}$$

(see, for instance, [52c]; also [63, Vol. III, Section 45]). But since $A(\varepsilon)$ is self-adjoint, all its eigenvalues (which are exactly the solution of (S6.1)) are real. This implies that in (S6.17) only terms with integral powers of $\varepsilon - \varepsilon_0$ can appear. Indeed, let b_m be the first nonzero coefficient in (S6.17): $b_1 = \cdots = b_{m-1} = 0$, $b_m \neq 0$. (If $b_i = 0$ for all $i = 1, 2, \ldots$, then our assertion is trivial.) Letting $\varepsilon - \varepsilon_0 \to 0$ through real positive values, we see that

$$b_m = \lim_{\varepsilon \to \varepsilon_0^+} \frac{\mu(\varepsilon) - \mu_0}{(\varepsilon - \varepsilon_0)^{m/p}}$$

is real, because $\mu(\varepsilon)$ is real for ε real. On the other hand, letting $\varepsilon - \varepsilon_0$ approach 0 through negative real numbers, we note that

$$(-1)^{m/p} b_m = \lim_{\varepsilon \to \varepsilon_0^-} \frac{\mu(\varepsilon) - \mu_0}{(\varepsilon_0 - \varepsilon)^{m/p}}$$

is also real. Hence $(-1)^{m/p}$ is a real number, and therefore m must be a multiple of p. We can continue this argument to show that only integral powers of $\varepsilon - \varepsilon_0$ can appear in (S6.17). In other words, the eigenvalues of $A(\varepsilon)$, when properly enumerated, are analytic functions of the real variable ε.

Let $\mu_1(\varepsilon)$ be one such analytic eigenvalue of $A(\varepsilon)$. We turn our attention to the eigenvectors corresponding to $\mu_1(\varepsilon)$. Put $B(\varepsilon) = A(\varepsilon) - \mu_1(\varepsilon)I$. Then $B(\varepsilon)$ is self-adjoint and analytic for real ε, and $\det B(\varepsilon) \equiv 0$. By Theorem S6.2 there is an analytic (for real ε) nonzero vector $x_1(\varepsilon)$ such that $B(\varepsilon)x_1(\varepsilon) \equiv 0$. In other words, $x_1(\varepsilon)$ is an eigenvector of $A(\varepsilon)$ corresponding to $\mu_1(\varepsilon)$.

Since $A(\varepsilon) = (A(\varepsilon))^*$, the orthogonal complement $\{x_1(\varepsilon)\}^\perp$ to $x_1(\varepsilon)$ in $\mathbb{C}^n$ is $A(\varepsilon)$-invariant. We claim that there is an analytic (for real ε) orthogonal basis in $\{x_1(\varepsilon)\}^\perp$. Indeed, let

$$x_1(\varepsilon) = [x_{11}(\varepsilon), \ldots, x_{1n}(\varepsilon)]^T.$$

By Theorem S6.2, there exists an analytic basis $y_1(\varepsilon), \ldots, y_{n-1}(\varepsilon)$ in $\operatorname{Ker}[x_{11}(\varepsilon), \ldots, x_{1n}(\varepsilon)]$. Then $\overline{y_1(\varepsilon)}, \ldots, \overline{y_{n-1}(\varepsilon)}$ is a basis in $\{x_1(\varepsilon)\}^\perp$ (the bar denotes complex conjugation). Since ε is assumed to be real, the basis $\overline{y_1(\varepsilon)}, \ldots, \overline{y_{n-1}(\varepsilon)}$ is also analytic. Applying to this basis the Gram–Schmidt orthogonalization process, we obtain an analytic orthogonal basis $z_1(\varepsilon), \ldots, z_{n-1}(\varepsilon)$ in $\{x_1(\varepsilon)\}^\perp$.

In this basis the restriction $A(\varepsilon)|_{\{x_1(\varepsilon)\}^\perp}$ is represented by an $(n-1) \times (n-1)$ self-adjoint matrix $\hat{A}(\varepsilon)$, which is also analytic (for real ε). Indeed, $\hat{A}(\varepsilon)$ is determined (uniquely) by the equality

$$A(\varepsilon)Z(\varepsilon) = Z(\varepsilon)\hat{A}(\varepsilon), \tag{S6.18}$$

where $Z(\varepsilon) = [z_1(\varepsilon), \ldots, z_{n-1}(\varepsilon)]$ is an $n \times (n-1)$ analytic matrix function with linearly independent columns, for every real ε. Fix $\varepsilon_0 \in \mathbb{R}$, and let $Z_0(\varepsilon)$ be an $(n-1) \times (n-1)$ submatrix in $Z(\varepsilon)$ such that $Z_0(\varepsilon_0)$ is nonsingular (we shall assume for simplicity of notation that $Z_0(\varepsilon)$ occupies the top $n-1$ rows of $Z(\varepsilon)$). Then $Z_0(\varepsilon)$ will be nonsingular in some real neighborhood of ε_0, and in this neighborhood $(Z_0(\varepsilon))^{-1}$ will be analytic. Now (S6.18) gives

$$\hat{A}(\varepsilon) = [(Z_0(\varepsilon))^{-1}, 0]A(\varepsilon)Z(\varepsilon);$$

so $\hat{A}(\varepsilon)$ is analytic in a neighborhood of ε_0. Since $\varepsilon_0 \in \mathbb{R}$ is arbitrary, $\hat{A}(\varepsilon)$ is analytic for real ε.

Now apply the above argument to find an analytic eigenvalue $\mu_2(\varepsilon)$ and corresponding analytic eigenvector $x_2(\varepsilon)$ for $\hat{A}(\varepsilon)$, and so on. Eventually we obtain a set of n analytic eigenvalues $\mu_1(\varepsilon), \ldots, \mu_n(\varepsilon)$ of $A(\varepsilon)$, with corresponding analytic eigenvectors $x_1(\varepsilon), \ldots, x_n(\varepsilon)$ and, moreover, these eigenvectors are orthogonal to each other. Normalizing them (this does not destroy analyticity because ε is real) we can assume that $x_1(\varepsilon), \ldots, x_n(\varepsilon)$ form an orthonormal basis in $\mathbb{C}^n$ for every $\varepsilon \in \mathbb{R}$. Now the requirements of Theorem S6.3 are satisfied with $U(\varepsilon) = [x_1(\varepsilon), \ldots, x_n(\varepsilon)]^{-1}$. □

Comments

The proofs in the first section originated in [37d]. They are based on the approach suggested in [74]. For a comprehensive treatment of perturbation problems in finite and infinite dimensional spaces see [48].

Theorem S6.1 can be obtained as a corollary from general results on analytic vector bundles; see [39].

References

1. S. Barnett, Regular polynomial matrices having relatively prime determinants, *Proc. Cambridge Philos. Soc.* **65**, 585–590 (1969).
2. H. Bart, Meromorphic operator-valued functions, Thesis, Free University, Amsterdam [Mathematical Centre Tracts 44, Mathematical Centre, Amsterdam, 1973].

3a. H. Bart, I. Gohberg, and M. A. Kaashoek, A new characteristic operator function connected with operator polynomials, *Integral Equations and Operator Theory* **1**, 1–12 (1978).

3b. H. Bart, I. Gohberg, and M. A. Kaashoek, Stable factorizations of monic matrix polynomials and stable invariant subspaces, *Integral Equations and Operator Theory* **1**, (4), 496–517 (1978).

3c. H. Bart, I. Gohberg, and M. A. Kaashoek, "Minimal Factorizations of Matrix and Operator Functions," Birkhaeuser, Basel, 1979.

4. H. Bart, I. Gohberg, M. A. Kaashoek, and P. van Dooren, Factorizations of transfer functions, *SIAM J. Control Optim.* **18** (6), 675–696 (1980).
5. H. Baumgärtel, "Endlich-dimensionale Analytische Störungstheorie," Akademie-Verlag, Berlin, 1972.
6. R. R. Bitmead, S. Y. Kung, B. D. O. Anderson and T. Kailath, Greatest common divisors via generalized Sylvester and Bezout matrices, *IEEE Trans. Automat. Control* **AC-23**, 1043–1047 (1978).
7. B. den Boer, Linearization of operator functions on arbitrary open sets, *Integral Equations Operator Theory* **1**, 19–27 (1978).
8. J. Bognar, "Indefinite Inner Product Spaces," Springer-Verlag, Berlin and New York, 1974.
9. M. S. Brodskii, "Triangular and Jordan representations of linear operators," Amer. Math. Soc., Transl. Math. Monographs, Vol. 32, Providence, Rhode Island, 1971.
10. R. S. Busby and W. Fair, Iterative solution of spectral operator polynomial equation and a related continued fraction, *J. Math. Anal. Appl.* **50**, 113–134 (1975).

11. S. L. Campbell, Linear systems of differential equations with singular coefficients, *SIAM J. Math. Anal.* **8**, 1057–1066 (1977).
12. S. Campbell and J. Daughtry, The stable solutions of quadratic matrix equations, *Proc. Amer. Math. Soc.* **74** (1), 19–23 (1979).
13. K. Clancey, private communication.
14. N. Cohen, On spectral analysis of matrix polynomials, M.Sc. Thesis, Weizmann Institute of Science, Israel, 1979.
15a. W. A. Coppel, Linear Systems, *Notes in Pure Mathematics*, Australian National University, Canberra, **6**, (1972).
15b. W. A. Coppel, Matrix quadratic equations, *Bull. Austral. Math. Soc.* **10**, 377–401 (1974).
16a. Ju. L. Daleckii and M. G. Krein, "Stability of Solutions of Differential Equations in Banach Space," Amer. Math. Soc., Transl. Math. Monographs, Vol. 43, Providence, Rhode Island, 1974.
16b. J. Daughtry, Isolated Solutions of Quadratic Matrix Equations, *Linear Algebra and Appl.* **21**, 89–94 (1978).
17. R. J. Duffin, A minimax theory for overdamped networks, *J. Rat. Mech. and Anal.* **4**, 221–233 (1955).
18. H. Flanders and H. K. Wimmer, On the matrix equations $AX - XB = C$ and $AX - YB = C$, *SIAM J. Appl. Math.* **32**, 707–710 (1977).
19. R. A. Frazer, W. J. Duncan and A. R. Collar, "Elementary matrices," 2nd ed., Cambridge Univ. Press, London and New York, 1955.
20. P. Fuhrmann, Simulation of linear system and factorization of matrix polynomials, preprint (1977).
21. P. A. Fuhrmann and J. C. Willems, Factorization indices at infinity for rational matrix functions, *Integral Equations Operator Theory*, **2**, 287–301 (1979).
22. F. R. Gantmacher, "The theory of matrices," Vols. I and II, Chelsea, Bronx, New York, 1959.
23. I. M. Gel'fand, "Lectures in linear algebra," Wiley (Interscience), New York, 1961.
24. I. Gohberg, On some problems in the spectral theory of finite-meromorphic operator functions, *Izv. Akad. Nauk, Armjan. SSR, Ser. Mat.*, VI, 160–181 (1971), (Russian); errata: VII, 152 (1972).
25. I. C. Gohberg and I. A. Feldman, "Convolution Equations and Projection Methods for their Solution," Amer. Math. Soc., Providence, Rhode Island (1974).
26. I. Gohberg and G. Heinig, The resultant matrix and its generalizations, I. The resultant operator for matrix polynomials, *Acta Sci. Math. (Szeged)* **37**, 41–61 (1975), (Russian).
27. I. Gohberg and M. A. Kaashoek, Unsolved problems in matrix and operator theory, II. Partial multiplicities for products, *Integral Equations Operator Theory* **2**, 116–120 (1979).
28. I. Gohberg, M. A. Kaashoek and D. Lay, Equivalence, linearization and decomposition of holomorphic operator functions, *J. Funct. Anal.* **28**, 102–144 (1978).
29a. I. Gohberg, M. A. Kaashoek, L. Lerer and L. Rodman, Common multiples and common divisors of matrix polynomials, I. Spectral method, *Ind. J. Math.* **30**, 321–356, (1981).
29b. I. Gohberg, M. A. Kaashoek, L. Lerer and L. Rodman, Common multiples and common divisors of matrix polynomials, II. Vandermonde and resultant, *Linear and Multilinear Algebra* (to appear).
30a. I. Gohberg, M. A. Kaashoek, and L. Rodman, Spectral analysis of families of operator polynomials and a generalized Vandermonde matrix, I. The finite dimensional case, in "Topics in Functional Analysis," (I. Gohberg and M. Kac, eds.), 91–128, Academic Press, New York, (1978).
30b. I. Gohberg, M. A. Kaashoek, and L. Rodman, Spectral analysis of families of operator polynomials and a generalized Vandermonde matrix, II. The infinite dimensional case, *J. Funct. Anal.* **30**, 358–389 (1978).

31. I. Gohberg, M. A. Kaashoek and F. van Schagen, Common multiples of operator polynomials with analytic coefficients, *Manuscripta Math.* **25**, 279–314 (1978).
32a. I. C. Gohberg and M. G. Krein, Systems of internal equation on a half line with kernels depending on the difference of arguments, *Amer. Math. Soc. Transl.* (2) **13**, 185–264 (1960).
32b. I. C. Gohberg and M. G. Krein, "Introduction to the Theory of Linear Nonself-adjoint Operators in Hilbert Space," Amer. Math. Soc., Transl. Math. Monographs, Vol. 24, Providence, Rhode Island, 1970.
32c. I. C. Gohberg and M. G. Krein, The basic proposition on defect numbers, root numbers and indices of linear operators, *Uspehi Mat. Nauk* **12**, 43–118 (1957); tranlation, *Russian Math. Surveys*, **13**, 185–264 (1960).
33. I. Gohberg and N. Krupnik, "Einführung in die Theorie der eindimensionalen singulären Integraloperatoren," Birkhäuser, Basel, 1979.
34a. I. Gohberg, P. Lancaster and L. Rodman, Spectral analysis of matrix polynomials – I. Canonical forms and divisors, *Linear Algebra Appl.* **20**, 1–44 (1978).
34b. I. Gohberg, P. Lancaster and L. Rodman, Spectral analysis of matrix polynomials – II. The resolvent form and spectral divisors, *Linear Algebra and Appl.* **21**, 65–88 (1978).
34c. I. Gohberg, P. Lancaster and L. Rodman, Representations and divisibility of operator polynomials, *Canad. J. Math.* **30**, 1045–1069 (1978).
34d. I. Gohberg, P. Lancaster and L. Rodman, On self-adjoint matrix polynomials, *Integral Equations Operator Theory* **2**, 434–439 (1979).
34e. I. Gohberg, P. Lancaster and L. Rodman, Perturbation theory for divisors of operator polynomials, *SIAM J. Math. Anal.* **10**, 1161–1183 (1979).
34f. I. Gohberg, P. Lancaster and L. Rodman, Spectral analysis of selfadjoint matrix polynomials, Research Paper No. 419, Dept. of Math. and Stat., University of Calgary, 1979.
34g. I. Gohberg, P. Lancaster and L. Rodman, Spectral analysis of selfadjoint matrix polynomials, *Ann. of Math.* **112**, 34–71 (1980).
34h. I. Gohberg, P. Lancaster and L. Rodman, Factorization of self-adjoint matrix polynomials with constant signature, *Linear and Multilinear Algebra* (1982) (to appear).
35a. I. C. Gohberg and L. E. Lerer, Resultants of matrix polynomials, *Bull. Amer. Math. Soc.* **82**, 565–567 (1976).
35b. I. Gohberg and L. Lerer, Factorization indices and Kronecker indices of matrix polynomials, *Integral Equations Operator Theory* **2**, 199–243 (1979).
36a. I. Gohberg, L. Lerer, and L. Rodman, Factorization indices for matrix polynomials, *Bull. Amer. Math. Soc.* **84**, 275–277 (1978).
36b. I. Gohberg, L. Lerer and L. Rodman, On canonical factorization of operator polynomials, spectral divisors and Toeplitz matrices, *Integral Equations Operator Theory* **1**, 176–214 (1978).
36c. I. Gohberg, L. Lerer, and L. Rodman, Stable factorization of operator polynomials and spectral divisors simply behaved at infinity, I, *J. Math. Anal. Appl.*, **74**, 401–431 (1980).
36d. I. Gohberg, L. Lerer and L. Rodman, Stable factorization of operator polynomials and spectral divisors simply behaved at infinity, II, *J. Math. Anal. Appl.*, **75**, 1–40 (1980).
37a. I. Gohberg and L. Rodman, On spectral analysis of nonmonic matrix and operator polynomials, I. Reduction to monic polynomials, *Israel J. Math.* **30**, 133–151 (1978).
37b. I. Gohberg and L. Rodman, On spectral analysis of nonmonic matrix and operator polynomials, II. Dependence on the finite spectral data, *Israel J. Math.* **30**, 321–334 (1978).
37c. I. Gohberg and L. Rodman, On the spectral structure of monic matrix polynomials and the extension problem, *Linear Algebra Appl.* **24**, 157–172 (1979).
37d. I. Gohberg and L. Rodman, Analytic matrix functions with prescribed local data, *J. Analyse Math.* (to appear).
38. I. Gohberg and E. I. Sigal, On operator generalizations of the logarithmic residue theorem and the theorem of Rouche, *Math. USSR-Sb.* **13**, 603–625 (1971).

39. H. Grauert, Analytische Fascrungen über holomorphvollständigen Raumen, *Math. Ann.* **135**, 263–273 (1958).
40. K. G. Guderley, On nonlinear eigenvalue problems for matrices, *SIAM J. Appl. Math.* **6**, 335–353 (1958).
41. G. Heinig, The concepts of a bezoutian and a resolvent for operator bundles, *Functional Anal. App.* **11**, 241–243 (1977).
42. P. Henrici, "Elements of Numerical Analysis," Wiley, New York, 1964.
43. I. S. Iohvidov and M. G. Krein, Spectral theory of operators in spaces with an indefinite metric, *Amer. Math. Soc. Translations* (2) **13**, 105–175 (1960); **34**, 283–373 (1963).
44. G. A. Isaev, Numerical range of operator pencils and Keldysh multiple completeness, *Functional Anal. Appl.* **9**, 27–30 (1975).
45. V. A. Jakubovič, Factorization of symmetric matrix polynomials, *Soviet Math. Dokl.* **11**, 1261–1264 (1970).
46. V. I. Kabak, A. S. Marcus, and V. I. Mereutsa, On a connection between spectral properties of a polynomial operator bundle and its divisors, in "Spectral Properties of Operators," pp. 29–57, Stiinca, Kishinev, 1977, (in Russian).
47. R. E. Kalman, Irreducible realizations and the degree of a rational matrix, *SIAM J. Control (Optim.)* **13**, 520–544 (1965).
48. T. Kato, "Perturbation theory for linear operators," Springer-Verlag, Berlin and New York, 1966.
49a. M. V. Keldysh, On eigenvalues and eigenfunctions of some classes of nonselfadjoint equations, *DAN SSSR* **77**, No. 1, 11–14 (1951), (in Russian).
49b. M. V. Keldysh, On the completeness of the eigenfunctions of some classes of nonselfadjoint linear operators, *Russian Math. Surveys*, **24**, No. 4, 15–44 (1971)
50. A. G. Kostyuchenko and M. B. Orazov, Certain properties of the roots of a selfadjoint quadratic bundle, *Functional Anal. Appl.* **9**, 295–305 (1975).
51. M. G. Krein and H. Langer, On some mathematical principles in the linear theory of damped oscillations of continua, I, II, *Integral Equations Operator Theory* **1**, 364–399, 539–566 (1978). (Translation).
52a. P. Lancaster, Inversion of lambda-matrices and applications to the theory of linear vibrations, *Arch. Rational Mech. Anal.* **6**, 105–114 (1960).
52b. P. Lancaster, "Lambda-matrices and Vibrating Systems," Pergamon, Oxford, 1966.
52c. P. Lancaster, "Theory of Matrices," Academic Press, New York, 1969.
52d. P. Lancaster, Some questions in the classical theory of vibrating systems, *Bul. Inst. Politehn. Iasi*, **17** (21), 125–132 (1971).
52e. P. Lancaster, A fundamental theorem of lambda-matrices with applications—I. Ordinary differential equations with constant coefficients, *Linear Algebra Appl.* **18**, 189–211 (1977).
52f. P. Lancaster, A fundamental theorem of lambda-matrices with applications—II. Difference equations with constant coefficients, *Linear Algebra Appl.* **18**, 213–222 (1977).
53. P. Lancaster and L. Rodman, Existence and uniqueness theorems for the algebraic Riccati equation, *Internat. J. Control* **32**, 285–309 (1980).
54. P. Lancaster and J. G. Rokne, Solution of nonlinear operator equations, *SIAM J. Math. Anal.* **8**, 448–457 (1977).
55. P. Lancaster and H. K. Wimmer, Zur Theorie der λ-Matrizen, *Math. Nachr.* **68**, 325–330 (1975).
56a. H. Langer, Spektraltheorie linearer Operatoren in J-Räumen und einige Anwendungen auf die Schar $L(\lambda) = \lambda^2 I + \lambda B + C$. Habilitationsschrift, Tehnischen Universität Dresden, 1964.
56b. H. Langer, Über Lancaster's Zerlegung von Matrizen-Scharen, *Arch. Rational Mech. Anal.* **29**, 75–80 (1968).

56c. H. Langer, Über eine Klasse nichtlinearer Eigenwertprobleme, *Acta Sci. Math.* (*Szeged*), **35**, 79–93 (1973).
56d. H. Langer, Factorization of operator pencils, *Acta Sci. Math.* (*Szeged*) **38**, 83–96 (1976).
57a. J. Leiterer, Local and global equivalence of Meromorphic Operator Functions; I, *Math. Nachr.* **83**, 7–29 (1978).
57b. J. Leiterer, Local and global equivalence of Meromorphic Operator Functions; II, *Math. Nachr.* **84**, 145–170 (1978).
58. Ya. B. Lopatinskii, Factorization of a polynomial matrix, *Nauch. Zap. L'vov. Politekh. Inst., Ser. Fiz.-Mat.* **38**, 3–9 (1956), (in Russian).
59. A. I. Mal'cev, "Foundations of Linear Algebra," Freeman, San Francisco, California, 1963.
60. A. S. Marcus, On homomorphic operator functions, *Dokl. Akad. Nauk SSSR* **119**, 1099–1102 (1958). (Russian).
61a. A. S. Marcus and V. I. Matsaev, On spectral properties of holomorphic operator functions in Hilbert space, *Mat. Issled.*, IX:4 (34), 79–91 (1974).
61b. A. S. Marcus and V. I. Matsaev, On spectral theory of holomorphic operator functions in Hilbert space, *Functional Anal. Appl.* **9**, 76–77 (1975), (in Russian).
61c. A. S. Marcus and V. I. Matsaev, On spectral theory of holomorphic operator functions, in: Research in Functional Analysis, Stiinca, Kishinev, 71–100 (1978), (in Russian).
62a. A. S. Marcus and I. V. Mereutsa, On the complete n-tuple of roots of the operator equation corresponding to a polynomial operator bundle, *Math. USSR Izv.* **7**, 1105–1128 (1973).
62b. A. S. Marcus and I. V. Mereutsa, On some properties of λ-matrices, *Mat. Issled.*, Kishinev **3**, 207–213 (1975), (in Russian).
63. A. I. Markushevich, "Theory of Functions of a Complex Variable," Vols. I–III, Prentice-Hall, Englewood Cliffs, New Jersey, 1965.
64a. I. V. Mereutsa, On the decomposition of an operator polynomial bundle into a product of linear factors, *Mat. Issled.* **8**, 102–114 (in Russian).
64b. I. V. Mereutsa, On decomposition of a polynomial operator bundle into factors, in "Research in Functional Analysis," Stiinca, Kishinev, pp. 101–119 (1978), (in Russian).
65. L. Mirsky, "An introduction to Linear Algebra," Oxford Univ. Press (Clarendon), London and New York, 1955.
66a. B. Mityagin, Linearization of holomorphic operator functions I, *Integral Equations Operator Theory*, **1**, 114–131 (1978).
66b. B. Mityagin, Linearization of holomorphic operator functions II, *Integral Equations Operator Theory* **1**, 226–249 (1978).
67. P. H. Müller, Über eine Klasse von Eigenwertaufgaben mit nichtlinearer Parameterähbangigkeit, *Math. Nachr.* **12**, 173–181 (1954).
68. N. I. Muskhelishvili, "Singular Integral Equations," Nordhoff, Groningen, 1953.
69. M. B. Orazov, On certain properties of analytic operator-functions of two arguments, *Uspehi Mat. Nauk*, **4**, 221–222 (1978), (in Russian).
70a. L. Rodman, On analytic equivalence of operator polynomials, *Integral Equations Operator Theory*, **1**, 400–414 (1978).
70b. L. Rodman, On existence of common multiples of monic operator polynomials, *Integral Equations Operator Theory*, **1**, 400–414 (1978).
70c. L. Rodman, Spectral theory of analytic matrix functions, Ph.D. Thesis, Tel-Aviv University, 1978.
70d. L. Rodman, On extremal solutions of the algebraic Riccati equation, *Amer. Math. Soc., Lectures in Appl. Math.* **18**, 311–327 (1980).
70e. L. Rodman, On connected components in the set of divisors of monic matrix polynomials, *Linear Algebra and Appl.*, **31**, 33–42 (1980).
70f. L. Rodman, On nonnegative invariant subspaces in indefinite scalar product spaces, *Linear and Multilinear Algebra* **10**, 1–14 (1981).

71. L. Rodman and M. Schaps, On the partial multiplicities of a product of two matrix polynomials, *Integral Equations Operator Theory* **2**, 565–599 (1979).
72. W. E. Roth, The equations $AX - YB = C$ and $AX - XB = C$ in matrices, *Proc. Amer. Math. Soc.* **3**, 392–396 (1952).
73. B. Rowley, Spectral analysis of operator polynomials, Ph.D. Thesis, McGill University, Montreal (1979).
74. Yu. L. Shmuljan, Finite dimensional operators depending analytically on a parameter, *Ukrainian Math. J.* IX (2), 195–204 (1957), (in Russian).
75. E. I. Sigal, Partial multiplicities of a product of operator functions, *Mat. Issled.* **8** (3), 65–79 (1973), (in Russian).
76a. G. P. A. Thijsse, On solutions of elliptic differential equations with operator coefficients, *Integral Equations Operator Theory* **1**, 567–579 (1978).
76b. G. P. A. Thijsse, Rules for the partial multiplicities of the product of holomorphic matrix functions, *Integral Equations Operator Theory* **3**, 515–528 (1980).
77. R. C. Thompson, The characteristic polynomial of a principal subpencil of a Hermitian matrix pencil, *Linear Algebra and Appl.* **14**, 135–177 (1976).
78. J. V. Uspensky, "Theory of Equations," McGraw-Hill, New York, 1978.
79a. H. K. Wimmer, A Jordan factorization theorem for polynomial matrices, *Proc. Amer. Math. Soc.*, 201–206 (1979).
79b. H. K. Wimmer, The structure of non-singular polynomial matrices, Research Paper 432, University of Calgary, 1979.
80. W. A. Wolovich, "Linear Multivariable Systems" (Applied Mathematical Sciences, Vol. 11), Springer-Verlag, Berlin and New York, 1974.
81. D. C. Youla and P. Tissi, An explicit formula for the degree of a rational matrix, Electrophysics Memo, PIBM RI-1273-65, Polytechnic Institute of Brooklyn, Electrophysics Department, 1965.

List of Notations and Conventions

Vectors, matrices and functions are assumed to be complex-valued if not stated otherwise. Linear spaces are assumed to be over the field $\mathbb{C}$ of complex numbers if not stated otherwise.

$\mathbb{C}^n$	linear space of all n-dimensional vectors written as columns
$\mathbb{R}$	the field of real numbers
I or I_m	identity linear transformation or identity matrix of size $m \times m$
0 or 0_m	zero linear transformation or zero matrix of size $m \times m$

For a constant matrix A,

$$\sigma(A) = \{\lambda \mid I\lambda - A \text{ is singular}\}$$

For a nonconstant matrix polynomial $L(\lambda)$,

$$\sigma(L) = \{\lambda \mid L(\lambda) \text{ is singular}\}$$

Matrices will often be interpreted as linear transformations (written in a fixed basis, which in $\mathbb{C}^n$ is generally assumed to be standard, i.e., formed by unit coordinate vectors), and vice versa.

$L^{(i)}(\lambda)$ ith derivative with respect to λ of matrix-valued function $L(\lambda)$

$\operatorname{Ker} T = \{x \in \mathbb{C}^n \mid Tx = 0\}$ for $m \times n$ matrix T

$\operatorname{Im} T = \{y \in \mathbb{C}^m \mid y = Tx \text{ for some } x \in \mathbb{C}^n\}$, for $m \times n$ matrix T

$A\mathcal{M}$	image of a subspace $\mathcal{M}$ under linear transformation A
$A\|_{\mathcal{M}}$	The restriction of linear transformation A to the domain $\mathcal{M}$.
$\mathcal{M} \dotplus \mathcal{N}$	direct sum of subspaces
$\mathcal{M} \oplus \mathcal{N}$	orthogonal sum of subspaces
$\operatorname{Span}\{x_1, \ldots, x_m\}$	subspace spanned by vectors $x_1, \ldots, x_m$

$\{0\}$	zero subspace
A^{T}	transposed matrix
A^*	the conjugate transpose of matrix A, or the adjoint linear transformation of A
$\|x\| = (\sum_{i=1}^{n} \|x_i\|^2)^{1/2}$	euclidean norm of $x = (x_1, \dots, x_n)^{\mathrm{T}} \in \mathbb{C}^n$.
$(x, y) = \sum_{i=1}^{n} x_i \bar{y}_i$	scalar product of vectors $x = (x_1, \dots, x_n)^{\mathrm{T}}$ and $y = (y_1, \dots, y_n)^{\mathrm{T}}$
$\|A\|$	norm of a matrix or linear transformation, generally given by

$$\|A\| = \max\{\|Ax\| \mid \|x\| = 1\}$$

$$\operatorname{row}(Z_i)_{i=1}^{m} = \operatorname{row}(Z_1, \dots, Z_m) = [Z_1 \quad Z_2 \quad \dots \quad Z_m]$$

block row matrix with blocks $Z_1, \dots, Z_m$

$$\operatorname{col}(Z_i)_{i=1}^{m} = \begin{bmatrix} Z_1 \\ Z_2 \\ \vdots \\ Z_m \end{bmatrix}$$ block column matrix with blocks $Z_1, \dots, Z_m$.

$$\operatorname{diag}[Z_i]_{i=1}^{m} = \operatorname{diag}[Z_1, \dots, Z_m] = Z_1 \oplus \cdots \oplus Z_m = \begin{bmatrix} Z_1 & 0 & \cdots & 0 \\ 0 & Z_2 & \cdots & 0 \\ \vdots & \vdots & \ddots & \vdots \\ 0 & 0 & \cdots & Z_m \end{bmatrix}$$

block diagonal matrix with blocks $Z_1, \dots, Z_m$

$\operatorname{ind}(X, T)$	index of stabilization of admissible pair (X, T)
$\operatorname{sig} H$	signature of hermitian matrix H
$\operatorname{sgn}(x, \lambda)$	a sign (+ or −) associated with eigenvalue λ and eigenvector x
$\mathscr{Im}\, \lambda$, $\mathscr{Re}\, \lambda$	imaginary and real parts of $\lambda \in \mathbb{C}$
$\bar{\lambda}$	complex conjugate of $\lambda \in \mathbb{C}$
$\delta_{ij} = \begin{cases} 1, & i = j \\ 0, & i \neq j \end{cases}$	Kronecker index
$GL(\mathbb{C}^n)$	group of all nonsingular $n \times n$ matrices
$\|X\|$	number of elements in a finite set X
$\mathscr{N} \setminus \mathscr{M}$	The complement of set $\mathscr{M}$ in set $\mathscr{N}$.
□	end of proof or example
$\stackrel{\text{def}}{=}$	equals by definition

Index

J

K

L

M

N

O

T

U

V

W

Computer Science and Applied Mathematics

A SERIES OF MONOGRAPHS AND TEXTBOOKS

Editor

Werner Rheinboldt

University of Pittsburgh

HANS P. KÜNZI, H. G. TZSCHACH, and C. A. ZEHNDER. Numerical Methods of Mathematical Optimization: With ALGOL and FORTRAN Programs, Corrected and Augmented Edition

AZRIEL ROSENFELD. Picture Processing by Computer

JAMES ORTEGA AND WERNER RHEINBOLDT. Iterative Solution of Nonlinear Equations in Several Variables

AZARIA PAZ. Introduction to Probabilistic Automata

DAVID YOUNG. Iterative Solution of Large Linear Systems

ANN YASUHARA. Recursive Function Theory and Logic

JAMES M. ORTEGA. Numerical Analysis: A Second Course

G. W. STEWART. Introduction to Matrix Computations

CHIN-LIANG CHANG AND RICHARD CHAR-TUNG LEE. Symbolic Logic and Mechanical Theorem Proving

C. C. GOTLIEB AND A. BORODIN. Social Issues in Computing

ERWIN ENGELER. Introduction to the Theory of Computation

F. W. J. OLVER. Asymptotics and Special Functions

DIONYSIOS C. TSICHRITZIS AND PHILIP A. BERNSTEIN. Operating Systems

ROBERT R. KORFHAGE. Discrete Computational Structures

PHILIP J. DAVIS AND PHILIP RABINOWITZ. Methods of Numerical Integration

A. T. BERZTISS. Data Structures: Theory and Practice, Second Edition

N. CHRISTOPHIDES. Graph Theory: An Algorithmic Approach

ALBERT NIJENHUIS AND HERBERT S. WILF. Combinatorial Algorithms

AZRIEL ROSENFELD AND AVINASH C. KAK. Digital Picture Processing

SAKTI P. GHOSH. Data Base Organization for Data Management

DIONYSIOS C. TSICHRITZIS AND FREDERICK H. LOCHOVSKY. Data Base Management Systems

JAMES L. PETERSON. Computer Organization and Assembly Language Programming

WILLIAM F. AMES. Numerical Methods for Partial Differential Equations, Second Edition

ARNOLD O. ALLEN. Probability, Statistics, and Queueing Theory: With Computer Science Applications

ELLIOTT I. ORGANICK, ALEXANDRA I. FORSYTHE, AND ROBERT P. PLUMMER. Programming Language Structures

ALBERT NIJENHUIS AND HERBERT S. WILF. Combinatorial Algorithms. Second edition.

JAMES S. VANDERGRAFT. Introduction to Numerical Computations

AZRIEL ROSENFELD. Picture Languages, Formal Models for Picture Recognition

ISAAC FRIED. Numerical Solution of Differential Equations

ABRAHAM BERMAN AND ROBERT J. PLEMMONS. Nonnegative Matrices in the Mathematical Sciences

BERNARD KOLMAN AND ROBERT E. BECK. Elementary Linear Programming with Applications

CLIVE L. DYM AND ELIZABETH S. IVEY. Principles of Mathematical Modeling

ERNEST L. HALL. Computer Image Processing and Recognition

ALLEN B. TUCKER, JR., Text Processing: Algorithms, Languages, and Applications

MARTIN CHARLES GOLUMBIC. Algorithmic Graph Theory and Perfect Graphs

GABOR T. HERMAN. Image Reconstruction from Projections: The Fundamentals of Computerized Tomography

WEBB MILLER AND CELIA WRATHALL. Software for Roundoff Analysis of Matrix Algorithms

ULRICH W. KULISCH AND WILLARD L. MIRANKER. Computer Arithmetic in Theory and Practice

LOUIS A. HAGEMAN AND DAVID M. YOUNG. Applied Interative Methods

I. GOHBERG, P. LANCASTER AND L. RODMAN. Matrix Polynomials.

In preparation

AZRIEL ROSENFELD AND AVINASH C. KAK. Digital Picture Processing, Second Edition, Vol. 1, Vol. 2

DIMITRI P. BERTSEKAS. Constrained Optimization and Lagrange Multiplier Methods

LEONARD UHR. Computer Arrays and Networks: Algorithm-Structured Parallel Architectures for Perception and Modeling